Lecture Notes in Physics

The Lecture Notes in Physics

The series Lecture Notes in Physics (LNP), founded in 1969, reports new developments in physics research and teaching – quickly and informally, but with a high quality and the explicit aim to summarize and communicate current knowledge in an accessible way. Books published in this series are conceived as bridging material between advanced graduate textbooks and the forefront of research and to serve three purposes:

- to be a compact and modern up-to-date source of reference on a well-defined topic
- to serve as an accessible introduction to the field to postgraduate students and nonspecialist researchers from related areas
- to be a source of advanced teaching material for specialized seminars, courses and schools

Both monographs and multi-author volumes will be considered for publication. Edited volumes should, however, consist of a very limited number of contributions only. Proceedings will not be considered for LNP.

Volumes published in LNP are disseminated both in print and in electronic formats, the electronic archive being available at springerlink.com. The series content is indexed, abstracted and referenced by many abstracting and information services, bibliographic networks, subscription agencies, library networks, and consortia.

Proposals should be sent to a member of the Editorial Board, or directly to the managing editor at Springer:

Christian Caron
Springer Heidelberg
Physics Editorial Department I
Tiergartenstrasse 17
69121 Heidelberg / Germany
christian.caron@springer.com

Sverre J. Aarseth
Christopher A. Tout
Rosemary A. Mardling (Eds.)

The Cambridge *N*-Body Lectures

Sverre J. Aarseth
University of Cambridge
Institute of Astronomy
Madingley Road
Cambridge CB3 0HA
United Kingdom
sverre@ast.cam.ac.uk

Christopher A. Tout
University of Cambridge
Institute of Astronomy
Madingley Road
Cambridge CB3 0HA
United Kingdom
cat@ast.cam.ac.uk

Rosemary A. Mardling
School of Mathematical Sciences
Monash University
Victoria 3800
Australia
mardling@sci.monash.edu.au

Aarseth, S. J. et al. (Eds.), *The Cambridge* N-*Body Lectures*, Lect. Notes Phys. 760 (Springer, Berlin Heidelberg 2008), DOI 10.1007/978-1-4020-8431-7

The Royal Astronomical Society Series. A series on Astronomy & Astrophysics, Geophysics, Solar and Solar-terrestrial Physics, and Planetary Sciences

ISBN: 978-1-4020-8430-0 e-ISBN: 978-1-4020-8431-7

DOI 10.1007/978-1-4020-8431-7

Lecture Notes in Physics ISSN: 0075-8450

Library of Congress Control Number: 2008929549

Cover design: Integra Software Services Pvt. Ltd.

Printed on acid-free paper

9 8 7 6 5 4 3 2 1

springer.com

Preface

This book gives a comprehensive introduction to the tools required for direct N-body simulations. The contributors are all active researchers who write in detail on their own special fields in which they are leading international experts. It is their previous and current connections with the Cambridge Institute of Astronomy, as staff or visitors, that gives rise to the title. The material is generally at a level suitable for a graduate student or postdoctoral worker entering the field.

The book begins with a detailed description of the codes available for N-body simulations. In a second chapter we find different mathematical formulations for special treatments of close encounters involving binaries or multiple systems, which have been implemented. The concept of chaos and stability plays a fundamental role in celestial mechanics and is highlighted here in a presentation of a new formalism for the three-body problem. The emphasis on collisional stellar dynamics enables the scope to be enlarged by including methods relevant for comparison purposes. Modern star cluster simulations include additional astrophysical effects by modelling real stars instead of point-masses. Several contributions cover the basic theory and comprehensive treatments of stellar evolution for single stars as well as binaries. Questions concerning initial conditions are also discussed in depth. Further connections with reality are established by an observational approach to data analysis of actual and simulated star clusters. Finally, important aspects of hardware requirements are described with special reference to parallel and GRAPE-type computers. The extensive chapters provide an essential framework for a variety of N-body simulations.

During an extensive summer school on astrophysical N-body simulations, held in Cambridge, www.cambody.org, the Royal Astronomical Society encouraged us to edit a volume on the topic, to be published in The Royal Astronomical Society Series. Subsequently, we collected the tutorial lecture notes assembled in this volume. We would like to take this opportunity to thank the Royal Astronomical Society for sponsoring the school and the Institute of Astronomy for provision of school facilities. We are grateful to all the authors who took time off from their busy schedules to deliver the manuscripts, which were then checked for both style and scientific content by the editors. This collection of topics, related to the gravitational N-body problem, will prove useful to both students and researchers in years to come.

Cambridge
May 2008

Sverre J. Aarseth
Christopher A. Tout
Rosemary A. Mardling

Contents

1

Direct N-Body Codes

Sverre J. Aarseth

University of Cambridge, Institute of Astronomy, Madingley Road, Cambridge CB3 0HA, UK
sverre@ast.cam.ac.uk

1.1 Introduction

The classical formulation of the gravitational N-body problem is deceptively simple. Given initial values of N masses, coordinates and velocities, the task is to calculate the future orbits. Although the motions are in principle completely determined by the underlying differential equations, accurate solutions can only be obtained by numerical methods. Self-gravitating stellar systems experience highly complicated interactions, which require efficient procedures for studying the long-term behaviour. In this chapter, we are concerned with describing aspects relating to direct summation codes that have been remarkably successful. This is the most intuitive approach and present-day technology allows surprisingly large systems to be considered for a direct attack. Astronomers and mathematicians alike are interested in many aspects of dynamical evolution, ranging from highly idealized systems to star clusters where complex astrophysical processes play an important role. Hence the need for modelling such behaviour poses additional challenges for both the numerical analyst and the code designer.

In the present chapter, we concentrate on describing some relevant procedures for star cluster simulation codes. Such applications are mainly directed towards studying large clusters. However, many techniques dealing with few-body dynamics have turned out to be useful here and their implementation will therefore be discussed too. At the same time, the GRAPE special-purpose supercomputers are increasingly being used for large-N simulations. Hence a diversity of tools are now employed in modern simulations and the practitioner needs to be versatile or part of a team. This development has led to complicated codes, which also require an effort in efficient utilization as well as interpretation of the results. It follows that designers of large N-body codes need to pay attention to documentation as well as the programming itself. Finally, bearing in mind the increasing complexity of challenging problems posed by new observations, further progress in software is needed to keep pace with the ongoing hardware developments.

Aarseth, S.J.: *Direct N-Body Codes*. Lect. Notes Phys. **760**, 1–30 (2008)
DOI 10.1007/978-1-4020-8431-7_1

1.2 Basic Features

Before delving more deeply into the underlying algorithms, it is desirable to define units and introduce the data structure that forms the back-bone of a general N-body code. From dimensional analysis we first construct fiducial velocity and time units by $\tilde{V}^* = 1 \times 10^{-5}(GM_\odot/L^*)^{1/2}\,\mathrm{km\,s^{-1}}$, $\tilde{T}^* = (L^{*3}/GM_\odot)^{1/2}\,\mathrm{s}$, with G the gravitational constant and $L^* = 3\times10^{18}$ cm as a convenient length unit. Given the length scale or virial radius R_V in pc and total mass NM_S in $M_\odot$, where M_S is the average mass specified as input, we can now write the corresponding values for a star cluster model as $V^* = 6.557 \times 10^{-2}(NM_\mathrm{S}/R_\mathrm{V})^{1/2}\,\mathrm{km\,s^{-1}}$and $T^* = 14.94(R_\mathrm{V}^3/NM_\mathrm{S})^{1/2}\,\mathrm{Myr}$. Hence scaled (or internal) N-body units of distance, velocity and time are converted to corresponding astrophysical units (pc, $\mathrm{km\,s^{-1}}$, Myr) by $\tilde{r} = R_\mathrm{V} r$, $\tilde{v} = V^* v$, $\tilde{t} = T^* t$. Finally, individual masses in $M_\odot$ are obtained from $\tilde{m} = M_\mathrm{S}\, m$ where M_S is now redefined in terms of the scaled mean mass.

As the next logical step on the road to an N-body simulation, we consider matters relating to the initial data. Let us assume that a complete set of initial conditions have been generated in the form $m_i, \tilde{\boldsymbol{r}}_i, \tilde{\boldsymbol{v}}_i$ for N particles, where the masses, coordinates and velocities can be in any units. A standard cluster model is essentially defined by $N, M_\mathrm{S}, R_\mathrm{V}$, together with a suitable initial mass function (IMF). After assigning the individual data, we evaluate the kinetic and potential energy, K and U, taking $U < 0$. The velocities are scaled according to the virial theorem by taking $\boldsymbol{v}_i = \mathrm{q}\,\tilde{\boldsymbol{v}}_i$, where $q = (Q_\mathrm{V}|U|/K)^{1/2}$ and Q_V is an input parameter (0.5 for overall equilibrium). Note that, in general, the virial energy should be used; however, the additional terms are not known ahead of the scaling. We now introduce so-called standard units by adopting the scaling $G = 1, \sum m_i = 1, E_0 = -0.25$, where E_0 is the new total energy (< 0). Here the energy condition is only applied for bound systems ($Q_\mathrm{V} < 1$), otherwise the convention $E_0 = 0.25$ is adopted. The final scaling is performed by $\hat{\boldsymbol{r}}_i = \tilde{\boldsymbol{r}}_i/S^{1/2}$, $\hat{\boldsymbol{v}}_i = \boldsymbol{v}_i S^{1/2}$ with $S = E_0/(q^2K + U)$. These variables define a standard crossing time $T_\mathrm{cr} = 2\sqrt{2}T^*$ Myr.

Many simulations include primordial binary stars for greater realism. Because of their internal binding energies, the above scaling cannot be implemented directly. Instead, the components of each binary are first combined into one object, whereupon the reduced population of N_s single stars and N_b binaries are subject to the standard scaling. It then remains for the internal two-body elements, such as semi-major axis, eccentricity and relevant angles to be assigned, together with the mass ratio. The choice of distributions is very wide, but should be motivated by astrophysical considerations. Of special interest here are the periods and mass ratios, which may well be correlated for luminous stars (e.g. spectroscopic binaries). More complicated ways of providing initial conditions with primordial binaries can readily be incorporated. Thus, for example, a consistent set of initial conditions that do not require scaling may be uploaded. Such a data set might in fact be acceptable by a well-written code, but this practice is not recommended.

1.3 Data Structure

The time has now come to introduce the data structure used in the Cambridge N-body codes. Complications of describing the quantities in a stellar system arise when some objects are no longer single stars. In the first instance, hard binaries are treated by two-body regularization (Kustaanheime & Stiefel 1965, hereafter KS). Now a convenient description refers to the relative motion as well as that of the centre of mass (c.m.). For the purposes of sequential predictions and force summations, it is natural to place the two KS components first in all relevant arrays, followed by single stars, with the c.m. last. Thus given $N_{\rm p}$ pairs, the type of object can be distinguished by its location i in the array, compared to $2N_{\rm p}$ and N. Likewise, for long-lived triples, where the inner binary of the hierarchy becomes the first member of the new KS pair and the outer component the second.

The new arrangement necessitates the introduction of so-called ghost stars, which retain the quantities associated with the outer component, except that the mass is temporarily set to zero. In other words, a ghost star is a dormant particle without any gravitational effect since it now forms part of the triple. Generalization to a quadruple consisting of two binaries forming a new KS follows readily. Note that in this case a ghost binary must be defined as well as a ghost c.m. particle. Higher-order systems of increasing complexity are defined in an analogous manner. The treatment of hierarchies continues as long as they are defined to be stable, as will be discussed in subsequent sections.

It now remains to introduce the final type of object in the form of a compact subsystem, which is treated by chain regularization (Mikkola & Aarseth 1993). Briefly, the idea here is to employ pairwise two-body regularization for the strongest interactions and include the other terms as perturbations. Such systems are invariably short-lived, but the special treatment is most conveniently carried out within the context of the standard data structure. At least two of the chain members are former components of a KS binary and the initial membership may be three or four. These systems are usually created following a strong interaction between a binary and another single particle or binary. Here one of the members is assigned to the role as the c.m. for the subsystem while the others become ghosts.

- Single stars $\quad 2\,N_{\rm p} < i \le N, \quad \mathcal{N}_i = i$
- KS pairs $\quad 1 \le i \le 2\,N_{\rm p}\,, \quad i_p = i_{\rm icm} - N$
- C.m. particles $\quad i > N, \quad \mathcal{N} = N_0 + \mathcal{N}_k\,, \quad k = 2i_p - 1$
- Stable triples $\quad$ KS + ghost, $\quad \mathcal{N}_{\rm cm} = -\,\mathcal{N}_k$
- Ghost particles $\quad \mathcal{N}_{\rm ghost} = \mathcal{N}_{2i_{\rm p}-1}\,, \quad m_{\rm ghost} = 0$
- Stable quadruples $\quad$ KS + KS ghost, $\quad \mathcal{N}_{\rm cm} = -\,\mathcal{N}_k$
- Higher orders $\quad$ T + KS, $\quad \mathcal{N}_{\rm cm} = -\,(2N_0 + \mathcal{N}_k)$
- Chain members $\quad 2\,N_{\rm p} < i_{\rm cm} \le N, \quad \mathcal{N}_{\rm cm} = 0$

The table summarizes the key features of the data structure. In order to keep track of the identity of the particles, we also assign a name to each, denoted by

$\mathcal{N}_i$. This quantity is useful for distinguishing the type of object, i.e. whether single, binary or even chain c.m. Thus the name of a binary c.m. is defined by $\mathcal{N}_{\rm cm} = N_0 + \mathcal{N}_k$, where N_0 is the initial particle number and $\mathcal{N}_k$ is the name of the first KS component. Likewise, the c.m. of hierarchical systems of different levels are identified by $\mathcal{N}_{\rm cm} < 0$ while $\mathcal{N}_i = 0$ for a chain c.m. with $i \le N$. Note that an arbitrary number of binaries can be accommodated, but only one chain. Given the location $i_{\rm cm}$ of any c.m., the corresponding KS pair index is obtained from $i_{\rm p} = i_{\rm cm} - N$, with the components at $2i_{\rm p} - 1$, $2i_{\rm p}$.

A new KS pair is created by exchanging the individual particle components with the two first single-particle arrays and introducing the corresponding c.m. at $N + N_{\rm p}$ after $N_{\rm p}$ has been updated. Conversely, termination of a KS solution requires the former components to be placed in the first available single-particle array (unless already in the correct location) and the c.m. to be eliminated. The case of terminating a hierarchical system is more complicated and will be considered later.

There are many advantages of having a clearly defined and simple data structure. The analogy with molecules is striking and this also extends to interactions since some objects may combine while others are disrupted in response to internal or external effects. On the debit side, all arrays of size $N + N_{\rm p}$ must be in correct sequential order after each creation or destruction of an object. Neighbour lists, to be discussed later, must also be updated consistently. However, the overheads still form a small fraction of the total CPU time. The same procedure applies when distant particles, known as escapers, are removed from the data set. Again, in the latter case, the name identifies the type of object involved.

1.4 *N*-Body Codes

A general N-body code consists of three main parts in the form of initial conditions, integration and run-time data analysis of the results. In the preceding sections we have discussed some relevant aspects dealing with the initial setup and data structure. Before attacking the next stage, it is useful to introduce the various algorithms that are used to advance the solutions. Ideally, different objects require a specially designed integration method in order to exploit the characteristic features. We start by considering single stars, which usually dominate by numbers and concentrate on the challenge of studying large systems. The first speed-up of such calculations can be obtained by assigning individual time-steps according to the local conditions. Since a Taylor series is used to describe the motion, we are concerned with relative convergence where smooth orbits in low-density regions may have longer steps.

From the N^2 nature of the gravitational problem, the calculation of the accelerations requires an increasing fraction of the total effort. Hence the simple approach of direct summation for each integration step is too expensive and restricts the type of problem for investigation. A second efficiency feature called a neighbour scheme (Ahmad & Cohen 1973, hereafter AC) enables consistent solutions to be obtained while still employing direct summation. The basic idea here is to introduce two time-scales for each particle, where contributions from close neighbours are evaluated frequently by direct summation while the more distant forces are included (and recalculated) on a longer time-scale. This two-polynomial scheme speeds up the calculation considerably at the expense of extra programming. Finally, we also mention the modern way to study large N and retain strict summation, namely special-purpose computers known as GRAPE (Makino et al. 1997).

Close encounters present another challenge that must be faced, either in the form of hyperbolic motion or as persistent binaries. Although the time-steps of two interacting bodies can be reduced accordingly, this may lead to significant accumulation of errors. A more elegant way, practised in the Cambridge codes, is to employ two-body regularization as mentioned above. Now the programming requirements are quite formidable. However, the payoff is that such solutions can be used with confidence since the equations of motion are linear for weak perturbations.

The next level of complexity arises when a regularized binary experiences a strong interaction with another object. A reliance on the two-body formulation makes for inefficient treatment during resonant interactions. Compact subsystems may instead be studied by three-body (Aarseth & Zare 1974) or chain regularization (Mikkola & Aarseth 1993). At present the former may be used if the external perturbations are small, while the latter takes account of perturbations and allows for up to six members. Once again the programming effort is substantial, but permits the study of extremely energetic interactions.

One more special procedure remains to be discussed. Although less spectacular, the treatment of long-lived hierarchies requires careful decision-making. A hierarchy is said to be stable if the orbital elements satisfy certain conditions. The main property of a stable system is that the inner semi-major axis should be secularly constant in the presence of an outer bound perturber. Essentially, the outer pericentre needs to exceed the inner semi-major axis by a factor depending on the orbital parameters (Mardling & Aarseth 1999). Once deemed to be stable, the closest perturber is regularized with respect to the inner binary c.m., which is now treated as a point-mass. However, the special configuration is terminated on large external perturbations or if the outer eccentricity increases sufficiently to violate the stability criterion.

The procedures outlined above constitute a veritable tool box for a wide variety of N-body simulations. Efficient use of these tools requires a complex network of decision-making. Moreover, it is desirable that the associated overheads should only represent a small proportion of the total CPU effort. Some of the relevant algorithms will be presented in later sections. Suffice it for now to state that this desirable requirement has been met, as can be ascertained by so-called run-time profiling.

In the following we shall concentrate on the code NBODY6, which combines all of the above features and is suitable for studying realistic star clusters as well as idealized systems on laptops and workstations. However, a section will be devoted to GRAPE procedures. With the above review as background, we now move to the next stage of presenting some of the main integration algorithms. In each case, further details are available elsewhere (Aarseth 2003).

1.5 Hermite Integration

Let us start by looking at the derivation of the Hermite scheme that has proved so successful in modern simulations. We expand Taylor series solution for the coordinates and velocities to fourth order in an interval Δt by

$$\begin{aligned} \boldsymbol{x}_1 &= \boldsymbol{x}_0 + \boldsymbol{v}_0 \Delta t + \frac{\boldsymbol{a}_0}{2}\Delta t^2 + \frac{\dot{\boldsymbol{a}}_0}{6}\Delta t^3 + \frac{\boldsymbol{a}_0^{(2)}}{24}\Delta t^4 + \alpha \frac{\boldsymbol{a}_0^{(3)}}{120}\Delta t^5 \\ \boldsymbol{v}_1 &= \boldsymbol{v}_0 + \boldsymbol{a}_0 \Delta t + \frac{\dot{\boldsymbol{a}}_0}{2}\Delta t^2 + \frac{\boldsymbol{a}_0^{(2)}}{6}\Delta t^3 + \frac{\boldsymbol{a}_0^{(3)}}{24}\Delta t^4 \,. \end{aligned} \tag{1.1}$$

Here $\boldsymbol{a}$ represents the acceleration, or force per unit mass, which will also be referred to as force for convenience, and α is an adjustable constant. The higher-order Newmark implicit method (Newmark 1959) takes the form

$$\begin{aligned} \boldsymbol{x}_1 &= \boldsymbol{x}_0 + \frac{1}{2}(\boldsymbol{v}_0 + \boldsymbol{v}_1)\Delta t - \frac{\alpha}{10}(\boldsymbol{a}_1 - \boldsymbol{a}_0)\Delta t^2 + \frac{6\alpha - 5}{120}(\dot{\boldsymbol{a}}_1 + \dot{\boldsymbol{a}}_0)\Delta t^3 \\ \boldsymbol{v}_1 &= \boldsymbol{v}_0 + \frac{1}{2}(\boldsymbol{a}_1 + \boldsymbol{a}_0)\Delta t - \frac{1}{12}(\dot{\boldsymbol{a}}_1 - \dot{\boldsymbol{a}}_0)\Delta t^2 \,. \end{aligned} \tag{1.2}$$

As can be verified by substitution for $\boldsymbol{v}_1$ into the first equation with $\alpha = 1$, the standard Taylor series is recovered after some simplification,

$$\begin{aligned} \boldsymbol{a}_1 &= \boldsymbol{a}_0 + \dot{\boldsymbol{a}}_0 \Delta t + \frac{1}{2}\boldsymbol{a}_0^{(2)}\Delta t^2 + \frac{1}{6}\boldsymbol{a}_0^{(3)}\Delta t^3 \\ \dot{\boldsymbol{a}}_1 &= \dot{\boldsymbol{a}}_0 + \boldsymbol{a}_0^{(2)}\Delta t + \frac{1}{2}\boldsymbol{a}_0^{(3)}\Delta t^2 \,. \end{aligned} \tag{1.3}$$

The subscripts 0, 1 can be reversed, hence the formulation is time-symmetric and consistent with the Hermite formulation. It has been shown (Kokubo & Makino 2004) that $\alpha = 7/6$ is the optimal choice for the leading term in the error of the longitude of the periapse. Moreover, secular errors in the elements a and e are removed by using constant time-steps (in the absence of encounters) for small eccentricities, $e \leq 0.1$. This makes it an efficient scheme for planetesimal dynamics (see below). It has been found that energy errors are improved by high-order prediction of the particle being advanced.

It is also instructive to present a traditional formulation of standard Hermite integration. We first write a Taylor series for the force per unit mass $\boldsymbol{F}$ and its explicit derivative $\boldsymbol{F}^{(1)}$ for a given particle i (with index suppressed) to be advanced by a time interval t as

$$
\begin{aligned}
\boldsymbol{F} &= \boldsymbol{F}_0 + \boldsymbol{F}_0^{(1)}\, t + \frac{1}{2}\boldsymbol{F}_0^{(2)}\, t^2 + \frac{1}{6}\boldsymbol{F}_0^{(3)}\, t^3 \\
\boldsymbol{F}^{(1)} &= \boldsymbol{F}_0^{(1)} + \boldsymbol{F}_0^{(2)}\, t + \frac{1}{2}\boldsymbol{F}_0^{(3)}\, t^2 \,.
\end{aligned}
\tag{1.4}
$$

After obtaining the initial values $\boldsymbol{F}_0$, $\boldsymbol{F}_0^{(1)}$ by summation, the coordinates and velocities of all particles are predicted to low order by

$$
\begin{aligned}
\boldsymbol{r}_j &= \left[\left(\frac{1}{6}\boldsymbol{F}_0^{(1)}\, \delta t'_j + \frac{1}{2}\boldsymbol{F}_0\right) \delta t'_j \, + \boldsymbol{v}_0\right] \delta t'_j + \boldsymbol{r}_0 \\
\boldsymbol{v}_j &= \left(\frac{1}{2}\boldsymbol{F}_0^{(1)}\, \delta t'_j + \boldsymbol{F}_0\right) \delta t'_j + \boldsymbol{v}_0 \,,
\end{aligned}
\tag{1.5}
$$

with $\delta t'_j = t - t_j$, where t_j is the time of the last force calculation. New values $\boldsymbol{F}$, $\boldsymbol{F}^{(1)}$ are now obtained in the usual way for the particle under consideration. This enables the higher derivatives to be constructed by inversion, which yields

$$
\begin{aligned}
\boldsymbol{F}_0^{(3)} &= [2(\boldsymbol{F}_0 - \boldsymbol{F}) + (\boldsymbol{F}_0^{(1)} + \boldsymbol{F}^{(1)})\, t]\, \frac{6}{t^3} \\
\boldsymbol{F}_0^{(2)} &= [-3(\boldsymbol{F}_0 - \boldsymbol{F}) - (2\boldsymbol{F}_0^{(1)} + \boldsymbol{F}^{(1)})\, t]\, \frac{2}{t^2} \,.
\end{aligned}
\tag{1.6}
$$

Consequently, the fourth-order corrector can be applied to the predicted solution of particle i by adding the contributions

$$
\begin{aligned}
\Delta \boldsymbol{r}_i &= \frac{1}{24}\boldsymbol{F}_0^{(2)} \Delta t^4 + \frac{1}{120}\boldsymbol{F}_0^{(3)}\, \Delta t^5 \\
\Delta \boldsymbol{v}_i &= \frac{1}{6}\boldsymbol{F}_0^{(2)} \Delta t^3 + \frac{1}{24}\boldsymbol{F}_0^{(3)}\, \Delta t^4 \,.
\end{aligned}
\tag{1.7}
$$

Before proceeding, we introduce so-called quantized time-steps according to the rule

$$
\Delta t_n = \left(\frac{s_{\max}}{2}\right)^{n-1} \,,
\tag{1.8}
$$

where $s_{\max}$ defines the maximum permitted value, usually taken as unity with standard scaling. Hence every time-step Δt_i should correspond to some value of n, which entails a slight reduction from a provisional choice. The reason for this novel procedure is to reduce the overheads involved in the predictions of all coordinates and velocities, namely once per step. Moreover, this prediction is made by hardware when using GRAPE. This procedure is referred to as a block-step scheme. Thus it requires truncation of the natural step to the nearest value of n. Moreover, time-steps can only be increased by a factor of 2 every other time to maintain synchronization of all $t_i + \Delta t_i$.

Here we also discuss a heliocentric formulation, which has proved efficient for planetesimal simulations (Kokubo, Yoshinaga & Makino 1998). In heliocentric coordinates, the equation of motion for a mass-point m_i is given by

$$
\ddot{\boldsymbol{r}}_i = - \sum_{j=1;\, j\neq i}^{N} m_j \left[\frac{\boldsymbol{r}_i - \boldsymbol{r}_j}{|\boldsymbol{r}_i - \boldsymbol{r}_j|^3} + \frac{\boldsymbol{r}_j}{r_j^3}\right] - \frac{M_0 + m_i}{r_i^3}\boldsymbol{r}_i \,,
\tag{1.9}
$$

where M_0 is the mass of the central star or dominant body. If the total mass in planetesimals is small (e.g. Saturn's ring), the indirect terms may be neglected.

In concise form, the following algorithm describes the essential steps involved in the integration itself for a group of selected particles.

- Determine members due for updating at new time t
- Predict all $\boldsymbol{r}$, $\dot{\boldsymbol{r}}$ to order $\dot{\boldsymbol{F}}$
- Improve $\boldsymbol{r}_i$, $\dot{\boldsymbol{r}}_i$ to order $\boldsymbol{F}^{(3)}$ for the first member
- Obtain $\boldsymbol{F}, \dot{\boldsymbol{F}}$ due to planetesimals
- Add optional gas drag or tidal damping
- Include the dominant force and first derivative
- Apply the Hermite corrector
- Perform a second iteration by the two last steps
- Specify provisional new time-step Δt_i
- Compare nearest neighbour step: $\Delta t_{\rm nb} = 0.1 R^2/\boldsymbol{R} \cdot \boldsymbol{V}$
- Check for close encounter: $R < R_{\rm cl}\,,\ \dot{R} < 0$
- Complete the cycle for any other $t_j + \Delta t_j = t$
- Include optional boundary crossings

Some comments on this scheme are in order. It is known as being time-symmetric Hermite of type $\texttt{P(EC)}^{\texttt{n}}$ (predict, evaluate, correct, etc.). The number of iterations n is usually chosen as 2, but $n = 3$ may also be worth while. Note that for large N, the expensive evaluation of the perturbations is not performed again because the two-body term dominates the errors. On GRAPE, the procedure for identifying close encounters is implemented by using the nearest-neighbour facility, which enables a suitable maximum time-step to be defined. In the alternative case of a standard calculation, the closest particle can readily be determined from the current neighbour list, which would usually be small.[1] Typically, a close encounter is defined by the distance $R_{\rm cl}$, which signals switching the solution method to regularization (if desired).

1.6 Ahmad–Cohen Neighbour Scheme

Most simulations aim for the largest systems that can be studied with a given resource. As already remarked, this invariably means the use of some kind of neighbour (or hybrid) procedure. In the following we summarize the salient features of the AC scheme since complete descriptions of the Hermite version are already available (Makino & Aarseth 1992; Aarseth 2003).

The basic idea is to split the total force acting on a particle into two parts, formally represented by

$$\boldsymbol{F}(t) = \sum_{j=1}^{n} \boldsymbol{F}_j + \boldsymbol{F}_{\rm d}(t)\,, \tag{1.10}$$

[1] A full-blown AC scheme might not satisfy the strict time-symmetry condition.

where the first term contains the contributions from the n nearest neighbours and $\boldsymbol{F}_{\mathrm{d}}$ represents the distant members as well as any external effects. Likewise, a similar equation can be written for the force derivative. The basic idea is to perform direct summation over the neighbours at suitably chosen small steps and add the *predicted* contributions from the distant particles, with fitting coefficients recalculated on a longer time-scale, Δt_{d}. This leads to a gain in performance provided that $N \gg n$ and $\Delta t_{\mathrm{d}} \gg \Delta t_n$ can be satisfied.

The total force used for the integration is obtained on the time-scale Δt_{d} when the neighbour list is also formed. At intermediate times, or so-called irregular time-steps, the total force and first derivative are evaluated by

$$\begin{aligned} \boldsymbol{F}(t) &= \boldsymbol{F}_n + \dot{\boldsymbol{F}}_{\mathrm{d}}(t - t_0) + \boldsymbol{F}_{\mathrm{d}}(t_0) \\ \dot{\boldsymbol{F}}(t) &= \dot{\boldsymbol{F}}_n + \dot{\boldsymbol{F}}_{\mathrm{d}} , \end{aligned} \tag{1.11}$$

where t_0 is the time of the last regular force calculation. For convenience, the two time-steps are commensurate but this is not a formal requirement, provided the total force is evaluated at the nearest irregular time. The determination of time-steps for each force polynomial will be discussed in the next section.

There are several possible strategies for neighbour selection. Essentially, the choice is between aiming for a constant value of n or adopt a more flexible approach depending on local conditions. Given that particles in the halo have smooth orbits as opposed to those in the core that are affected by strong interactions, it seems appropriate to employ a criterion depending on the density. The neighbour radius itself is updated according to the relation

$$R_{\mathrm{s}}^{\mathrm{new}} = R_{\mathrm{s}}^{\mathrm{old}} \left(\frac{n_{\mathrm{p}}}{n} \right)^{1/3} . \tag{1.12}$$

Here the predicted neighbour number n_{p} is expressed in terms of the density contrast $C \propto n/R_{\mathrm{s}}^3$ as

$$n_{\mathrm{p}} = n_{\max}(0.04C)^{1/2} , \tag{1.13}$$

subject to an upper limit. Again the choice of $n_{\max}$ is a matter of taste, but a value near $2N^{1/2}$ has proved itself for large N. In fact, there are compensating factors affecting code performance such that smaller n requires more frequent updating of the neighbours. The neighbour selection is made during the total force calculation using $|\boldsymbol{r}_i - \boldsymbol{r}_j| < R_{\mathrm{s}}$ and is essentially free since all distances are calculated in any case.

The combination of two-force polynomials requires some care when there is a change in the neighbour population. In general, there is a flux across the neighbour sphere, which must be accounted for in the higher derivatives. To do this we evaluate the explicit derivatives $\boldsymbol{F}_{ij}^{(2)}$, $\boldsymbol{F}_{ij}^{(3)}$ from the corresponding members j and add or subtract the corrections to the higher derivatives that are kept separately. However, this extra cost may be avoided by performing the energy check and result analysis at times commensurate with $s_{\max}$ since all the solutions are then known to highest order. This is possible because only predictions up to $\boldsymbol{F}_i^{(1)}$ are used in the general integration.

As regards performance, the neighbour scheme is comparable to a single-force polynomial code for $N \simeq 50$ and speeds up as $N^{1/4}$. Moreover, a comparison with the GRAPE-6A (so-called micro-Grape) with the same host shows the latter being faster by a factor of 11 for $N = 25\,000$. Finally, we emphasize that neighbour lists are also very useful for identifying other close members in connection with regularization and for estimating the density contrast.

1.7 Time-Step Criteria

Any integration method based on individual time-steps tries to employ an appropriate criterion, which optimizes the overall solution accuracy. At the simplest level are expressions of the type

$$\Delta t = \frac{\alpha |\boldsymbol{r}|}{|\boldsymbol{v}|}, \quad \Delta t = \frac{\beta |\boldsymbol{F}|}{|\boldsymbol{F}^{(1)}|}, \tag{1.14}$$

where α and β are suitable dimensionless constants. However, such simple forms invariably cause numerical problems, mainly because close encounters are not detected in time for step reduction. Since we are dealing with a Taylor series for the force, it is natural to look for a relative criterion involving higher derivatives. The most convenient simple time-step can be constructed from

$$\Delta t = \left(\frac{\eta |\boldsymbol{F}|}{|\boldsymbol{F}^{(2)}|} \right)^{1/2}, \tag{1.15}$$

where $\eta \simeq 0.02$ would give reasonable behaviour. For many years this relation was used with success.

The idea of relative convergence can be extended to take into account all the force derivatives. Consequently, we write a general expression in the form

$$\Delta t = \left(\frac{\eta (|\boldsymbol{F}||\boldsymbol{F}^{(2)}| + |\boldsymbol{F}^{(1)}|^2)}{|\boldsymbol{F}^{(1)}||\boldsymbol{F}^{(3)}| + |\boldsymbol{F}^{(2)}|^2} \right)^{1/2}. \tag{1.16}$$

This criterion has several useful properties. Compared to (1.15) it gives a well-defined large value when the force is small, as is the case near a tidal boundary. Moreover, two bodies with different masses will tend to have similar time-steps during close encounters, which facilitates decision-making. In fact, after the truncation according to (1.8) the two steps are often identical, but this cannot be assumed. It is worth emphasizing that a relative time-step criterion of the above type is independent of the (non-zero) mass.

From past experience, it seems most efficient to assign slightly different values for the dimensionless accuracy factors. Hence in most practical work, regardless of N, the respective values $\eta_I = 0.02$, $\eta_R = 0.03$ for the irregular and regular time-steps have been adopted. For $N \simeq 1000$, typical time-step ratios of about 6 are seen; this increases slowly as N is increased.

In the case of planetesimal simulations, special care is needed to ensure detection of close encounters and physical collisions. We therefore employ an additional criterion based on the nearest neighbour,

$$\Delta t = \frac{\beta R^2}{|\boldsymbol{R} \cdot \boldsymbol{V}|}, \tag{1.17}$$

where $\beta = 0.1$ has proved sufficient. The different strategies for GRAPE and conventional computers in this problem were commented on in a previous section.

For completeness, we also include KS regularization in this discussion since it has relevance for the general time-step criterion. Briefly, for the unperturbed case the equation governing the relative motion is given by

$$\boldsymbol{F}_{\rm u} = \frac{1}{2} h \boldsymbol{u}, \tag{1.18}$$

where h is the specific two-body energy and $\boldsymbol{u}$ the generalized coordinates, which have the useful property $\boldsymbol{u} \cdot \boldsymbol{u} = R$. Since $h < 0$ for a binary, we define the constant time-step in terms of the frequency as

$$\Delta \tau = \frac{\eta_u}{(2|h|)^{1/2}}, \tag{1.19}$$

with $\eta_u = 0.2$ for accurate solution (Mikkola & Aarseth 1998). Substitution into (1.16) by carrying out explicit differentiation (with $h' = 0$) simplifies to the adopted form, thereby giving some support for this apparently complicated expression. Note that the basic time-step (1.19) is reduced appropriately in the presence of significant perturbations.

1.8 Two-Body Regularization

Regularization plays an important part in the codes under discussion. In the following we outline some of the main aspects of the KS method and describe various relevant algorithms. The latter can be divided into a purely local part involved with studying the relative motion and a global part that forms an interface with the whole system. Let us begin with a summary of the well-known classical formulation (Kustaanheimo & Stiefel 1965) for the 3D treatment, which is described in more detail elsewhere (Aarseth 2003).

New coordinates in 4D are introduced by the condition

$$R = u_1^2 + u_2^2 + u_3^2 + u_4^2. \tag{1.20}$$

As usual in regularization, a time transformation is also needed and we choose the simplest differential relation,

$$\mathrm{d}t = R\,\mathrm{d}\tau\,, \tag{1.21}$$

or $t' = R$. It turns out that the coordinate transformation

$$\boldsymbol{R} = \mathcal{L}(\boldsymbol{u})\,\boldsymbol{u} \tag{1.22}$$

is satisfied by the Levi-Civita matrix

$$\mathcal{L}(\boldsymbol{u}) = \begin{bmatrix} u_1 & -u_2 & -u_3 & u_4 \\ u_2 & u_1 & -u_4 & -u_3 \\ u_3 & u_4 & u_1 & u_2 \end{bmatrix} \tag{1.23}$$

as can be verified by substitution into the equation for $\boldsymbol{R}$. For completeness, we also include the appropriate relations for the relative velocity. Thus the regularized velocities are obtained by

$$\boldsymbol{u}' = \frac{1}{2}\mathcal{L}^T(\boldsymbol{u})\dot{\boldsymbol{R}}\,, \tag{1.24}$$

while the physical values are recovered from

$$\dot{\boldsymbol{R}} = 2\mathcal{L}(\boldsymbol{u})\boldsymbol{u}'/R\,. \tag{1.25}$$

Starting from the perturbed two-body problem for m_k and m_l,

$$\ddot{\boldsymbol{R}} = -\frac{m_k + m_l}{R^3}\boldsymbol{R} + \boldsymbol{P}\,, \tag{1.26}$$

with $\boldsymbol{P}$ the tidal perturbation, the equations of relative motion can be derived. The complete set is given by

$$\begin{aligned} \boldsymbol{u}'' &= \frac{1}{2}\,h\,\boldsymbol{u} + \frac{1}{2}\,R\,\mathcal{L}^T\,\boldsymbol{P} \\ h' &= 2\,\boldsymbol{u}'\cdot\mathcal{L}^T\,\boldsymbol{P} \\ t' &= \boldsymbol{u}\cdot\boldsymbol{u}\,, \end{aligned} \tag{1.27}$$

where $\mathcal{L}^T$ represents the transpose matrix.

The 10 equations describing the relative motion in the presence of external perturbations are regular in the sense that the solutions are well defined for $R \to 0$. In order to describe the actual orbit in a stellar system, we introduce the associated c.m. by

$$\boldsymbol{r}_{\rm cm} = \frac{m_k\,\boldsymbol{r}_k + m_l\boldsymbol{r}_l}{m_k + m_l}\,. \tag{1.28}$$

Likewise, the c.m. force is obtained from

$$\ddot{\boldsymbol{r}}_{\rm cm} = \frac{m_k\,\boldsymbol{P}_k + m_l\,\boldsymbol{P}_l}{m_k + m_l}\,. \tag{1.29}$$

Hence the c.m. is added to the system of N particles as a fictitious member, to be advanced in time. Individual coordinates are obtained by combining the two motions, which yields

$$\begin{aligned} \boldsymbol{r}_k &= \boldsymbol{r}_{cm} + \mu \boldsymbol{R}/m_k \\ \boldsymbol{r}_l &= \boldsymbol{r}_{cm} - \mu \boldsymbol{R}/m_l \,, \end{aligned} \tag{1.30}$$

where $\mu = m_k m_l/(m_k + m_l)$ is the reduced mass, and similarly for the global velocities.

Given the regularized time-step defined above, the equations for the relative motion are advanced by an efficient Hermite method (Mikkola & Aarseth 1998). Although this formulation is fairly complicated, the KS equations can also be written in standard Hermite form by including the terms $\boldsymbol{F}_u^{'}$ and h''.

Implementation of two-body regularization has many practical benefits. First, the equations of motion take the form of a perturbed harmonic oscillator and are therefore regular. This treatment permits a constant time-step for small perturbations while for direct integration, $\Delta t \propto R^{3/2}$ which can be troublesome when treating very eccentric binaries. Moreover, with linearized equations the accuracy per step is higher and only about 30 steps are needed for an orbit. Integration of relative motion also permits a faster force calculation because $P \propto 1/R^3$ for tidal perturbation. Finally, on the credit side, unperturbed two-body motion is justified in case there are no perturbers within a distance $d = \lambda a(1 + e)$, with $\lambda \simeq 100$. Likewise, if $d > \lambda R$, the c.m. approximation can be used in force calculations with binaries.

The price to pay for all the advantages comes in the form of coordinate and velocity transformations at the interface between relative and global motion. However, these operations are fast and do not involve the square root. As for simulations using GRAPE, there is a further cost due to differential force corrections since the hardware is based on point-mass interactions.

Several optional features are worth mentioning. For small perturbations, the principle of adiabatic invariance can be used to slow down the motion by scaling the perturbation (Mikkola & Aarseth 1996). So-called energy rectification improves the solutions of $\boldsymbol{u}, \boldsymbol{u}'$ by scaling to the explicit value of h, which is integrated independently. The availability of completely regular two-body elements like the semi-major axis (a) and eccentricity (e) can also be beneficial when employing averaged expressions to model secular evolution of stable triples or tidal circularization (Mardling & Aarseth 2001).

1.9 KS Decision-Making

A variety of algorithms are involved in the overall management of the regularization scheme. Broadly speaking, we may distinguish between aspects of initialization, integration and termination and these will be covered in turn.

The first question which presents itself is when to choose two particles for regularization treatment. A close encounter is traditionally defined by the two main parameters

$$R_{\rm cl} = \frac{4\,r_{\rm h}}{N\,C^{1/3}}, \quad \Delta t_{\rm cl} = \beta\left(\frac{R_{\rm cl}^3}{\bar{m}}\right)^{1/2}, \tag{1.31}$$

where $r_{\rm h}$ is the half-mass radius, C is the central density contrast and β a dimensionless constant determined by experimentation. Thus a particle with time-step $\Delta t_k < \Delta t_{\rm cl}$ needs to have a close neighbour inside the distance $R_{\rm cl}$. Further conditions of negative radial velocity and dominant two-body motion must also be satisfied. The latter is ensured by comparing the two-body terms due to any other members identified in the close encounter search. In the case of GRAPE, a list of particles with small time-steps is maintained and updated during the force calculation when the host computer is idle.

The principle of initializing KS polynomials is the same as for single particles, except that time derivatives must also be obtained. By employing explicit differentiation, the latter terms are readily constructed from the available data involving $\boldsymbol{u}$ and its derivatives. A conversion by Taylor series expansion for $\Delta\tau$ finally gives the time-step in physical units, which is used for the scheduling of regularized solutions. Thus any KS pair which needs to be advanced during the next block-step is treated first.

Initially and during the integration, a consistent perturber list must also be available. The perturber search is carried out after each apocentre passage, $R_{\rm ap} = a(1+e)$, using the tidal limit approximation. Particles inside a distance

$$r_{\rm p} = \left(\frac{2m_{\rm p}}{m_{\rm b}\gamma_{\rm min}}\right)^{1/3} a\,(1+e) \tag{1.32}$$

are selected from the neighbour list, where $m_{\rm b}$ is the mass of the binary and $\gamma_{\rm min}$ is a small dimensionless perturbation, usually taken as 10^{-6}. An extra procedure is included to increase the neighbour list for c.m. particles if $R_{\rm s} < \lambda a(1+e)$.

A useful quantity for many purposes is the dimensionless relative perturbation, defined by

$$\gamma = \frac{|\boldsymbol{P}_k - \boldsymbol{P}_l| R^2}{m_k + m_l}\,. \tag{1.33}$$

If evaluated in the apocentre region, this dimensionless quantity is a measure of dominant two-body motion. In general, it is advantageous to initiate regularization if $\gamma \simeq 0.1$, but larger values are acceptable during the treatment.

The KS integration itself begins with the prediction of $\boldsymbol{u}$ and $\boldsymbol{u}'$ to highest order, $\boldsymbol{u}^{(5)}$, while h is predicted to order $h^{(2)}$. As usual in the Hermite scheme, perturbers are predicted to low order. Transformations yield global coordinates and velocities, $\boldsymbol{r}_k, \boldsymbol{r}_l, \dot{\boldsymbol{r}}_k, \dot{\boldsymbol{r}}_l$, which are needed for the force calculation. The physical perturbation $\boldsymbol{P} = \boldsymbol{P}_k - \boldsymbol{P}_l$ and $\dot{\boldsymbol{P}}$ can now be obtained.

By virtue of the time transformation we have $\boldsymbol{P}' = R\dot{\boldsymbol{P}}$. This enables the corrector to be applied, with new values $\boldsymbol{u}, \boldsymbol{u}'$ to order $\boldsymbol{u}^{(5)}$ and h to $h^{(4)}$. An iteration without recalculation of the perturbations improves the final solution.

The conversion to physical time must also be carried out to highest order. Taylor series expansion yields the desired terms by successive explicit differentiation, beginning with $t'' = 2\boldsymbol{u} \cdot \boldsymbol{u}'$ and continued up to $t^{(6)}$ using known terms. This permits the corresponding physical time-step to be obtained by

$$\Delta t = \sum_{k=1}^{6} \frac{1}{k!} t_0^{(k)} \Delta\tau^k . \tag{1.34}$$

Time inversion is required when calculating the force on single particles. Given a physical interval δt, this is achieved by expanding $\dot{\tau} = 1/R$ to sufficient order. Note that division by R is not dangerous here since the c.m. approximation is used for small values.

Conditions for unperturbed motion have been alluded to above. By careful analysis of the velocity distribution of nearby particles, it is possible to extend the analytical solution to many Kepler periods. This is achieved by identifying the particles that provide the maximum force as well the smallest time of minimum approach. If there are no perturbers, we estimate the minimum time to reach the boundary $\gamma \simeq \gamma_{\rm min}$ as well as the free fall time of the nearest particle. Depending on the remaining time, a number of unperturbed orbits may be adopted and the KS motion will remain dormant until the next time for checking. Several extra conditions are also included in order to avoid premature interactions inside the unperturbed boundary.

Following the general exposition, we now comment on the final stage of the KS cycle. Termination of hard binaries is appropriate for strong perturbation, say $\gamma \geq 0.5$, which would most likely result in switching to another dominant pair (temporary capture or so-called resonance) or chain regularization. For softer binaries, a smaller perturbation limit is called for. After termination, standard force polynomials are initialized for the two single particles.

As a technical point, except for collisions, termination is delayed until the end of the block-step; i.e. until the remaining interval $\delta t = T_{\rm block} - t$ falls below the physical step Δt converted from $\Delta\tau$. A final iteration to the exact value can then readily be performed, with $\Delta\tau$ obtained from $\dot{\tau}$, $\ddot{\tau}$ and δt.

1.10 Hierarchical Systems

Long-lived triples or even quadruples form an important constituent in N-body simulations. Typically, a triple is formed through a strong interaction between two hard binaries, where the weakest binary is disrupted and one component is ejected. The other component may then be captured into an orbit around the inner binary because of energy and angular momentum conservation. Such systems may have long life-times and their treatment by direct integration poses very severe numerical problems (or even code crash) by loss of accuracy as well as greater effort.

Over the years there has been a quest for stability criteria, which would allow the description of hierarchies to be simplified by assuming the inner semi-major axis to be constant, permiting the c.m. approximation to be used. In the absence of secular changes, the outer component (a single particle or another binary) may then be regularized with respect to the inner binary c.m., thereby speeding up the calculation by a large factor. For this purpose we have employed a stability criterion that has been tested successfully for a limited range of parameters (Mardling & Aarseth 1999, 2001). A sharper stability criterion has been developed recently for the general three-body problem, based on first principles. The underlying theory is discussed in Chap. 3, together with a practical algorithm that has been implemented in NBODY4/6. Given all the elements describing the inner and outer orbit, this algorithm defines stability or otherwise for a hierarchical configuration, instead of estimating the distance from the stability boundary. Consequently, the stability test needs to be re-assessed during the subsequent evolution.

The identification of a hierarchical candidate system involves checking many conditions. In the first instance, a search is initiated after each apocentre turning point, provided the c.m. step is sufficiently small; in other words if $\Delta t_{\rm cm} < \Delta t_{\rm cl}$. This condition implies that the new hierarchy is likely to form a hard outer binary. However, it should be stated that the same test is also performed for a new chain regularization, which again involves strong interactions. After identifying the two most dominant neighbours, the outer two-body elements are constructed for the main perturber. Among further conditions to be checked are the perturbation on the outer orbit as well as the requirement of a new hard binary. Moreover, extra tests are performed if the outer component is another binary, in which case a modified criterion is used depending on the ratio of semi-major axes.

Acceptance of the stability condition entails a considerable programming effort in order to maintain a consistent data structure, as discussed in an earlier section. The relevant algorithmic steps are set out in the following table and are mostly self-explanatory.

- Increase the control index for decision-making
- Save relevant masses m_k, m_l in a hierarchy table
- Copy c.m. neighbour list for later corrections
- Terminate KS solution and update $N_{\rm p}$ and arrays
- Evaluate potential energy of components and old neighbours
- Record $\boldsymbol{R} = r_k - \boldsymbol{r}_l$, $\boldsymbol{V} = v_k - \boldsymbol{v}_l$ and h in the special table
- Form binary c.m. in location of the primary, $j = 2N_{\rm p} + 1$
- Define ghost ($m = 0,\ x = 10^6$) and initialize prediction variables
- Obtain potential energy of inner c.m. body and neighbours
- Remove ghost from neighbour and perturber lists
- Initialize new KS for outer component in $l = k + 1$

- Specify c.m. and ghost names: $\mathcal{N}_{\rm cm} = -\mathcal{N}_k$, $\mathcal{N}_{\rm ghost} = \mathcal{N}_l$
- Set pericentre stability limit in $R_0(N_{\rm p})$ for termination test
- Update the internal and differential energy: $\Delta E = \mu h_0 + \Delta\Phi$

Integration of hierarchical systems proceeds in the usual way, except that the stability condition needs to be checked. This is done at each apocentre turning point, using the property $\mathcal{N}_{\rm cm} < 0$ for identification. One way in which the stability test may no longer apply is when the outer eccentricity increases due to perturbations, otherwise similar termination criteria are used as for hard binaries. For completeness, we also give the algorithm dealing with the main points of termination.

- Locate current position in the hierarchy table: $\mathcal{N}_i = \mathcal{N}_{\rm cm}$
- Save c.m. neighbours for correction procedure
- Terminate the outer KS solution (k, l) and update $N_{\rm p}$
- Evaluate potential energy of c.m. wrt neighbours & l
- Determine location of ghost: $\mathcal{N}_j = \mathcal{N}_{\rm ghost}$, $j = 1, \ldots, N + N_{\rm p}$
- Restore inner binary components from saved quantities
- Add l to neighbour lists containing first component k
- Initialize force polynomials for outer component
- Copy basic KS variables h, $\boldsymbol{u}$, u' from the table
- Re-activate inner binary as new KS solution
- Obtain potential energy of inner components and perturbers
- Update internal energy for conservation: $\Delta E = \Delta\Phi - \mu h$
- Reduce control index and compress tables (including escapers)

1.11 Three-Body Regularization

More than 30 years ago a break-through in regularization theory made it possible to study the strong interactions of three particles (Aarseth & Zare 1974). The basic idea is simple, namely to employ two different KS solutions of m_1 and m_2 separately with respect to the so-called reference body m_3. It is also instructive to review this development because of its connection with the subsequent chain regularization mentioned above.

In the following we summarize the key points of the formulation. The initial conditions are first expressed in the local c.m. frame, with coordinates $\boldsymbol{r}_i$ and momenta $\boldsymbol{p}_i$. Given the three respective distances R_1, R_2, R, with R the distance between m_1 and m_2, and $\boldsymbol{p}_3 = -(\boldsymbol{p}_1 + \boldsymbol{p}_2)$ as the momentum of m_3, the basic Hamiltonian can be written as

$$H = \sum_{k=1}^{2} \frac{1}{2\mu_{k3}} \boldsymbol{p}_k^2 + \frac{1}{m_3} \boldsymbol{p}_1^T \cdot \boldsymbol{p}_2 - \frac{m_1 m_3}{R_1} - \frac{m_2 m_3}{R_2} - \frac{m_1 m_2}{R} \,, \qquad (1.35)$$

with $\mu_{k3} = m_k m_3/(m_k + m_3)$. As can be seen, the kinetic energy is expressed by the momenta of m_1 and m_2 together with a cross product, which represents the mutual interaction of m_1 and m_2. Likewise, the potential energy is a sum of the three relevant terms. Thus omitting any references to m_2 reduces to the familiar form of the two-body problem.

In analogy with standard KS we introduce a coordinate transformation for the distances R_1 and R_2 by

$$\boldsymbol{Q}_k^2 = R_k\,. \quad (k = 1, 2). \tag{1.36}$$

Several alternative time transformations are available. Here we adopt the original choice, which is the most intuitive but not necessarily the best, giving the differential relation between physical and regularized time

$$\mathrm{d}t = R_1 R_2 \,\mathrm{d}\tau\,. \tag{1.37}$$

This enables a regularized Hamiltonian to be formed as $\Gamma^* = R_1 R_2\,(H - E_0)$, where E_0 is the initial energy. By construct Γ^* should be zero along the solution path. Making use of the KS property $\boldsymbol{p}_k^2 = \boldsymbol{P}_k^2/4R_k$, where $\boldsymbol{P}_k$ now is the regularized momentum, the new Hamiltonian becomes

$$\begin{aligned}\Gamma^* &= \sum_{k=1}^{2} \frac{1}{8\mu_{k3}} R_l\,\boldsymbol{P}_k^2 + \frac{1}{16m_3} \boldsymbol{P}_1^T \boldsymbol{A}_1 \cdot \boldsymbol{A}_2^T \boldsymbol{P}_2 \\ &\quad - m_1 m_3 R_2 - m_2 m_3 R_1 - \frac{m_1 m_2 R_1 R_2}{|\boldsymbol{R}_1 - \boldsymbol{R}_2|} - E_0 R_1 R_2\,,\end{aligned} \tag{1.38}$$

where $l = 3 - k$. For historical reasons, $\boldsymbol{A}_i$ is taken as twice the transpose Levi-Civita matrix of (1.23). Finally, the equations of motion are given by

$$\frac{\mathrm{d}\boldsymbol{Q}_k}{\mathrm{d}\tau} = \frac{\partial \Gamma^*}{\partial \boldsymbol{P}_k}\,, \qquad \frac{\mathrm{d}\boldsymbol{P}_k}{\mathrm{d}\tau} = -\frac{\partial \Gamma^*}{\partial \boldsymbol{Q}_k}\,. \tag{1.39}$$

It can be seen from inspection of the Hamiltonian that the solutions are regular for $R_1 \to 0$ or $R_2 \to 0$. Moreover, the singular terms are numerically smaller than the regular terms, provided $|\boldsymbol{R}_1 - \boldsymbol{R}_2| > \max\,(R_1,\, R_2)$. Hence a switch to another reference body can be made when R is no longer the largest (or second largest) distance, which usually ensures a regular behaviour. Full details of the transformations can be found in the original publication.

So far three-body regularization has only been used in unperturbed form within the N-body codes when chain regularization is not available, which is quite rare. However, it can be quite efficient as a stand-alone code for scattering experiments. In particular, the simplicity of decision-making as well as the ability to achieve accurate results by a high-order integrator makes it a good choice for such problems (Aarseth & Heggie 1976).

1.12 Wheel-Spoke Regularization

The recent interest in massive objects in the form of black holes has inspired a closer look at alternative regularization methods. The so-called wheel-spoke formulation is a direct generalization of three-body regularization to include more members (Zare 1974). Such a configuration may be appropriate if the reference body dominates the mass, in which case the need for switching is no longer an issue and leads to further simplification. The scheme is outlined here in the expectation that it will prove a popular tool since its effectiveness has been demonstrated recently (Aarseth 2007).

Let us consider a subsystem of n single particles of mass m_i and a dominant body of mass m_0 where the initial conditions $\tilde{\boldsymbol{q}}_i$, $\tilde{\boldsymbol{p}}_i$ are expressed in the local c.m. frame. Introducing relative coordinates $\boldsymbol{q}_i$ with respect to m_0, we write the Hamiltonian as

$$H = \sum_{i=1}^{n} \frac{\boldsymbol{p}_i^2}{2\mu_i} + \frac{1}{m_0} \sum_{i<j}^{n} \boldsymbol{p}_i^{\mathrm{T}} \cdot \boldsymbol{p}_j - m_0 \sum_{i=1}^{n} \frac{m_i}{R_i} - \sum_{i<j}^{n} \frac{m_i m_j}{R_{ij}} , \tag{1.40}$$

where $\mu_i = m_i m_0/(m_i + m_0)$ and $R_i = |\boldsymbol{q}_i|$. As can be seen, this is a direct generalization of (1.35) to $n > 2$, where m_0 plays the role of reference body. This implies that the technical treatment will also be similar. However, the original time transformation is now replaced by the inverse Lagrangian energy as $t' = 1/L$ since a multiple product would be cumbersome and might not work for critical cases. This choice has many advantages and would also be suitable for three-body regularization.

The use of a fixed reference body, albeit with dominant mass, raises a technical problem of dealing with close encounters between two light bodies. Thus for small separations, the last term of (1.40) may become arbitrarily large if $R_{ij} \to 0$. At present this difficulty is overcome by introducing a small softening in these terms while still retaining the conservative nature of the Hamiltonian. It turns out that the powerful integrator (Bulirsch & Stoer 1966) is able to handle quite small values of non-regularized distances so that the essential dynamics is preserved.

The regularized coordinates and momenta $\boldsymbol{Q}_i$, $\boldsymbol{P}_i$ are obtained in the usual way. Conversely, the physical values are recovered from the inverse transformations by

$$\boldsymbol{q}_i = \frac{1}{2}\boldsymbol{A}_i^{\mathrm{T}}\boldsymbol{Q}_i, \quad \boldsymbol{p}_i = \frac{1}{4}\boldsymbol{A}_i^{\mathrm{T}}\boldsymbol{P}_i/R_i . \tag{1.41}$$

For completeness, we also give the full set of transformations to the final values in the local c.m. system, corrected for a sign error,

$$\begin{aligned} \tilde{\boldsymbol{q}}_i &= \tilde{\boldsymbol{q}}_0 + \boldsymbol{q}_i , \quad \tilde{\boldsymbol{q}}_0 = -\sum_{i=1}^{n} m_i \boldsymbol{q}_i \Big/ \sum_{i=0}^{n} m_i \\ \tilde{\boldsymbol{p}}_i &= \boldsymbol{p}_i , \quad (i = 1, \ldots, n) \quad \tilde{\boldsymbol{p}}_0 = -\sum_{i=1}^{n} \boldsymbol{p}_i . \end{aligned} \tag{1.42}$$

The method presented here may also be used for more conventional calculations involving comparable masses without the restriction of a fixed reference body or softening. This would be a simpler alternative to chain regularization, but would at most be effective for four or five members.

1.13 Post-Newtonian Treatment

The wheel-spoke formulation is particularly suited to studying a compact subsystem containing a massive object inside a star cluster. Especially attractive is the possibility of including relativistic terms in the most dominant two-body motion. The corresponding post-Newtonian equation of motion can be written in the convenient form (Blanchet & Iyer 2003; Mora & Will 2004)

$$\frac{\mathrm{d}^2\boldsymbol{r}}{\mathrm{d}t^2} = \frac{m_i + m_0}{r^2}\left[(-1+A)\frac{\boldsymbol{r}}{r} + B\boldsymbol{v}\right], \tag{1.43}$$

where the dimensionless quantities A and B represent relativistic effects. Here the two-body term is contained in the regularized Hamiltonian with the remaining contributions added as a perturbation.

The coefficients A, B can be expanded as functions of v/c, with c the speed of light. Using the current notation, this gives rise to the perturbing *force*

$$\boldsymbol{P}_{GR} = \frac{m_i m_0}{c^2 r^2}\left[\left(A_1 + \frac{A_2}{c^2} + \frac{A_{5/2}}{c^3}\right)\frac{\boldsymbol{r}}{r} + \left(B_1 + \frac{B_2}{c^2} + \frac{B_{5/2}}{c^3}\right)\boldsymbol{v}\right]. \tag{1.44}$$

Here the first-order precession is described by

$$A_1 = 2(2+\eta)\frac{m_i + m_0}{r} - (1+3\eta)v^2 + \frac{3}{2}\eta\dot{r}^2, \quad B_1 = 2(2-\eta)\dot{r}, \tag{1.45}$$

with $\eta = m_i m_0/(m_i + m_0)^2$. Next comes the second-order precession terms A_2, B_2, which are somewhat more complicated. Of most interest is the energy loss by gravitational radiation, represented by $A_{5/2}$, $B_{5/2}$.

For energy conservation purposes, an extra equation for the relativistic contribution is integrated according to

$$\Delta E_{GR} = \int \boldsymbol{P}_{GR} \cdot \boldsymbol{v}\,\mathrm{d}t. \tag{1.46}$$

In order to carry out the treatment in regularized time, the right-hand side is converted into an expression analogous to h' in (1.27). Also note that derivative evaluations of the physical perturbation are not required for solution of first-order equations. The associated time-scale for shrinkage employed in the decision-making is given by (Peters 1964)

$$\tau_{GR} = \frac{5a^4c^5}{64 m_i m_0^2}\frac{(1-e^2)^{7/2}}{g(e)}, \tag{1.47}$$

where $g(e)$ is a known function and standard N-body units apply.

Implementation of the wheel-spoke scheme into a large N-body code presents many interesting aspects. To begin with, a suitably compact subsystem is chosen from a binary containing the heavy body if there is at least one close perturber inside R_{cl}. The subsystem is initialized in the usual way, including transformations to KS-type variables $\boldsymbol{Q}$, $\boldsymbol{P}$. The perturber list is again constructed according to (1.32), which now yields a smaller mass factor and hence requires less effort in coordinate prediction.

Although the innermost binary is invariably long-lived, the question of membership changes must be considered. Decisions of addition or removal are based on the central distance and radial velocity of perturbers or existing members, respectively. Simple criteria including a combination of an appropriate perturbation (say $\gamma > 0.05$) and distance ($r_p < \sum R_k$) are used in the former case while removal is controlled by $\dot{R}^2 > 2m_0/R$ and $R_k > R_{\rm cl}$. In analogy with the integration of KS binaries, the c.m. force is obtained by vectorial summation over the components.

The addition of post-Newtonian terms necessitates the introduction of physical units. This is achieved by specifying the total mass and half-mass radius as well as the speed of light. From $NM_{\rm S}$ and $r_{\rm h}$, we have $c = 3\times10^5/V^*$, with the velocity scaling factor V^* expressed in $\rm km\,s^{-1}$. This enables the coalescence distance to be defined as three Schwarzschild radii by

$$r_{\rm coal} = \frac{6(m_i + m_0)}{c^2} \,. \tag{1.48}$$

Alternatively, a disruption distance may be defined for white dwarfs. An experimental scheme has been adopted where the different GR terms are activated progressively, depending on the value of the time-scale (1.47). Thus the radiation term is included first on the supposition that precession does not play an important role during the early stages. However, due care must be exercised if the innermost binary is subject to Kozai cycles (Kozai 1962).

Simulations of centrally concentrated cluster models have been made with a GRAPE code for $m_0 = N^{1/2}M_{\rm S}$ and $N = 10^5$ equal-mass stars. Here the innermost binary shrank by a significant factor and also developed very high eccentricity by the Kozai resonance. In some cases, the resulting pericentre distance was sufficiently small for stars with white dwarf radii to be affected by further gravitational radiation shrinkage before disruption (Aarseth 2007).

1.14 Chain Regularization

This contribution would not be complete without a discussion of chain regularization, which has proved to be a powerful tool in star cluster simulations. In the following we shall review some of the essential features as well as the main algorithms since the relevant details can be found elsewhere (Mikkola & Aarseth 1993; Aarseth 2003).

The basic idea takes its cue from three-body regularization. A system is suitable for special treatment if one hard binary has a close perturber in the form of a single particle or another binary. Upon termination of the KS binary, the coordinates and momenta are expressed in the local c.m. frame. Thus $N-1$

chain vectors connect the particles experiencing the strongest pair-wise forces and are defined in terms of the coordinates $\boldsymbol{q}_k$ by

$$\boldsymbol{R}_k = \boldsymbol{q}_{k+1} - \boldsymbol{q}_k \,, \quad k = 1, \ldots, N-1 \,. \tag{1.49}$$

In Hamiltonian theory, the generating function,

$$S = \sum_{k=1}^{N-1} \boldsymbol{W}_k \cdot (\boldsymbol{q}_{k+1} - \boldsymbol{q}_k), \tag{1.50}$$

connects the old momenta with the new ones by $\boldsymbol{p}_k = \partial S / \partial \boldsymbol{q}$. The relative physical momenta $\boldsymbol{W}_k$ can then be obtained by the recursion

$$\boldsymbol{W}_k = \boldsymbol{W}_{k-1} - \boldsymbol{p}_k \,, \quad k = 2, \ldots, N-2 \,, \tag{1.51}$$

with $\boldsymbol{W}_1 = -\boldsymbol{p}_1$ and $\boldsymbol{W}_{N-1} = -\boldsymbol{p}_N$ due to the c.m. condition. Substitution into a Hamiltonian of the type (1.40) yields

$$H = \frac{1}{2} \sum_{k=1}^{N-1} \left(\frac{1}{m_k} + \frac{1}{m_{k+1}} \right) \boldsymbol{W}_k^2 - \sum_{k=2}^{N-1} \frac{1}{m_k} \boldsymbol{W}_{k-1} \cdot \boldsymbol{W}_k$$
$$- \sum_{k=1}^{N-1} \frac{m_k m_{k+1}}{R_k} - \sum_{1 \le i \le j-2}^{N} \frac{m_i m_j}{R_{ij}}, \tag{1.52}$$

where the first momentum term contains the reduced mass. In spite of the similarity with (1.40), the formalism differs in some important respects, mainly because there is no reference body.

As stated earlier, the inverse Lagrangian energy is a good choice for the time transformation. Multiplication by $t' = 1/L$ gives the regularized Hamiltonian $\Gamma^* = t'(H - E_0)$, which can be differentiated in the usual way to yield the equations of motion. Note that for technical reasons, the differentiation of the product $t'H$ is done explicitly. This procedure enables the term $H - E_0$ (which should be zero) to be retained for stabilizing the solutions. It can be seen that the two-body solutions are regular for any individual $R_k \to 0$ at separate times. As usual, the KS relations can be used to recover the physical variables via the standard transformations

$$\boldsymbol{R}_k = \mathcal{L}_k \boldsymbol{Q}_k \,, \quad \boldsymbol{W}_k = \mathcal{L}_k \boldsymbol{P}_k / 2 Q_k^2, \tag{1.53}$$

from which the momenta $\boldsymbol{p}_k$ are readily derived.

The implementation of chain regularization into an N-body code contains many algorithms, some of which will be described briefly. Following initialization in the c.m. frame and evaluation of the total energy E_0, the chain vectors must be constructed. The selection of the corresponding chain indices presents a considerable algorithmic challenge if (as may occur later) there are more than four members (cf. Mikkola & Aarseth 1993). Thus the scheme may not work efficiently if the chain vectors fail to connect the dominant two-body forces. The canonical variables $\boldsymbol{Q}$, $\boldsymbol{P}$ are introduced as before and the integration can begin after specifying a suitably small time-step.

Several quantities are useful for the decision-making. Among these are the characteristic external perturbation $\gamma_{\rm ch}$ and gravitational radius $R_{\rm grav}$, where the latter represents the effective size of the subsystem. Thus a perturber is considered for chain membership if $\gamma_{\rm ch}$ is significant, provided certain other conditions are fulfilled. The perturber list is updated at appropriate times by (1.32), with $R_{\rm grav}$ replacing the apocentre distance. Likewise, an existing member with positive radial velocity is a candidate for removal if we have

$$\dot{R}_k^2 > \frac{2\sum m_k}{R_k}\,, \quad R_k > 3R_{\rm grav}\,. \tag{1.54}$$

Here the former condition requires transformation to the local c.m. system. The chain integration is continued as long as there are at least three members, with re-initialization after any changes. Note that the membership procedure also allows for a hard binary to be added or removed.

It turns out that the chain structure is a convenient tool for checking the dynamical state. Thus any escaping single particle or binary can readily be identified by considering the distances at the beginning and end of the chain if $N > 3$. As in the case of two-body regularization, the internal integration is continued up to the next block-step time. This entails inverting the integral of Ldt for an upper limit to ensure that the block-step is not exceeded. Note that here we do not have a Taylor series expansion for the time derivatives.

In general, termination is carried out if $\max\{R_k\} > 3R_{\rm cl}$ for three particles or two hard binaries. Provisions are also included for termination of a stable hierarchy, followed by switching to the more efficient KS treatment. As discussed previously, one way in which this can occur is after a strong interaction of two binaries. Finally, procedures for physical collisions or tidal circularization are also included, albeit with considerable programming effort.

1.15 Astrophysical Procedures

A star cluster simulation code should include a wide range of astrophysical processes for a realistic treatment. In the following we touch briefly on some of the most relevant aspects of the Cambridge codes. By now the addition of synthetic stellar evolution has enabled the introduction of many interesting features that pose numerical challenges. The simulation of realistic star clusters requires an IMF containing a significant proportion of heavy stars, as discussed in Sect. 7.4. It has been known for a long time that a few heavy bodies exert an unduly large influence on the dynamics of stellar systems. Such a distribution also leads to mass segregation on a short time-scale, which may be comparable to the main-sequence life-time for typical cluster parameters. Mass loss from evolving stars is therefore important for all but the youngest clusters and its inclusion in a simulation code is essential for observational interpretation.

Since the basic ingredients of the stellar evolution scheme are discussed at length in Chaps. 10 and 12, we concentrate on some of the related algorithms here. The primary quantities associated with each star are updated at sufficiently frequent intervals for a smooth representation. For dynamical purposes, only the process of mass loss requires special treatment. It is usually confined to a small fraction of all stars. The main procedures can be summarized under the following headings.

- Mass loss from single stars and binaries
- Roche-lobe mass transfer and common-envelope evolution
- Magnetic braking and spin-orbit coupling
- Inspiralling of compact binaries
- Supernova explosions and neutron star kicks
- Physical collisions (KS or chain regularization)

In the case of significant mass loss, $\Delta m > 0.1 M_{\odot}$, force polynomials for the nearest neighbours are re-initialized in order to reduce discontinuity effects. Likewise, appropriate corrections are made to ensure overall energy conservation. This entails knowledge of the potential since we assume that the ejected mass escapes rapidly from the cluster. When using GRAPE, the cost of a full N summation can be avoided in most cases (except small Δt_i and large Δm) by employing the available potential, corrected for the net force contribution up to the current time,

$$\Delta \phi = -\boldsymbol{v}_i \cdot (\boldsymbol{F}_i - \boldsymbol{F}_{\rm tide})(t - t_i) \,. \tag{1.55}$$

Close binaries undergoing general mass loss on a slow time-scale also require updating of their KS elements. Consequently, the orbital parameters are modified at constant eccentricity, based on the adiabatic approximation $M_{\rm b} a = $ const. A corresponding correction for the inner binary elements of a hierarchical triple can be carried out explicitly. Here it is necessary to re-assess the stability condition because the inner orbit expands more than the outer one.

A realistic period distribution invariably includes binaries that experience Roche-lobe mass-transfer after the primary leaves the main sequence. This stage is initiated by tidal circularization or the formation of a circular binary following common envelope evolution. Since the complicated astrophysical modelling is discussed in Chap. 12, we limit our comments to some computational aspects for completeness. For practical reasons, the continuous process of mass transfer is divided into an active and a coasting phase, where the latter is updated at frequent intervals. The duration of the active phase is restricted to the c.m. time-step for consistency with the dynamics. After the internal adjustment of the essentially circular orbit has been completed, any system mass loss is corrected for in the same way as for single stars.

Magnetic braking and inspiralling of compact binaries by gravitational radiation are catered for both within the Roche process as well as for certain non-interacting binaries. In either case, changes in the rotational spin of the components are treated according to the recipes outlined in Chap. 12. We note that these processes themselves do not involve any mass loss.

Stars above about $8M_\odot$ undergo supernova explosions and eject a significant amount of mass during the transition to neutron stars. In the absence of a consensus on neutron star kicks, we have adopted a Maxwellian distribution with large dispersion; hence practically all the neutron stars escape from the cluster. Now the correction procedure includes the increased kinetic energy as well as the potential energy contribution of the expelled mass. Since the ejection of high-velocity members is also a feature of stellar systems containing binaries, we have implemented an algorithm for preventing discontinuous changes in the neighbour force for large time-steps.

The determination and implementation of collisions in chain regularization require special care and have been discussed elsewhere in considerable detail (Aarseth 2003). For highly eccentric binaries, the KS solution facilitates a check on the pericentre distance, provisionally identified by a negative product of the old and new radial velocity $R' = 2\boldsymbol{u} \cdot \boldsymbol{u}'$ and $R < a$. The outcome of a collision depends on the stellar types so that a variety of remnants may be produced (see Chap. 12). Here we note that the device of ghost stars can be used when two stars are replaced by one non-zero mass.

Tidal fields represent another important feature of star cluster simulations. Two different types of external effects are catered for. Most open clusters in the solar neighbourhood move in nearly circular orbits, which admit a linearized tidal force to be included in the equations of motion. This simple representation gives rise to an energy integral and imposes a tidal boundary that is useful for defining escape. The tidal radius is given by

$$r_{\text{tide}} = \left(\frac{GM}{4A(A-B)} \right)^{1/3} , \tag{1.56}$$

where A and B are the classical rotation constants. Traditionally, stars outside $2r_{\text{tide}}$ are removed from the calculation since their subsequent effect on bound cluster members is negligible.

The general case of 3D motion requires a full galactic model, with explicit expressions for the force and its derivative. The equations of motion are now most conveniently expressed in a non-rotating coordinate system (Aarseth 2003). It is still possible to have an approximate energy integral by monitoring the accumulated work done by the perturbing force $\boldsymbol{P}_i$ during each (regular) time-step. Expanding the integrated contribution to *third* order in terms of the *initial* values and expressing the result at the end of the time-step, we obtain

$$\Delta E_i = m_i \left(\frac{1}{2} \ddot{W}_i \Delta t_i^2 - \dot{W}_i \Delta t_i \right) , \tag{1.57}$$

where $\dot{W}_i = \boldsymbol{v}_i \cdot \boldsymbol{P}_i$. Knowledge of $\dot{\boldsymbol{P}}_i$ enables the second order to be included in the expansion, and the resulting conservation is satisfactory. Although distant stars are usually removed from the active data structure using a nominal value of the tidal radius, their orbits in the galactic potential can still be integrated. Hopefully, these recent code innovations will encourage more comprehensive studies of eccentric globular cluster orbits and associated tidal tails.

1.16 GRAPE Implementations

Since the use of GRAPE-type special-purpose computers is gaining more widespread use, it may be of interest to describe some of the procedures in the simulation code NBODY4. In particular, it should be emphasized that the internal GRAPE data structure differs from the host in several important respects, which calls for additional software.

We take advantage of the work-sharing facility to speed up the calculation by carrying out some operations on the host while GRAPE is busy. In general for large N, many particles are due to be advanced at the same time but the number may also be quite small during episodes of strong multiple interactions. After prediction of the first 48 block members $n_{\rm block}$, the relevant procedures can be summarized as follows.

- Begin force calculation for the first block-step members
- Predict the next 48 members (if any) while GRAPE is busy
- Predict $\boldsymbol{r}_i, \boldsymbol{v}_i$ of c.m. and perturbed KS components (first time)
- Form a list of small time-steps (first time, $n_{\rm block} \leq 32$)
- Correct the previous block members and specify new time-steps
- Copy the force and force derivatives from GRAPE
- Correct the last block members after repeating the above
- Send all the corrected $\boldsymbol{r}_i, \boldsymbol{v}_i$ and also $\boldsymbol{F}_i, \dot{\boldsymbol{F}}_i$ to GRAPE

The scheduling of particles to be advanced is essentially the same as in NBODY6. However, coordinate and velocity predictions on the host are now restricted to block-step members since a fast prediction of all particles are carried out on the GRAPE hardware. When these quantities are copied to the corresponding GRAPE variables for data transfer, an optional prediction to second order in the force derivative may be included for increased accuracy. With regularized binaries present, the data structure on GRAPE consists of single particles and the c.m. of each KS pair. Consequently, the force acting on a binary is in the first instance obtained by direct summation from $2N_{\rm p}+1$ to $N + N_{\rm p}$, where a c.m. is treated as a single particle. Differential force corrections are then applied for each binary perturber to be consistent with the c.m. force and likewise for any perturber forces. These corrections involve subtracting the c.m. terms before adding the vectorial contributions due to the two components. Any particles which are not on the block-step must therefore be predicted on the host before these corrections are performed. Note that the subtraction procedure invariably introduces small errors due to the lower precision of the GRAPE hardware.

Another aspect of the prediction strategy concerns the indirect terms in the heliocentric formulation (1.9). Again the coordinates and velocities of any significant members for which $t_j + \Delta t_j > t$ need to be predicted first. This can most readily be achieved by maintaining a list of any important planetesimal perturber, which is updated following changes in the data structure. In order to check energy conservation in the heliocentric case, the expression for kinetic energy takes the form

$$K = \frac{1}{2} \sum_{i=1}^{N} m_i \boldsymbol{v}_i^2 - \frac{1}{2} \left(M_0 - \sum m_i \right) \boldsymbol{v}_0^2 \,, \tag{1.58}$$

where $\boldsymbol{v}_0 = -\sum m_i \boldsymbol{v}_i / M_0$ is the velocity of the dominant body of mass M_0 and the second sum in (1.58) refers to the heavy perturbers.

As mentioned in Sect. 1.5, the determination of a maximum time-step also differs when using a GRAPE in connection with (1.9). We employ a special function that supplies the index of the closest neighbour at no extra cost during the force evaluation. The current relative coordinates and velocity $\boldsymbol{R}, \boldsymbol{V}$ define an appropriate time-step $\Delta t_{\rm nb} = 0.1 R^2 / \boldsymbol{R} \cdot \boldsymbol{V}$, which may be smaller than the standard value. Another point to note is that the direct force summation does not include the dominant body whose effect is added in the iteration. Since provisional values of $\boldsymbol{F}_i, \dot{\boldsymbol{F}}_i$ for each member on the block-step are supplied to GRAPE for scaling purposes, it is necessary to subtract the dominant contributions first. On the other hand, decisions on new regularizations or terminations are made during the time-step determination and executed in the usual way at the end of the block-step.

Procedures for wheel-spoke regularization have also been combined with the GRAPE code NBODY4 making a separate version, NBODY7. A new feature here is how to recognize a compact subsystem suitable for special treatment. Given the presence of a massive binary together with the conditions

$$R < 2R_{\rm cl} \,, \quad r_{\rm p} < \frac{1}{4} R_{\rm cl} \, (m_0/\bar{m})^{1/2} \,, \quad \dot{r}_p < 0 \,, \tag{1.59}$$

with $r_{\rm p}$ the distance to the closest perturber, this system is initialized and additional perturbers are selected as for chain regularization. A list of neighbours is updated on the local crossing time, from which significant perturbers are selected. Frequent checks are made on membership changes of the subsystem, taking care to avoid near-collisions in the overlap region although no direct test is made at present.[2]

The post-Newtonian algorithms discussed above have also been implemented. Again these procedures are carried out on the host computer. Several

[2]Interactions between subsystem members and perturbers are not softened, hence the use of an overall perturbation with respect to the c.m. only acts as a guide.

models where the relativistic terms become important have been studied for centrally concentrated systems with $N = 10^5$ equal-mass particles and one massive black hole of mass $m_0 = 300\,\bar{m}$ (Aarseth 2007). A typical simulation over 100 time units and including GR coalescence can be done in a few days. Experience shows that the less powerful GRAPE-6A is well suited for this purpose since for much of the time the host constitutes the computational bottleneck, especially during relativistic episodes. Because the central subsystem is now advanced by the accurate but more expensive Bulirsch–Stoer method, the overall energy conservation is somewhat better than for standard cluster simulations.

When using GRAPE, all regularization procedures are treated in essentially the same way as in NBODY6. Depending on the requirements, there is a choice of chain regularization, time-transformed leapfrog (see Chap. 2) or wheel-spoke method for studying three different types of problems, but only one scheme is chosen for a given calculation. Some of these procedures are distinguished by options and there are also different directories containing routines of the same name. In conclusion, this GRAPE software package has already yielded some interesting results that open up new avenues for future exploration.

1.17 Practical Aspects

In the preceding sections we have described the main procedures of the code NBODY6 and also NBODY4, which is similar. The actual use of these codes involves many additional considerations. Here we attempt a general summary of some practical features that play a key role.

To begin with, the code needs to be installed and tested. This necessitates downloading the software and extracting the relevant files.[3] Certain parameters governing maximum array sizes should be checked, otherwise the (generous) defaults will be adopted. It is expected that the code will compile successfully on most conventional computers. Likewise, results of the test input should be examined before any further work is attempted. When trying out a new code, it is of interest to evaluate the performance by so-called profiling as explained in the manual which can also be downloaded.

A versatile code requires a number of input parameters, especially if there are many alternative procedures. To facilitate explanation, we distinguish between different types of input. In the first group are the particle number N, maximum neighbour membership $n_{\rm max}$, as well as the number of primordial binaries $n_{\rm bin}$. The second set of parameters, η_I, η_R, η_u, are concerned with the integration itself and are dimensionless, i.e. the same for most problems.

Initial conditions may be generated internally or uploaded from a file. In the former case there is a choice of IMF distributions with upper and lower

[3]See `http://www.ast.cam.ac.uk/research/nbody`.

mass limits. The main scaling parameters are the length unit R_V in pc and mean mass M_S in solar units, as well as the virial theorem ratio Q_V discussed earlier. The network of 40 options are defined in a table and allows a variety of tasks to be considered. However, the choice must be consistent, which requires due care. All the close encounter parameters have been discussed in the KS section. Special input templates are also available for simulations with primordial binaries or cluster orbits in a 3D galactic potential.

An example of typical input parameters is given for illustration purposes, where the main categories are placed together.

- $N = 1000$, $n_{\max} = 70$, $\eta_I = 0.02$, $\eta_R = 0.03$
- $S_0 = 0.3$, $\Delta T = 2$, $T_{\rm crit} = 100$
- $Q_E = 2 \times 10^{-5}$, $R_V = 2$, $M_S = 0.5$
- # 1, 2, 5, 7, 14, 16, 20, 23
- $\Delta t_{\rm cl} = 10^{-4}$, $R_{\rm cl} = 0.001$, $\eta_u = 0.2$, $\gamma_{\min} = 10^{-6}$
- $\alpha = 2.3$, $m_1 = 10.0$, $m_N = 0.2$

In the second line S_0 is an initial guess for the neighbour sphere, the output interval is ΔT and $T_{\rm crit}$ gives the termination time. Moreover, the relative energy tolerance Q_E is used for automatic error control. The line of options contains some useful suggestions but is by no means complete. Finally, the IMF is defined by the classical Salpeter exponent α, together with the upper and lower mass limits in terms of the average mass. More detailed information on the full set of input parameters can be found in the manual. Thus for example, there are options for external perturbations or stellar evolution. Taking into account the wide range of available procedures, the complete input file is quite compact in comparison with many other large codes.

Presentation of results constitutes another challenge for code development. It also requires an effort by the practitioner to extract the available data in a suitable form. Here we may distinguish between result summaries and detailed information. To elucidate the possibilities, the table summarizes some of the main optional procedures with a brief explanation.

Procedure	Explanation
Cluster core	N^2 algorithm for core radius and density centre
Lagrangian radii	Percentile mass radii and half-mass radius
Error control	Automatic error check and restart from last time
Escape	Removal of distant members and table updates
Time offset	Rescaling of all global times for large values
Event counters	Stellar types and remnant statistics
Binary analysis	Regularized binary histograms and energy budget
Binary data bank	Characteristic parameters for regularized binaries
HR diagram	Evolutionary state of single stars and binaries
General data bank	Detailed snapshots for data analysis

Each of these procedures is activated by specifying a non-zero option, as defined in the manual. There is also a facility for changing any option at later times. Many of the result summaries are self-explanatory and will not be reviewed here. Likewise, the manual illustrates the principle of adding new variables to the code while preserving the total size of the common blocks.

We conclude by commenting on the way in which the total energy is obtained. Thus rather than evaluating the kinetic and potential energies directly, the different contributions are derived consistently according to the calculation method. For example, the binding energies of KS pairs are given by $\sum \mu_i h_i$, where h_i is predicted to highest order. Monitoring the internal energies of hierarchical systems and collisions events enable a conservation scheme to be maintained at high accuracy because dissipative processes are also accounted for.

References

Aarseth S. J., 2003, Gravitational N-Body Simulations, Cambridge University Press, Cambridge

Aarseth S. J., 2007, MNRAS, 378, 285

Aarseth S. J., Heggie D. C., 1976, A&A, 53, 259

Aarseth S. J., Zare K., 1974, Celes. Mech., 10, 185

Ahmad A., Cohen L., 1973, J. Comput. Phys., 12, 389

Blanchet L., Iyer B., 2003, Class. Quantum Grav., 20, 755

Bulirsch R., Stoer J., 1966, Num. Math., 8, 1

Kokubo E., Makino J., 2004, PASJ, 56, 861

Kokubo E., Yoshinaga K., Makino J., 1998, MNRAS, 297, 1967

Kozai Y., 1962, AJ, 67, 591

Kustaanheimo P., Stiefel E., 1965, J. Reine Angew. Math., 218, 204

Makino J., Aarseth S. J., 1992, PASJ, 44, 141

Makino J., Taiji M., Ebisuzaki T., Sugimoto, D., 1997, ApJ, 480, 432

Mardling R. A., Aarseth S. J., 1999, in Steves B. A., Roy A. E., eds, The Dynamics of Small Bodies in the Solar System, Kluwer, Dordrecht, p. 385

Mardling, R., Aarseth S., 2001, MNRAS, 321, 398

Mikkola S., Aarseth S. J., 1993, Celes. Mech. Dyn. Ast., 57, 439

Mikkola S., Aarseth S. J., 1996, Celes. Mech. Dyn. Ast., 64, 197

Mikkola S., Aarseth S. J., 1998, New Astron., 3, 309

Mora T., Will C. M., 2004, Phys. Rev. D 69, 104021 (gr-qc/0312082)

Newmark N. M., 1959, J. Eng. Mech., 85, 67

Peters P. C., 1964, Phys. Rev., 136, B1224

Zare K., 1974, Celes. Mech., 10, 207

2

Regular Algorithms for the Few-Body Problem

Seppo Mikkola

Tuorla Observatory, University of Turku, Finland
mikkola@utu.fi

2.1 Introduction

In N-body simulations the most common strong interactions are due to close encounters of just two bodies. Most classical numerical integration methods lose precision for such situations due to the $1/r^2$ singularity of the mutual force of the two bodies. In a close encounter the relative motion of the participating bodies is so fast that, for a brief moment, the rest of the system can be considered frozen. Consequently, the most important feature of a regularizing algorithm must be that it can handle reliably the perturbed two-body problem. There are two basically different types of methods available: Coordinate and time transformations and algorithms that produce regular results without coordinate transformation.

The first coordinate-transformation method was that of Levi-Civita (1920), but the method works only in two dimensions. Later, Kustaanheimo & Stiefel (1965) generalized this by applying a transformation (*KS-transformation*) from four dimensions to three dimensions (see also Aarseth 2003). More recently, two versions of *algorithmic regularization* have been proposed. These are the *logarithmic Hamiltonian* (LogH), suggested by Mikkola & Tanikawa (1999a, b) and independently by Preto & Tremaine (1999).

A further development, the *Time Transformed Leapfrog* (TTL), was presented by Mikkola & Aarseth (2002). Finally, Mikkola & Merritt (2006, 2008) combined the LogH and TTL as well as a generalized midpoint method to modify the algorithmic regularization such that it can handle the case of velocity dependent perturbations, which are important in, for example, post-Newtonian dynamics (Soffel 1989).

2.2 Hamiltonian Manipulations

All known regularization methods require the introduction of a new independent variable. Due to the importance of the Hamiltonian formalism, this is

Mikkola, S.: *Regular Algorithms for the Few-Body Problem*. Lect. Notes Phys. **760**, 31–58 (2008)
DOI 10.1007/978-1-4020-8431-7_2

often done by transforming the Hamiltonian. Let $\boldsymbol{q}$ and $\boldsymbol{p}$ be the coordinates and momenta, $T = T(\boldsymbol{p})$ the kinetic energy and $U = U(\boldsymbol{r}, t)$ the potential. Then $H(\boldsymbol{p}, \boldsymbol{q}, t) = T(\boldsymbol{p}) - U(\boldsymbol{q}, t)$ is the Hamiltonian. If one defines a new independent variable s by the differential equation

$$\mathrm{d}t = g(p, q, t)\mathrm{d}s, \tag{2.1}$$

the equations of motion can be derived from the extended phase space Hamiltonian Γ (Poincaré's transformation)

$$\Gamma = g(p, q, t)(H(p, q, t) + B), \tag{2.2}$$

where B is the momentum of time and initially

$$B(0) = -H(p(0), q(0), t_0). \tag{2.3}$$

Time is now a coordinate and one notes that the Poincaré transformation makes the new Hamiltonian Γ conservative, since it does not depend explicitly on the new independent variable. Due to this and the choice of the initial value for B, the numerical values are $\Gamma = 0$ and $B = -H$ (binding energy) along the trajectory.

One often uses

$$\Gamma = (H + B)/L; \text{ or } \Gamma = (H + B)/U. \tag{2.4}$$

Here U is the potential energy and $L = T + U$ the Lagrangian. The equations of motion take the form

$$t' = \frac{\partial \Gamma}{\partial B} = g; \quad q' = \frac{\partial \Gamma}{\partial p} = +g\frac{\partial H}{\partial p} + \frac{\partial g}{\partial p}(H + B) \tag{2.5}$$
$$B' = -\frac{\partial \Gamma}{\partial t} = -g\frac{\partial H}{\partial t} - \frac{\partial g}{\partial t}(H + B); \quad p' = -\frac{\partial \Gamma}{\partial q} = -g\frac{\partial H}{\partial q} - \frac{\partial g}{\partial q}(H + B),$$

which is correct because $H + B = 0$ along the orbit. However, *this does not mean that the latter terms can be dropped.* The reason for this will become clear in the example in Sect. 2.3.

Another way to manipulate the Hamiltonian is the use of the functional Hamiltonian (Preto & Tremaine 1999)

$$\Lambda = f(T + B) - f(U), \tag{2.6}$$

where $f(z)$ is any function that satisfies $f'(z) \geq 0$. A most interesting function is $f(z) = \log(z)$ (Mikkola & Tanikawa 1999a, b; Preto & Tremaine 1999), which gives $t' = \partial\Lambda/\partial B = 1/(T + B)$. Along the correct trajectory we also have $1/(T + B) = 1/U$, and thus the time transformation is essentially the same as $g = 1/U$. A special feature of the functional Hamiltonian is that it allows the use of the (symplectic) leapfrog algorithm because the equations of motion

$$\dot{\boldsymbol{r}} = \frac{\partial \Lambda}{\partial \boldsymbol{p}} = f'(T+B)\frac{\partial T}{\partial \boldsymbol{p}}; \quad \dot{\boldsymbol{p}} = -\frac{\partial \Lambda}{\partial \boldsymbol{r}} = f'(U)\frac{\partial U}{\partial \boldsymbol{r}} \tag{2.7}$$

are such that the right-hand sides do not depend on variables on the left-hand side.

2.3 Coordinate Transformations

2.3.1 One-Dimensional Case

A simple example is provided by the one-dimensional two-body problem. The Keplerian Hamiltonian $H = p^2/2 - M/q$ may be transformed by the point-transformation $q = Q^2$, $p = P/(2Q)$ into the form $H = P^2/(8Q^2) - M/Q^2$. Using $g = q = Q^2$, one obtains

$$\Gamma = Q^2\left(\frac{P^2}{8Q^2} - \frac{M}{Q^2} + B\right) = \frac{1}{8}P^2 + BQ^2 - M, \tag{2.8}$$

and the equations of motion are

$$Q' = \frac{1}{4}P, \ \ P' = -2BQ; \quad \text{or} \quad Q'' = -\frac{B}{2}Q, \tag{2.9}$$

which is a harmonic oscillator because $B = -H = \text{constant}$.

Note that, had we dropped the $(H+B)$ factored terms in (2.5), we would have had

$$Q' = \frac{1}{4}P, \ \ P' = -2\left(\frac{1}{8}P^2 - M\right)/Q; \quad \text{or} \quad Q'' = -\frac{1}{2}\left(\frac{1}{8}P^2 - M\right)/Q, \tag{2.10}$$

which is singular (but still analytically regular, due to energy conservation i.e., because $\frac{1}{8}P^2 - M = BQ^2$).

2.3.2 Three-Dimensional Case: KS-Transformation

The KS-transformations (Kustaanheimo & Stiefel 1965) between the three-dimensional position and momentum $\boldsymbol{r}$ and $\boldsymbol{p}$ and the corresponding four-dimensional KS-variables $\boldsymbol{Q}$ and $\boldsymbol{P}$ may be written

$$\boldsymbol{r} = \widehat{\mathbf{Q}}\boldsymbol{Q}; \ \ \boldsymbol{p} = \widehat{\mathbf{Q}}\boldsymbol{P}/(2Q^2). \tag{2.11}$$

Here $\widehat{\mathbf{Q}}$ is the KS-matrix (Stiefel & Scheifele 1971, p. 24)

$$\widehat{\mathbf{Q}} = \begin{pmatrix} Q_1 & -Q_2 & -Q_3 & Q_4 \\ Q_2 & Q_1 & -Q_4 & -Q_3 \\ Q_3 & Q_4 & Q_1 & Q_2 \\ Q_4 & -Q_3 & Q_2 & -Q_1 \end{pmatrix}. \tag{2.12}$$

Another way to write the transformation is

$$x = Q_1^2 - Q_2^2 - Q_3^2 + Q_4^2; \;\; y = 2(Q_1Q_2 - Q_3Q_4); \;\; z = 2(Q_1Q_3 + Q_2Q_4). \quad (2.13)$$

Note that the fourth components of $\boldsymbol{r}$ and $\boldsymbol{p}$ that (2.11) produces are zeros due to the structure and properties of the transformation.

Due to increased number of variables, the Q's corresponding to given physical coordinates are not unique. However, one may choose any solution, for example, with $\boldsymbol{r} = (x, y, z)^t$, $r = |\boldsymbol{r}|$ we calculate

$$\begin{aligned} u_1 &= \sqrt{\tfrac{1}{2}(r + |x|)} \\ u_2 &= \;\; Y/(2u_1) \\ u_3 &= \;\; Z/(2u_1) \\ u_4 &= \;\; 0, \end{aligned} \quad (2.14)$$

and the components of $\boldsymbol{Q}$ are

$$\boldsymbol{Q} = \begin{cases} (u_1, u_2, u_3, u_4)^t \; ; & X \geq 0 \\ (u_2, u_1, u_4, u_3)^t \; ; & X < 0 \end{cases} . \quad (2.15)$$

(This algorithm is used to avoid round-off error.)

Initial values for the KS momenta are given by

$$\boldsymbol{P} = 2\widehat{\mathbf{Q}}^t \boldsymbol{p}. \quad (2.16)$$

For the two-body problem $H = \frac{1}{2}\boldsymbol{p}^2 - M/r$, the time-transformed Hamiltonian Γ in (2.2) takes the form

$$\Gamma = \frac{1}{8}\boldsymbol{P}^2 - M + B\boldsymbol{Q}^2, \quad (2.17)$$

i.e. a harmonic oscillator, in complete analogy with the one-dimensional case.

When regularized by the KS-transformation, the equations of motion for a perturbed binary

$$\ddot{\boldsymbol{r}} + M\boldsymbol{r}/r^3 = \boldsymbol{F} \quad (2.18)$$

take the explicit form

$$\begin{aligned} \boldsymbol{Q}'' &= -\frac{1}{2}B\boldsymbol{Q} + \frac{1}{2}r\widehat{\mathbf{Q}}^t\boldsymbol{F} \\ B' &= -2\boldsymbol{Q}' \cdot \widehat{\mathbf{Q}}^t\boldsymbol{F} \\ t' &= r = \boldsymbol{Q} \cdot \boldsymbol{Q}. \end{aligned} \quad (2.19)$$

Here $\boldsymbol{F}$ is the physical perturbation exerted by other particles (or any other physical effect) and

$$B = \frac{M}{r} - \frac{\boldsymbol{p}^2}{2}$$

is the two-body binding (Kepler-)energy. Since the equations are regular, they can be solved with any reasonable numerical method.

2.4 KS-Chain(s)

When the KS-transformation is applied in N-body systems, one does not obtain a harmonic oscillator, but close approaches can still be regularized. First, one forms a chain of particles such that all the small critical distances are included in the chain and then one applies the KS-transformation to the chain vectors. For details of the chain selection procedure see Sect. 2.7.1.

Let a time-transformed multiparticle Hamiltonian be

$$\Gamma = (T - U + B)/(T + U),$$

where

$$T = \sum_{\nu} \boldsymbol{p}_{\nu}^{2}/(2m_{\nu}); \quad U = \sum_{i<j} m_i m_j / r_{ij}.$$

Let us introduce new coordinates

$$\boldsymbol{X}_k = \boldsymbol{r}_{i_k} - \boldsymbol{r}_{j_k},$$

then we can use the generating function

$$S = \sum_k \boldsymbol{W}_k \cdot \boldsymbol{X}_k = \sum_k \boldsymbol{W}_k \cdot (\boldsymbol{r}_{i_k} - \boldsymbol{r}_{j_k}). \tag{2.20}$$

In terms of the new momenta $\boldsymbol{W}$, the old ones are

$$\boldsymbol{p}_{\nu} = \frac{\partial S}{\partial \boldsymbol{r}_{\nu}} = \sum_k \boldsymbol{W}_k \cdot (\delta_{\nu i_k} - \delta_{\nu j_k}), \tag{2.21}$$

where the δ's are the Kronecker symbols. Thus we have

$$T = \frac{1}{2} \sum_{\alpha\beta} T_{\alpha\beta} \boldsymbol{W}_{\alpha} \cdot \boldsymbol{W}_{\beta} \tag{2.22}$$

$$U = \sum_k \frac{m_{i_k} m_{j_k}}{|\boldsymbol{X}_k|} + \sum_{i<j,\ (i,j)\notin\{i_k,j_k\}} \frac{m_i m_j}{r_{ij}}, \tag{2.23}$$

where

$$T_{\alpha\beta} = \sum_{\nu} \frac{1}{m_{\nu}} (\delta_{\nu i_\alpha} - \delta_{\nu j_\alpha})(\delta_{\nu i_\beta} - \delta_{\nu j_\beta}),$$

and the second potential energy term

$$\sum_{i<j,\ (i,j)\notin\{i_k,j_k\}} \frac{m_i m_j}{r_{ij}}$$

contains all the distances $r_{ij} = r_{ij}(X_1, X_2 \ldots)$ that are not included among the vectors $\boldsymbol{X}_k$.

After application of the KS transformation by (2.11) to every momentum-coordinate pair by

$$\boldsymbol{W},\ \boldsymbol{X} \rightarrow \boldsymbol{P},\ \boldsymbol{Q},$$

one can obtain the regularized Hamiltonian

$$\Gamma(\boldsymbol{P}, \boldsymbol{Q}) = (T - U + B)/(T + U)$$

and form the canonical equations of motion,

$$B' = -\frac{\partial \Gamma}{\partial t}; \quad \boldsymbol{P}' = -\frac{\partial \Gamma}{\partial \boldsymbol{Q}} \tag{2.24}$$

$$t' = \frac{\partial \Gamma}{\partial B}; \quad \boldsymbol{Q}' = \frac{\partial \Gamma}{\partial \boldsymbol{P}}. \tag{2.25}$$

Note that the number of new variables may exceed the number of the old ones. This, however, is not a problem: all the physical results remain correct (Heggie 1974).

The above formulation is completely general at least to the point that all the well-known methods, the Zare (1974) method in which all particles are regularized with respect to a central body, Heggie's global regularization (Heggie 1974) (in which all the interparticle vectors are taken as new variables and collisions are regularized by the KS transformation) and the chain method (Mikkola & Aarseth 1993), are included. The vectors $\boldsymbol{X}$ of these methods are schematically illustrated in Fig. 2.1.

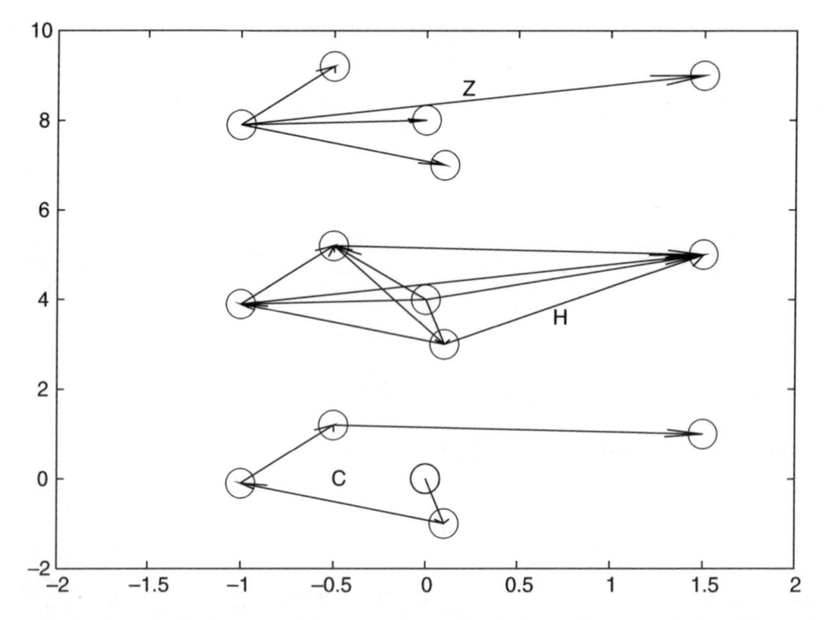

Fig. 2.1. Regularized interactions (schematically) in Zare method (Z), global method of Heggie (H) and chain method (C)

In fact, one can regularize any interparticle vector. Thus any kind of branching and looping chains can be handled. This could be seen as an intermediate form between the Heggie method and the chain. However, it is not clear if such alternatives are actually more useful than the simple chain. Comprehensive instructions for use of the KS-chain can be found in Mikkola & Aarseth (1993) and Aarseth (2003).

2.5 Algorithmic Regularization

The algorithmic regularization, contrary to KS regularization, does not use coordinate transformation but only a time transformation and a suitable algorithm that produces regular results despite the singularity in the force. The first such methods were invented in 1999 independently in two places (Mikkola & Tanikawa 1999a, b; Preto & Tremaine 1999).

2.5.1 The Logarithmic Hamiltonian (LogH)

Let $\boldsymbol{p}$ be the momenta and $\boldsymbol{q}$ the coordinates, $T(\boldsymbol{p})$ the kinetic energy and $U(\boldsymbol{q}, t)$ the force function. Then the Hamiltonian in extended phase-space is

$$H = T + B - U. \tag{2.26}$$

Here B is the momentum of time (which is now a coordinate: $\dot{t} = \frac{\partial H}{\partial B} = 1$).

If $B(0) = -H(0)$, then the function

$$\Lambda = \log(T + B) - \log(U) \tag{2.27}$$

can be used as a Hamiltonian in the extended phase space.

Demonstration

The equations of motion derivable from Λ read

$$\boldsymbol{p}' = -\frac{\partial \Lambda}{\partial \boldsymbol{q}} = \frac{\partial U}{\partial \boldsymbol{q}}/U; \quad B' = -\frac{\partial \Lambda}{\partial t} = \frac{\partial U}{\partial t}/U \tag{2.28}$$

$$\boldsymbol{q}' = \frac{\partial \Lambda}{\partial \boldsymbol{p}} = \frac{\partial T}{\partial \boldsymbol{p}}/T_e; \quad t' = \frac{\partial \Lambda}{\partial B} = 1/T_e, \tag{2.29}$$

where $T_e = T + B$ and a prime denotes differentiation with respect to the (new) independent variable s.

Since Λ does not depend explicitly on s, the value of Λ is constant. Thus $T + B = U$ due to choice of initial value for B. Using this and dividing the equations of motion by the equation for time (2.29), we get for the time derivatives

$$\dot{\boldsymbol{p}} = \frac{\partial U}{\partial \boldsymbol{q}}, \quad \dot{B} = \frac{\partial U}{\partial t} \quad \text{and} \quad \dot{\boldsymbol{q}} = \frac{\partial T}{\partial \boldsymbol{p}}, \tag{2.30}$$

i.e. the normal Hamiltonian equations.

LogH for Two bodies

To introduce the method we first consider the simple case of two-body motion $H = \boldsymbol{p}^2/2 - M/r$, which gives

$$\Lambda = \log(\boldsymbol{p}^2/2 + B) + \log(r), \tag{2.31}$$

after dropping $\log(M)$.
Thus the time transformation is

$$\mathrm{d}t = \mathrm{d}s\,\frac{\partial\Lambda}{\partial B} = \frac{\mathrm{d}s}{(\boldsymbol{p}^2/2 + B)}. \tag{2.32}$$

B remains constant, $B = -(\boldsymbol{p}^2/2 - M/r)$. The new independent variable s is

$$s = \int^t (\boldsymbol{p}^2/2 + B)\,\mathrm{d}t = \int^t \frac{M}{r}\,\mathrm{d}t, \tag{2.33}$$

i.e. a quantity proportional to the eccentric anomaly increment.

With stepsize h and initial values $\boldsymbol{p}_0$, $\boldsymbol{r}_0$, t_0, the leapfrog algorithm takes the form (illustration in Fig. 2.2)

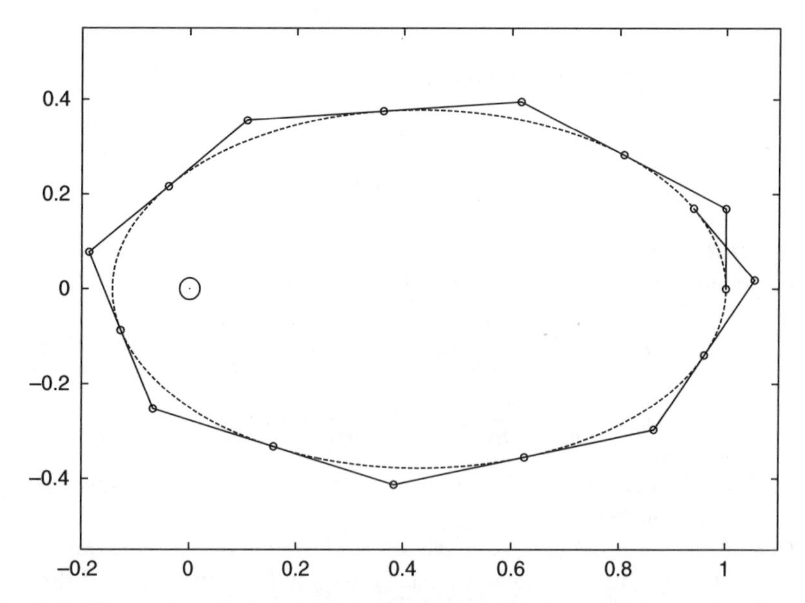

Fig. 2.2. Illustration of the working of the algorithmic regularization in the case of an elliptic two-body motion. The points on the ellipse are the starting and end points in a leapfrog step, while those outside the ellipse are the $\boldsymbol{r}_{\frac{1}{2}}$-points

$$\boldsymbol{r}_{\frac{1}{2}} = \boldsymbol{r}_0 + \frac{h}{2}\boldsymbol{p}_0/(\frac{\boldsymbol{p}_0^2}{2} + B) \tag{2.34}$$

$$\boldsymbol{p}_1 = \boldsymbol{p}_0 - h\,\boldsymbol{r}_{\frac{1}{2}}/r_{\frac{1}{2}}^2 \tag{2.35}$$

$$\boldsymbol{r}_1 = \boldsymbol{r}_{\frac{1}{2}} + \frac{h}{2}\boldsymbol{p}_1/(\frac{\boldsymbol{p}_1^2}{2} + B) \tag{2.36}$$

$$t_1 = t_0 + \frac{h}{2}\left[\frac{1}{(\frac{\boldsymbol{p}_0^2}{2} + B)} + \frac{1}{(\frac{\boldsymbol{p}_1^2}{2} + B)}\right]. \tag{2.37}$$

This algorithm produces correct positions and momenta on the associated Keplerian ellipse (Mikkola & Tanikawa 1999a, b; Preto & Tremaine 1999); however, time is not correct and the method thus has phase errors. This result applies even for collision orbits where the eccentricity $e = 1$.

Although the singularity when $r \to 0$ is not removed, one expects the algorithm to be applicable for the N-body problem since the functions are not evaluated precisely at $r = 0$.

2.5.2 Time-Transformed Leapfrog (TTL)

Consider the general system

$$\dot{\boldsymbol{r}} = \boldsymbol{v}, \quad \dot{\boldsymbol{v}} = \boldsymbol{F}(\boldsymbol{r}), \tag{2.38}$$

where $\boldsymbol{r}$ and $\boldsymbol{v}$ are position and velocity vectors of arbitrary dimension. We now introduce a time transformation

$$\mathrm{d}s = \Omega(\boldsymbol{r})\,\mathrm{d}t, \tag{2.39}$$

where $\Omega(\boldsymbol{r}) > 0$ is arbitrary.

If $W = \Omega$, then one may write

$$\boldsymbol{r}' = \boldsymbol{v}/W, \quad t' = 1/W, \quad \boldsymbol{v}' = \boldsymbol{F}/\Omega,$$

where a prime means $\frac{\mathrm{d}}{\mathrm{d}s}$. If W is obtained from the differential equation

$$\dot{W} = \boldsymbol{v}\cdot\frac{\partial\Omega}{\partial\boldsymbol{r}} \quad \text{or } W' = \boldsymbol{v}\cdot\frac{\partial\Omega}{\partial\boldsymbol{r}}/\Omega, \tag{2.40}$$

instead of $W = \Omega$ directly, we have

$$\begin{pmatrix}\boldsymbol{r}'\\ t'\\ \boldsymbol{v}'\\ W'\end{pmatrix} = \begin{pmatrix}\boldsymbol{v}/W\\ 1/W\\ \boldsymbol{0}\\ 0\end{pmatrix} + \begin{pmatrix}\boldsymbol{0}\\ 0\\ \boldsymbol{F}(\boldsymbol{r})/\Omega(\boldsymbol{r})\\ \boldsymbol{v}\cdot\partial\ln(\Omega)/\partial\boldsymbol{r}\end{pmatrix}. \tag{2.41}$$

This allows the Time-Transformed Leapfrog (TTL):

$$\boldsymbol{r}_{\frac{1}{2}} = \boldsymbol{r}_0 + \frac{h}{2}\frac{\boldsymbol{v}_0}{W_0} \tag{2.42}$$

$$t_{\frac{1}{2}} = t_0 + \frac{h}{2}\frac{1}{W_0} \tag{2.43}$$

$$\boldsymbol{v}_1 = \boldsymbol{v}_0 + h\frac{\boldsymbol{F}(\boldsymbol{r}_{\frac{1}{2}})}{\Omega(\boldsymbol{r}_{\frac{1}{2}})} \tag{2.44}$$

$$W_1 = W_0 + h\frac{\boldsymbol{v}_0 + \boldsymbol{v}_1}{2\Omega(\boldsymbol{r}_{\frac{1}{2}})} \cdot \frac{\partial\Omega(\boldsymbol{r}_{\frac{1}{2}})}{\partial\boldsymbol{r}_{\frac{1}{2}}} \tag{2.45}$$

$$\boldsymbol{r}_1 = \boldsymbol{r}_{\frac{1}{2}} + \frac{h}{2}\frac{\boldsymbol{v}_1}{W_1} \tag{2.46}$$

$$t_1 = t_{\frac{1}{2}} + \frac{h}{2}\frac{1}{W_1}. \tag{2.47}$$

A Simple Fortran Code for Two Bodies (LogH)

```
      implicit real*8 (a-h,m,o-z)
      read(5,*)h,tmx,mass ! read stepsize, maximum time & mass
      read(5,*)x,y,z,vx,vy,vz   ! read initial coords/vels

c     initializations
      t=0
      r=sqrt(x*x+y*y+z*z) !distance
      vv=vx*vx+vy*vy+vz*vz       !v-square
      B=mass/r-vv/2                          !binding-E
c
c     Integration of the two-body motion

1     continue
      dt=h/(vx*vx+vy*vy+vz*vz+2*B)  ! time increment
      x=x+dt*vx
      y=y+dt*vy
      z=z+dt*vz
      t=t+dt
      dtc=h/(x*x+y*y+z*z)
      vx=vx-x*dtc
      vy=vy-y*dtc
      vz=vz-z*dtc
      dt=h/(vx*vx+vy*vy+vz*vz+2*B)  ! new time increment
      x=x+dt*vx
      y=y+dt*vy
      z=z+dt*vz
      t=t+dt ! time has an O(h^3) error
c     diagnostics:  time, coords & error
      write(6,2)t,x,y,z,
     & (B+(vx*vx+vy*vy+vz*vz)/2)-mass/sqrt(x**2+y**2+z**2)
      if(t.lt.Tmx)goto 1
```

```
2          format(1x,1p,5g12.4)
           end
```

If one takes

$$\Omega = 1/r, \tag{2.48}$$

the increment of W in one step is

$$\Delta W = -h\frac{\boldsymbol{r}}{r^3} \cdot \frac{\boldsymbol{v}_1 + \boldsymbol{v}_0}{2} \tag{2.49}$$

and

$$\Delta\frac{1}{2}\boldsymbol{v}^2 = \frac{1}{2}(\boldsymbol{v}_1^2 - \boldsymbol{v}_0^2) = \frac{1}{2}(\boldsymbol{v}_1 - \boldsymbol{v}_0)\cdot(\boldsymbol{v}_1 + \boldsymbol{v}_0) = -h\frac{\boldsymbol{r}}{r^3} \cdot \frac{\boldsymbol{v}_1 + \boldsymbol{v}_0}{2},$$

which means that, for the unperturbed two-body problem, this algorithm is mathematically equivalent to the LogH-method (more generally this is the case if $\Omega = U$). Numerically, however, this does not apply. The reason is that in case of a close approach W first increases, then decreases fast. This means that the increments are large numbers and there is considerable cancellation and possible round-off error. Combined with the extrapolation method, this alternative leapfrog can be a powerful integrator for some systems.

Remark: Especially interesting is the fact that the method can be efficient for potentials that differ from the Newtonian $1/r$ behaviour at small distances. One notes that both the LogH and TTL are useful for the soft potential

$$U \propto 1/\sqrt{r^2 + \epsilon^2},$$

which cannot be regularized with the KS-transformation.

Remark: If $\Omega = 1/r$, the (numerical) relation $W = 1/r$ remains valid after every step and, somewhat surprisingly, this is true for any radial force field $\boldsymbol{F} = f(r)\boldsymbol{r}/r$.

A Simple Fortran Code for Two Bodies (TTL)

```
           implicit real*8 (a-h,m,o-z)
           read(5,*)h,tincr,tmx,mass ! read step,tincr, maxtime, mass
           read(5,*)x,y,z,vx,vy,vz   ! read initial coords/vels

           tnext=0
c          initializations
           t=0
           r=sqrt(x*x+y*y+z*z) !distance
           vv=vx*vx+vy*vy+vz*vz        !v-square
           E0=vv/2-mass/r
           W=mass/r
c
c          Integration of two-body motion
```

```
1       continue
        dt=h/W/2  ! time increment
        t=t+dt
        x=x+dt*vx
        y=y+dt*vy
        z=z+dt*vz
c
        dtc=h/(x*x+y*y+z*z)
        dw=  -(x*vx+y*vy+z*vz)*dtc/2
        vx=vx-x*dtc
        vy=vy-y*dtc
        vz=vz-z*dtc
        W=W+dw-(x*vx+y*vy+z*vz)*dtc/2
c
        dt=h/W/2  ! new time increment
        t=t+dt                          ! this has an O(h^3) error
        x=x+dt*vx
        y=y+dt*vy
        z=z+dt*vz
c       diagnostics
        if(t.lt.tnext)goto 1
        tnext=tnext+tincr
        r=sqrt(x*x+y*y+z*z)
        err=-E0+(vx*vx+vy*vy+vz*vz)/2-mass/r
        write(6,2)t,x,y,z,err*r ,W*r-mass ! time, coords & error
        if(t.lt.Tmx)goto 1
2       format(1x,1p,10g12.4)
        end
```

2.5.3 A Simple LogH Algorithm for the Three-Body Problem

The three-body problem is still one of the most studied problems in few-body dynamics. Therefore, it may be of interest to consider in more detail a simple regular three-body algorithm. This also serves as further illustration of the use of the algorithmic regularization.

Following Heggie (1974), we use the three interparticle vectors (see Fig. 2.3)

$$\boldsymbol{X}_1 = \boldsymbol{r}_3 - \boldsymbol{r}_2; \quad \boldsymbol{X}_2 = \boldsymbol{r}_1 - \boldsymbol{r}_3; \quad \boldsymbol{X}_3 = \boldsymbol{r}_2 - \boldsymbol{r}_1 \tag{2.50}$$

as new coordinates. Let the corresponding velocities be $\boldsymbol{V}_k = \dot{\boldsymbol{X}}_k$, then the kinetic and potential energies (in c.m. system) can be written

$$T = \frac{1}{2M} \sum_{i<j} m_i m_j \boldsymbol{V}^2_{k_{ij}}; \quad U = \sum_{i<j} \frac{m_i m_j}{|X_{k_{ij}}|}, \tag{2.51}$$

where $M = \sum_k m_k$ is the total mass and $k_{ij} = 6 - i - j$. The equations of motion are

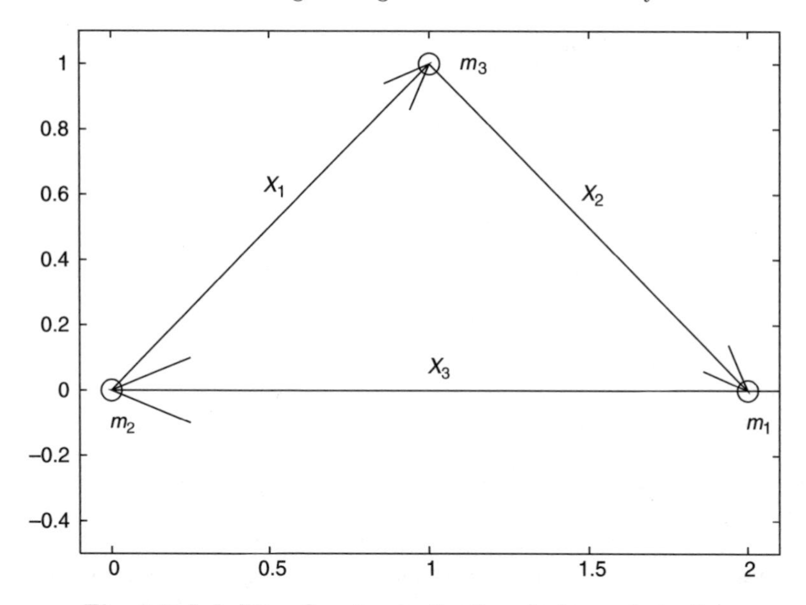

Fig. 2.3. Labelling of vectors in the three-body regularization

$$\dot{\boldsymbol{X}}_k = \boldsymbol{V}_k; \quad \dot{\boldsymbol{V}}_k = -M\frac{\boldsymbol{X}_k}{|\boldsymbol{X}_k|^3} + m_k \sum_{\nu} \frac{\boldsymbol{X}_\nu}{|\boldsymbol{X}_\nu|^3}, \tag{2.52}$$

and after the application of the logarithmic Hamiltonian modification they read

$$t' = 1/(T+B); \quad \boldsymbol{X}'_k = \dot{\boldsymbol{X}}_k/(T+B); \quad \boldsymbol{V}'_k = \dot{\boldsymbol{V}}_k/U, \tag{2.53}$$

which are suitable for the leapfrog algorithm, given in (2.58) and (2.59), as well as for Yoshida's (1990) higher-order leapfrogs.

The usage of the relative vectors, instead of some inertial coordinates, is advantageous in attempting to avoid large round-off effects. One could also integrate only two of the triangle sides, obtaining the remaining one from the conditions

$$\sum_k \boldsymbol{X}_k = \boldsymbol{0}; \quad \sum_k \boldsymbol{V}_k = \boldsymbol{0}.$$

However, this hardly reduces the computational effort required by the method. Instead one may, occasionally, compute the longest side, and the corresponding velocity, from the above triangle conditions. Note, however, that the sums of the sides are not only integrals of the exact solution but are also exactly conserved by the leapfrog mapping.

The transformation from the variables $\boldsymbol{X}$ to centre-of-mass coordinates $\boldsymbol{r}$ can be done as

$$\boldsymbol{r}_1 = \frac{(m_3\boldsymbol{X}_2 - m_2\boldsymbol{X}_3)}{M}; \quad \boldsymbol{r}_2 = \frac{(m_1\boldsymbol{X}_3 - m_3\boldsymbol{X}_1)}{M}; \quad \boldsymbol{r}_3 = \frac{(m_2\boldsymbol{X}_1 - m_1\boldsymbol{X}_2)}{M}, \tag{2.54}$$

and the velocities obey the same rule.

2.6 *N*-Body Algorithms

In an N-body system the *Logarithmic Hamiltonian (LogH)*

$$\Lambda = \ln(T + B) - \ln(U), \tag{2.55}$$

gives the equations of motion

$$t' = \frac{\partial \Lambda}{\partial B} = 1/(T+B); \quad \boldsymbol{r}_k' = \boldsymbol{v}_k/(T+B); \quad \boldsymbol{v}_k' = \boldsymbol{A}_k/U, \tag{2.56}$$

where $\boldsymbol{v}_k = \dot{\boldsymbol{r}_k}$ and $\boldsymbol{A}_k = \frac{\partial U}{\partial \boldsymbol{r}_k}/m_k$ are the velocity and acceleration correspondingly.

It is important to note that the derivatives of coordinates only depend on velocities and vice versa. This makes a simple leapfrog algorithm possible (see below). The most important feature is that, as discussed in Sect. 2.5.1, the resulting leapfrog is exact for two-body motion, except for a phase error, and thus *regularizes close approaches.*

The *Time-Transformed Leapfrog* (TTL) method is a generalization of this idea (Mikkola & Aarseth 2002). In the time transformation, one chooses some other function $\Omega(\boldsymbol{r})$ in place of the potential U and defines an auxiliary quantity W by the differential equation $\dot{W} = \dot{\Omega} = \frac{\partial \Omega}{\partial \boldsymbol{r}} \cdot \boldsymbol{v}$.

The resulting TTL equations read

$$t' = 1/W; \quad \boldsymbol{r}_k' = \frac{1}{W}\frac{\partial T}{\partial \boldsymbol{p}_k}; \quad \boldsymbol{v}_k' = \frac{1}{\Omega}\boldsymbol{A}_k; \quad W' = \sum_k \frac{\partial \Omega}{\partial \boldsymbol{r}_k} \cdot \boldsymbol{v}_k/\Omega, \tag{2.57}$$

and these can also be used to construct a leapfrog-like mapping which, for suitable functions Ω, are asymptotically exact for two-body motion near collision. It can be shown that TTL is mathematically equivalent to LogH if one takes $\Omega = U$.

2.6.1 LogH Leapfrog

First, one computes the constant $B = -T + U$ from initial values. The equations of motion can be used to define the basic mappings $\boldsymbol{X}(s)$ and $\boldsymbol{V}(s)$ as

$$\begin{aligned} &\boldsymbol{X}(\mathrm{s}): \ \delta t = s/(T+B); \quad t \to t + \delta t; \quad \boldsymbol{r}_k \to \boldsymbol{r}_k + \delta t\, \boldsymbol{v}_k \\ &\boldsymbol{V}(\mathrm{s}): \ \widetilde{\delta t} = s/U; \quad \boldsymbol{p} \to \boldsymbol{p}_k + \widetilde{\delta t}\boldsymbol{A}_k, \end{aligned} \tag{2.58}$$

which can be evaluated in a sequence

$$\boldsymbol{X}(\mathrm{h}/2)\boldsymbol{V}(\mathrm{h})\mathrm{X}(\mathrm{h}/2),$$

using always the most recent results as input for the next operation.

2.6.2 TTL

Here one first evaluates the initial value of $W = \Omega$, then uses the leapfrog mappings

$$\boldsymbol{X}(\mathrm{s}): \quad \delta t = s/W; \quad t \to t + \delta t; \quad \boldsymbol{r}_k \to \boldsymbol{r}_k + \delta t\, \boldsymbol{v}_k \tag{2.59}$$

$$\boldsymbol{V}(\mathrm{s}): \quad \widetilde{\delta t} = s/\Omega; \quad \delta \boldsymbol{v}_k = \widetilde{\delta t}\boldsymbol{A}_k; \quad W \to W + \widetilde{\delta t} \sum_k \frac{\partial \Omega}{\partial \boldsymbol{r}_k} \cdot \left(\boldsymbol{v}_k + \frac{1}{2}\delta \boldsymbol{v}_k\right)$$

$$\boldsymbol{v}_k \to \boldsymbol{v}_k + \delta \boldsymbol{v}_k, \tag{2.60}$$

to advance the coordinates and velocities using the operation sequence

$$\boldsymbol{X}(\mathrm{h}/2)\boldsymbol{V}(\mathrm{h})\boldsymbol{X}(\mathrm{h}/2)$$

repeatedly.

For Ω one may use any suitable function, but usually it is advantageous to take

$$\Omega = \sum_{i<j} \frac{\Omega_{ij}}{r_{ij}},$$

where

$$\Omega_{ij} = 1, \qquad \text{or} \qquad \Omega_{ij} = m_i m_j,$$

the latter choice being recommended if the masses are comparable.

The leapfrog alone is, however, in many cases not accurate enough. The accuracy can be improved, e.g. by using the higher-order leapfrog algorithms of Yoshida (1990). Alternatively, one may use the extrapolation method (Bulirsch & Stoer 1966; Press et al. 1986).

2.7 AR-Chain

First of all it is necessary to emphazise the importance of the chain structure, not only in the KS-chain method but also when one uses one of the algorithmic regularizations. The reason is round-off errors. If one uses centre-of-mass coordinates, the relative coordinates of a distant close pair are differences of large numbers and there is considerable cancellation of significant figures, leading to irrecoverable errors.

This section discusses a new code that uses the chain structure and a mixture of the LogH and TTL-methods.

2.7.1 Finding and Updating the Chain

We begin by finding the shortest interparticle vector for the first part of the chain. Next we search for the particle closest to one or the other end of the presently known part of the chain. This particle is added to the closest end

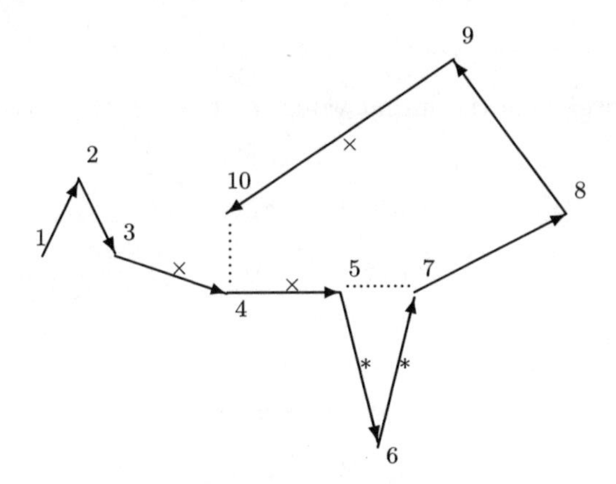

Fig. 2.4. Illustration of the chain and the checking of switching conditions. Distances like $R_{5,7}$ are compared with the smaller of the two distances $R_{5,6}$ and $R_{6,7}$ (marked by *). Interparticle distances like $R_{4,10}$ are compared with the smallest of those in contact with the considered distance (marked by $\times$)

of the already existing chain. This is repeated until all particles are included in the chain. The particles are then re-numbered along the chain as $1, 2, \ldots N$ for ease of programming.

After every integration step, we check for the need of updating the chain. Figure 2.4 illustrates the case of a 10-particle chain. To avoid some potential round-off problems, it is advantageous to carry out the transformation from the old chain vectors $\boldsymbol{X}_k$ to the new ones directly by expressing the new chain vectors as sums of the old ones.

Let the actual "physical" names of the chain particles $1, \ldots, N$ (as defined above) be $I_1,\ I_2, \ldots, I_N$ and let us use the notation I_k^{old} and I_k^{new} for the names in the old and new chains. Then we may write

$$\boldsymbol{r}_{I_k^{\text{old}}} = \sum_{\nu=1}^{k-1} \boldsymbol{X}_\nu^{\text{old}} \tag{2.61}$$

$$\boldsymbol{X}_\mu^{\text{new}} = \boldsymbol{r}_{I_{\mu+1}^{\text{new}}} - \boldsymbol{r}_{I_\mu^{\text{new}}}. \tag{2.62}$$

Thus we need to use the correspondence between the old and the new indices to express the new chain vectors $\boldsymbol{X}$ in terms of the old ones. One finds that if k_0 and k_1 are two indices such that $I_{k_0}^{\text{old}} = I_\mu^{\text{new}}$ and $I_{k_1}^{\text{old}} = I_{\mu+1}^{\text{new}}$, then

$$\boldsymbol{X}_\mu^{\text{new}} = \sum_{\nu=1}^{N-1} B_{\mu\nu} \boldsymbol{X}_\nu^{\text{old}}, \tag{2.63}$$

where $B_{\mu\nu} = +1$ if$(k_1 > \nu \;\&\; k_0 \le \nu)$ and $B_{\mu\nu} = -1$ if$(k_1 \le \nu \;\&\; k_0 > \nu)$, otherwise $B_{\mu\nu} = 0$.

2.7.2 Transformations

After selecting the chain, and renaming the particles as $1, 2, \ldots, N$ along the chain, one can evaluate the initial values for the chain vectors and velocities as

$$\boldsymbol{X}_k = \boldsymbol{r}_{k+1} - \boldsymbol{r}_k \tag{2.64}$$

$$\boldsymbol{V}_k = \boldsymbol{v}_{k+1} - \boldsymbol{v}_k, \tag{2.65}$$

where $\boldsymbol{v}_k = \dot{\boldsymbol{r}}_k$. At the same time one may evaluate the centre-of-mass quantities

$$M = \sum_k m_k \tag{2.66}$$

$$\boldsymbol{r}_{cm} = \sum_k m_k \boldsymbol{r}_k / M \tag{2.67}$$

$$\boldsymbol{v}_{cm} = \sum_k m_k \boldsymbol{v}_k / M. \tag{2.68}$$

The transformation back to $\boldsymbol{r}, \boldsymbol{v}$ can be done by simple summation

$$\widetilde{\boldsymbol{r}}_1 = \boldsymbol{0} \tag{2.69}$$

$$\widetilde{\boldsymbol{v}}_1 = \boldsymbol{0} \tag{2.70}$$

$$\widetilde{\boldsymbol{r}}_{k+1} = \widetilde{\boldsymbol{r}}_k + \boldsymbol{X}_k \tag{2.71}$$

$$\widetilde{\boldsymbol{v}}_{k+1} = \widetilde{\boldsymbol{v}}_k + \boldsymbol{V}_k, \tag{2.72}$$

followed by reduction to the centre of mass

$$\widetilde{\boldsymbol{r}}_{cm} = \sum_k m_k \widetilde{\boldsymbol{r}}_k / M \tag{2.73}$$

$$\widetilde{\boldsymbol{v}}_{cm} = \sum_k m_k \widetilde{\boldsymbol{v}}_k / M \tag{2.74}$$

$$\boldsymbol{r}_k = \widetilde{\boldsymbol{r}}_k - \widetilde{\boldsymbol{r}}_{cm} \tag{2.75}$$

$$\boldsymbol{v}_k = \widetilde{\boldsymbol{v}}_k - \widetilde{\boldsymbol{v}}_{cm}. \tag{2.76}$$

However, it is not always necessary to reduce the coordinates to the centre-of-mass system since accelerations only depend on the differences.

2.7.3 Equations of Motion and the Leapfrog

The equations of motion read

$$\dot{\boldsymbol{X}}_k = \boldsymbol{V}_k \tag{2.77}$$

$$\dot{\boldsymbol{V}}_k = \boldsymbol{A}_{k+1} - \boldsymbol{A}_k, \tag{2.78}$$

where the accelerations $\boldsymbol{A}_k$, with possible external effects $\boldsymbol{f}_k$, are

$$\boldsymbol{A}_k = -\sum_{j\neq k} m_j \frac{\boldsymbol{r}_{jk}}{|\boldsymbol{r}_{jk}|^3} + \boldsymbol{f}_k, \tag{2.79}$$

and, for $j < k$

$$\boldsymbol{r}_{jk} = \begin{cases} \boldsymbol{r}_k - \boldsymbol{r}_j, & \text{if } k > j+2 \\ \boldsymbol{X}_j, & \text{if } k = j+1\ . \\ \boldsymbol{X}_j + \boldsymbol{X}_{j+1}, & \text{if } k = j+2 \end{cases} \tag{2.80}$$

For $k > j$ one uses the fact that $\boldsymbol{r}_{jk} = -\boldsymbol{r}_{kj}$. The use of $\boldsymbol{X}_j$ and $\boldsymbol{X}_j + \boldsymbol{X}_{j+1}$ reduces the round-off effect significantly. More generally, one could also use

$$\boldsymbol{r}_{kj} = \sum_{\nu=j}^{k-1} \boldsymbol{X}_\nu, \tag{2.81}$$

but for many bodies it is faster to use the above recipe (2.80) and the latter alternative seems not to improve the results.
The kinetic energy is

$$T = \frac{1}{2}\sum_k m_k \boldsymbol{v}_k^2, \tag{2.82}$$

and the potential energy

$$U = \sum_{i<j} \frac{m_i m_j}{|\boldsymbol{r}_{ij}|}, \tag{2.83}$$

which is evaluated along with the accelerations according to (2.80). We introduce further a time transformation function

$$\Omega = \sum_{i<j} \frac{\Omega_{ij}}{|\boldsymbol{r}_{ij}|}, \tag{2.84}$$

where Ω_{ij} are some selected coefficients (to be discussed below).

Now one may define the two time transformations

$$t' \;=\; 1/(\alpha(T+B) + \beta\omega + \gamma) = \;1/(\alpha U + \beta\Omega + \gamma), \tag{2.85}$$

where α, β and γ are adjustable constants, $B = U - T$ is the N-body binding energy and ω is defined by the differential equation

$$\dot{\omega} = \sum_k \frac{\partial\Omega}{\partial\boldsymbol{r}_k}\cdot\boldsymbol{v}_k, \tag{2.86}$$

and the initial value $\omega(0) = \Omega(0)$. The binding energy B changes according to

$$\dot{B} = -\sum_k m_k \boldsymbol{v}_k \cdot \boldsymbol{f}_k. \tag{2.87}$$

The equations of motion that can be used to construct the leapfrog which provides algorithmic regularization are, for time and coordinates respectively,

$$t' = 1/(\alpha(T+B) + \beta\omega + \gamma) \tag{2.88}$$

$$\boldsymbol{r}'_k = t'\boldsymbol{v}_k, \tag{2.89}$$

and for velocities B and ω,

$$\tau' = 1/(\alpha U + \beta\Omega + \gamma) \tag{2.90}$$

$$\boldsymbol{v}'_k = \tau'\boldsymbol{A}_k \tag{2.91}$$

$$B' = \tau'\sum_k(-m_k\boldsymbol{v}_k \cdot \boldsymbol{f}_k) \tag{2.92}$$

$$\omega' = \tau'\sum_k \frac{\partial\Omega}{\partial\boldsymbol{r}_k} \cdot \boldsymbol{v}_k. \tag{2.93}$$

To account for the $\boldsymbol{v}$-dependence of B' and ω', one must follow Mikkola & Aarseth (2002), i.e. first the $\boldsymbol{v}_k$ are advanced and then the average $< \boldsymbol{v}_k >= (\boldsymbol{v}_k(0) + \boldsymbol{v}_k(h))/2$ is used to evaluate B' and ω'.

The leapfrog for the chain vectors $\boldsymbol{X}_k$ and $\boldsymbol{V}_k$ can be written most easily in terms of the two mappings

$\boldsymbol{X}(s)$:

$$\delta t = s/(\alpha(T+B) + \beta\omega + \gamma) \tag{2.94}$$

$$t = t + \delta t \tag{2.95}$$

$$\boldsymbol{X}_k \to \boldsymbol{X}_k + \delta t \boldsymbol{V}_k \tag{2.96}$$

$$\tag{2.97}$$

$\boldsymbol{V}(s)$:

$$\widetilde{\delta t} = s/(\alpha U + \beta\Omega + \gamma) \tag{2.98}$$

$$\boldsymbol{V}_k \to \boldsymbol{V}_k + \widetilde{\delta t}(\boldsymbol{A}_{k+1} - \boldsymbol{A}_k) \tag{2.99}$$

$$B \to B + \widetilde{\delta t}\sum_k(-m_k < \boldsymbol{v}_k > \cdot \boldsymbol{f}_k) \tag{2.100}$$

$$\omega \to \omega + \widetilde{\delta t}\sum_k \frac{\partial\Omega}{\partial\boldsymbol{r}_k} \cdot < \boldsymbol{v}_k >, \tag{2.101}$$

where $< \boldsymbol{v}_k >$ is the average of the initial and final $\boldsymbol{v}$'s here. Note that it is also necessary to evaluate the individual velocities $\boldsymbol{v}_k$, because the expression for B' and ω' would otherwise (in terms of the chain vector velocities $\boldsymbol{V}_k$) become rather cumbersome.

One leapfrog step can then be written simply as

$$\boldsymbol{X}(h/2)\boldsymbol{V}(h)\boldsymbol{X}(h/2),$$

and a longer sequence of n steps reads

$$\boldsymbol{X}(h/2)\left[\Pi_{\nu=1}^{n-1}(\boldsymbol{V}(h)\boldsymbol{X}(h))\right]\boldsymbol{V}(h)\boldsymbol{X}(h/2).$$

This is the formulation to be used with the extrapolation method when proceeding over a total time interval of length nh.

2.7.4 Alternative Time Transformations

If one takes

$$\Omega_j = m_i m_j, \tag{2.102}$$

then $\alpha = 0,\ \beta = 1,\ \gamma = 0$ is mathematically equivalent to $\alpha = 1,\ \beta = \gamma = 0$, as was shown in Mikkola & Aarseth (2002). However, numerically these are not equivalent, and the LogH alternative is much more stable. On the other hand, as noted above, it is desirable to get stepsize shortening (and thus regularization) also for encounters of small bodies and thus some function Ω should also be included.

To increase the numerical stability for strong interactions of big bodies and smooth the encounters of small bodies, one may use $\alpha = 1,\ \beta \neq 0$ and

$$\Omega_{ij} = \begin{cases} \widetilde{m}^2, & \text{if } m_i m_j < \epsilon \widetilde{m}^2 \\ 0, & \text{otherwise} \end{cases}, \tag{2.103}$$

where $\widetilde{m}^2 = \sum_{i<j} m_i m_j/(N(N-1)/2)$ is the mean mass product and ϵ an adjustable parameter ($\epsilon \sim 10^{-3}$ may be a good guess). It is sometimes advantageous to integrate (2.86) for ω even if $\beta = 0$. This is because the integrator (extrapolation method!) is forced to use short steps where $\dot{\omega}$ is large, thus giving higher precision when required.

Remarks

1. If $(\alpha, \beta, \gamma) \propto (1, 0, 0)$, the method is the logarithmic Hamiltonian method (LogH) of Mikkola & Tanikawa (1999a).
2. If $(\alpha, \beta, \gamma) \propto (0, 1, 0)$, the method is the transformed leapfrog (TTL) (Mikkola & Aarseth 2002).
3. If $(\alpha, \beta, \gamma) \propto (0, 0, 1)$, the method is the normal basic leapfrog.
4. Which combination of the numbers (α, β, γ) is best cannot be answered in general. For N-body systems with very large mass ratios, one must have $\beta \neq 0$, but some small value is advantageous. This is because low-mass bodies do not contribute significantly to the energies and if $\beta = 0$, the stepsize is not reduced sufficiently during a close encounter.

2.8 Basic Algorithms for the Extrapolation Method

2.8.1 Leapfrog

The extrapolation method (Gragg 1964, 1965; Bulirsch & Stoer 1966), which extrapolates results from a simple basic integrator to zero stepsize, is one of the most efficient methods to convert results of low-order basic integrators into highly accurate final outcomes. Often such an integrator can be conveniently chosen to be a composite integrator, like the leapfrog. Let the differential equations to be

$$\dot{\boldsymbol{x}} = \boldsymbol{f}(\boldsymbol{y}); \quad \dot{\boldsymbol{y}} = \boldsymbol{g}(\boldsymbol{x}), \tag{2.104}$$

then one can construct the the simple leapfrog algorithm

$$\boldsymbol{x}_{\frac{1}{2}} = \boldsymbol{x}_0 + \frac{h}{2}\boldsymbol{f}(\boldsymbol{y}_0) \tag{2.105}$$

$$\boldsymbol{y}_1 = \boldsymbol{y}_0 + h\boldsymbol{g}(\boldsymbol{x}_{\frac{1}{2}}) \tag{2.106}$$

$$\boldsymbol{x}_1 = \boldsymbol{x}_{\frac{1}{2}} + \frac{h}{2}\boldsymbol{f}(\boldsymbol{y}_1). \tag{2.107}$$

One notes that this is a slightly generalized formulation of the very basic leapfrog, which is obtained if $\boldsymbol{f}(\boldsymbol{y}) = \boldsymbol{y}$. In this case therefore $\boldsymbol{x}$ would be the coordinate vector, $\boldsymbol{y}$ the velocity vector and $\boldsymbol{g}(\boldsymbol{x})$ the acceleration.

Let us introduce the two mappings (or "subroutines")

$$\boldsymbol{X}(s): \quad \boldsymbol{x} \to \boldsymbol{x} + s\boldsymbol{f}(\boldsymbol{y}) \tag{2.108}$$

and

$$\boldsymbol{Y}(s): \quad \boldsymbol{y} \to \boldsymbol{y} + s\boldsymbol{g}(\boldsymbol{x}), \tag{2.109}$$

with which the above leapfrog can be symbolized as $\boldsymbol{X}(h/2)\boldsymbol{Y}(h)\boldsymbol{X}(h/2)$. When we want to compute n steps of stepsize $= h/n$, we can write

$$\boldsymbol{X}\left(\frac{h}{2n}\right)\left[\boldsymbol{Y}\left(\frac{h}{n}\right)\boldsymbol{X}\left(\frac{h}{n}\right)\right]^{n-1}\boldsymbol{Y}\left(\frac{h}{n}\right)\boldsymbol{X}\left(\frac{h}{2n}\right). \tag{2.110}$$

This advances the system over the time interval h.

The final results can now be considered to be a function of h/n and thus it is possible to extrapolate to zero stepsize. Due to the time symmetry of the leapfrog, the error has an (asymptotic) expansion of the form

$$a_2(h/n)^2 + a_4(h/n)^4 + \ldots,$$

i.e. the expansion contains only even powers of h. This makes the extrapolation process particularly efficient.

2.8.2 Midpoint Method

In addition to the leapfrog algorithm, commonly used in connection with the extrapolation method, we have the so-called modified midpoint method. This algorithm can also be formally written as a leapfrog. Let the differential equation be

$$\dot{\boldsymbol{z}} = \boldsymbol{f}(\boldsymbol{z}), \tag{2.111}$$

and let us split this into two parts as

$$\dot{\boldsymbol{x}} = \boldsymbol{f}(\boldsymbol{y}); \quad \dot{\boldsymbol{y}} = \boldsymbol{f}(\boldsymbol{x}). \tag{2.112}$$

If this pair of equations is solved using the initial conditions $\boldsymbol{x}(0) = \boldsymbol{y}(0) = \boldsymbol{z}(0)$, the solution is simply $\boldsymbol{x}(t) = \boldsymbol{y}(t) = \boldsymbol{z}(t)$. On the other hand, (2.112) is of the same form as (2.104) except that $\boldsymbol{g} = \boldsymbol{f}$, and it is possible to construct the leapfrog algorithm

$$\boldsymbol{x}_{\frac{1}{2}} = \boldsymbol{x}_0 + \frac{h}{2}\boldsymbol{f}(\boldsymbol{y}_0) \tag{2.113}$$

$$\boldsymbol{y}_1 = \boldsymbol{y}_0 + h\boldsymbol{f}(\boldsymbol{x}_{\frac{1}{2}}) \tag{2.114}$$

$$\boldsymbol{x}_1 = \boldsymbol{x}_{\frac{1}{2}} + \frac{h}{2}\boldsymbol{f}(\boldsymbol{y}_1), \tag{2.115}$$

the results of which can also be used for extrapolation to zero stepsize. Note that it is the vector $\boldsymbol{x}$ that is extrapolated while here $\boldsymbol{y}$ is just an auxiliary quantity. If one defines the mapping

$$\boldsymbol{A}(\boldsymbol{y}, \boldsymbol{x}, s): \quad \boldsymbol{x} \to \boldsymbol{x} + s\boldsymbol{f}(\boldsymbol{y}), \tag{2.116}$$

then, similar to (2.110), one can write for the results with stepsize $= h/n$

$$\boldsymbol{A}\left(\boldsymbol{y}, \boldsymbol{x}, \frac{h}{2n}\right)\left[\boldsymbol{A}\left(\boldsymbol{x}, \boldsymbol{y}, \frac{h}{n}\right)\boldsymbol{A}\left(\boldsymbol{y}, \boldsymbol{x}, \frac{h}{n}\right)\right]^{n-1}\boldsymbol{A}\left(\boldsymbol{x}, \boldsymbol{y}, \frac{h}{n}\right)\boldsymbol{A}\left(\boldsymbol{y}, \boldsymbol{x}, \frac{h}{2n}\right), \tag{2.117}$$

where $\boldsymbol{x} = \boldsymbol{z}(0)$, $\boldsymbol{y} = \boldsymbol{z}(0)$ initially.

2.8.3 Generalized Midpoint Method

Here we introduce a generalization of the well-known modified midpoint method. In this algorithm, the basic approximation to advance the solution is not just the evaluation of the derivative at the midpoints, but any method to approximate the solution. Thus e.g. the algorithmic regularization by the leapfrog can be used even when there are additional forces depending on velocities. This provides a regular basic algorithm, which is made suitable for the extrapolation method by means of the generalized midpoint method.

The starting point in this algorithm (Mikkola & Merritt 2006, 2008) is the same as in the previous (midpoint method) section, i.e. the problem considered is

$$\dot{\boldsymbol{z}} = \boldsymbol{f}(\boldsymbol{z}), \quad \boldsymbol{z}(0) = \boldsymbol{z}_0, \tag{2.118}$$

and it is split into two as $\dot{\boldsymbol{x}} = \boldsymbol{f}(\boldsymbol{y})$, $\dot{\boldsymbol{y}} = \boldsymbol{f}(\boldsymbol{x})$ and the leapfrog-like algorithm (the modified midpoint method) is

$$\boldsymbol{x}_{\frac{1}{2}} = \boldsymbol{x}_0 + \frac{h}{2}\boldsymbol{f}(\boldsymbol{y}_0), \quad \boldsymbol{y}_1 = \boldsymbol{y}_0 + hf(\boldsymbol{x}_{\frac{1}{2}}), \quad \boldsymbol{x}_1 = \boldsymbol{x}_{\frac{1}{2}} + \frac{h}{2}\boldsymbol{f}(\boldsymbol{y}_1).$$

A new interpretation of the above can be obtained by first rewriting it in the form

$$\boldsymbol{x}_{\frac{1}{2}} = \boldsymbol{x}_0 + \left[+\frac{h}{2}\boldsymbol{f}(\boldsymbol{y}_0)\right] \tag{2.119}$$

$$\boldsymbol{y}_{\frac{1}{2}} = \boldsymbol{y}_0 - \left[-\frac{h}{2}f(\boldsymbol{x}_{\frac{1}{2}})\right] \tag{2.120}$$

$$\boldsymbol{y}_1 = \boldsymbol{y}_{\frac{1}{2}} + \left[+\frac{h}{2}f(\boldsymbol{x}_{\frac{1}{2}})\right] \tag{2.121}$$

$$\boldsymbol{x}_1 = \boldsymbol{x}_{\frac{1}{2}} - \left[-\frac{h}{2}\boldsymbol{f}(\boldsymbol{y}_1)\right]. \tag{2.122}$$

In (2.119) the bracketed term is an (Euler-method) approximation to the increment of $\boldsymbol{x}$ over the time interval $h/2$ with the initial value $\boldsymbol{y}_0$, while in (2.120) the initial value is $\boldsymbol{x}_{\frac{1}{2}} \approx \boldsymbol{x}(h/2)$ and the time interval is $-h/2$. Finally, this increment is added – with a minus sign – to $\boldsymbol{y}_0$ to obtain an approximation for $\boldsymbol{y}(h/2)$. In the remaining formulae (2.121) and (2.122), the idea is the same but the roles of $\boldsymbol{x}$ and $\boldsymbol{y}$ have been changed.

A generalization of this follows readily. Let $\mathbf{d}(\boldsymbol{z}_0, \Delta t)$ be an increment for $\boldsymbol{z}$, such that

$$\boldsymbol{z}(\Delta t) \approx \boldsymbol{z}_0 + \mathbf{d}(\boldsymbol{z}_0, \Delta t) \tag{2.123}$$

is an approximation to the solution of (2.118) over a time interval Δt. One step in the generalized midpoint method can now be written

$$\boldsymbol{x}_{\frac{1}{2}} = \boldsymbol{x}_0 + \mathbf{d}\left(\boldsymbol{y}_0, +\frac{h}{2}\right) \tag{2.124}$$

$$\boldsymbol{y}_{\frac{1}{2}} = \boldsymbol{y}_0 - \mathbf{d}\left(\boldsymbol{x}_{\frac{1}{2}}, -\frac{h}{2}\right) \tag{2.125}$$

$$\boldsymbol{y}_1 = \boldsymbol{y}_{\frac{1}{2}} + \mathbf{d}\left(\boldsymbol{x}_{\frac{1}{2}}, +\frac{h}{2}\right) \tag{2.126}$$

$$\boldsymbol{x}_1 = \boldsymbol{x}_{\frac{1}{2}} - \mathbf{d}\left(\boldsymbol{y}_1, -\frac{h}{2}\right), \tag{2.127}$$

or, if we define the mapping (or "subroutine")

$$\boldsymbol{A}(\boldsymbol{x}, \boldsymbol{y}, h): \quad \boldsymbol{x} \to \boldsymbol{x} + \mathbf{d}\left(\boldsymbol{y}, +\frac{h}{2}\right) \tag{2.128}$$

$$\boldsymbol{y} \to \boldsymbol{y} - \mathbf{d}\left(\boldsymbol{x}, -\frac{h}{2}\right), \tag{2.129}$$

we can write the algorithm with many (n) steps as

$$\begin{aligned} &1.\ \text{Initialize } \boldsymbol{y} = \boldsymbol{x};\\ &2.\ \text{Repeat } \boldsymbol{A}(\boldsymbol{x},\boldsymbol{y},h)\boldsymbol{A}(\boldsymbol{y},\boldsymbol{x},h)\ \ n \text{ times};\\ &3.\ \text{Take } \boldsymbol{x} \text{ as the final result.} \end{aligned} \tag{2.130}$$

Thus one simply calls the subroutine $\boldsymbol{A}$ alternately with arguments $(\boldsymbol{x},\boldsymbol{y})$ and $(\boldsymbol{y},\boldsymbol{x})$ such that the sequence is time-symmetric (starts and stops with $\boldsymbol{x}$ in (2.130)).

This basic algorithm has the correct symmetry – because it was derived from a leapfrog-like treatment and thus the Gragg-Bulirsch-Stoer extrapolation method can be used to obtain high accuracy.

This generalized midpoint algorithm may be especially useful if one employs a special method, well-suited to the particular problem at hand, to obtain the increment $\boldsymbol{d}$. For the few-body problem, with velocity-dependent external perturbations, such a method is the algorithmic regularization leapfrog. The external perturbation (with possible dependence on velocities) can be added to the increment as

$$\mathbf{d} \rightarrow \mathbf{d} + \Delta t \boldsymbol{f}(\boldsymbol{v},..), \tag{2.131}$$

where $\boldsymbol{f}$ is the external perturbation and $\boldsymbol{v}$ is the most recent velocity value available. Further on, the leapfrog can be replaced by any other method that is not necessarily time-symmetric since the algorithm generates the right kind of symmetry.

2.8.4 Lyapunov Exponents

When the Lyapunov exponents (usually the largest one is sufficient) are required, the normal practice is that one derives the variational equations and then programs the integration of those equations. In practice there exists another, simpler, way to do the necessary programming:

1. First one writes the code to integrate the basic problem. It is a good idea to use rather simple program statements.
2. One differentiates the resulting (and tested!) code, line by line, adding the necessary lines for evaluation of the variations.
3. This is the simplest way to write the code for the variations, since there is no reason to consider the variational equations at all. Instead, one mechanically differentiates every program statement, thus getting the exact variations of the algorithm.
4. That is the best one can do!

Perhaps, the best way to clarify the above is to give a simple example. Here is a leapfrog algorithm for the harmonic oscillator. First is shown the pure harmonic oscillator code, then the version with variations. The differentiated lines that evaluate the variations are marked as "var".

```
 c Leapfrog code for a harmonic oscillator:
c-----------------------------------------------
      implicit real*8 (a-h,o-z)
      x=1
      p=0
      h=0.01d0
      E0=(p*p+x*x)/2
      t=0
1     continue
      x=x+h/2*p ! this is
      p=p-h*x   ! a leapfrog
      x=x+h/2*p ! step
      t=t+h
c     diagnostics
      E=(p*p+x*x)/2
      write(6,*)t,x,p,E-E0
      if(t.lt.100.)goto 1 ! max time=100.
      end


c     Differentiated  leapfrog for harmonic oscillator
c----------------------------------------------
      implicit real*8 (a-h,o-z)
      x=1
         dx=1             ! var
      p=0
         dp=0             ! var
      E0=(p*p+x*x)/2
         dE0=p*dp+x*dx    ! var
      t=0
      h=0.01d0 ! stepsize
1     continue
      x=x+h/2*p ! this is
        dx=dx+h/2*dp      ! var
      p=p-h*x   ! a leapfrog
        dp=dp-h*dx        ! var
      x=x+h/2*p ! step
        dx=dx+h/2*dp      ! var
      t=t+h
c     diagnostics
      E=(p*p+x*x)/2
        dE=p*dp+x*dx      ! var (this should be constant!)
      write(6,*)t,x,p,E-E0,dE-dE0
      if(t.lt.100.)goto 1 ! max time=100.
      end
```

The harmonic oscillator example is almost trivial but explains anyway how the variations can be obtained by differentiating the original code mechanically, without any need to consider the variational equations. The same technique is useful for almost any algorithm, however complicated. One easy check to implement for the the variations is based on the fact that the differentials of constants of motion are also constants of motion. Above there is only one integral, the total energy. The differential should thus remain (approximately) constant. In the few-body problem this applies to the components of angular momentum also. Finally, in terms of the variations δq, the Lyapunov exponents (approximations for) can be obtained as

$$\lambda \approx \ln(|\delta q|)/t \tag{2.132}$$

when the time t is sufficiently large.

In time-transformed systems all the variables, including the time t, have variations. Often the results are wanted in the "physical" system where time is the independent variable. One must thus eliminate the time-variation effect. If f is any function of the system variables and time, the physical system variation Δf and the time-transformed system variation δf are related by

$$\Delta f = \delta f - \delta t\, \dot{f}, \tag{2.133}$$

where $\dot{f}$ is the total time derivative of f.

2.9 Accuracy of the AR-Chain

To demonstrate the ability of the AR-chain code to handle large mass ratios, we plot in Fig. 2.5 the energy and angular momentum errors in a system with a wide range of masses (two masses $m_1 = m_2 = 1$, and the rest were assigned values $0.1, 0.01, 0.001, \ldots, 10^{-8}$. Due to the large range of masses, the KS-chain cannot integrate the motions in this system satisfactorily, but AR-chain is fast and accurate.

The system evolves by ejecting most of the small masses in the time interval illustrated. The energy errors in this example are shown in two ways: the uppermost curve gives the relative error in energy computed as $1-E/E_0$ while the lowermost curve is the value of the logarithmic Hamiltonian (essentially the same as $(E - E_0)/U$. The absolute error of the angular momentum is also illustrated in the figure. Somewhat surprisingly, the relative error of the energy fluctuates considerably, while the value of the logarithmic Hamiltonian evolves much more slowly. The reason for this is that, since the Hamiltonian is $\log((T-E/U))$, the algorithm attempts to keep this quantity constant (and not the energy E). In fact, it is inevitable that integration errors give a small non-zero value for the logarithmic Hamiltonian $\log((T - E)/U) = \epsilon$, from which we can derive the energy error

$$\delta E = \epsilon U, \tag{2.134}$$

assuming the logarithmic Hamiltonian remains constant. Thus it is essentially the variation of the potential energy U that causes the fluctuation of the energy error in the above figure. We conclude that all the illustrated errors are sufficiently small, of the order of magnitude of round-off error effects.

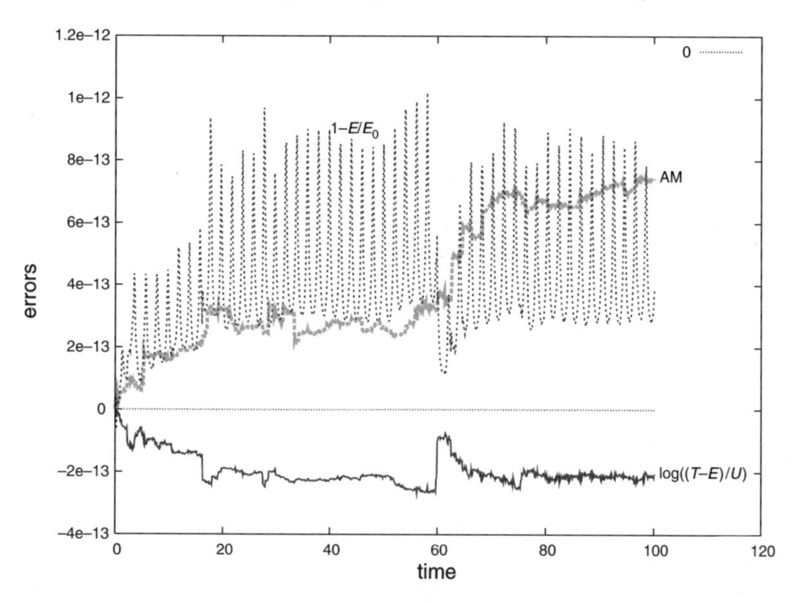

Fig. 2.5. Errors in a 10-body problem integrated with the AR-chain code. The system consists of a heavy binary (component masses = 1, eccentricity $e = 0.5$) and the other particles have masses 10^{-n} for $n = 1, 2, 3, \ldots, 8$. Uppermost curve: relative error of energy ($= 1 - E/E_0$); lowermost curve: $\log((T - E)/U)$, which is the value of the logarithmic Hamiltonian; the thick curve (AM): absolute error in the angular momentum

2.10 Conclusions

Experience has shown that generally the AR-chain is comparable in accuracy with the KS-chain in most practical problems (the one-dimensional N-body problem being an exception). With the modified midpoint method, AR-chain is efficient also in problems with velocity-dependent external forces. A further advantage is the fact that, contrary to KS-chain, soft potentials can readily be treated without any problem. Also the differentiation of the algorithms is sufficiently simple, especially for the three-body algorithm, discussed in Sect. 2.5.3, so that one can evaluate the Lyapunov exponents.

In summary:

1. KS-chain is the most efficient KS-regularized code, but restricted to comparable masses (say mass ratios of $\sim 10^4$). A possible drawback for some problems is that a soft potential cannot be used.
2. LogH is a good alternative for comparable masses.
3. TTL can handle large mass ratios, but may suffer from round-off errors.
4. AR-chain can handle large mass ratios and soft potential. With the generalized midpoint method, velocity-dependent external forces can also be

included with no problem. Consequently, AR-chain is a good alternative to the KS-chain and in many problems the best method.

5. For all the algorithms discussed here, use of the extrapolation method (Bulirsch & Stoer 1966; Press et al. 1986) is necessary to improve the leapfrog results to high accuracy.

Finally, it is necessary to stress that the codes discussed here are stand-alone few-body codes requiring additional programming when implementing them for large N-body systems.[1]

References

Aarseth S. J., 2003, Gravitational N-Body Simulations. Cambridge University Press, Cambridge
Bulirsch R., Stoer J., 1966, Num. Math., 8, 1
Gragg W. B., 1964, Ph.D. thesis, University of California, Los Angeles
Gragg W. B., 1965, SIAM J. Numer. Anal., 2, 384
Heggie D. C., 1974, Celes. Mech., 10, 217
Kustaanheimo P., Stiefel E., 1965, J. Reine Angew. Math., 218, 204
Levi-Civita T., 1920, Acta Math., 42, 99
Mikkola S., Aarseth S. J., 1993, Celes. Mech. Dyn. Astron., 57, 439
Mikkola S., Aarseth S., 2002, Celes. Mech. Dyn. Astron., 84, 343
Mikkola S., Merritt D., 2006, MNRAS, 372, 219
Mikkola S., Merritt D., 2008, AJ, 135, 2398
Mikkola S., Tanikawa K., 1999a, MNRAS, 310, 745
Mikkola S., Tanikawa K., 1999b, Celes. Mech. Dyn. Astron., 74, 287
Press W. H., Flannery B. P.,Teukolsky S. A., Wetterling W. T., 1986, Numerical Recipes. Cambridge University Press, Cambridge
Preto M., Tremaine S., 1999, AJ, 118, 2532
Soffel M. H., 1989, Relativity in Astrometry, Celestial Mechanics and Geodesy. Springer-Berlin, p. 141
Stiefel E. L., Scheifele G., 1971, Linear and Regular Celestial Mechanics. Springer, Berlin
Yoshida H., 1990, Phys. Lett. A, 150, 262
Zare K., 1974, Celes. Mech., 10, 207

[1]Some source codes can be found on `http://www.cambody.org/codes.php`.

3

Resonance, Chaos and Stability: The Three-Body Problem in Astrophysics

Rosemary A. Mardling

School of Mathematical Sciences, Monash University, Victoria 3800, Australia, mardling@sci.monash.edu.au

3.1 Introduction

In his Oppenheimer lecture entitled "Gravity is cool, or, why our universe is as hospitable as it is", Freeman Dyson discusses how time has two faces: the quick violent face and the slow gentle face, the face of the destroyer and the face of the preserver (Dyson 2000). He entirely attributes these two faces to gravity and the ease with which gravitational energy can change irreversibly into other forms of energy. The simplest system exhibiting these two faces is that of three gravitating bodies; for most configurations, the slow gentle face is the norm, while for a very important subset, violence is the order of the day. In fact it is this violence, resulting in one of the bodies being ejected from the system, which is responsible for much of the structure we see in the universe, from planets to giant elliptical galaxies.

The simplest example of a quiescent gravitating system is that of two bodies orbiting each other at a distance large enough that their potentials are essentially those of point masses. Their paths about the common centre of mass are simple ellipses, and these paths do not change from orbit to orbit; their shapes (eccentricities) are preserved as are their sizes (semi-major axes) and orientations in space (inclination and longitudes of periastron and ascending nodes measured with respect to some reference set of axes; see Fig. 3.1). However, add one more body to the system and this wealth of symmetry is lost, at least to some extent. In the simplest case, if the binary components have equal mass and the third body orbits the binary in the same plane and is "sufficiently distant", the original binary will simply rotate about its centre of mass: this is apsidal motion. Its eccentricity and semi-major axis will not be affected, and the third body will orbit the centre of mass of the binary as if the latter were a single body with mass equal to the sum of the component masses. No *net* energy or angular momentum is exchanged between the inner and outer orbits in this simple case. If the inner binary components have different masses, some angular momentum *is* exchanged between the orbits, with the result that the eccentricities oscillate about some mean values.

Mardling, R.A.: *Resonance, Chaos and Stability: The Three-Body Problem in Astrophysics*. Lect. Notes Phys. **760**, 59–96 (2008)
DOI 10.1007/978-1-4020-8431-7_3

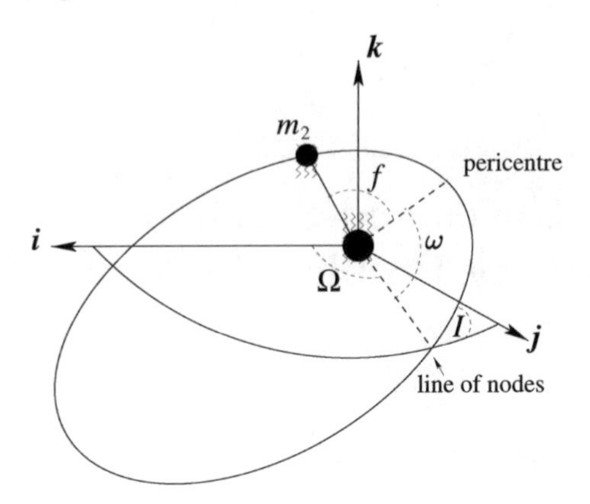

Fig. 3.1. Orbital elements specifying the orientation and phase of a binary relative to a fixed coordinate system: ω is the argument of periastron; Ω is the longitude of the ascending node; I is the orbital inclination and f is the true anomaly, the latter being one of several ways of specifying the orbital phase

This is most pronounced when one body is much more massive than the other two, as is the case in a planetary system, because very close stable systems can exist.

If the orbit of the third body is out of the plane of the binary, in addition to apsidal motion both orbits will rock (nutate) up and down, that is, their relative inclination will oscillate about some mean value and the planes of their orbits will rotate about the direction defined by the total angular momentum of the system (precession).[1] No energy and very little angular momentum is exchanged between the orbits of such a system,[2] even though the eccentricity of the inner binary may oscillate substantially about some mean value, a phenomenon called the *Kozai effect* (Kozai 1962).

These variations of the elements generally occur on time-scales much longer than the component orbital periods and are referred to as *secular variations*. They are characterized by zero energy exchange between the orbits, which manifests itself in the constancy of the semi-major axes of both the inner and the outer orbits.[3] In contrast to this, *unstable* systems, defined as those for which one body eventually escapes to infinity, necessarily must exchange energy between the orbits in order for this to occur. If one makes a plot in the parameter space of initial conditions associated with secular and unstable behaviour, one finds a very sharp boundary between the two.

I was led to the study of stability in the three-body problem after discovering that the energy exchange process between the tides and the orbit in a close binary system can be chaotic (Mardling 1995a,b). One day Sverre

[1]Note that apsidal motion is often mistakenly referred to as precession.

[2]Again, except if the system is a very close planetary-like system.

[3]Except for stable *resonant* systems; see later.

Aarseth was looking at my stability plots and commented that they reminded him of some plots made by Peter Eggleton and Luda Kiseleva for three-body hierarchies (Eggleton & Kiseleva 1995). He wondered whether or not the two problems might be linked. It turns out that they are; much of the analysis presented in this chapter can equally be applied to the binary-tides problem.

Throughout this chapter I will refer to five intimately related works submitted or in progress: M1a (Mardling 2008a) and M1b discuss stability in the three-body problem, the former coplanar systems and the latter inclined, M2 discusses the resonant structure of eccentric planetary systems, M3 (Mardling 2008b) presents a simple formalism for studying the secular evolution of arbitrary triple configurations,[4] while M4 presents a new formalism for studying strong three-body interactions.

3.2 Resonance in Nature

The most familiar example of resonance in action is a parent pushing a child on a swing. The only way to increase the amplitude of the swing consistently is to push it at its natural frequency. But if you think about it, the "natural frequency" varies depending on the amplitude of the swing; while it is pretty much constant over the range of amplitudes tolerated by most children, for the intrepid child who prefers heights substantially more than that of the parent's, one needs to wait considerably longer for her to complete a full swing before she gets her next push! This *amplitude dependence of the frequency* is a characteristic of non-linear oscillators of which the pendulum is one example, and we will see that it is fundamental to understanding stability in the three-body problem.

Resonance is responsible for both structure and destruction in Nature, and not just via gravity. *It is Nature's way of moving energy around in bulk.* For example, molecular structure depends on resonance between internal electronic states; the formation of carbon in stars via the triple-alpha process relies on a resonant reaction between an alpha particle and a very short-lived beryllium nucleus, leading to the formation of an excited state of the carbon nucleus; even the Archimedes spiral of a sunflower relies on resonance for its formation [see Reichl (1992) for a discussion of the golden mean as the "most irrational number"]. But when gravity is involved, resonance plays a role on every astrophysical scale through the dynamics of three-body instability.

3.2.1 Three-Body Processes in Astrophysics

Three-body processes are at the heart of structure on all astrophysical scales, from planet formation via the accumulation of planetesimals to giant elliptical galaxies through the forced collisions of smaller galaxies. Processes occurring

[4] Some animations of stable and unstable triples may be found at `http://users.monash.edu.au/~ro`.

in star clusters include binary–single star scattering in the cores of globular clusters, a process largely responsible for the prevention of total core collapse (Aarseth 1971), the formation of X-ray binaries in globular cluster cores through binary–single and binary–binary collisions (Hills 1976), the formation of massive stars that almost certainly occasionally (if not exclusively) form through collisions induced in small-N systems, the building of intermediate-mass black holes through the so-called Kozai mechanism (Aarseth 2007), the formation of close binaries through the Kozai mechanism (Eggleton & Kiseleva 2001; Fabrycky & Tremaine 2007), the stability or otherwise of planetary systems in star clusters (Spurzem et al. 2006), and hypervelocity stars originating from galactic centre (Hills 1976). In addition, many objects thought to be binary stars are revealing themselves to be triple or higher-order configurations (Tokovinin et al. 2006); such systems may well be the remnants of even higher-order systems that have decayed since their birth in the natal star cluster (Reipurth & Clarke 2001).

To understand all these processes it is necessary to understand how energy and angular momentum move around inside a triple, and under what circumstances a given configuration is stable. The rest of this chapter is devoted to this question through a study of resonance in the three-body problem.

3.3 The Mathematics of Resonance

3.3.1 The Pendulum

Before we discuss resonance, it is necessary to review the mechanics of a pendulum. As we will show, pendulum-like behaviour is fundamental to an understanding of the three-body problem.

The equation governing the motion of a pendulum of length l in a uniform gravitational field g is

$$\ddot{\phi} + \omega_0^2 \sin\phi = 0, \tag{3.1}$$

where $\omega_0^2 = g/l$. Clearly for $\max(\phi) \ll 1$, (3.1) reduces to the equation for simple harmonic motion with natural frequency ω_0. We will refer to ω_0 as the *small angle frequency*, and to the associated libration period the *small angle libration period*. Figure 3.2(a) plots ϕ against time, the latter measured in units of small angle libration periods for $\phi(0) = 0$ and various values of $\dot{\phi}(0)$, while Fig. 3.1(b) plots solutions in phase-space, that is, $\dot{\phi}$ against ϕ. Solutions that oscillate between fixed values of $\phi < \pi$ are referred to as *libratory* and those for which ϕ is unbounded are called *circulatory*. These two kinds of motion are separated in phase space by the *separatrix*, the two branches of which are indicated by the dashed curves in each panel. Clearly, the libration period increases from $2\pi/\omega_0$ for small maximum $\phi \equiv \phi_m$ to infinity for $\phi_m = \pi$. Note in particular the so-called hyperbolic fixed points on the separatrix $(\phi, \dot{\phi}) = (\pm\pi, 0)$ in panel (b): these play a vital role in unstable triples as we will demonstrate.

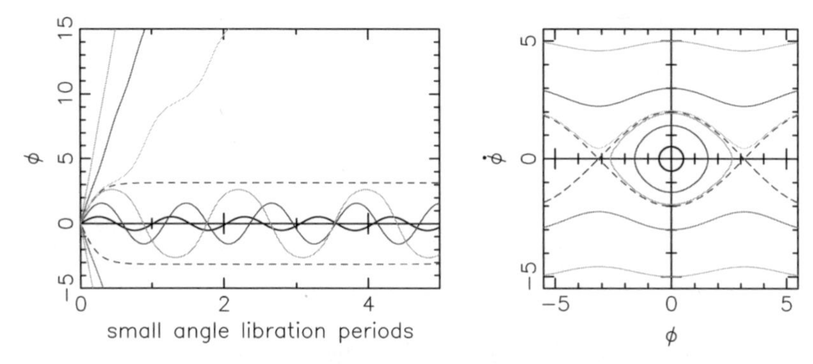

Fig. 3.2. Libration versus circulation of a pendulum. Corresponding curves in (**a**) and (**b**) have the same colour. The dashed curves correspond to the separatrix; after starting at $\phi(0) = 0$ the system takes an infinite amount of time to reach the unstable equilibrium points $(\phi, \dot{\phi}) = (\pm\pi, 0)$ (also known as hyperbolic fixed points)

Equation (3.1) has an integral of the motion, which we refer to as the *pendulum energy*:

$$E = \frac{1}{2}\dot{\phi}^2 - \omega_0^2(\cos\phi + 1), \tag{3.2}$$

where we have chosen the zero of E to correspond to the separatrix, that is, the curve which passes through $(\phi, \dot{\phi}) = (\pi, 0)$. The equation for the separatrix is therefore

$$\dot{\phi} = \pm 2\,\omega_0 \cos(\phi/2). \tag{3.3}$$

For systems with $E < 0$, the libration period, $T_{\rm lib}$, is given by

$$T_{\rm lib} = \int_0^{T_{\rm lib}} {\rm d}t = 4\int_0^{\phi_m} \frac{{\rm d}\phi}{\dot{\phi}} = \frac{2\sqrt{2}}{\omega_0}\int_0^{\phi_m} \frac{{\rm d}\phi}{\sqrt{\cos\phi - \cos\phi_m}}, \tag{3.4}$$

where again ϕ_m is the maximum value of ϕ, therefore corresponding to $\dot{\phi} = 0$. Note that for $\phi_m \ll 1$, $T_{\rm lib} \simeq 2\pi/\omega_0$.

For systems with $E > 0$, the circulation period, $T_{\rm circ}$, is given by

$$T_{\rm circ} = 2\int_0^{\pi} \frac{{\rm d}\phi}{\dot{\phi}} = 2\int_0^{\pi} \frac{{\rm d}\phi}{\sqrt{\dot{\phi}_0^2 + 2\,\omega_0^2(\cos\phi - 1)}}, \tag{3.5}$$

where $\dot{\phi}_0$ is the value of $\dot{\phi}$ corresponding to $\phi = 0$. Note that for $\dot{\phi}_0 \gg 2\omega_0$, $T_{\rm circ} \simeq 2\pi/\dot{\phi}_0$.

The libration and circulation frequencies, $\omega_{\rm lib} \equiv 2\pi/T_{\rm lib}$ and $\omega_{\rm circ} \equiv 2\pi/T_{\rm circ}$, respectively, are plotted in Fig. 3.3. Note the steep dependence of $\omega_{\rm lib}$ on ϕ_m near $\phi_m = \pi$ and $\omega_{\rm circ}$ on $\dot{\phi}_0$ near $\dot{\phi}_0 = 0$. As we will now demonstrate, it is this steep dependence which is responsible for chaos in weakly coupled non-linear systems.

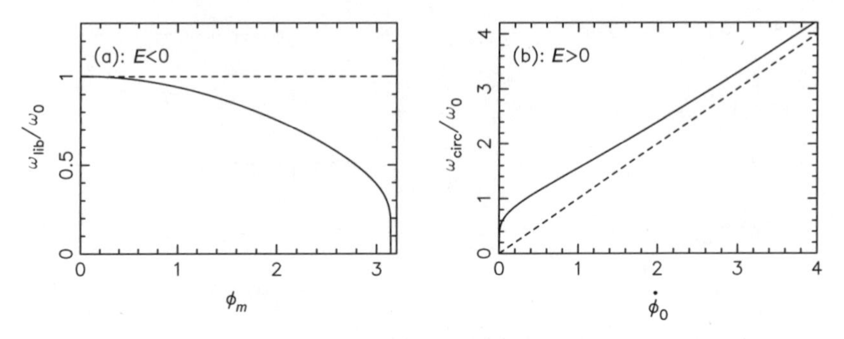

Fig. 3.3. Amplitude dependence of pendulum libration and circulation frequencies. Note the extremely steep dependence of $\omega_{\rm lib}$ on ϕ_m near π – one of the secrets to understanding chaos in weakly interacting systems. The dashed curves correspond to (**a**) the small angle frequency and (**b**) $\dot\phi_0 = 2\omega_0$

3.3.2 Linear Versus Non-Linear Resonance

Consider a simple undamped spring with natural frequency ω which is forced at the frequency Ω. If ϕ is the displacement away from equilibrium, then given the initial conditions $\phi(0) = \dot\phi(0) = 0$, the solution to the equation of motion

$$\ddot\phi + \omega^2\phi = A\sin\Omega t \tag{3.6}$$

is

$$\phi(t) = \frac{A}{\Omega^2-\omega^2}\left[(\Omega/\omega)\sin\omega t - \sin\Omega t\right] \tag{3.7}$$

when $\Omega \neq \omega$, and

$$\phi(t) = \frac{A}{2\omega^2}\left[\sin\omega t - \omega t\cos\omega t\right] \tag{3.8}$$

when $\Omega = \omega$. These two types of solution are plotted in Fig. 3.4(a) and (b) respectively. In the first case, a near-resonant value of $\Omega = 0.9\omega$ produces the phenomenon called *beating*, where the frequency of the envelope of the solution is $|\Omega - \omega|$. The maximum value attained is approximately $(A/\omega)/|\Omega - \omega|$. However, when $\Omega = \omega$, the envelope is given by $\phi(t) = \pm At/2\omega$ and the solution grows without bound. This is *linear* resonance.

Unlike a simple spring whose natural oscillation frequency is independent of the amplitude, the libration frequency of a pendulum is amplitude-dependent except when the libration angle is small. Consider a pendulum which is forced at a constant frequency Ω, and let its small angle frequency be ω_0. Its equation of motion is almost identical to (3.6) except that ϕ is replaced by $\sin\phi$:

$$\ddot\phi + \omega_0^2\sin\phi = A\sin\Omega t. \tag{3.9}$$

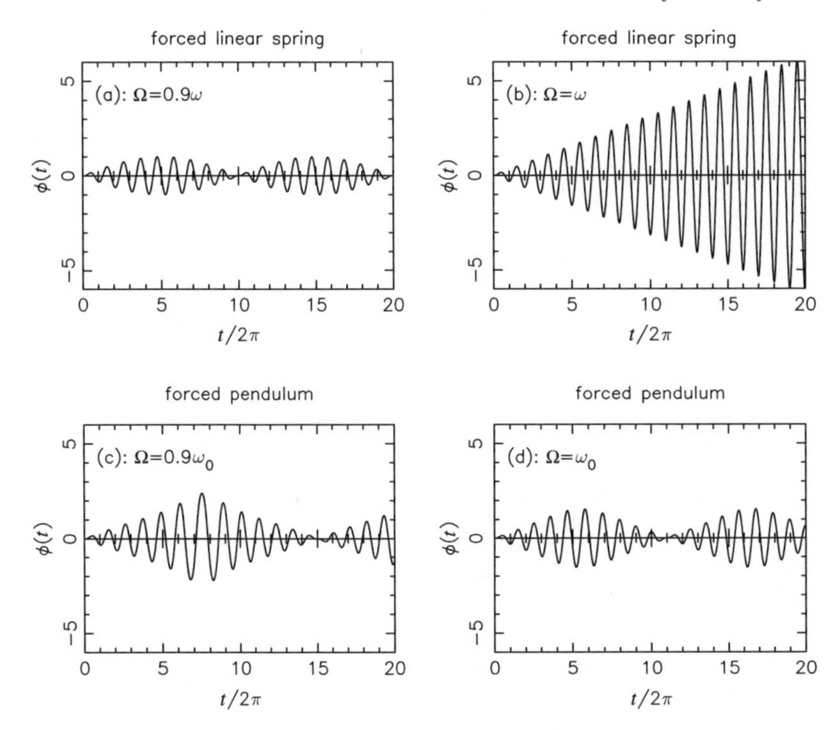

Fig. 3.4. Forced linear spring vs forced pendulum. Linear spring: (**a**) beating with $\Omega \lesssim \omega$ and (**b**) linear resonance with $\Omega = \omega$. Pendulum: (**c**) and (**d**). Both solutions exhibit beating but the system which is forced with a frequency *less* than the small-angle frequency attains a larger amplitude because, as the amplitude increases, the libration frequency *decreases* moving it closer to the forcing frequency. In contrast, system (**d**) moves away from the forcing frequency from the start and therefore does not attain as large an amplitude. For all four systems $A = 0.1$ and $\phi(0) = \dot{\phi}(0) = 0$

Now there is no closed-form solution; in fact, this differential equation admits chaotic solutions. In order to understand how such solutions arise (and ultimately, to understand why the three-body problem admits chaotic solutions), consider solutions to (3.9) with the same initial conditions as for the forced spring; these are shown in Fig. 3.4(c) and (d). Both solutions exhibit beating but the system which is forced with a frequency *less* than the small angle frequency attains a larger amplitude because, as the amplitude increases, the libration frequency *decreases* moving it closer to the forcing frequency (see Fig. 3.3). In contrast, system (b) moves away from the forcing frequency from the beginning.

What happens if A is increased in (3.9)? While doing this merely scales the amplitude for a linear spring, the response is quite different for a forced pendulum because the response frequency actually depends on the amplitude. Figure 3.5 shows solutions for various values of $\overline{A} \equiv A/\omega_0^2$ for $\phi(0) = \dot{\phi}(0) = 0$

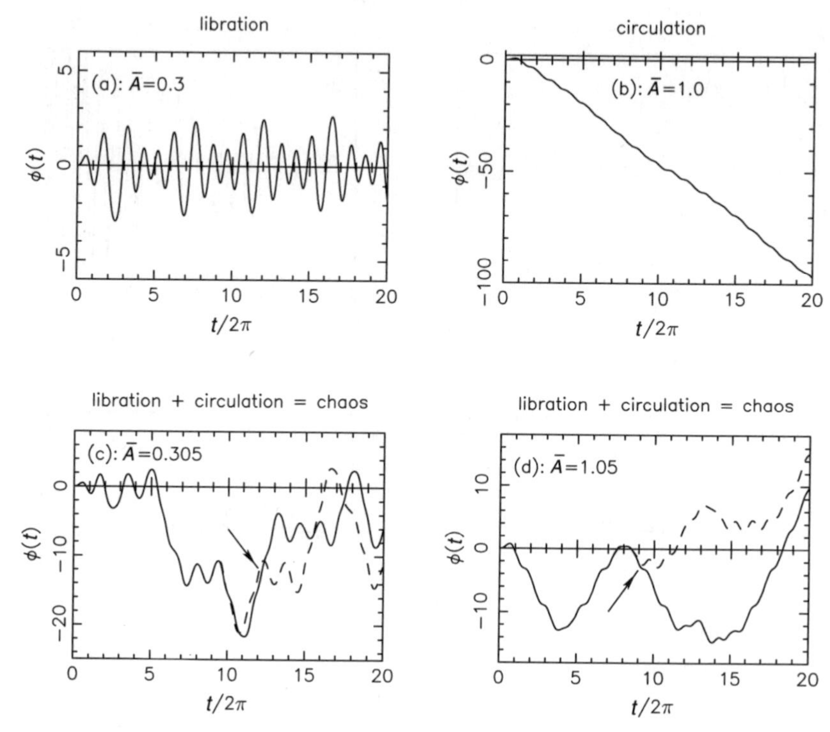

Fig. 3.5. Strong forcing of a pendulum. All systems have $\Omega = 0.9\,\omega_0$ and $\phi(0) = \dot{\phi}(0) = 0$, except for the dashed curves for which $\phi(0) = 10^{-6}$. (**a**) $\overline{A} \equiv A/\omega_0^2 = 0.3$: *libration*. Here the pendulum frequency drops further below the forcing frequency and beating is less pronounced. Note especially that the amplitude gets dangerously close to π, that is, the separatrix. (**b**) $\overline{A} = 1.0$: *circulation*. Safely past the separatrix, the system is sufficiently forced to simply circulate. (**c**) $\overline{A} = 0.305$ and (**d**) $\overline{A} = 1.05$: *chaos*. The system is forced sufficiently strongly to show a mixture of libration and circulation. The dashed curves illustrate the sensitivity of chaotic systems to initial conditions. In fact, both (**a**) and (**b**) are also chaotic but these systems do not come sufficiently close to the separatrix during this time interval. Note that the values of $\overline{A}$ in (**c**) and (**d**) are only slightly different to those in (**a**) and (**b**), respectively, suggesting that the time at which obvious divergence of nearby trajectories takes place is statistical. Note also that different scales have been used for each panel

and $\Omega = 0.9\,\omega_0$. In (a), $\overline{A} = 0.3$, the motion remains libratory over this time interval ($E < 0$), but the amplitude comes close to π (maximum 2.6). In (b), $\overline{A} = 1.0$ and the stronger forcing allows the system to be completely circulatory with $E > 0$ at all times shown. Panels (c) and (d) exhibit sensitivity to initial conditions, a diagnostic of chaos, even though their values for $\overline{A}$ are only slightly different to those in (a) and (b). This is demonstrated by plotting trajectories with the same initial conditions except for the initial values for ϕ, which differ by 10^{-6}. Note that for longer integration times (a) and

(b) also display similar sensitivity to initial conditions, including a mixture of libration and circulation.

3.3.3 The Butterfly Effect Explained

When a system is near the separatrix, a small difference in ϕ can correspond to at least an order of magnitude difference in the pendulum frequency $\omega_{\rm lib}$ or $\omega_{\rm circ}$ (see Fig. 3.3). Since the libration amplitude depends sensitively on the current value of $\omega_{\rm lib}$ relative to the forcing frequency [for example, compare Fig. 3.4(c) and (d)], such differences can eventually lead to a significant divergence of initially nearby solutions *as long as the system is not periodic or quasi-periodic* (see below).[5] A system that is sufficiently strongly forced may even cross the separatrix and begin to circulate; this almost never happens at the same time as a neighbouring trajectory because of the differences in their pendulum frequencies at the time. The situation is indicated by arrows in Fig. 3.5(c) and (d). *This behaviour is the essence of chaos in weakly interacting systems.*

Let us consider the situation more closely. Given the values of ϕ and $\dot\phi$ at any time t, one can define the instantaneous (or osculating) pendulum frequency ω to be such that

$$\omega(t) = \begin{cases} \omega_{\rm lib}, & E<0 \\ -\omega_{\rm circ}, & E>0, \end{cases} \tag{3.10}$$

where again, $\omega_{\rm lib} = 2\pi/T_{\rm lib}$ and $\omega_{\rm circ} = 2\pi/T_{\rm circ}$ with $T_{\rm lib}$ and $T_{\rm circ}$ defined in (3.4) and (3.5). These latter quantities depend on knowing ϕ_m and $\dot\phi_0$, that is, respectively, ϕ at $\dot\phi = 0$ for a librating system and $\dot\phi$ at $\phi = 0$ for a circulating system. The instantaneous values of these can be *defined* via the pendulum energy E (which is now not conserved). Thus from (3.2),

$$\dot\phi^2 - \omega_0^2(1+\cos\phi) = -\omega_0^2(1+\cos\phi_m) \tag{3.11}$$

and

$$\dot\phi^2 - \omega_0^2(1+\cos\phi) = \dot\phi_0^2 - 2\omega_0^2. \tag{3.12}$$

Note that defining the pendulum frequency to be negative when $E>0$ simply ensures that $d\omega/dt$ is continuous through $\omega = 0$, that is, for the purpose of graphical representation there is a smooth transition from libration to circulation. More importantly, it allows for a meaningful measure of the "distance" between neighbouring trajectories (see discussion below).

Figure 3.6(b) plots $\omega(t)$ for the stable case shown in panel (a) of the same figure for which $\overline{A} = 0.1$, $\Omega = 0.9\,\omega_0$. The pendulum frequency is clearly

[5] A system is *N-fold quasi-periodic* if it can be represented as the product of N Fourier series with associated frequencies ω_i, $i = 1,\dots,N$, such that the ω_i are not commensurate. If the ω_i *are* commensurate, the system is *periodic.*

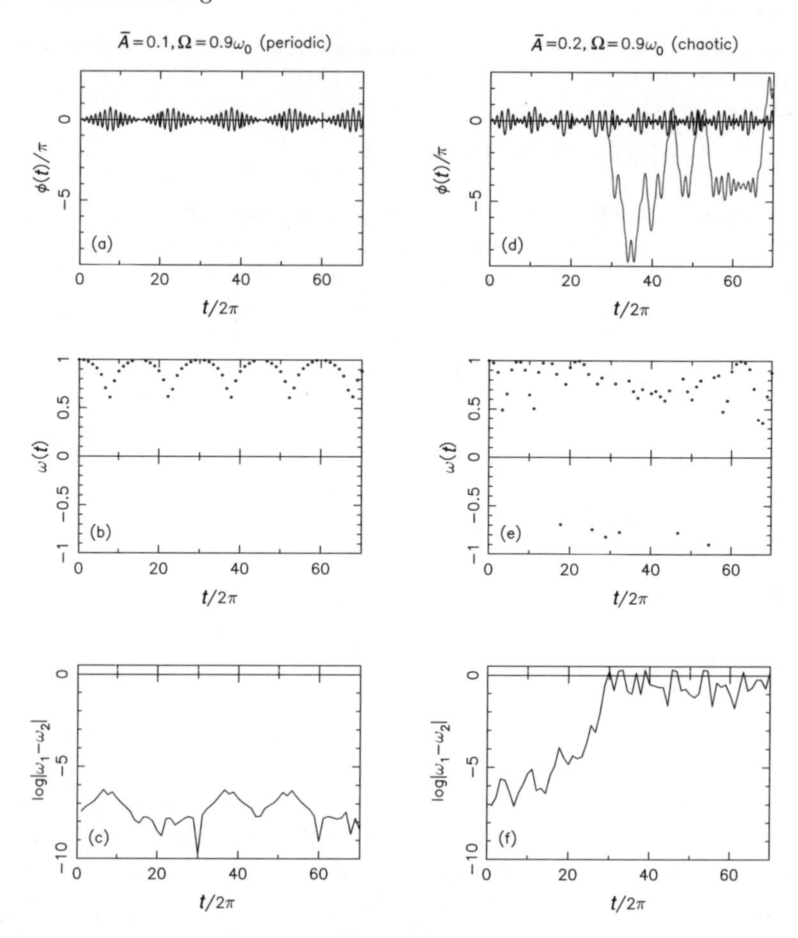

Fig. 3.6. Exponential divergence of chaotic trajectories. Panel (**a**) shows the evolution of ϕ (in units of π) for two initially close trajectories ($\delta\phi(0) = 10^{-6}$) for $\overline{A} = 0.1$ and $\Omega/\omega_0 = 0.9$. No unstable behaviour is indicated, and this is supported by panel (**c**), which plots the logarithm of the difference in the pendulum frequencies. Panel (**b**) shows the evolution of the pendulum frequency $\omega(t)$ ((3.10)) for the system with $\phi(0) = 0$. Points are plotted only when the forcing is zero, that is, when the pendulum is "free". Since ϕ is quasi-periodic (in fact for this example, it is actually periodic because ω_0 and Ω are commensurate), the pendulum frequencies come in and out of step over time and their differences, therefore, never build up. Panels (**d**), (**e**) and (**f**) show the evolution of these quantities for the chaotic system $\overline{A} = 0.2$ and $\Omega/\omega_0 = 0.9$. The initially close trajectories diverge strongly around $t/2\pi = 30$, even though the system appears to be stable before then. However, it is clearly not even quasi-periodic, and panel (**f**) reveals that the trajectories are in fact exponentially diverging because $|\phi|$ comes close enough to π for ω_1 to be significantly different to ω_2 at those times. In particular, notice how individual peaks in panel (**f**) correspond to minimum values of $|\omega(t)|$. The forcing is strong enough to allow the system to cross the separatrix and occasionally circulate. Since ϕ is not periodic, differences in ω accumulate and remain $\mathcal{O}(|\omega|)$

periodic, with minima corresponding to maximum forcing (notice in (a) how the response "stretches" at maximum amplitude; this is seen in more detail in Fig. 3.4(c)). Panel (c) plots the logarithm of the difference between the pendulum frequencies, ω_1 and ω_2, of two initially close systems for which the difference in $\phi(0)$ is again $10^{-6} \equiv \epsilon$. The difference remains of the order or less than ϵ for the time shown here, and for longer times grows linearly before turning over when $|\omega_1 - \omega_2| \simeq 0.01$. This behaviour is common to quasi-periodic (and periodic) systems for which accumulation of differences in ω is limited to how out of phase the two systems become.

In contrast, the right-hand panels (d), (e) and (f) show $\phi(t)$, $\omega(t)$ and $\log|\omega_1 - \omega_2|$ for the chaotic system $\overline{A} = 0.2$ and $\Omega = 0.9\,\omega_0$. Unlike the stable system, this one is not periodic or quasi-periodic, and the consequence is that differences in ω do accumulate. These differences are a maximum when $|\omega(t)|$ is a minimum because of its steep dependence on ϕ_0 as $\phi_0 \to \pi$, and this can be seen if one compares panels (e) and (f). Eventually, $|\omega_1 - \omega_2| = \mathcal{O}(|\omega|)$ when one of the systems is sufficiently forced to start circulating. Note that system 1 first circulates at $t/2\pi \simeq 84$.

The slope of the curve in panel (f) indicates the time-scale τ on which exponential trajectory divergence takes place. This is normally associated with the largest Lyapunov exponent λ, which is related to τ such that $\lambda \sim 1/\tau$.

The following questions arise: how strong does the forcing have to be (how large should A be) and/or how close should the forcing frequency Ω be to ω_0 in order that the system is not exclusively libratory? Are all systems which do not circulate quasi-periodic or periodic (i.e. do all chaotic systems involve circulation)? These and other related questions have been studied extensively in the context of *conservative Hamiltonian systems*, of which the general three-body problem is an example. In fact, the three-body problem (or simplified versions of it) motivated Poincaré to invent the modern theory of dynamical systems and chaos (Barrow-Green 1997) and led to the famous *Kolmogorov–Arnol'd–Moser* or KAM theory of weakly interacting Hamiltonian systems (see below).

3.3.4 Pendulums, the Three-Body Problem and Resonance Overlap

The previous examples demonstrate how springs and pendulums respond to fixed forcing. How are these related to the three-body problem? Most three-body configurations can be regarded as being composed of an "inner binary" and an "outer binary", the latter being composed of the inner binary and the third body; this is referred to as a three-body hierarchy (see Fig. 3.8). When a system is stable (or at least, close to stable), these two binaries constitute a weakly interacting conservative system with each binary forcing the other.

Figure 3.7 shows the evolution of the semi-major axis, a_i, of the inner binary of (a) a stable triple and (b) an unstable triple. The behaviour of the stable system is very similar to the forced pendulum in Figs. 3.4(c) and

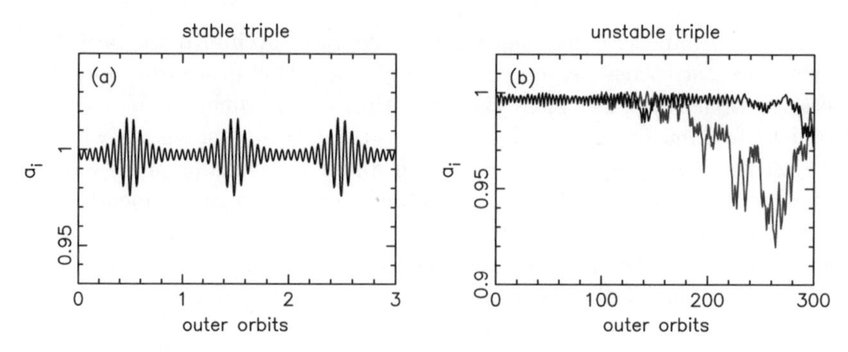

Fig. 3.7. Evolution of the semi-major axis, a_i, of the inner binary of a stable triple (**a**) and an unstable triple (**b**). The initial conditions are such that for both (**a**) and (**b**) the ratio of the outer periastron distance to the inner semi-major axis is 3.6 and the inner binary is circular, while the outer eccentricity is 0.3 and 0.5 for (**a**) and (**b**), respectively. In (**b**) we also show the evolution of an almost identical configuration for which the initial inner eccentricities differ by 10^{-6}

3.6(a); here the forcing is provided by the third body, with outer periastron passage occurring at 0.5 phase. The chaotic system in (b) is reminiscent of Fig. 3.6(d), in this case with a mixture of oscillation between two fixed values ("libration") and approximately steady increase or decrease ("circulation") of a_i. In fact, the inner and outer orbits exchange energy via an interaction potential or *disturbing function*, which can be written as an infinite series of *resonance angles*, each a linear combination of all the angles in the system and each obeying a forced pendulum equation. The forcing of each individual "pendulum" is provided by all the other "pendulums", and when the system is stable the forcing is negligible (in fact, exponentially small). For almost all stable systems the pendulum motions are circulatory with exponentially small amplitudes; however, some stable systems exist in a *resonant* state in which case one resonance angle librates.[6] In order for stability to be maintained, the forcing of such an angle must remain small in the sense discussed in the previous section. When the forcing is such that the pendulum libration amplitude (i.e. the single resonance angle that is librating) comes close to π, the system is unstable, again in the same sense as discussed in the previous section. However, here the forcing is provided by another "pendulum" with almost the same frequency, i.e. by another resonance angle. In order for the forcing to be sufficiently strong, it turns out that such a resonance angle (in general) must also be librating and we have the situation where the system exists in two "neighbouring" resonant states: this is referred to as *resonance overlap*. Thus the diagnostic for instability is simply that two neighbouring resonances be librating; *this is the resonance overlap stability criterion.*

[6] In fact, the stable resonant state actually consists of a *superposition* of resonance angles (M2), but this is usually only important for extreme mass-ratio systems that have stable low-order resonances.

The reader is referred to the original paper by Walker & Ford (1969) in which this idea is discussed in a clear and straightforward way, while Chirikov (1979) provides a deeper and more extensive analysis. The concept of resonance in weakly interacting conservative systems originates in a theorem proposed and partially proved by Kolmogorov (1954), itself inspired by the work of Poincaré (1993). This theorem was fully proved by Arnol'd (1963) and independently by Moser (1962). The three papers constitute the famous Kolmogorov–Arnol'd–Moser, or KAM, theorem, which would provide a proof that "stable" triple systems are formally stable for all time were it not for the fact that one of the assumptions made in the proof of the theorem is violated! The aim of the KAM theorem is to show that if one perturbs an *integrable* Hamiltonian system sufficiently weakly,[7] then some of the KAM tori on which solutions were originally quasi-periodic will be only slightly distorted and quasi-periodicity will be preserved. Although not a conservative Hamiltonian system, we see this behaviour in going from the forced spring in Fig. 3.4(a) to the forced pendulum in panel (c) of the same figure; a pendulum can be regarded as a linear spring with a non-linear perturbation. However, if the perturbation is too strong, quasi-periodicity is lost and the motion becomes unpredictable. If the KAM theorem applied to the three-body problem it would prove that a large subset of configurations exists whose members remain stable for all time (because they are stuck on KAM tori). But the catch is that one requires the characteristic frequencies of the decoupled system to be non-commensurate and this is not the case because the apsidal motion and precession frequencies are equal (in fact, equal to zero).

So a formal proof of the ultimate stability of general three-body configurations remains elusive, although it can be proved in some restricted cases, for example, when the eccentricities and inclinations are small so that the secular theory of Laplace applies and can be used as the underlying "unperturbed" system; see Arnol'd (1978) p. 414. We must therefore (at least for now) be content with our observation that apparently stable systems seem to mimic quasi-periodic systems for which the KAM theorem *does* apply, and proceed to use the tools of the theorem (in particular, the resonance overlap stability criterion) to predict, albeit approximately, the boundary between stable and unstable behaviour.

---

[7]An integrable Hamiltonian system that is a function of N coordinate and N momentum variables is one which has N integrals of the motion. For such systems one can then find a coordinate transformation such that the new momenta are the integrals themselves and the new coordinates q_i, $i = 1, \ldots, N$ are linear functions of time, $q_i(t) = \omega_i t + C_i$, where the ω_i are the characteristic frequencies of the system and the C_i are constants. If the ω_i are not commensurate, that is, there exists no integers k_i such that $\sum k_i \omega_i = 0$, the solutions are restricted to and densely cover so-called *KAM tori* and the motion is quasi-periodic. If the ω_i are commensurate, the motion is periodic.

3.4 The Three-Body Problem

The three-body problem is famously easy to formulate and impossible to solve – at least analytically. Newton is said to have suffered from sleeplessness and headaches trying to find closed-form solutions after having had such an easy time with the two-body problem. After many attempts by the best mathematicians of their time, Poincaré noticed that perturbation techniques unavoidably involved singularities associated with resonances and concluded that the three-body problem has solutions that cannot be represented by convergent series.

In order to appreciate fully the dynamics of the three-body problem, we begin by reviewing some aspects of the two-body problem, in particular, its integrals of the motion. These express various symmetries inherent in the equations of motion, one (sometimes more) of which survives when a third body is added and the system is stable (the *total* energy and linear and angular momenta are still conserved).

3.4.1 Symmetries in the Two-Body Problem

The equations of motion of two bodies with masses m_1 and m_2 acting under the influence of each other's gravity are

$$m_1\ddot{\mathbf{r}}_1 = \frac{Gm_1m_2}{r_{12}^2}\hat{\mathbf{r}}_{12} \tag{3.13}$$

$$m_2\ddot{\mathbf{r}}_2 = -\frac{Gm_1m_2}{r_{12}^2}\hat{\mathbf{r}}_{12}, \tag{3.14}$$

where $\mathbf{r}_{12} = \mathbf{r}_2 - \mathbf{r}_1$. Equations (3.13) and (3.14) constitute a twelfth-order system of differential equations. However, it has *eight* independent integrals of the motion and, as is well known, this restricts the motion to a simple curve in space as we now show. Three of the integrals of motion are the components of the *total linear momentum* $\mathbf{P}$, which one obtains by adding (3.13) and (3.14) together and integrating, that is,

$$m_1\dot{\mathbf{r}}_1 + m_2\dot{\mathbf{r}}_2 \equiv \mathbf{P}. \tag{3.15}$$

Dividing through by the masses, subtracting (3.13) from (3.14) and defining $\mathbf{r}$ to be the position vector of m_2 *relative to* m_1, that is, $\mathbf{r} \equiv \mathbf{r}_{12}$, we reduce the system to sixth order:

$$\ddot{\mathbf{r}} = -\frac{Gm_{12}}{r^2}\hat{\mathbf{r}}, \tag{3.16}$$

where $r = |\mathbf{r}|$ and $m_{12} = m_1 + m_2$. Taking the cross product of each side with $\mu\mathbf{r}$ and integrating we get another three integrals of the motion; these are the components of the *total angular momentum* $\mathbf{J}$:

$$\mu \mathbf{r} \times \dot{\mathbf{r}} \equiv \mathbf{J}, \tag{3.17}$$

where $\mu = m_1 m_2 / m_{12}$ is the reduced mass of the system. A seventh integral of the motion is the *total energy*; this is obtained by taking the dot product of (3.16) with $\mu \dot{\mathbf{r}}$ and integrating:

$$\frac{1}{2}\mu \dot{\mathbf{r}} \cdot \dot{\mathbf{r}} - \frac{G m_1 m_2}{r} \equiv E, \tag{3.18}$$

where we have used the chain rule

$$\frac{\mathrm{d}}{\mathrm{d}t} = \frac{\partial}{\partial t} + \dot{\mathbf{r}} \cdot \frac{\partial}{\partial \mathbf{r}}, \tag{3.19}$$

with $\partial/\partial \mathbf{r} \equiv \nabla$. The seven integrals reflect natural symmetries of isolated conservative mechanical systems: the conservation of energy and linear momentum reflect the fact that the equations of motion are independent of the origin of time and space, respectively, while the conservation of angular momentum reflects the fact that the solution is independent of the orientation of the system. For all these symmetries, there is no external landmark which could be used to distinguish one system from another under such transformations.

What symmetry does the eighth integral correspond to? It is well known that solutions to (3.13) and (3.14) are conic sections. In particular, these curves are fixed in space, that is, their *orientation is invariant*, a fact peculiar to the two-body problem (see Goldstein (1980) p. 104 for a discussion of this). This is normally expressed as the invariance of the *Runge–Lenz* vector (also called the Laplace vector), a vector which points in the direction of periastron and is defined by

$$\mathbf{e} = \dot{\mathbf{r}} \times (\mathbf{r} \times \dot{\mathbf{r}})/G m_{12} - \hat{\mathbf{r}}, \tag{3.20}$$

and whose magnitude is the orbital eccentricity e. But this appears to add *three* extra integrals; in fact one can show that only one is independent of the other seven (Goldstein 1980).

The two-body problem has six degrees of freedom and hence one only needs six integrals of the motion in order that the system be completely integrable (in the sense discussed in the footnote on p. 71). The fact that we have eight restricts the motion to closed curves *in the frame of reference of the centre of mass of the system*. Solution curves are the conic sections (see Goldstein (1980) for a method of solution).

3.4.2 The Three-Body Problem

The equations of motion of three bodies with masses m_1, m_2 and m_3 acting under the influence of each other's gravity are

$$m_1\ddot{\mathbf{r}}_1 = \frac{Gm_1m_2}{r_{12}^2}\hat{\mathbf{r}}_{12} + \frac{Gm_1m_3}{r_{13}^2}\hat{\mathbf{r}}_{13} \tag{3.21}$$

$$m_2\ddot{\mathbf{r}}_2 = -\frac{Gm_1m_2}{r_{12}^2}\hat{\mathbf{r}}_{12} + \frac{Gm_2m_3}{r_{23}^2}\hat{\mathbf{r}}_{23} \tag{3.22}$$

$$m_3\ddot{\mathbf{r}}_3 = -\frac{Gm_1m_3}{r_{13}^2}\hat{\mathbf{r}}_{13} - \frac{Gm_2m_3}{r_{23}^2}\hat{\mathbf{r}}_{23}, \tag{3.23}$$

where the vectors $\mathbf{r}_i$, $i = 1, 2, 3$ are referred to the centre of mass of the system (see Fig. 3.8), and $\mathbf{r}_{ij} = \mathbf{r}_j - \mathbf{r}_i$ with $r_{ij} = |\mathbf{r}_{ij}|$. The differential equations (3.21), (3.22) and (3.23) constitute an 18th-order system. While it again yields the seven integrals of total energy, linear momentum and angular momentum, there is no analogue of the Runge–Lenz integral. Thus we are two integrals short of a totally integrable system. This fact results in the possibility of the system admitting *chaotic* solutions, that is, solutions that are exquisitely sensitive to the initial conditions and are hence unpredictable. In fact for some systems with negative total energy, it allows for infinite separation of one body from the other pair. These are systems referred to as *Lagrange unstable*, which in general do not rely on the close approach of two of the bodies (such systems are referred to as *Hill unstable*).

We thus ask the general question: given a particular three-body configuration, how can we determine whether or not it is (Lagrange) stable for all time? As discussed in Sect. 3.3.4, there is no rigorous answer to this question. However, there is no doubt that there exists a sharp (albeit fractal-like) boundary in parameter space between unstable systems, which decay on a relatively short time-scale and those which *appear* to remain intact (are stable) indefinitely. It is this boundary that is approximately delineated in this chapter using the so-called *resonance overlap criterion*, which itself involves identifying internal resonances in the system. In order to do this, we begin by introducing *Jacobi* or *hierarchical coordinates* $\mathbf{r}$ and $\mathbf{R}$, which together with

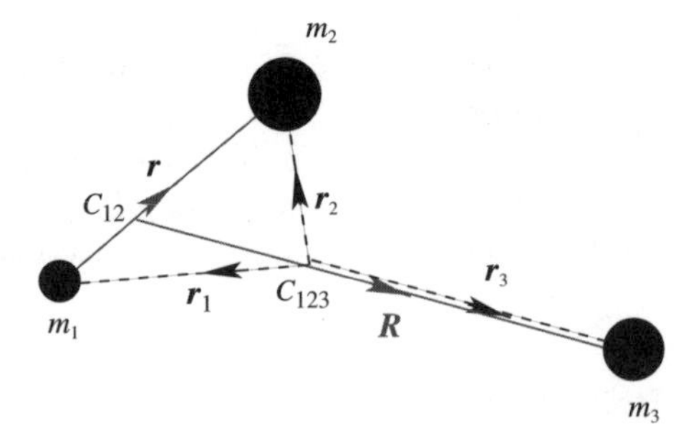

Fig. 3.8. Centre of mass coordinates $\mathbf{r}_i$ and Jacobi coordinates $\mathbf{r}$ and $\mathbf{R}$. C_{12} is the centre of mass of bodies 1 and 2 while C_{123} is the centre of mass of the whole system

conservation of linear momentum, replace the *centre-of-mass coordinates* $\mathbf{r}_1$, $\mathbf{r}_2$ and $\mathbf{r}_3$ (see Fig. 3.8).

3.4.3 Equations of Motion in Jacobi Coordinates

Intuitively, it seems reasonable that three-body configurations are more likely to be stable the further one of the bodies (let us take this to be body 3) is separated from the other two. In fact, a very distant third body will orbit the other two as if they were almost a single body. Thus we can conceive of an "inner binary" composed of bodies 1 and 2, and an "outer binary" composed of bodies (1+2) and body 3. Jacobi coordinates conveniently express this arrangement. Just as for the two-body problem, $\mathbf{r}$ is defined to be the position vector of m_2 relative to m_1, that is, $\mathbf{r} = \mathbf{r}_2 - \mathbf{r}_1$, while $\mathbf{R}$ is the position vector of m_3 relative to the *centre of mass* of m_1 and m_2. In fact, it turns out that $\mathbf{R}$ passes through the centre of mass of the system and as such is in the same direction as $\mathbf{r}_3$ with $\mathbf{R} = (m_{123}/m_{12})\,\mathbf{r}_3$ (Fig. 3.8), where $m_{123} = m_1 + m_2 + m_3$. Using these definitions, we can reduce the 18th-order system (3.21), (3.22) and (3.23) to the 12th-order system

$$\mu_i \ddot{\mathbf{r}} + \frac{G m_1 m_2}{r^2}\hat{\mathbf{r}} = \frac{\partial \mathcal{R}}{\partial \mathbf{r}} \tag{3.24}$$

$$\mu_o \ddot{\mathbf{R}} + \frac{G m_{12} m_3}{R^2}\hat{\mathbf{R}} = \frac{\partial \mathcal{R}}{\partial \mathbf{R}}, \tag{3.25}$$

where $R = |\mathbf{R}|$, $\mu_i = m_1 m_2/m_{12}$ and $\mu_o = m_{12} m_3/m_{123}$ are the reduced masses associated with the inner and outer orbits, respectively, and

$$\mathcal{R} = -\frac{G m_{12} m_3}{R} + \frac{G m_2 m_3}{|\mathbf{R} - \alpha_1 \mathbf{r}|} + \frac{G m_1 m_3}{|\mathbf{R} + \alpha_2 \mathbf{r}|} \tag{3.26}$$

is the *disturbing function*[8] with $\alpha_i = m_i/m_{12}$, $i = 1, 2$. As $r/R \to 0$ and/or $m_3/m_{12} \to 0$, $\mathcal{R} \to 0$ and the inner and outer orbits decouple. In fact, the disturbing function contains *all* the information about how the inner and outer orbits exchange energy and angular momentum. Since we are interested in determining which configurations are unstable, that is, which allow the escape to infinity of one of the bodies, and this necessarily generally involves a substantial exchange of energy between the orbits, our focus for the rest of this chapter will be on the disturbing function: *it contains all the secrets of the three-body problem!*

Before we proceed, we need to define the orbital elements of the inner and outer binaries in terms of which the stability boundary will be expressed. Using

[8]Note that, as a quantity introduced to study the *restricted* three-body problem, the disturbing function has historically been defined to have units of energy per unit mass. Here it has units of energy.

subscripts i and o to denote the inner and outer orbits respectively,[9] these are the semi-major axes a_i and a_o, the eccentricities e_i and e_o, the orientation angles ω_i, Ω_i, I_i and ω_o, Ω_o, I_o, which are respectively the arguments of periastron, the longitudes of the ascending node and the inclinations (see Fig. 3.1), and the phase angles f_i, M_i, λ_i, ϵ_i and f_o, M_o, λ_o, ϵ_o, which are respectively the true anomaly, the mean anomaly, the mean longitude and the mean longitude at epoch (Murray & Dermott 2000). Note that *longitude* angles are measured with respect to a fixed direction (which here we take to be the **i** direction in Figs. 3.1 and 3.9); we will use longitudes when we construct the *resonance angle* in the next section. Thus rather than $\omega_{i,o}$ we will use the *longitudes* of periastron, defined to be $\varpi_i = \omega_i + \Omega_i$ and similarly for ϖ_o. From Fig. 3.1 we see that for inclined orbits this is a dog-leg angle! The phase angles $f_{i,o}$, $M_{i,o}$ and $\lambda_{i,o} \equiv M_{i,o} + \varpi_{i,o}$ are used to express the angular positions of the bodies in the two-body orbit, the choice of which depends on the application (there are at least another two phase angles in use: the *true longitude* $\equiv f + \varpi$ and *eccentric anomaly*, neither of which we will use here). The mean longitude at epoch is the mean longitude at $t = 0$ ((3.45)). See Murray & Dermott (2000) for a more detailed discussion of the various orbital elements.

3.4.4 Spherical Harmonic Expansions

Since our aim is to determine which configurations are stable, it is useful to write the disturbing function in terms of the orbital elements of the inner and outer binaries. To do this we somehow need to separate information about the inner orbit from that of the outer orbit. The form of the second and third terms in (3.26) suggest using a Legendre expansion:

$$\frac{1}{|\mathbf{b}-\mathbf{a}|} = \sum_{l=0}^{\infty} \left(\frac{a^l}{b^{l+1}}\right) P_l(\cos\gamma), \tag{3.27}$$

where $b = |\mathbf{b}|$, $a = |\mathbf{a}|$ with $a < b$, $P_l(\cos\gamma)$ is a Legendre polynomial of degree l and $\cos\gamma = \hat{\mathbf{a}} \cdot \hat{\mathbf{b}}$. However, for us, this involves the angle between $\mathbf{r}$ and $\mathbf{R}$: information about the two orbits is still "tangled". We can go one step further and use something called the *addition theorem* (Jackson 1975), which expresses a Legendre polynomial of order l in terms of spherical harmonics, Y_{lm}, whose arguments are the spherical polar coordinate angles of the vectors $\mathbf{r}$ and $\mathbf{R}$, both referred to a fixed coordinate system (Fig. 3.9):

$$P_l(\cos\gamma) = \frac{4\pi}{2l+1} \sum_{m=-l}^{l} Y_{lm}(\theta,\varphi) Y^*_{lm}(\Theta,\Psi). \tag{3.28}$$

[9] When no subscript is used, the elements refer to any (or either) two-body orbit.

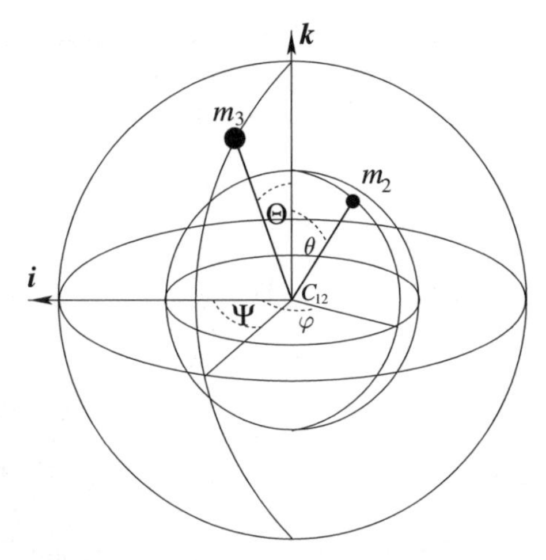

Fig. 3.9. Spherical polar angles associated with $\mathbf{r}$ (θ, φ) and $\mathbf{R}$ (Θ, Ψ). The origin corresponds to the centre of mass of m_1 and m_2, C_{12}

Spherical harmonics are defined in terms of *associated Legendre functions*, $P_l^m(\cos\theta)$, and trigonometric functions (see Jackson (1975) for an extensive discussion of their properties):

$$Y_{lm}(\theta,\varphi) = \sqrt{\frac{2l+1}{4\pi}\frac{(l-m)!}{(l+m)!}}\,P_l^m(\cos\theta)\,e^{im\varphi}, \tag{3.29}$$

where the numerical coefficient is chosen so that the spherical harmonics have a particularly simple orthogonality relation:

$$\int_0^{2\pi}\int_0^{\pi} Y_{lm}(\theta,\varphi)\,Y^*_{l'm'}(\theta,\varphi)\sin\theta\,\mathrm{d}\theta\,\mathrm{d}\varphi = \delta_{ll'}\delta_{mm'}. \tag{3.30}$$

Spherical harmonics are especially important in quantum mechanics. Combining (3.27) and (3.28), the disturbing function (3.26) becomes

$$\mathcal{R} = G\mu_i m_3 \sum_{l=2}^{\infty}\sum_{m=-l}^{l}\left(\frac{4\pi}{2l+1}\right)\mathcal{M}_l\left(\frac{r^l}{R^{l+1}}\right)Y_{lm}(\theta,\varphi)\,Y^*_{lm}(\Theta,\psi), \tag{3.31}$$

where

$$\mathcal{M}_l = \frac{m_1^{l-1} + (-1)^l m_2^{l-1}}{m_{12}^{l-1}}. \tag{3.32}$$

Notice how the sum over l begins at $l = 2$ and not $l = 0$; this is because the $l = 0$ term is cancelled by the first term in (3.26), while the $l = 1$ term (the *dipole* term) is zero because $\mathcal{M}_1 = 0$. Thus the leading term is proportional to r^2/R^3 so that $\mathcal{R}$ provides a *perturbation* to the inner and outer orbits for small r/R. The $l = 2$ contribution is called the *quadrupole* term while the $l = 3$ contribution is called the *octopole* term. Notice also that $\mathcal{M}_2 = 1$, and that when $m_1 = m_2$, $\mathcal{M}_l = 0$ for l odd.

Since the focus of classical treatments of the three-body problem has been the Solar System in which mass ratios, eccentricities and inclinations are generally small, these elements have been used as expansion parameters. The so-called *literal expansion* (Murray & Dermott 2000) involves *Laplace coefficients*, which are functions of the ratio of semimajor axes, and *is valid for orbits which cross*, an example of which is the Neptune–Pluto pair. Apart from being restricted to small eccentricities and inclinations, it also assumes that one of the participating orbits is not affected by the presence of the third body: this is the *restricted* three-body problem. The formulation presented here is, instead, restricted by the condition $r/R < 1$ for the spherical harmonic expansion (3.31) to be valid. Note that it is similar to the (rather tedious to follow) formulation of Kaula (1961); however, the latter is also based on the restricted three-body problem.

Our aim here is to identify internal resonances so that we can apply the resonance overlap criterion and determine stability boundaries. The two most fundamental frequencies in the system are the inner and outer orbital frequencies, ν_i and ν_o, respectively, and these are the only frequencies present when the orbits are not coupled. For example, recall that the orientation of a two-body orbit remains fixed in space, and this is expressed by the constancy of the Runge–Lenz vector. However, when a third body is introduced, this symmetry is broken and the original orbit rotates in space, in a manner similar to a spinning top acting under the applied torque of the Earth. As discussed in the Introduction, the presence of a third body introduces four new frequencies (apsidal advance and precession of the inner and outer orbits), which are usually much slower than the orbital frequencies. Resonances will, in general, involve linear combinations of all six frequencies. Our next task, then, is to express the disturbing function in terms of six angles associated with these frequencies, and as discussed earlier, these are chosen to be longitudes. The mean longitudes $\lambda_{i,o}$ are associated with $\nu_{i,o}$ while the angles associated with apsidal motion and precession are the longitudes of periastron, $\varpi_{i,o}$, and the longitudes of the ascending node, $\Omega_{i,o}$, respectively.

For clarity and simplicity, the rest of the chapter will assume coplanar motion; see M1a and M3 for the general analysis involving inclined systems. Taking the plane of the orbits to be the x–y plane, the polar angles are then $\theta = \Theta = \pi/2$ so that from (3.29),

$$Y_{lm}(\pi/2,\varphi) = \sqrt{\frac{2l+1}{4\pi}\frac{(l-m)!}{(l+m)!}}\,P_l^m(0)\,e^{im\varphi} \equiv \sqrt{\frac{2l+1}{4\pi}}\,c_{lm}\,e^{im\varphi} \qquad (3.33)$$

and similarly for $Y_{lm}(\pi/2,\Psi)$. Values for c_{lm}^2 for some values of l and m are listed in Table 3.1.

Table 3.1. Spherical harmonic constants

l	m	c_{lm}^2
2	2	3/8
	0	1/4
3	3	5/16
	1	3/16

Referring to Figs. 3.1 and 3.9 and recalling that we are working in the plane ($I = 0$), we have $\varphi = f_i + \omega_i + \Omega_i = f_i + \varpi_i$ and $\Psi = f_o + \varpi_o$. Substituting these together with (3.33) into (3.31) gives

$$\mathcal{R} = G\mu_i m_3 \sum_{l=2}^{\infty} \sum_{m=-l}^{l} c_{lm}^2 \, \mathcal{M}_l \left(r^l e^{imf_i}\right) \left(\frac{e^{-imf_o}}{R^{l+1}}\right) e^{im(\varpi_i - \varpi_o)}, \tag{3.34}$$

where we have collected together plane polar variables associated with each orbit in the two pairs of large brackets. For uncoupled orbits these are periodic functions with frequencies ν_i and ν_o. Since we are interested in weak interaction between the orbits, it makes sense to expand these expressions in Fourier series in these frequencies. Using the familiar two-body expressions

$$r = \frac{a_i(1-e_i^2)}{1+e_i \cos f_i} \quad \text{and} \quad R = \frac{a_o(1-e_o^2)}{1+e_o \cos f_o}, \tag{3.35}$$

we have

$$\left(\frac{r}{a_i}\right)^l e^{imf_i} = \sum_{n'=-\infty}^{\infty} s_{n'}^{(lm)}(e_i)\, e^{in' M_i} \tag{3.36}$$

and

$$\frac{e^{-imf_o}}{(R/a_o)^{l+1}} = \sum_{n=-\infty}^{\infty} F_n^{(lm)}(e_o)\, e^{-inM_o}, \tag{3.37}$$

where

$$s_{n'}^{(lm)}(e_i) = \frac{1}{2\pi} \int_{-\pi}^{\pi} \left(\frac{r}{a_i}\right)^l e^{imf_i} e^{-in' M_i} \, \mathrm{d}M_i \tag{3.38}$$

and

$$F_n^{(lm)}(e_o) = \frac{1}{2\pi} \int_{-\pi}^{\pi} \frac{e^{-imf_o}}{(R/a_o)^{l+1}} e^{inM_o} \, \mathrm{d}M_o. \tag{3.39}$$

Note that the mean anomalies are related to the orbital frequencies by

$$M_i(t) = \nu_i t + M_i(0) \quad \text{and} \quad M_o(t) = \nu_o t + M_o(0). \tag{3.40}$$

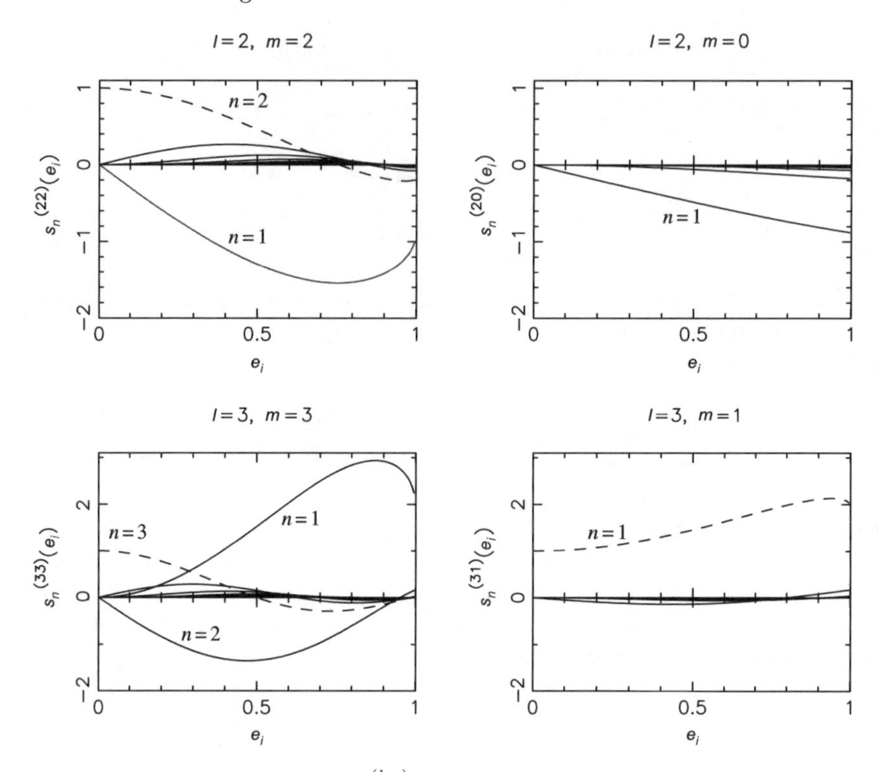

Fig. 3.10. Fourier coefficients $s_{n'}^{(lm)}(e_i)$ for various values of l, m and $n' = 1, 2, \ldots, 10 (= n$ in figure). Dashed curves correspond to $n' = m$. The most important coefficient for the stability analysis of similar-mass systems is $s_1^{(22)}(e_i)$ (shown in red (grey); note that it is negative for all values of e_i)

The *real* eccentricity-dependent Fourier coefficients $s_{n'}^{(lm)}(e_i)$ and $f_n^{(lm)}(e_o) = (1-e_o)^{l+1} F_n^{(lm)}(e_o)$ are plotted in Figs. 3.10 and 3.11 for some values of l, m, n and n'. In Sect. 3.4.7 we present approximations to the functions used in our stability analysis. Substituting (3.36) and (3.37) into the disturbing function (3.34) gives

$$\mathcal{R} = G\mu_i m_3 \sum_{l=2}^{\infty} \sum_{m=-l,2}^{l} \sum_{n'=-\infty}^{\infty} \sum_{n=-\infty}^{\infty} c_{lm}^2 \mathcal{M}_l \left(\frac{a_i^l}{a_o^{l+1}}\right) s_{n'}^{(lm)}(e_i) F_n^{(lm)}(e_o) e^{i\phi_{mnn'}}$$
$$= 2G\mu_i m_3 \sum_L \zeta_m c_{lm}^2 \, \mathcal{M}_l \left(\frac{a_i^l}{a_o^{l+1}}\right) s_{n'}^{(lm)}(e_i) \, F_n^{(lm)}(e_o) \, \cos\left(\phi_{mnn'}\right), \quad (3.41)$$

where

$$\begin{aligned}\phi_{mnn'} &= n'M_i - nM_o + m(\varpi_i - \varpi_o) \\ &= n'\lambda_i - n\lambda_o + (m-n')\varpi_i - (m-n)\varpi_o \end{aligned} \quad (3.42)$$

is called a *resonance angle*. Here $\zeta_m = 1/2$ if $m = 0$ and is 1 otherwise, and

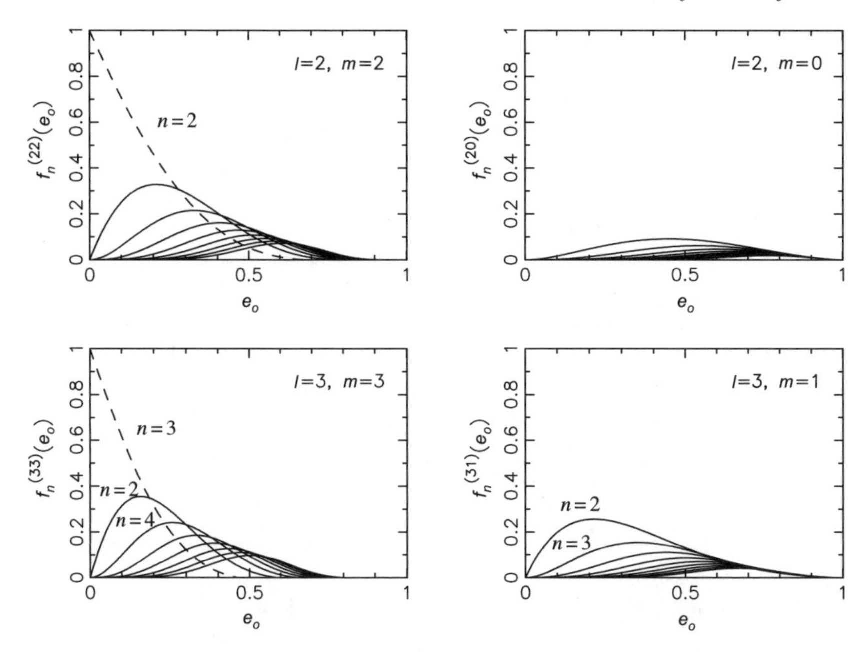

Fig. 3.11. Fourier coefficients $f_n^{(lm)}(e_i) = (1-e_o)^{l+1}F_n^{(lm)}$ for various values of l, m and $n = 2, \ldots, 10$. Dashed curves correspond to $n = m$. The most important coefficients for the stability analysis of similar-mass systems are $f_n^{(22)}$

$$\sum_L \equiv \sum_{l=2}^{\infty} \sum_{m=m_{\min},2}^{l} \sum_{n'=-\infty}^{\infty} \sum_{n=-\infty}^{\infty}, \tag{3.43}$$

where the sum over m is in steps of two for coplanar systems (M1a) and $m_{\min} = 0$ or 1 if l is even or odd, respectively.

We now have the disturbing function expressed in terms of all the relevant orbital elements including the four angles λ_i, λ_o, ϖ_i and ϖ_o, which appear in linear combination in the resonance angle (for coplanar systems the ascending node longitudes do not appear explicitly).

3.4.5 Energy Transfer Between Orbits

The defining characteristic of (most) stable hierarchical systems is that (essentially) no net energy is exchanged between the orbits over one outer orbital period. The usual way to show this is via orbit-averaging over the inner orbit. This involves a time-average over one entire orbit, assuming that all the orbital variables except the inner orbital phase remain constant on this short time-scale. The form of (3.41) makes this extremely easy to perform; but first we need an expression for the rate of change of the orbital energy. The simplest way to obtain such an expression is to use *Lagrange's planetary equation* for the rate of change of the semi-major axis.

Lagrange's Planetary Equations

Lagrange's planetary equations express the rates of change of all the elements of a two-body orbit which is being perturbed by some external potential. No assumption is made about the smallness of mass ratios (or any other parameters) so that it is perfectly well applicable to the general three-body problem, the results of which are meaningful as long as the inner and outer orbits retain their identities. The derivation of these equations can be found in Brouwer & Clements (1961) and is based on the method of variation of parameters. The parameters in this case are the orbital element which remain constant when the orbit is unperturbed, that is, e, a, ϖ, Ω, I and $\epsilon = M(0)+\varpi$. The Lagrange equation relevant to us here is that for the rate of change of the semi-major axis. For the inner and outer orbits of a triple this is

$$\frac{\mathrm{d}a_i}{\mathrm{d}t} = \frac{2}{\mu_i \nu_i a_i}\frac{\partial \mathcal{R}}{\partial \epsilon_i} \quad \text{and} \quad \frac{\mathrm{d}a_o}{\mathrm{d}t} = \frac{2}{\mu_o \nu_o a_o}\frac{\partial \mathcal{R}}{\partial \epsilon_o}, \tag{3.44}$$

respectively, where $\mathcal{R}$ is given by (3.41) (recall that our disturbing function has dimensions of energy).

Now the usual definition of the mean longitude is

$$\lambda = M + \varpi = \nu t + M(0) + \varpi = \nu t + \epsilon. \tag{3.45}$$

But this assumes that the orbital frequency (and hence the semi-major axis by Kepler's law, and also the orbital energy) is constant, something we certainly do not wish to assume once we consider unstable systems. A more general definition is

$$\lambda = \int_0^t \nu(t')\,\mathrm{d}t' + \epsilon^*, \tag{3.46}$$

where ϵ^* is a generalization of ϵ, which takes into account the variation of ν (Brouwer & Clements (1961), p. 286, and Murray & Dermott (2000), p. 252; we do not need the precise definition here). It turns out that, using this definition of λ, one can replace ϵ_i and ϵ_o with λ_i and λ_o in (3.44) so that the rates of change of the semi-major axes become

$$\frac{\mathrm{d}a_i}{\mathrm{d}t} = \frac{2}{\mu_i \nu_i a_i}\frac{\partial \mathcal{R}}{\partial \lambda_i} \quad \text{and} \quad \frac{\mathrm{d}a_o}{\mathrm{d}t} = \frac{2}{\mu_o \nu_o a_o}\frac{\partial \mathcal{R}}{\partial \lambda_o}. \tag{3.47}$$

Writing the inner orbital energy, E_i, in terms of inner semimajor axis, $E_i = -Gm_1m_2/2a_i$, the rate of change of E_i is then

$$\begin{aligned}\frac{1}{E_i}\frac{\mathrm{d}E_i}{\mathrm{d}t} &= -\frac{1}{a_i}\frac{\mathrm{d}a_i}{\mathrm{d}t} \\ &= 4\nu_i\left(\frac{m_3}{m_{12}}\right)\sum_L n'\zeta_m c_{lm}^2 \mathcal{M}_l \left(\frac{a_i}{a_o}\right)^{l+1} s_{n'}^{(lm)}(e_i)F_n^{(lm)}(e_o)\sin\left(\phi_{mnn'}\right) \\ &\equiv \sum_L n'\,C_{lmnn'}\sin(\phi_{mnn'}).\end{aligned} \tag{3.48}$$

Performing a time-average over the inner orbit assuming all elements except λ_i are constant (including a_i, i.e., putting $\lambda_i = \nu_i t + \epsilon_i$) gives

$$\left\langle \frac{1}{E_i}\frac{\mathrm{d}E_i}{\mathrm{d}t} \right\rangle = \sum_L \frac{n' C_{lmnn'}}{T_i} \int_0^{T_i} \sin\phi_{mnn'} \mathrm{d}t$$
$$= \sum_L n' C_{lmnn'} \sin(\phi_{mnn'})\, \delta_{n'0} = 0, \tag{3.49}$$

where $T_i = 2\pi/\nu_i$ is the outer orbital period. A simpler way to look at this is to ask for the contributions to (3.48) which are not rapidly varying (i.e. terms which do not depend on λ_i and λ_o), that is, to retain only the "secular" (slowly varying) terms by putting $n' = n = 0$. This automatically gives $<\dot{E}_i/E_i>= 0$ due to the factor n' in (3.48). This simple approach also yields the secular rates of change of the other orbital elements via the Lagrange equations (M3).

Resonance

How do we reconcile (3.49) with the fact that significant energy transfer is needed for escape of one body to occur? It seems that the assumption that elements other than λ_i hardly change over an inner orbital period must be wrong in such cases. In fact, it is not so much that the other elements do not change much, but rather that in some circumstances certain *combinations* of angles vary slowly, and this can result in significant energy transfer. For example, imagine a system for which the outer orbital period is almost exactly two times the inner orbital period, that is,

$$\nu_i - 2\nu_o \simeq 0. \tag{3.50}$$

Noting from (3.42) and (3.46) that

$$\dot{\phi}_{mnn'} = n'\nu_i - n\nu_o + [n'\dot{\epsilon}_i - n\dot{\epsilon}_o + (m-n')\dot{\varpi}_i - (m-n)\dot{\varpi}_o]$$
$$\simeq n'\nu_i - n\nu_o, \tag{3.51}$$

where the frequencies in square brackets are generally much smaller than the orbital frequencies, (3.50) is simply $\dot{\phi}_{m21} \simeq 0$ for any m. In practice, it is terms with $m = 2$ which contribute the most to energy transfer because these involve the quadrupole $l = 2$ terms (note the power of a_i/a_o in (3.48) and recall that the summation over l begins at 2). A system for which (3.50) holds is referred to as *resonant* for obvious reasons. In fact, except for systems for which $m_2, m_3 \ll m_1$, e.g. star–planet–planet systems or intermediate/massive black hole–star–star systems, the so-called 2:1 resonance is unstable because adjacent resonances overlap and produce instability. However, there are now several stable 2:1 planetary systems known. One example is GJ 876 (Rivera et al. 2005) whose orbital periods are 30.34 days and 60.935 days with masses $m_1 = 0.3M_\odot$, $m_2 = 0.62M_J$ and $m_3 = 1.93M_J$, where M_J is the mass of

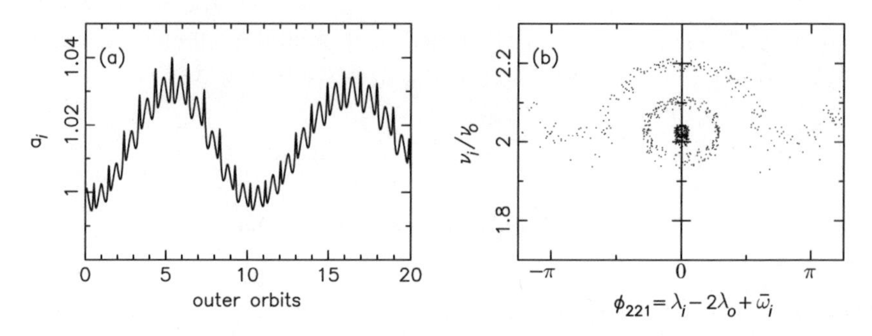

Fig. 3.12. The 2:1 resonance in the GJ 876 planetary system. (**a**): the evolution of the inner semi-major axis for $\max(\nu_i/\nu_o) = 2.1$. The small wiggles correspond to energy exchange during periastron passage of the outer planet (two peaks per passage corresponding to superior and inferior conjunction). (**b**): libration and circulation: $\nu_i/\nu_o \equiv \sigma$ vs the resonance angle ϕ_{221} for (*from centre*) $\sigma = 2.008$, 2.1 and 2.2

Jupiter. This period ratio is such that $\nu_i/\nu_o = 2.008$, that is, the system is very close to exact resonance. In order to demonstrate clearly the resonant variation of a_i, Fig. 3.12(a) plots its evolution for a slightly larger value of σ ($\sigma = 2.1$), while Fig. 3.12(b) plots $\nu_i/\nu_o \equiv \sigma$ vs the resonance angle ϕ_{221} for $\sigma = 2.008$ (the innermost set of points), $\sigma = 2.1$ (the librating set of points forming a fuzzy circle) and $\sigma = 2.2$ (the circulating set of points). The fact that a_i varies significantly in Fig. 3.12(a) indicates that a substantial amount of energy is exchanged between the orbits (when the inner orbit shrinks, the outer orbit expands due to conservation of energy). Resonant orbits are also associated with libration of one or more resonance angles. The *width* of a resonance is the "distance" from exact resonance to the separatrix, calculated at $\phi_{mnn'} = 0$; if this separatrix overlaps the separatrix of a neighbouring resonance, we have instability. Thus our task is to determine the width of resonances and to ask for what orbital parameters are these wide enough to overlap neighbouring resonances.

Before we leave this section on energy exchange and resonance, we quote a result from M4 which gives approximately the energy exchanged between the inner and outer orbits over one outer orbital period (from apastron to apastron):

$$\frac{\Delta E_i}{E_i} \simeq \mathcal{I}_{22}^2 + 2\, e_i(0)\, \mathcal{I}_{22} \sin\left[\phi(0)\right], \tag{3.52}$$

where $e_i(0)$ is the inner eccentricity at $t = 0$ and

$$\mathcal{I}_{22} = \frac{9}{4}\left(\frac{m_3}{m_{12}}\right)\left(\frac{a_i}{a_o}\right)^3 \mathcal{E}_{22}(e_o, \sigma), \tag{3.53}$$

with an asymptotic expression for the "overlap integral"

$$\mathcal{E}_{22}(e_o, \sigma) = \nu_i e^{-i\sigma\pi} \int_0^{T_o} \frac{e^{-2if_o}}{(R/a_o)^3} e^{i\nu_i t} \, dt \tag{3.54}$$

$$\simeq \frac{4\sqrt{2\pi}}{3} \frac{(1-e_o^2)^{3/4}}{e_o^2} \sigma^{5/2} e^{-\sigma\xi(e_o)} \tag{3.55}$$

(M1a). Here T_o is the outer orbital period and $\xi(e_o) = \cosh^{-1}(1/e_o) - \sqrt{1-e_o^2}$. Also,

$$\phi(0) = M_i(0) + \sigma\pi + 2(\varpi_i - \varpi_o) \simeq \phi_{2n1}(0), \tag{3.56}$$

that is, $\phi(0)$ is approximately the value of the resonance angle ϕ_{2n1} when the outer body is at apastron (see (3.42)), exact equality holding when $\sigma = n$. The expression (3.55) includes only quadrupole $l = 2$, $m = 2$ terms and is obtained using an asymptotic method similar to that of Heggie (1975), which gives the energy exchanged during the flyby of a binary by a third body. Note that $\lim_{e_o \to 0} \mathcal{E}_{22} = 0$ for $\sigma > 2$, is finite for $\sigma = 2$ and is not defined for $\sigma < 2$, and that $\lim_{e_o \to 1}(1-e_o)^3 \mathcal{E}_{22}$ is finite.

The form of (3.55) shows that the amount of energy transferred during one outer orbit of a bound triple is exponentially small except when $\sigma\xi(e_o)$ is small. This is consistent with the orbit-averaging result $\langle \dot{E}_i/E_i \rangle = 0$, and it strongly suggests that "stable" systems are stable *for all time*, although as previously discussed a proof is not yet available.

3.4.6 A Pendulum Equation for the Resonance Angle

Figure 3.12(b) illustrates how a resonance angle librates when the orbital frequencies are near-commensurate. This suggests that resonance angles should satisfy a pendulum-like equation; the ability to write down such an equation would then give us the full machinery outlined in Sect. 3.3.1 for pendulums. In particular, we could calculate the distance from exact resonance to the separatrix, that is, the resonance width; recall that we need this in order to determine when neighbouring resonances overlap and hence when a system is unstable.

Referring to (3.1), we see the second time derivative of the resonance angle is required. Starting from (3.51):

$$\dot{\phi}_{mnn'} = n'\nu_i - n\nu_o, \tag{3.57}$$

where we have replaced the approximation symbol with equality, we then have

$$\ddot{\phi}_{mnn'} = n'\dot{\nu}_i - n\dot{\nu}_o. \tag{3.58}$$

Relating the rates of change of the orbital frequencies to the rates of change of the semi-major axes:

$$\frac{\dot{\nu}_i}{\nu_i} = -\frac{3}{2}\frac{\dot{a}_i}{a_i} \quad \text{and} \quad \frac{\dot{\nu}_o}{\nu_o} = -\frac{3}{2}\frac{\dot{a}_o}{a_o}, \tag{3.59}$$

we can again make use of Lagrange's planetary equation for the rate of change of the semi-major axis, (3.47), together with (3.48) and its equivalent for $\dot{a}_o$. Substituting these into (3.58) *and assuming that the resonance is isolated (not forced), that is, that the only significant terms in the summations are those with the same values of* m, n *and* n', we get

$$\ddot{\phi}_{mnn'} = -n'^2 \nu_o^2 \mathcal{A}_{mnn'} \sin(\phi_{mnn'}), \tag{3.60}$$

where

$$\begin{aligned}\mathcal{A}_{mnn'} &\equiv -6\,\zeta_m \sum_{l=l_{min},2}^{\infty} c_{lm}^2 s_{n'}^{(lm)}(e_i) F_n^{(lm)}(e_o) \\ &\quad \cdot \left[M_i^{(l)}\, \sigma^{-(2l-4)/3} + M_o^{(l)} (n/n')^2 \sigma^{-2l/3} \right] \\ &\simeq -6\,\zeta_m \sum_{l=l_{min},2}^{\infty} c_{lm}^2 s_{n'}^{(lm)}(e_i) F_n^{(lm)}(e_o)\,(n/n')^{-(2l-4)/3} \\ &\quad \cdot \left[M_i^{(l)} + M_o^{(l)} (n/n')^{2/3} \right],\end{aligned} \tag{3.61}$$

and we have put $\sigma \simeq n/n'$ in the last step. Here $l_{min} = 2$ if m is even and $l_{min} = 3$ if m is odd. The dependence on the masses is solely through the functions

$$M_i^{(l)} = \mathcal{M}_l \left(\frac{m_3}{m_{12}}\right) \left(\frac{m_{12}}{m_{123}}\right)^{(l+1)/3} \quad \text{and} \quad M_o^{(l)} = \mathcal{M}_l \left(\frac{m_1 m_2}{m_{12}^2}\right) \left(\frac{m_{12}}{m_{123}}\right)^{l/3}. \tag{3.62}$$

Except for very low values of n corresponding to planetary-like problems, it is usually adequate to include only the first term in the summation over l.

Comparing (3.60) with (3.1), we have that the "small angle frequency" ω_0 is $n'\nu_o|\mathcal{A}_{mnn'}|^{1/2}$. When $\mathcal{A}_{mnn'} > 0$ we have libration around zero, and when $\mathcal{A}_{mnn'} < 0$ we have libration around π. It turns out that for systems for which at least two of the masses are reasonably similar (this is quantified in Sect. 3.4.10), the dominant resonances are those with $m = 2$ and $n' = 1$. Using the notation introduced in M1a, these are the $[n:1](2)$ resonances. Referring to Figs. 3.10 and 3.11 and recalling that we only need include $l = 2$ when $m = 2$, we see that $s_1^{(22)}(e_i) < 0$ for all $0 \le e_i \le 1$, and that $f_n^{(22)}(e_o) > 0$ for $0 \le e_o \le 1$ so that $\mathcal{A}_{2n1} > 1$ for all n. Thus libration is around zero for all resonances of interest here. Putting $n' = 1$ and $m = 2$ in (3.61), retaining only the $l = 2$ term and setting $\phi_{2n1} \equiv \phi_n$ and $\mathcal{A}_{2n1} \equiv \mathcal{A}_n$, the resonances of interest to us are governed by

$$\ddot{\phi}_n = -\nu_o^2 \mathcal{A}_n \sin\phi_n, \tag{3.63}$$

where

$$\mathcal{A}_n = -\frac{9}{2}\, s_1^{(22)}(e_i) F_n^{(22)}(e_o) \left[M_i^{(2)} + M_o^{(2)} n^{2/3} \right], \tag{3.64}$$

with

$$M_i^{(2)} = \frac{m_3}{m_{123}} \quad \text{and} \quad M_o^{(2)} = \left(\frac{m_1 m_2}{m_{12}^2}\right)\left(\frac{m_{12}}{m_{123}}\right)^{2/3}, \tag{3.65}$$

and we have used $c_{22}^2 = 3/8$ from Table 3.1. In Sect. 3.4.5 p. 84, we defined the width of a resonance to be the distance from exact resonance and the separatrix, calculated at $\phi_{mnn'} = 0$. Equation (3.3) gives an expression for the separatrix so that the width of a resonance is

$$\Delta\dot{\phi} = 2\omega_0 = 2\nu_o\sqrt{\mathcal{A}_n} \tag{3.66}$$

for the $[n:1](2)$ resonances of interest here. It is usually more convenient to define the width of a resonance in terms of the change in σ. Since

$$\dot{\phi}_n = \nu_i - n\,\nu_o = \nu_o(\sigma - n), \tag{3.67}$$

we can define the width of the $[n:1](2)$ resonance to be

$$\Delta\sigma_n = 2\sqrt{\mathcal{A}_n}. \tag{3.68}$$

We can associate an "energy" E_n with the pendulum-like motion of a resonance such that $E_n < 0$ for libration and $E_n > 0$ for circulation of ϕ_n. Following (3.2) we then have

$$E_n = \frac{1}{2}\dot{\phi}_n^2 - \nu_o^2 \mathcal{A}_n(\cos\phi_n + 1). \tag{3.69}$$

It is useful to define a dimensionless version of this such that $E_n = \nu_o \overline{E}_n$, that is,

$$\overline{E}_n = \frac{1}{2}[\delta\sigma_n]^2 - \mathcal{A}_n(\cos\phi_n + 1), \tag{3.70}$$

where $\delta\sigma_n = \sigma - n$ is the "distance" from exact resonance corresponding to ϕ_n. Note that $\delta\sigma_n$ is a maximum when $\phi_n = 0$ (for libration around $\phi_n = 0$). We will use (3.70) in a simple algorithm to determine the stability of any given configuration (Sect. 3.4.10).

The form of (3.64) makes it relatively easy to see how resonance widths depend on the various parameters. Before we make use of (3.68) to determine the stability boundary, it is necessary to discuss evaluation of the eccentricity functions $s_1^{(22)}(e_i)$ and $F_n^{(22)}(e_o)$.

3.4.7 Eccentricity Functions

Since the eccentricity functions $s_1^{(22)}(e_i)$ and $F_n^{(22)}(e_o)$ are integrals with no closed form expressions (except for $n=0$: see M3), it is of interest to find approximations. A simple Taylor expansion of the integrand of $s_1^{(22)}(e_i)$ about $e_i=0$ allows for the integral to be performed, and, if one expands up to $\mathcal{O}(e_i^7)$, allows for the function to be well represented for all $e_i \le 1$. This procedure gives

$$s_1^{(22)}(e_i) \simeq -3e_i + \frac{13}{8}e_i^3 + \frac{5}{192}e_i^5 - \frac{227}{3072}e_i^7. \tag{3.71}$$

If ϵ_i is the difference between the exact and approximate expression, $|\epsilon_i| < 0.001$ for $e_i < 0.63$, $|\epsilon_i| < 0.01$ for $e_i < 0.79$ and $|\epsilon_i| < 0.1$ for $e_i < 1$.

While it is possible to find Taylor series approximations to $F_n^{(22)}(e_o)$, we would need hundreds of these for a general stability algorithm since systems with very high outer eccentricity can involve very high values of n (since $\sigma = \nu_i/\nu_o \simeq n/n' = n$). Instead, we make use of the asymptotic expression (3.54) to evaluate (3.39). Making the substitution $M_o = \nu_o t - \pi$ in (3.54) (since the outer orbit starts at $-\pi$, that is, $M_o(0) = -\pi$) so that $\nu_i t = \sigma(M_o + \pi)$ and $\nu_i dt = \sigma dM_o$, the integral becomes

$$\mathcal{E}_{22}(e_o, \sigma) = \sigma \int_{-\pi}^{\pi} \frac{e^{-2if_o}}{(R/a_o)^3} e^{i\sigma M_o}\, \mathrm{d}M_o. \tag{3.72}$$

Comparing this with (3.39) we see that

$$F_n^{(22)}(e_o) \simeq \mathcal{E}_{22}(e_o, n)/2\pi n. \tag{3.73}$$

Thus we have the beautiful result that the resonance widths are exponentially small when $\sigma\xi(e_o)$ is small, consistent with the fact that an exponentially small amount of energy is exchanged between the orbits in such circumstances.

3.4.8 Induced Eccentricity and Secular Effects

The expression for the resonance width, (3.68), together with (3.64) and (3.71) suggest that systems whose inner binary is circular have zero resonance widths (since $s_1^{(22)}(0) = 0$). But this surely is *not* true! Figure 3.13 plots the evolution of the inner eccentricity for an equal mass three-body system whose initial eccentricities are $e_i(0) = 0$ and $e_o(0) = 0.5$, and for which (a) $\sigma = 10$ and (b) $\sigma = 8$. Both systems start at outer apastron, and significant eccentricity is induced when they pass through outer periastron. The formalism used to estimate the energy transferred between orbits (see Sect. 3.4.5 and (3.52)) can also be used to estimate the induced inner eccentricity. This is given by

$$e_i(T_o) = \left[e_i(0)^2 - 2\, e_i(0)\, \mathcal{I}_{22} \sin[\phi(0)] + \mathcal{I}_{22}^2\right]^{1/2}, \tag{3.74}$$

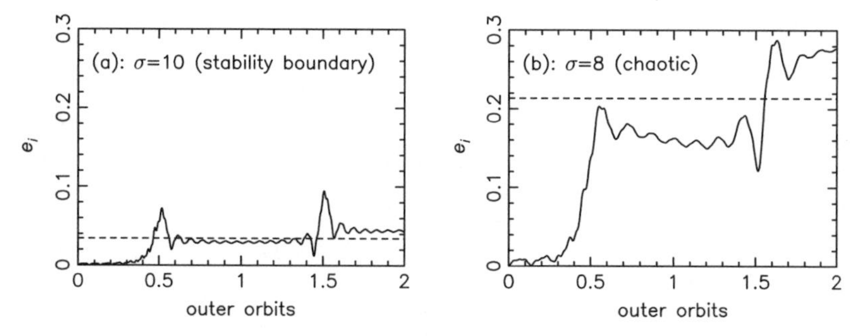

Fig. 3.13. Induced inner eccentricity of a circular binary. (**a**): $\sigma = 10$ and (**b**): $\sigma = 8$. In both cases $e_o = 0.5$ and the system is started at outer apastron with $M_i(0) = 0$ and $\varpi_i - \varpi_o = 0$. Both systems are chaotic but (**a**) is on the stability boundary while (**b**) is deep inside the unstable region. The dashed lines correspond to the estimated induced eccentricity ((3.74)), following the first outer periastron passage

where $e_i(0)$ and $e_i(T_o)$ are the inner eccentricity at initial and final outer apastron, and $\mathcal{I}_{22}$ and $\phi(0)$ are given by (3.53) and (3.56), respectively. The dashed curves in Fig. 3.13 indicate these estimates.

It turns out that using $e_i(T_o)$ instead of $e_i(0)$ in the expression for the resonance width quite accurately predicts the stability boundary when octopole effects are unimportant (see Fig. 3.15).

Octopole Variations for Coplanar Systems

For systems with $m_1 \neq m_2$, secular octopole contributions to the disturbing function (terms with $n = n' = 0$) can cause the inner eccentricity to vary considerably on time-scales of thousands of inner orbits (Murray & Dermott 2000, M3). This is especially important for close planetary systems. While the outer eccentricity also varies, the main effect on the resonance widths comes from the variation of $s_1^{(22)}(e_i)$, which is a maximum at the maximum of the octopole cycle in e_i. Referring to this maximum as $e_i^{(\mathrm{oct})}$, it is given approximately by (Mardling 2007, M1a)

$$e_i^{(\mathrm{oct})} = \begin{cases} (1+\alpha)e_i^{(\mathrm{eq})}, & \alpha \leq 1 \\ e_i(0) + 2e_i^{(\mathrm{eq})}, & \alpha > 1, \end{cases} \tag{3.75}$$

where $\alpha = |1 - e_i(0)/e_i^{(\mathrm{eq})}|$ and $e_i^{(\mathrm{eq})}$ is the "equilibrium" or "fixed point" eccentricity, which is the root of the eighth-order polynomial $\sum_{n=1}^{8} a_n x^n$ in [0,1], where the a_n are given by

$$a_0 = -B^2$$
$$a_1 = 2AB$$
$$a_2 = B^2 + C^2 - A^2$$

$$\begin{aligned} a_3 &= -2(AB + 4CD) \\ a_4 &= A^2 + 3C^2 + 16D^2 \\ a_5 &= -18CD \\ a_6 &= \frac{9}{4}C^2 + 24D^2 \\ a_7 &= -9CD \\ a_8 &= 9D^2, \end{aligned} \tag{3.76}$$

with

$$\begin{aligned} A &= \frac{3}{4}\left(\frac{m_3}{m_{12}}\right)\left(\frac{a_i}{a_o}\right)^3 \varepsilon_o^{-3} \\ B &= \frac{15}{64}\left(\frac{m_3}{m_{12}}\right)\left(\frac{m_1 - m_2}{m_{12}}\right)\left(\frac{a_i}{a_o}\right)^4 \varepsilon_o^{-5} \\ C &= \frac{3}{4}\left(\frac{m_1 m_2}{m_{12}^2}\right)\left(\frac{a_i}{a_o}\right)^2 \varepsilon_o^{-4} \\ D &= \frac{15}{64}\left(\frac{m_1 m_2}{m_{12}^2}\right)\left(\frac{m_1 - m_2}{m_{12}}\right)\left(\frac{a_i}{a_o}\right)^3 \left(\frac{1 + 4e_o^2}{e_o \varepsilon_o^6}\right), \end{aligned} \tag{3.77}$$

and $\varepsilon_o = \sqrt{1 - e_o^2}$. In the limit $e_i \ll 1$, the equilibrium eccentricity reduces to

$$e_i^{(eq)} = \frac{(5/4) e_o m_3 (m_1 - m_2)(a_i/a_o)^2 \sigma \varepsilon_o^{-1}}{|m_1 m_2 - m_{12} m_3 (a_i/a_o) \varepsilon_o \sigma|}. \tag{3.78}$$

Note that even though (3.78) is not accurate away from the stability boundary where $e_i(T_o)$ is large, it can be used to determine the boundary if $e_i(0)$ is small because $e_i(T_o)$ tends to be small there in that case (see Fig. 3.13).

3.4.9 Resonance Overlap and the Stability Boundary

The stability of any given coplanar configuration depends on the values of the eight parameters m_2/m_1, m_3/m_{12}, σ, e_i, e_o, $\varpi_i - \varpi_o$, $M_i(0)$ and $M_o(0)$. In order to represent the stability boundary in two dimensions, we need to fix the values of six of these and vary the other two. Here we choose to plot e_o against σ for $\varpi_i - \varpi_o = M_i(0) = 0$ and $M_o = -\pi$, and for a selection of mass ratios and $e_i(0)$.

For a given value of n and for fixed values of $e_i(0)$, m_2/m_1 and m_3/m_{12}, the two boundaries of the $[n:1](2)$ resonance are given by

$$\sigma(e_o) = n \pm \Delta\sigma_n(e_o) = n \pm 2\left[\mathcal{A}_n(e_o)\right]^{1/2}, \tag{3.79}$$

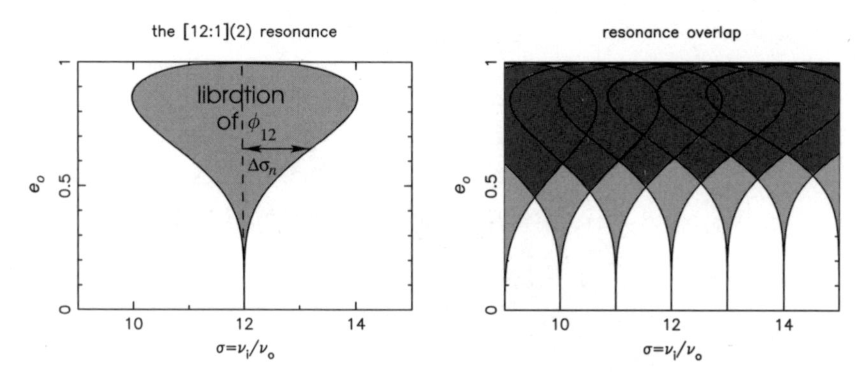

Fig. 3.14. (**a**): The [12:1](2) resonance. (**b**): Resonance overlap. This example corresponds to $m_2/m_1 = m_3/m_1 = 0.01$ and $e_i(0) = 0.5$ (see Fig. 3.16). See text for discussion

where $\mathcal{A}_n(e_o)$ is given by (3.64). Note that this assumes exact resonance occurs when

$$\dot{\phi}_n = \nu_i - n\nu_o = 0, \tag{3.80}$$

that is when $\sigma = \nu_i/\nu_o = n$; however, if $\dot{\varpi}_i/\nu_o$ is significant, it will shift exact resonance away from this (recall the precise expression (3.51) for $\dot{\phi}_n$; see also Fig. 3.15). Figure 3.14(a) plots e_o against σ for the [12:1](2) resonance for a particular set of initial conditions, with the shaded region corresponding to libration of the resonance angle ϕ_{12},[10] while panel (b) shows the overlap of the resonances $[n:1](2)$, $n = 9, 10, \ldots, 15$ for the same initial conditions. The lower (green)-shaded regions in panel (b) formally correspond to stable libration of the resonance angles ϕ_n, while the unshaded regions correspond to stable circulation for which the inner and outer orbits have constant semi-major axes. The upper (red)-shaded region corresponds to the overlap of neighbouring resonances (as well as more distant resonances), so that a system with initial conditions corresponding to any point in this region is predicted by the resonance overlap stability criterion to be unstable.

How does this compare with direct numerical experiments? Figure 3.15(a) shows a stability map for equal-mass configurations with initially circular inner binaries, for various initial period ratios and outer eccentricities. A dot corresponding to the initial values of σ and e_o is plotted if a direct numerical integration of the three-body equations of motion results in an unstable system. Rather than integrating the system until one of the bodies escapes, two almost identical systems (the given system and its "ghost") are integrated in parallel and the difference in the inner semi-major axes at outer apastron is monitored (because this variable is approximately constant for non-resonant systems). Taking advantage of the sensitivity of a chaotic system to initial conditions, this difference will grow in proportion to the initial difference between

[10]Even though (3.79) gives σ as a function of e_o, it seems more natural to plot the resonance boundaries with σ as the independent variable.

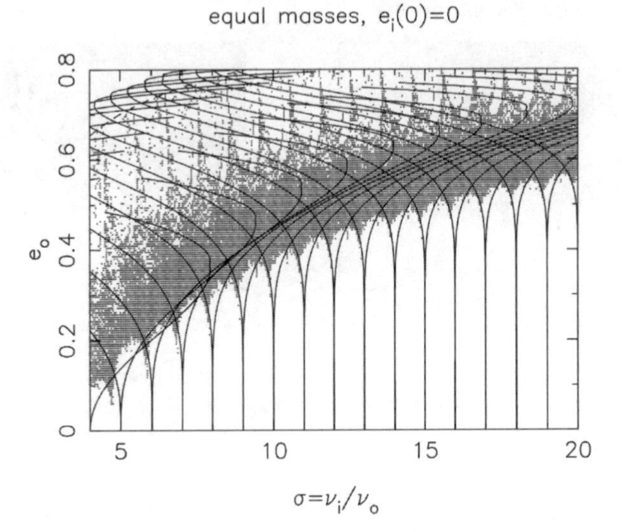

Fig. 3.15. Experimental vs theoretical stability boundary. The position of each red (grey) dot in $(\sigma - e_o)$ space corresponds to the initial conditions of an unstable system for which the masses are equal, $e_i(0) = 0$, and $M_i(0) = 0$ and $M_o(0) = -\pi$. The black curves are the resonance boundaries given by (3.79), which terminate at points for which $e_i(T_o) = 1$. Notice the structure of the distribution of dots near these termination points; this reflects the process of exchange of m_3 into the inner binary (consistent with $e_i(T_o) > 0$). Systems deemed stable (see text for how this decision is made) are those for which exchange occurs rapidly. While the resonance overlap stability criterion predicts the stability boundary fairly accurately, some of the red dots fall inside single-resonance regions which ought to be stable according to the criterion. But the criterion assumes that when only one resonance angle is librating, the forcing is negligible; this clearly is not true at these points. Also notice how the red dots trace the separatrix at the left-hand boundaries, and in particular notice the offset which is prominent for the 5:1 resonance; this is analogous to spectral line splitting by a magnetic field and is a result of the influence of $\dot{\varpi}_i$, which has been neglected in (3.79)

two systems (10^{-7} in the inner eccentricity) for a stable system, but will grow exponentially for an unstable system as discussed in Sect. 3.3.3. The actual stability boundary fairly accurately follows the points at which neighbouring resonances overlap; however, the stability criterion does not predict the unstable nature of some systems inside single-librating regions (corresponding to the green regions in Fig. 3.14(b)) because it assumes that forcing is negligible there.

Figure 3.16 shows stability maps for a variety of initial conditions. Each map has $m_1 = 1$, $M_i(0) = 0$ and $M_o(0) = -\pi$ and aligned periastra except for panel (f). Consider the systems (a), (c) and (e) for which $e_i(0) \neq 0$ and $\eta = \varpi_i - \varpi_o = 0$. The librating regions for which there is no overlap with a neighbouring resonance are relatively free of unstable systems, while those for

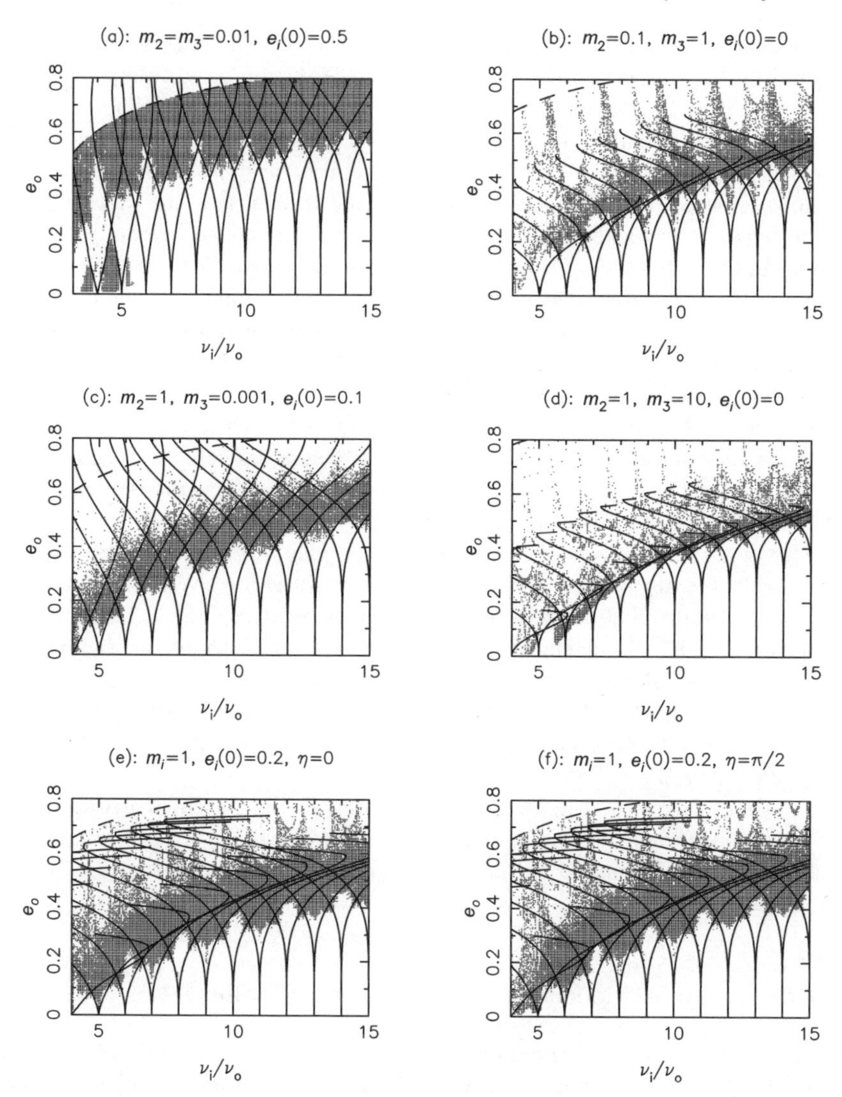

Fig. 3.16. Stability maps for a variety of initial conditions ($m_1 = 1$). Notice how resonance shapes vary significantly from panel to panel, but the resonance overlap stability criterion is still successful at predicting the stability boundary (except for the single-librating regions). The *dashed curve* in the top left-hand corner of each panel corresponds to $R_p/a_i = 1$, where $R_p = a_o(1 - e_o)$ is the outer periastron distance (data were not collected beyond this curve). (**a**): planetary-like system with significant inner eccentricity; (**b**): low-mass secondary with zero initial inner eccentricity; (**c**): Jupiter-like outer body orbiting an equal-mass eccentric binary; (**d**): "binary" consisting of a heavy body and an equal-mass binary; (**e**) and (**f**): equal-mass system with $e_i(0) = 0.2$. Here $\eta = \varpi_i - \varpi_o$, the two plots demonstrating the effect of rotating the orbits relative to each other. Notice that even resonances are more stable than odd in (**a**) while the opposite is true in (**b**) (see text for discussion)

odd resonances tend to be full down to near the resonance cross-over points. The reason for this is as follows. Referring to (3.42) on p. 80, we see (putting $n' = 1$ and $m = 2$) that for these initial conditions, $\phi_n(0) = n\pi$. Since libration is around zero (because $\mathcal{A}_n > 0$), a system starting at exact resonance, that is, with $\sigma = n$, will stay there if n is even because it is at the very centre of the resonance (see Fig. 3.12 on p.84), while if n is odd, the system starts at the hyperbolic fixed point on the separatrix! An odd-n system for which $\sigma \neq n$ (and is indicated on the stability map to be inside a resonance) actually begins *outside* the librating region; recall that the definition of the resonance boundary uses the value of the separatrix at $\phi_n = 0$. However, it will still be strongly forced and its proximity to the separatrix will cause it to be unstable. A more detailed analysis can be found in M1a.

We should expect from this discussion that a system for which $\eta = \varpi_i - \varpi_o \neq 0$ will exhibit different behaviour, and this is indeed the case as panel (f), for which $\eta = \pi/2$, reveals. In this case $\phi_n = (n+1)\pi$ and we see that it is the *even* resonances that are now more unstable.

The fact that $e_i(0) \neq 0$ for the examples just discussed means that the inner orbit begins with a definite periastron direction. What about when $e_i(0) = 0$? Figure 3.15 as well as panels (b) and (d) in Fig. 3.16 show that points on the left-hand sides of the resonances tend to be unstable while points on the right-hand side are stable up to where the resonances overlap. We interpret this as indicating that the induced periastron direction associated with the induced eccentricity tends to be such that $\eta(T_o) \simeq \pi/4$ so that $\phi_n = (2n+1)\pi/2$.

Another feature of Fig. 3.16 worth noting is the patch of instability at the lower-left corner of panel (a). This is common for low-order resonances in planetary-like systems and actually corresponds to libration around π (this is discussed in detail in M2).

3.4.10 A Simple Algorithm for Predicting Stability

For most applications one needs to know the stability characteristics of single systems. Thus rather than give a formula for the stability boundary, we end this chapter by presenting an algorithm for testing the stability of individual configurations. Note that it only holds for coplanar systems[11] and is restricted to systems for which the $[n:1](2)$ resonances dominate. These are such that either *both* $m_2/m_1 > 0.01$ and $m_3/m_1 > 0.01$ *or at least one of* $m_2/m_1 > 0.05$ or $m_3/m_1 > 0.05$. The algorithm is as follows:

1. Identify which $[n:1](2)$ resonance the system is near and calculate the distance $\delta\sigma_n$ from that resonance: $\delta\sigma_n = \sigma - n$, where $n = \lfloor\sigma\rfloor$ (the nearest integer for which $n \leq \sigma$),

[11] A Fortran routine for arbitrarily inclined systems is available from the author.

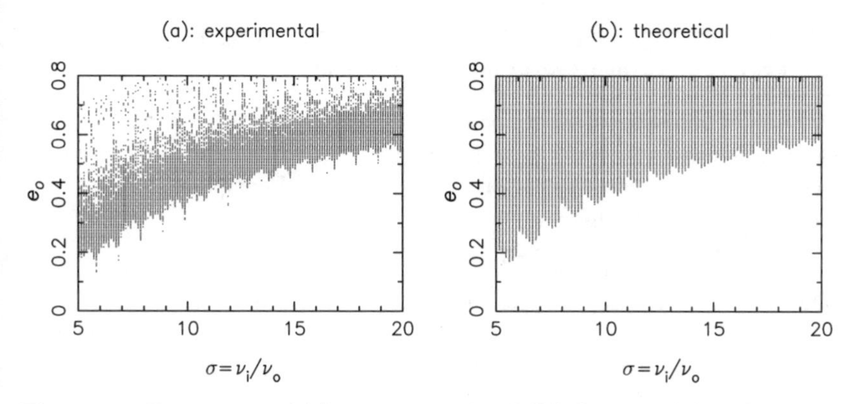

Fig. 3.17. Comparison of (**a**) experimental and (**b**) theoretical data for equal mass coplanar systems with $e_i(0) = 0$

2. Take the associated resonance angle to be zero rather than the definition (3.42) (see discussion below), $\phi_n = 0$;
3. Calculate the induced eccentricity from (3.74) and (if $m_1 \neq m_2$) the maximum octopole eccentricity from (3.75). Determine $e_i = \max[e_i(T_o), e_i^{(\rm oct)}]$ for use in $s_1^{(22)}(e_i)$,
4. Calculate $\mathcal{A}_n$ from (3.64);
5. Calculate $\overline{\mathcal{E}}_n$ and $\overline{\mathcal{E}}_{n+1}$ from (3.70) and deem the system unstable if $\overline{\mathcal{E}}_n < 0$ *and* $\overline{\mathcal{E}}_{n+1} < 0$.

Figure 3.17 compares the experimental data shown in Fig. 3.15 with data generated using the algorithm above. A dot is plotted if a system is deemed to be unstable. The boundary structure is reproduced reasonably well, although the boundary itself should be slightly lower, a result of the fact that the resonance overlap criterion does not recognize the unstable nature of points near to but outside the separatrix. This is also the reason for taking $\phi_n = 0$ for all initial conditions (recall the discussion in the previous section on odd and even resonances).

References

Aarseth S. J., 1971, Ap&SS, 13, 324
Aarseth S. J., 2007, MNRAS, 378, 285
Arnol'd V. I., 1963, Russian Mathematical Surveys, 18, 9
Arnol'd V. I., 1978, Mathematical Methods of Classical Mechanics. Springer-Verlag, New York
Barrow-Green J., 1997, Poincare and the Three Body Problem (History of Mathematics, V. 11). American Mathematical Society
Brouwer D., Clements G. M., 1961, Methods of Celestial Mechanics. Academic Press, New York and London
Chirikov B. V., 1979, Phys. Rep., 52, 263

Dyson F. J., 2000, Oppenheimer Lecture, University of California, Berkeley. http://www.hartford-hwp.com/archives/20/035.html
Eggleton P., Kiseleva L., 1995, ApJ, 455, 640
Eggleton P. P., Kiseleva-Eggleton L., 2001, ApJ, 562, 1012
Fabrycky D., Tremaine, S., 2007, ApJ, 669, 1298
Goldstein H., 1980, Classical Mechanics. Addison-Wesley, Philippines
Heggie D. C., 1975, MNRAS, 173, 729
Hills J. G., 1976, MNRAS, 175, 1P
Hills J. G., 1988, Nature, 331, 687
Jackson J. D., 1975, Classical Electrodynamics, Wiley, New York, 2nd ed
Kaula W. M., 1961, Geophys. J. Roy. Astr. Soc., 5, 104
Kolmogorov A. N., 1954, Dokl. Akad. Nauk, 98, 527
Kozai Y., 1962, AJ, 67, 591
Mardling R. A., 1995a, ApJ, 450, 722
Mardling R. A., 1995b, ApJ, 450, 732
Mardling R. A., 2007, MNRAS, 382, 1768
Mardling R. A., 2008a, submitted to MNRAS
Mardling R. A., 2008b, submitted to MNRAS
Moser J., 1962, Nachr. Akad. Wiss. Gottingen II, Math. Phys. KD, 1, 1
Murray C. D., Dermott S. F., 2000, Solar System Dynamics. Cambridge Univ. Press, Cambridge
Poincaré H., 1993, New Methods of Celestial Mechanics (Vol. 1). Goro D. L., ed., AIP, New York, I23, 22
Reichl L. E., 1992, The Transition to Chaos in Conservative Classical Systems: Quantum Manifestations, Springer-Verlag, New York
Reipurth, B., & Clarke, C. 2001, AJ, 122, 432
Rivera E. J., et al., 2005, ApJ, 634, 625
Spurzem R., Giersz M., Heggie D. C., Lin D. N. C., 2006, astro-ph/0612757
Tokovinin, A., Thomas, S., Sterzik, M., & Udry, S. 2006, A&A, 450, 681
Walker G. H., Ford J., 1969, Physical Review, 188, 416

4

Fokker–Planck Treatment of Collisional Stellar Dynamics

Marc Freitag

University of Cambridge, Institute of Astronomy, Madingley Road, Cambridge CB3 0HA, UK,
marc.freitag@gmail.com

4.1 Introduction

In this chapter, I explain how the evolution of an N-body system can be described using a formalism explicitly based on the distribution function in phase space. Such an approach can be contrasted with direct N-body simulations in which the trajectories of a large number of particles are integrated. Because trajectories with close initial conditions diverge exponentially in gravitational N-body systems (Goodman et al. 1993; Hemsendorf & Merritt 2002, and references therein), most results of N-body simulations must be interpreted statistically. It is therefore interesting to consider the simulation methods that treat the gravitational system in an explicitly statistical way.

Since the early 1980s, the numerical solution of the Fokker–Planck (FP) equation has been the technique of choice for a statistical treatment of collisional systems such as globular clusters or dense galactic nuclei. In its basic version, on which I focus, this equation (combined with the Poisson equation) describes the evolution of a stellar system in dynamical equilibrium; but evolving slowly through the effects of two-body relaxation. In this chapter, I further restrict myself to spherically symmetric configurations with no net rotation, as most researchers in the field have done, to make the problem easier to tackle. As far as relaxation is concerned, the Monte-Carlo numerical scheme, presented in Chap. 5, is essentially equivalent to solving the FP equation using a particle-based representation of the distribution function instead of tabulated data. Therefore, the assumptions and limitations inherent in the FP description of relaxation, which are described in detail in this chapter, also apply to Monte-Carlo techniques.

A note of caution is required here. The dynamics of a gravitational N-body system is highly non-linear, with the possibility that small differences in the "microscopic" conditions (such as the existence and properties of a binary star) can lead to rather large macroscopic differences in evolution. The FP approach does not provide a statistical description of the various macroscopically distinct possible evolutions. When such divergences are expected to

Freitag, M.: *Fokker–planck Treatment of Collisional Stellar Dynamics*. Lect. Notes Phys. **760**, 97–121 (2008)
DOI 10.1007/978-1-4020-8431-7_4

occur, such as in the process of collisional runaway or post-collapse core oscillations (see Sect. 4.5), the only way to capture them in a satisfying way by means of FP simulations is probably by including some explicit stochastic process and repeat the simulation several times with different random sequences (see Takahashi & Inagaki (1991) for an example in the case of core oscillations).

In the last decade or so, FP codes have lost some ground to direct N-body and Monte-Carlo codes. Indeed, these particle-based methods make it easier to include a variety of physical effects thought to play an important role in real systems and faster computers enable the use of higher and higher particle numbers. Nevertheless, because FP computations are very fast and produce data that are much smoother, less memory-consuming and easier to manipulate than particle-based simulations, they are an invaluable tool for exploring large volumes of parameter space. They also help in gaining a better understanding of "macroscopic" collisional stellar dynamics by providing a description at a level more suitable than that of "microscopic" point-mass particles attracting each other.

In Sect. 4.2, I present the Boltzmann equation which is at the heart of the statistical description of an N-body system. In Sect. 4.3, I give an outline of the derivation of the main forms of the FP equation used to simulate the effects of relaxation in spherical stellar systems. Finally, Sect. 4.5 is a quick overview of the applications of the FP approach in stellar dynamics with a focus on the additional physics that can be incorporated into that framework.

4.2 Boltzmann Equation

4.2.1 Notation

The following notations are in use in this section. Position and velocity in 3D space are denoted by

$$\boldsymbol{x} = (x, y, z) = (x_1, x_2, x_3)\,,$$

and

$$\boldsymbol{v} = (v_x, v_y, v_z) = (v_1, v_2, v_3)\,.$$

For a point in the 6D phase space, I use the notation

$$\underline{w} = (\boldsymbol{x}, \boldsymbol{v}).$$

The gradient of a field u in 3D space is written

$$\boldsymbol{\nabla} u \equiv \frac{\partial u}{\partial \boldsymbol{x}} = \left(\frac{\partial u}{\partial x}, \frac{\partial u}{\partial y}, \frac{\partial u}{\partial z}\right),$$

and the gradient in the 6D phase space is

$$\underline{\nabla} u \equiv \frac{\partial u}{\partial \underline{w}} = \left(\frac{\partial u}{\partial x}, \frac{\partial u}{\partial y}, \frac{\partial u}{\partial z}, \frac{\partial u}{\partial v_x}, \frac{\partial u}{\partial v_y}, \frac{\partial u}{\partial v_z}\right).$$

4.2.2 Collisionless System

In this section, I follow mostly the treatment presented in Sect. 4.1 of Binney & Tremaine (1987, hereafter BT87).

We consider a large number N_* of bodies moving under the influence of a *smooth* gravitational potential $\Phi(\boldsymbol{x}, t)$. Here smooth means essentially that Φ does not change much over distances of the order of (a few times) the average inter-particle distance $n^{-1/3}$, where n is the particle number density. No other forces affect the motion of these objects. The potential Φ may be the gravitational field created by these bodies themselves or an external field. The system of particles is described through the one-particle phase-space distribution function (DF for short) $f(\boldsymbol{x}, \boldsymbol{v}, t)$. A useful interpretation of f is as a probability density if it is normalised to 1. Then $f(\boldsymbol{x}, \boldsymbol{v}, t)\mathrm{d}^3x\,\mathrm{d}^3v$ is the probability of finding at time t, any given particle within a volume of phase space $\mathrm{d}^3x\,\mathrm{d}^3v$ around the 6D phase-space point $\underline{w} = (\boldsymbol{x}, \boldsymbol{v})$. The mean number of particles in this volume is $N_* f(\boldsymbol{x}, \boldsymbol{v}, t)\mathrm{d}^3x\,\mathrm{d}^3v$.

From the knowledge of the initial conditions $f_0(\boldsymbol{x}, \boldsymbol{v}) \equiv f(\boldsymbol{x}, \boldsymbol{v}, t_0)$, we want to predict $f(\boldsymbol{x}, \boldsymbol{v}, t)$ at some future time $t > t_0$. We define the velocity in the 6D phase-space

$$\underline{\dot{w}} = (\dot{\boldsymbol{x}}, \dot{\boldsymbol{v}}) = (\boldsymbol{v}, -\boldsymbol{\nabla}\Phi). \tag{4.1}$$

As long as Φ is sufficiently smooth, the particles evolve in a smooth, continuous way in the phase-space. Therefore, f must satisfy a continuity equation

$$\frac{\partial f}{\partial t} + \underline{\nabla} \cdot (f\underline{\dot{w}}) = \frac{\partial f}{\partial t} + \sum_{i=1}^{3} \frac{\partial (f v_i)}{\partial x_i} - \sum_{i=1}^{3} \frac{\partial (f \partial_{x_i}\Phi)}{\partial v_i} = 0. \tag{4.2}$$

This equation can be simplified using the fact that, in the phase-space representation, the x_i and v_i are independent variables ($\partial v_i/\partial x_j = 0$), and that Φ does not depend on the velocities so that $\partial\Phi/\partial v_i = 0$. Therefore, we have

$$\frac{\partial f}{\partial t} + \sum_{i=1}^{3} v_i \frac{\partial f}{\partial x_i} - \sum_{i=1}^{3} \partial_{x_i}\Phi \frac{\partial f}{\partial v_i} = \frac{\partial f}{\partial t} + \boldsymbol{v} \cdot \boldsymbol{\nabla} f - \boldsymbol{\nabla}\Phi \cdot \frac{\partial f}{\partial \boldsymbol{v}} = 0. \tag{4.3}$$

This is the collisionless Boltzmann equation. It can be written simply as

$$D_t f = 0, \tag{4.4}$$

where D_t is a notation for the "Lagrangian" or advective rate of change of f. This equation means that, if we follow the trajectory of a (real or imaginary) particle in the phase-space, the number density around it does not change. In other words, the flow in phase-space is incompressible.

We note that there is an equation which is equivalent, but more general (and of less practical use) for the distribution function in the N_*-particle phase-space, in which a point represents all the positions and velocities of the N_* bodies of the system. It is Liouville's theorem (BT87, Sect. 8.2). The collisionless Boltzmann equation follows from Liouville's theorem and the assumptions that the number of particles is very large and that there are no two-particle correlations. In other words, the probability of finding particle 1 at $\underline{w_1}$ and particle 2 at $\underline{w_2}$ is simply given by the product $f(\underline{w_1}, t) f(\underline{w_2}, t) \mathrm{d}^6 w_1 \, \mathrm{d}^6 \underline{w_2}$ (BT87, Sect. 8.3). While the first approximation is certainly valid in many astrophysical situations such as galaxies and globular clusters (but see comments below about multi-component systems), the second is violated by two-body effects such as mutual deflections or the existence of small bound sub-groups, in particular binaries. In fact, as long as they do not interact closely with other objects and are themselves numerous enough, binaries can in principle be treated as just a special component, for which a particle is really a binary. Two-particle effects such as deflection due to close encounters are called *collisional effects*, and the Fokker–Planck treatment, described below, is an approximate but manageable way to take them into account.

The Boltzmann equation is valid whether f is interpreted as a number, mass, luminosity or probability density. The distribution function f does not need to represent a system of objects with identical physical properties (stellar masses, radii, etc.) but may be used, globally, for a mixed population. As long as all sub-populations share the same f_0, or if we are not interested in distinguishing between them, and the system is collisionless, a unique f is enough to describe the system and its evolution. If there are different sub-populations with initially distinct distribution functions (as would be the case for a globular cluster with primordial mass segregation), each population (index α) can be assigned its own DF f_α. In the absence of collisional terms, the only coupling between the evolution of the various f_α is through the fact that they move in the same global potential Φ to which each component contributes, unless it is treated as a mass-less tracer. Specifically, Φ is obtained from the f_α's and a possible external potential Φ_{ext} through the Poisson equation,

$$\Phi(\boldsymbol{x}) = \Phi_{\mathrm{self}} + \Phi_{\mathrm{ext}} \quad \text{with} \quad \boldsymbol{\nabla}^2 \Phi_{\mathrm{self}} = 4\pi G \sum_{\alpha=1}^{N_{\mathrm{comp}}} \underbrace{M_\alpha \int \mathrm{d}^3 v \, f_\alpha(\boldsymbol{x}, \boldsymbol{v})}_{\rho_\alpha}, \tag{4.5}$$

where N_{comp} is the number of components and M_α the total mass in component α (with the normalisation $\int \mathrm{d}^3 v \, \mathrm{d}^3 x \, f_\alpha = 1$). In the following we will generally assume a fully self-gravitating system, $\Phi(\boldsymbol{x}) = \Phi_{\mathrm{self}}$.

Because the Boltzmann equation simply states conservation of the phase-space density along physical trajectories, it keeps the same form if another coordinate system is used instead of the Cartesian (x, y, z) as long as f still represents the number density per unit volume of the (x, y, z, v_x, v_y, v_z) phase-space.

4.2.3 Collision Terms

When particles are subject to forces other than those produced by the smooth Φ, the convective derivative of f does not vanish anymore. In particular, in a real self-gravitating N-particle system the potential cannot be smooth on small scales. Instead, it exhibits some graininess, i.e. short-term, small-scale fluctuations, $\Phi_{\text{real}} = \Phi + \Delta\Phi_{\text{grainy}}$. Here I call relaxation the effects of these fluctuations on the evolution of the system described by f. Schematically, they are due to the fact that a given particle does not see the rest of the system as a smooth mass distribution but as a collection of point-masses. Relaxational effects, also known (somewhat confusingly) as collisional effects, can therefore be seen as particles influencing each other individually as opposed to collectively. To allow for these effects, a right-hand collision term Γ has to be introduced into the Boltzmann equation,

$$D_t f = \Gamma[f]. \tag{4.6}$$

We now develop an expression for Γ. Let $\Psi(\underline{w}, \underline{\Delta w})\mathrm{d}^6(\Delta w)\mathrm{d}t$ be the probability that a particle at the phase-space position $\underline{w}$ is perturbed (through forces not derived from Φ) to $\underline{w} + \underline{\Delta w}$ during $\mathrm{d}t$. In general, Ψ is also a function of t but I drop this dependence here to simplify notation. Stars are scattered out of an element of phase space around $\underline{w}$ at a rate

$$\Gamma_- = -f(\underline{w}) \int \mathrm{d}^6(\Delta w)\Psi(\underline{w}, \underline{\Delta w}), \tag{4.7}$$

while stars from other phase-space positions $(\underline{w} - \underline{\Delta w})$ are scattering into this element at a rate

$$\Gamma_+ = \int \mathrm{d}^6(\Delta w) f(\underline{w} - \underline{\Delta w})\Psi(\underline{w} - \underline{\Delta w}, \underline{\Delta w}). \tag{4.8}$$

The collision term is thus $\Gamma = \Gamma_+ + \Gamma_-$ and the Boltzmann equation with such a collision term is called the master equation.

4.3 Fokker–Planck Equation

4.3.1 Fokker–Planck Equation in Position-Velocity Space

Theoretically, the master equation is of very general applicability because very few simplifying assumptions have been made so far. Unfortunately, it is of little practical use unless some explicit expression for the transition probability Ψ is known. The Fokker–Planck treatment is based on the assumption that Ψ is sufficiently smooth and that typical changes $\underline{\Delta w}$ are small. We can then develop Ψ and f around $\underline{w}$ in a Taylor series to second order in $\underline{\Delta w}$. Specifically, in the term Γ_+, we write

$$f(\underline{w}-\underline{\Delta w})\Psi(\underline{w}-\underline{\Delta w},\underline{\Delta w}) = f(\underline{w})\Psi(\underline{w},\underline{\Delta w}) - \sum_{i=1}^{6}\Delta w_i \frac{\partial}{\partial w_i}\left[\Psi(\underline{w},\underline{\Delta w})f(\underline{w})\right] + \frac{1}{2}\sum_{i,j=1}^{6}\Delta w_i \Delta w_j \frac{\partial^2}{\partial w_i \partial w_j}\left[\Psi(\underline{w},\underline{\Delta w})f(\underline{w})\right] + \mathcal{O}((\Delta w)^3). \tag{4.9}$$

Defining the diffusion coefficients (DCs)

$$\begin{aligned}\langle \Delta w_i \rangle &\equiv \int \mathrm{d}^6(\Delta w)\Delta w_i \Psi(\underline{w},\underline{\Delta w}),\\ \langle \Delta w_i \Delta w_j \rangle &\equiv \int \mathrm{d}^6(\Delta w)\Delta w_i \Delta w_j \Psi(\underline{w},\underline{\Delta w}),\end{aligned} \tag{4.10}$$

and plugging the development (4.9) into the collision term of the master equation, we obtain the general Fokker–Planck (FP) equation,

$$D_t f = -\sum_{i=1}^{6}\frac{\partial}{\partial w_i}\left[f(\underline{w})\langle \Delta w_i \rangle\right] + \frac{1}{2}\sum_{i,j=1}^{6}\frac{\partial^2}{\partial w_i \partial w_j}\left[f(\underline{w})\langle \Delta w_i \Delta w_j \rangle\right]. \tag{4.11}$$

Here $\langle \Delta w_i \rangle$ is the mean change in w_i per unit time due to collisional effects. These diffusion coefficients are generally functions of $\underline{w}$ and t, but I have not written these dependencies explicitly.

Now, in the case of stellar dynamics, we identify the collisional changes $\underline{\Delta w}$ with the effect of Keplerian, hyperbolic, uncorrelated two-body encounters and assume that they occur instantaneously, i.e. on a time-scale much shorter than the dynamical time-scale $t_{\mathrm{dyn}} \equiv R_{\mathrm{cl}}^{3/2}(GM_{\mathrm{cl}})^{-1/2}$, where M_{cl} is the total mass of the system and R_{cl} is some typical length scale such as the half-mass radius. In this local approximation, we neglect the change in position and only consider changes in velocity. This means that $\Psi(\underline{w},\underline{\Delta w}) = 0$ if $\Delta \boldsymbol{x} \neq 0$ and the Fokker–Planck equation reads

$$D_t f = -\sum_{i=1}^{3}\frac{\partial}{\partial v_i}\left[f(\boldsymbol{x},\boldsymbol{v})\langle \Delta v_i \rangle\right] + \frac{1}{2}\sum_{i,j=1}^{3}\frac{\partial^2}{\partial v_i \partial v_j}\left[f(\boldsymbol{x},\boldsymbol{v})\langle \Delta v_i \Delta v_j \rangle\right]. \tag{4.12}$$

4.3.2 Diffusion Coefficients and Approximations for Relaxation

Let us sketch the computation of the velocity diffusion coefficients. In practice, we do not need to compute the transition probability Ψ. Instead we use the fact that, for instance, $\langle \Delta v_i \rangle$ is the mean rate of change of the component i of the velocity of a given particle (called the test particle) as it is perturbed by all other particles (the field particles). To carry out the computations, we have to adopt the following set of approximations, usually referred to as "Chandrasekhar theory of relaxation" (Chandrasekhar 1943, 1960. See for instance Hénon 1973; Saslaw 1985; Spitzer 1987; Binney & Tremaine 1987; Heggie & Hut 2003):

1. *Local approximation.* The collisional perturbations to the motion of the test particle are assumed to take place on a scale much smaller than the size of its orbit. Formally, this holds if perturbations from distant stars with a long time-scale are negligible.
2. *Small perturbations approximation.* We assume that on time-scales of order $t_{\rm dyn}$ (or shorter), the "collisions" produce only a small change in the orbital parameters of a particle; for the diffusion coefficients this translates into $t_{\rm dyn}\langle \Delta v_i \rangle \ll v$, $t_{\rm dyn}\langle \Delta v_i \Delta v_j \rangle \ll v^2$. This is an extension of the FP approximation, which will make it possible to average the FP equation over the orbit of stars. Most importantly for the time being, it justifies the assumption that perturbations are two-body effects only and that they add linearly. In other words, to this level of approximation, the combined effect of two field particles on a test particle are the same as the sum of the effects of each taken independently. In particular, the interaction between both field particles can be neglected. Hence, we are only considering the so-called two-body relaxation. This simplification only holds if perturbations from very close stars (leading to large changes in $\boldsymbol{v}$) are negligible.
3. *Homogeneity approximation.* This is sometimes considered part of the local approximation. We assume that the cumulative effects of the perturbations on the test object are as if the properties of the field particles (density, velocity distribution) were the same in the whole system and equal to what they are in the vicinity of the test object. In other words, the local conditions are representative of the global ones. This arguably looks like an unjustified assumption, given how heterogeneous stellar systems are (for instance, the density in a globular cluster or galactic nucleus decreases by many orders of magnitude from the centre to the half-mass radius) and the long-range, unshielded nature of the gravitational force. We will see as we proceed why it may be a reasonable simplification, but we note that it can only work if distant perturbations do not dominate.

To sum up, the standard theory of relaxation is based on the assumptions that relaxation can be reduced to the cumulative effects of a large number of uncorrelated two-body encounters that can be treated like (local) Keplerian small-angle hyperbolic velocity deflections due to objects with a density and velocity distribution identical to the local ones.

All these approximations are shared by other explicitly statistical methods used to follow the long-term evolution of stellar clusters, such as the Monte-Carlo scheme (see Chap. 5) and the gaseous model (Bettwieser & Spurzem 1986; Louis & Spurzem 1991; Giersz & Spurzem 1994; Spurzem & Takahashi 1995; Amaro-Seoane et al. 2004), but some approximations can be improved on. In particular, large velocity changes (due to close encounters) can be included (Goodman 1983a; Freitag et al. 2006a).

To compute the diffusion coefficients we start by looking at the hyperbolic Keplerian encounter between the test particle, with velocity $\boldsymbol{v}$ and mass m and a field particle with velocity $\boldsymbol{v}_\mathrm{f}$ and mass m_f. We only consider field particles of a given mass, possibly different from m. Standard numerical methods based on the FP equation require that the mass spectrum is discretised. Hence, we assume there are N_f particles of mass m_f described by the distribution function f_f, now with the normalisation $\int \mathrm{d}^3x\mathrm{d}^3v f_\mathrm{f} = N_\mathrm{f}$.

Using the local approximation, we can assume that the encounter takes place in a vacuum. In other words, the orbits are straight lines at large separation ("infinity"). The relative velocity at infinity is $\boldsymbol{v}_\mathrm{rel} = \boldsymbol{v} - \boldsymbol{v}_\mathrm{f}$ and the velocity of the centre-of-mass (CM) of the pair $\boldsymbol{v}_\mathrm{cm} = \mu\boldsymbol{v} + (1-\mu)\boldsymbol{v}_\mathrm{f}$ with $\mu = m/(m+m_\mathrm{f})$. If b is the impact parameter, the effect of the encounter is simply to rotate the relative velocity by an angle

$$\tan\left(\frac{\theta}{2}\right) = \frac{b_0}{b} \quad \text{with} \quad b_0 = \frac{G\,(m+m_\mathrm{f})}{v_\mathrm{rel}^2}. \tag{4.13}$$

The value b_0 is the impact parameter leading to a deflection angle $\pi/2$ (in the CM frame). We decompose the change of velocity $\boldsymbol{\Delta v}$ into components parallel and perpendicular to the initial relative velocity $\boldsymbol{v}_\mathrm{rel}$,

$$\Delta v_\perp = 2(1-\mu)v_\mathrm{rel}\frac{b}{b_0}\left(1+\frac{b^2}{b_0^2}\right)^{-1}, \quad \Delta v_\parallel = 2(1-\mu)v_\mathrm{rel}\left(1+\frac{b^2}{b_0^2}\right)^{-1}. \tag{4.14}$$

We then transform from the reference frame aligned with $\boldsymbol{v}_\mathrm{rel}$ (dependent on $\boldsymbol{v}_\mathrm{f}$) to the external frame to get the Δv_i's. The next step is to average over all (equally probable) possible orientations of the impact parameter vector around the direction of $\boldsymbol{v}_\mathrm{rel}$. This gives values of $\langle\Delta v_i\rangle$ and $\langle\Delta v_i \Delta v_j\rangle$ for fixed $\boldsymbol{v}_\mathrm{f}$ and b. Now we sum the effects of all the encounters with field stars having this velocity. The number density of such objects is $f_\mathrm{f}\mathrm{d}^3v_\mathrm{f}$ (considered independent of the position, owing to the homogeneity approximation) and the rate of encounters with an impact parameter between b and $b+db$ is $2\pi b\,\mathrm{d}b v_\mathrm{rel} f_\mathrm{f}\mathrm{d}^3v_\mathrm{f}$. We have to integrate over all possible impact parameters. This involves the integrals

$$\begin{aligned}
\int_0^{b_\mathrm{max}} \Delta v_\parallel b\,\mathrm{d}b &= v_\mathrm{rel}(1-\mu)b_0^2\ln(1+\Lambda^2)\\
\int_0^{b_\mathrm{max}} (\Delta v_\parallel)^2 b\,\mathrm{d}b &= 2v_\mathrm{rel}^2(1-\mu)^2 b_0^2\left(1-\frac{1}{1+\Lambda^2}\right)\\
\int_0^{b_\mathrm{max}} (\Delta v_\perp)^2 b\,\mathrm{d}b &= 2v_\mathrm{rel}^2(1-\mu)^2 b_0^2\left(\ln(1+\Lambda^2)-1+\frac{1}{1+\Lambda^2}\right).
\end{aligned} \tag{4.15}$$

In these relations, $\Lambda = b_\mathrm{max}/b_0$ where b_max is the ill-defined maximum impact parameter. For a system that is not too centrally concentrated, we can set $b = R_\mathrm{cl}$. In most cases, $\Lambda \gg 1$ so the integrals can be approximated by

$$\int_0^{b_{\rm max}} \Delta v_{\parallel} b\, {\rm d}b \simeq 2v_{\rm rel}(1-\mu)b_0^2 \ln \Lambda,$$
$$\int_0^{b_{\rm max}} (\Delta v_{\parallel})^2 b\, {\rm d}b \simeq 0, \tag{4.16}$$
$$\int_0^{b_{\rm max}} (\Delta v_{\perp})^2 b\, {\rm d}b \simeq 4v_{\rm rel}^2(1-\mu)^2 b_0^2 \ln \Lambda.$$

Hence, the cut-off $b_{\rm max}$ only enters the computation of the diffusion coefficients through the multiplicative Coulomb logarithm $\ln \Lambda$. Due to the very weak logarithmic dependency, we can replace m and $m_{\rm f}$ in b_0 by the mean value $M_{\rm cl}/N_*$ and $v_{\rm rel}$ by the 1D velocity dispersion σ_v measured, for example, at the half-mass radius, unless σ_v is a very steep function of the position such as around a massive black hole. Further, for a self-gravitating system in virial equilibrium, $\sigma_v^2 \approx GM_{\rm cl}/R_{\rm cl}$ so that Λ must be of order N_*. Putting $\Lambda = \gamma_{\rm c} N_*$, direct N-body experiments indicate that $\gamma_{\rm c} \approx 0.1$ for single-mass systems and $\gamma_{\rm c} \approx 0.01$ (with considerable uncertainty) if objects have a realistic mass spectrum (See Hénon 1975 for theoretical estimates and Giersz & Heggie 1994, 1996, amongst others, for the determinations based on N-body simulations).

Although the above integrals are carried out from $b = 0$, remember that the FP approximation requires small changes in $\boldsymbol{v}$. This suggests that encounters with b smaller than a few b_0 (causing deflection angles not small compared to $\pi/2$) cannot be taken into account. But truncating the integrations at $b_{\rm min}$ = a few b_0 would just bring in terms smaller than those in (4.16) by a factor $\ln \Lambda$. This is reflected by the fact that the typical time-scale for an encounter within kb_0, with k some numerical coefficient, is

$$t_{\rm la} = \left[n\sigma_v \pi (kb_0)^2 \left(1 + \frac{2Gm}{kb_0\sigma_v} \right) \right]^{-1} \approx \left(n\sigma_v \pi (kb_0)^2 \right)^{-1} \approx \frac{\sigma_v^3}{k\, G^2 m^2 n}, \tag{4.17}$$

where n is the number density, σ_v the velocity dispersion and m the (mean) mass of a particle. For $k \approx 1$ this large-angle deflection time-scale is of order $\ln \Lambda$ longer than the relaxation time (see (4.24)). However, from these considerations, it does not follow that large-angle deflection cannot play an important role in some circumstances; while the standard two-body relaxation, by definition, leads to gradual changes in orbital properties, a single large-angle encounter causes sudden orbit modifications, which may have very different consequences. This may produce ejections or lead to strong interactions between stars and a central massive black hole in a galactic nucleus (Hénon 1960; Lin & Tremaine 1980; Freitag et al. 2006a. See also Chap. 5).

The contribution to the relaxation of encounters with b between b_1 and b_2 with $b_2 > b_1 \gg b_0$ is proportional to $\ln(b_1/b_2)$. This explains why the structure of the stellar system at large distances from the test particle has little importance in practice. The average inter-particle distance is

$$\bar{d} \equiv n^{-1/3} = \left(\frac{m}{\rho} \right)^{1/3} \approx \left(\frac{mR_{\rm cl}^3}{M_{\rm cl}} \right)^{1/3} = N_*{}^{-1/3} R_{\rm cl}, \tag{4.18}$$

while $b_0 \approx N_*^{-1} R_{\rm cl}$. So, somewhat surprisingly, about two thirds of the contribution to two-body relaxation come from encounters with impact parameters smaller than $\bar{d}$. This is why the homogeneity approximation is a good one.

Carrying out the computation of the diffusion coefficients using (4.16), we arrive at

$$\begin{aligned} \langle \Delta v_i \rangle &= 4\pi \ln \Lambda \, G^2 m_{\rm f} (m + m_{\rm f}) \frac{\partial h(\boldsymbol{v})}{\partial v_i}, \\ \langle \Delta v_i \Delta v_j \rangle &= 4\pi \ln \Lambda \, G^2 m_{\rm f}^2 \frac{\partial^2 g(\boldsymbol{v})}{\partial v_i \partial v_j}, \end{aligned} \tag{4.19}$$

where $h(\boldsymbol{v})$ and $g(\boldsymbol{v})$ are the Rosenbluth potentials (Rosenbluth et al. 1957),

$$h(\boldsymbol{v}) = \int \mathrm{d}^3 u f_{\rm f}(\boldsymbol{u}) \, |\boldsymbol{v} - \boldsymbol{u}|^{-1} \quad \text{and} \quad g(\boldsymbol{v}) = \int \mathrm{d}^3 u f_{\rm f}(\boldsymbol{u}) \, |\boldsymbol{v} - \boldsymbol{u}| \,. \tag{4.20}$$

Recall that all these quantities have an implicit $\boldsymbol{x}$-dependence.

If the velocity distribution is isotropic, we can go further in the computation of the diffusion coefficients for the velocity. We find (e.g. Spitzer 1987)

$$\begin{aligned} \langle \Delta v_{\parallel} \rangle &= -4\pi \lambda m_{\rm f}^2 \left(1 + \frac{m}{m_{\rm f}}\right) E_2^{<}(V), \\ \langle \Delta v_{\perp} \rangle &= 0, \\ \langle (\Delta v_{\parallel})^2 \rangle &= \frac{8\pi}{3} \lambda m_{\rm f}^2 v (E_4^{<}(v) + E_1^{>}(v)), \\ \langle (\Delta v_{\perp})^2 \rangle &= \frac{8\pi}{3} \lambda m_{\rm f}^2 v (3E_4^{<}(v) - E_4^{<}(v) + 2E_1^{>}(v)), \\ \langle \Delta v_{\parallel} \Delta v_{\perp} \rangle &= 0, \end{aligned} \tag{4.21}$$

where $\lambda \equiv 4\pi G^2 \ln \Lambda$,

$$E_n^{<}(v) = \int_{u=0}^{v} \left(\frac{u}{v}\right)^n f_{\rm f}(u) \mathrm{d}u, \quad \text{and } E_n^{>}(v) = \int_{u=v}^{\infty} \left(\frac{u}{v}\right)^n f_{\rm f}(u) \mathrm{d}u. \tag{4.22}$$

We see that the mass of the test particle m only appears in the coefficient $\langle \Delta v_{\parallel} \rangle$ for dynamical friction. From this, the diffusion coefficients for the energy can be computed using $\Delta E = v \Delta v_{\parallel} + \frac{1}{2}(\Delta v_{\perp})^2 + \frac{1}{2}(\Delta v_{\parallel})^2$, which gives

$$\begin{aligned} \langle \Delta E \rangle &= 4\pi \lambda m_{\rm f}^2 v \left(E_1^{>}(v) - \frac{m}{m_{\rm f}} E_2^{<}(v) \right), \\ \langle (\Delta E)^2 \rangle &= \frac{8\pi}{3} \lambda m_{\rm f}^2 v^3 \left(E_4^{<}(v) + E_1^{>}(v) \right). \end{aligned} \tag{4.23}$$

We can write $E_n^{>,<} = \xi_n^{>,<} n \sigma_v^{-3}$, where $\xi_n^{>,<}$ are dimensionless, order-of-unity (and position-dependent) numbers, n is the local number density of field stars and σ_v their local 1D velocity dispersion. The time-scale $t_{\rm rlx}$ over which the direction of the velocity of a typical star (with $v = \bar{v} \equiv 3^{1/2}\sigma_v$) has changed completely due to relaxation can be estimated, using (4.23) and the definition $\langle(\Delta v_\perp)^2\rangle_{\bar{v}} t_{\rm rlx} \equiv \sigma_v^2$. We find $t_{\rm rlx}^{-1} \approx \ln \Lambda G^2 m_{\rm f}^2 n \sigma_v^{-3}$. A conventional definition of the local relaxation time is obtained by assuming that the velocity distribution is isotropic and Maxwellian and using the mean stellar mass m (Spitzer 1987),

$$t_{\rm rlx} \equiv 0.339 \frac{\sigma_v^3}{\ln \Lambda\, G^2 m^2 n}. \quad (4.24)$$

In the case of a system with objects of different masses, the relaxational effect of a species α is proportional to $n_\alpha m_\alpha^2$ rather than its density (e.g. Perets et al. 2007). On the other hand, dynamical friction, corresponding to the second, negative term for $\langle\Delta E\rangle$ (see (4.23)), has a time-scale proportional to $\rho = mn$, the total mass density of the field, irrespective of the individual masses of the stars (for more on dynamical friction, see Chap. 7).

This is as far as we can go without further restriction on the distribution function $f_{\rm f}$. If there is a single species of particles, $f_{\rm f} = f$ and the FP equation, consisting of (4.12) with the above diffusion coefficients (4.19), together with the Poisson equation, determine the evolution of the DF in a self-contained way. Unfortunately, the FP equation is a very intricate integro-differential equation which, at this point, cannot be solved in whole generality.

Furthermore, realistic stellar systems are composed of objects with a range of properties (in particular masses). We can assume that there is a discrete set of populations orbiting in their common total potential and influencing each other through two-body relaxation. Each component k is described by DF $f_{\rm k}$, which follows an FP equation, but the diffusion coefficients are now a sum of contributions from each component

$$\langle\Delta v_i\rangle_k = 4\pi \ln \Lambda\, G^2 \times \sum_{l=1}^{N_{\rm comp}} \left[m_l(m_k + m_l) \frac{\partial}{\partial v_i} \left(\int {\rm d}^3 u f_l(\boldsymbol{u})\, |\boldsymbol{v} - \boldsymbol{u}|^{-1} \right) \right]. \quad (4.25)$$

4.4 Orbit-Averaged Fokker–Planck Equation

4.4.1 General Considerations

To go further and obtain more easily usable versions of the FP equation, we need to restrict ourselves to stellar systems that are spherically symmetric in all their properties.[1] The use of the FP equation to study the structure and

[1]This does not imply that the velocity distribution is isotropic, meaning that it is spherically symmetric in velocity space, but that the local velocity distribution depends only on the moduli of the components of the velocity parallel and perpendicular to the radius-vector.

evolution of stellar clusters was pioneered by Hénon (1961), who derived the FP equation for an isotropic (but multi-mass) cluster and found an analytical, self-similar solution for the single-mass case, assuming the existence of a central energy source. The first numerical codes producing general time-dependent solutions were written by Cohn (1979, 1980) and, to this day, most of the work in this field is based on the formalism and numerical methods developed by this author (but see Takahashi 1995 and references therein for a finite-element scheme to solve the FP equation, based on a variational principle).

The FP equation can also be used for systems with axial symmetry, such as globular clusters or galactic nuclei with global rotation, but we will not treat this approach here (see Goodman 1983b; Einsel 1996; Einsel & Spurzem 1999; Kim et al. 2002, 2004; Fiestas 2006; Fiestas et al. 2006; Kim et al. 2008 for this original line of research under active development).

We also assume that the stellar system is in (quasi-)dynamical equilibrium. In other words, it evolves very little over dynamical timescales, $\left|f/\dot{f}\right| \ll t_{\rm dyn}$. If evolution is only due to two-body relaxation and the system is fully self-gravitating, this assumption holds provided N_* is sufficiently large because $\left|f/\dot{f}\right| \approx t_{\rm rlx} \approx N_*(\ln \Lambda)^{-1} t_{\rm dyn}$ with $\ln \Lambda = \ln(\gamma_{\rm c} N_*) \approx 5 - 15$. For single-mass systems with $N_* \lesssim 10^3$, the distinction between dynamical and relaxational effects (or between the smooth and grainy parts of the potential) becomes blurred. When stars have a broad mass spectrum, a larger number of stars is required for a clear distinction between dynamical and relaxational regimes.

From Jeans' theorem (Jeans 1915; Merritt 1999) for a spherical system in dynamical equilibrium, the DF f can depend on the phase-space coordinates $(\boldsymbol{x}, \boldsymbol{v})$ only through the (specific) orbital energy E and modulus of the angular momentum J,

$$f(\boldsymbol{x}, \boldsymbol{v}) = F(E(\boldsymbol{x}, \boldsymbol{v}), J(\boldsymbol{x}, \boldsymbol{v})) \quad \text{with} \quad E = \phi(r) + \frac{1}{2}v^2, \quad J = r\, v_{\rm t}, \tag{4.26}$$

where $r = |\boldsymbol{x}|$, $v = |\boldsymbol{v}|$ in a system of reference centred on the cluster centre,[2] ϕ is the spherically symmetric smooth gravitational potential so that $\Phi(\boldsymbol{x}) = \phi(r)$, and $v_{\rm t}$ is the modulus of the component of the velocity perpendicular to the radius-vector $\boldsymbol{x}$.

4.4.2 Isotropic Spherical Cluster

We first consider the simpler case of a cluster with isotropic velocity dispersion, where F depends on E only. We also assume only one component. Let $N(E)\mathrm{d}E$ be the number of stars with energy between E and $E+\mathrm{d}E$. The transformation from F to N is found by integrating over the phase-space accessible to orbits

[2] I use the word "cluster" to designate all (spherically) symmetric stellar systems including galactic nuclei.

with energy between E and $E + \delta E$ and then letting δE be an infinitesimal $\delta E \to \mathrm{d}E$,

$$N(E)\delta E = \int_{[E,E+\delta E]} \mathrm{d}^3x\, \mathrm{d}^3v F(E) = 16\pi^2 \int_r \mathrm{d}r\, r^2 \left[\int_v \mathrm{d}v\, v^2 F(E) \right]. \quad (4.27)$$

We bring $F(E)$ out of the integrals because it is nearly constant in the integration domain (by definition). We first realise the v-integration, at fixed r, which runs from $v = \sqrt{2(E - \phi(r))}$ to $v + \delta v$ with $\delta v \simeq \delta E/v$ giving $\int_v \mathrm{d}v\, v^2 \simeq \sqrt{2(E - \phi(r))}\delta E$. Finally remains the integration over r which runs from 0 to $r_{\mathrm{max}}(E)$ defined such that $\phi(r_{\mathrm{max}}) = E$. We neglect the small part of the integration domain with r between $r_{\mathrm{max}}(E)$ and $r_{\mathrm{max}}(E + \delta E)$ because its contribution is of higher order in δE. Once we replace δE by $\mathrm{d}E$, we find

$$N(E) = 16\pi^2 p(E) F(E), \quad (4.28)$$

with

$$p(E) = \int_0^{r_{\mathrm{max}}} r^2 v\, \mathrm{d}r = \int_0^{r_{\mathrm{max}}} r^2 \sqrt{2(E - \phi(r))}\mathrm{d}r. \quad (4.29)$$

Note that the quantity $p(E)$ is proportional to the radial orbital period averaged in J space (isotropised orbital period),

$$p(E) = \frac{1}{2} \int_0^{J_c^2(E)} \mathrm{d}(J^2) P_{\mathrm{orb}}(E, J) \quad \text{with} \quad P_{\mathrm{orb}}(E, J) = 2 \int_{r_{\mathrm{min}}}^{r_{\mathrm{max}}} \frac{\mathrm{d}r}{v_{\mathrm{r}}}, \quad (4.30)$$

where $J_c(E)$ is the angular momentum of a circular orbit of energy E.

We could transform the FP equation in $(\boldsymbol{x}, \boldsymbol{v})$-space (4.12) into an equation for the rate of change of $N(E)$, but it is much simpler to start over from scratch. The collisional term of an FP equation for $N(E)$ simply reads

$$\left.\frac{\mathrm{d}N}{\mathrm{d}t}\right|_{\mathrm{coll}} = -\frac{\partial}{\partial E}\left[\{\Delta E\} N(E)\right] + \frac{1}{2}\frac{\partial^2}{\partial E^2}\left[\{(\Delta E)^2\} N(E)\right]. \quad (4.31)$$

Here the computation of the diffusion coefficients involve averaging over the volume of space accessible to a particle of energy E, reflecting the transformation from $F(E)$ to $N(E)$ (4.28) and (4.29),

$$\{\Delta E\} = p(E)^{-1} \int_0^{r_{\mathrm{max}}} r^2 v \langle \Delta E \rangle \mathrm{d}r, \quad (4.32)$$

where $\langle \Delta E \rangle$ is the local diffusion coefficient for the kinetic energy. In other words, the mean rate of change of $\frac{1}{2}v^2$ for a particle at position r with velocity $v = \sqrt{2(E - \phi(r))}$.

The smooth potential ϕ may change slowly as a result of the relaxational evolution of the cluster itself, or because of an external influence. In any case, this will induce a change in the energy not accounted for by the collisional term (4.31). So, if we write $D_t N(E)$ for the "Lagrangian" rate of change of density in energy space following the ϕ-induced change in E, we obtain the right-hand side of the FP for $N(E)$,

$$D_t N(E) = \frac{\partial N}{\partial t} + \frac{\partial N}{\partial E}\frac{\mathrm{d}E}{\mathrm{d}t}\bigg|_{\phi} = \frac{\mathrm{d}N}{\mathrm{d}t}\bigg|_{\mathrm{coll}}, \tag{4.33}$$

where $\mathrm{d}E/\mathrm{d}t|_{\phi}$ is the change in energy due to the evolution of the potential. It can be shown that it is equal to the phase-space averaged value of $\partial\phi/\partial t$,

$$\frac{\mathrm{d}E}{\mathrm{d}t}\bigg|_{\phi} = p(E)^{-1}\int_0^{r_{\max}} \frac{\partial\phi(r)}{\partial t} r^2 v \mathrm{d}r. \tag{4.34}$$

We see that the FP equation for $N(E)$ as well as its generalisation to the anisotropic case (see Sect. 4.4.3) are orbit-averaged. Again, the condition for this averaging to be valid is that the system evolves only very little over one dynamical time, staying close to dynamical equilibrium.

To solve numerically the FP equation, it is usual to write it in a flux-conservation form,

$$D_t N(E) = -\frac{\partial \mathcal{F}_E}{\partial E} \quad \text{with} \quad \mathcal{F}_E = m\mathcal{D}_E F - \mathcal{D}_{EE}\frac{\partial F}{\partial E}. \tag{4.35}$$

Using (4.23), it can be shown that the flux coefficients are

$$\begin{aligned} \mathcal{D}_E =& 16\pi^3 \lambda m_{\mathrm{f}} \int_{\phi(0)}^{E} \mathrm{d}E' p(E') F_{\mathrm{f}}(E'), \\ \mathcal{D}_{EE} =& 16\pi^3 \lambda m_{\mathrm{f}}^2 \left[q(E) \int_E^0 \mathrm{d}E' F_{\mathrm{f}}(E') + \int_{\phi(0)}^{E} \mathrm{d}E' q(E') F_{\mathrm{f}}(E') \right], \end{aligned} \tag{4.36}$$

where

$$q(E) = \int_{\phi(0)}^{E} \mathrm{d}E' p(E') = \frac{1}{3}\int_0^{r_{\max}} r^2 v^3 \,\mathrm{d}r. \tag{4.37}$$

Here $q(E)$ is the volume of phase-space accessible to particles with energies lower than E, and $p(E)$ is the area of the hypersurface bounding this volume, that is, $p(E) = \partial q/\partial E$ (Goodman 1983a). $q(E)$ is also proportional to the isotropised radial action,

$$q(E) = \frac{1}{4}\int_0^{J_{\mathrm{c}}^2(E)} \mathrm{d}(J^2) Q(E,J), \quad \text{with } Q(E,J) = 2\int_{r_{\min}}^{r_{\max}} \mathrm{d}r\, v_{\mathrm{r}}. \tag{4.38}$$

We have used an index "f" for "field" to distinguish the mass and DF of the population we follow (test stars) from the "field" objects. This distinction does not apply to a single-component system, but makes it very easy to generalise to a multi-component situation by summing over components to get the total flux coefficient,

$$\mathcal{D}_E = \sum_{l=1}^{N_{\rm comp}} \mathcal{D}_{E,l}, \quad \mathcal{D}_{EE} = \sum_{l=1}^{N_{\rm comp}} \mathcal{D}_{EE,l}, \tag{4.39}$$

where the flux coefficient for component l can be written by replacing the subscript "f" by "l" in (4.36) (e.g. Murphy & Cohn 1988).

We now explain schematically how the FP equation is used numerically to follow the evolution of star clusters. A more detailed description can be found in, for example, Chernoff & Weinberg (1990). In the most common scheme, pioneered by Cohn (1980), two types of steps are realised in alternation.

1. *Diffusion step.* The change in the distribution function F for a discrete time step Δt is computed by use of the FP equation *assuming the potential* ϕ *is fixed*, setting $D_t N = \frac{\partial N}{\partial t} = \frac{\partial N}{\partial t}\big|_{\rm coll}$. The FP equation, written as a flux-conserving equation, is discretised on an energy grid. The flux coefficients are computed using the DF(s) of the previous step; this makes the equations linear in the values of F on the grid points. The finite-differentiation scheme is the implicit Chang & Cooper (1970) algorithm, which is first-order in time and energy.
2. *Poisson step.* Now the change of potential resulting from the modification in F is computed and F is modified to account for the term $\mathrm{d}E/\mathrm{d}t|_\phi$, assuming $D_t N = \frac{\partial N}{\partial t} + \frac{\partial N}{\partial E}\frac{\mathrm{d}E}{\mathrm{d}t}\big|_\phi = 0$. This is done implicitly by using the fact that as long as the change in ϕ over Δt is very small, the actions of each orbit are adiabatic invariants. Hence, during the Poisson step, the distribution function, expressed as a function of the actions, does not change. Using the isotropised radial action $q(E)$ defined above, $\tilde{F}(q)\mathrm{d}q = F(E)p(E)\mathrm{d}E$ with $\tilde{F}(q) = F(E(q))$. In other words, the modified $F(E)$ is obtained by recomputing the relation $q(E)$ in the modified potential. In practice, an iterative scheme is used to compute the modified potential, determined implicitly by the modified DF through the relation

$$\phi(r) = -4\pi G\left[\frac{1}{r}\int_0^r \mathrm{d}s s^2\rho(s) + \int_r^\infty \mathrm{d}s s\rho(s)\right], \tag{4.40}$$

with

$$\rho(r) = 4\pi m \int_{\phi(r)}^{E_{\rm max}} \mathrm{d}E\sqrt{2(E-\phi(r))}F(E), \tag{4.41}$$

for one component. The iteration is started with the values of ϕ, ρ, etc. computed before the previous diffusion step.

4.4.3 Anisotropic Spherical Cluster

The anisotropic FP treatment was already used to study some aspects of the structure of globular clusters by Spitzer & Shapiro (1972). This type of approach was then applied to the distribution of stars around a massive black hole (assuming $\phi = -GM_{\rm BH}/r$ where $M_{\rm BH}$ is the mass of the black hole) by Lightman & Shapiro (1977) and Cohn & Kulsrud (1978). Although the first self-consistent FP simulations by Cohn (1979) made use of an anisotropic code, further work on such models was relatively limited in comparison to the isotropic case because the Chang & Cooper (1970) discretisation scheme, which proved so useful for getting good energy conservation when the DF depended only on E (and t), has no exact equivalent for the case of a 2D (E, J) dependence. Also, in most circumstances, it seems that forcing isotropy does not affect the results much and allows a substantial reduction in the computational burden. Cohn (1985) first presented results of anisotropic FP models based on an extension of the Chang–Cooper scheme. Since then, Takahashi (1995, 1996, 1997) and Drukier et al. (1999) have developed FP codes for spherical clusters with anisotropic velocity distributions.

Let $F(E(\boldsymbol{x},\boldsymbol{v}), J(\boldsymbol{x},\boldsymbol{v}))\mathrm{d}^3x\,\mathrm{d}^3v$ be the number of stars with position within a volume d^3x around $\boldsymbol{x}$ and velocity within d^3v around $\boldsymbol{v}$. Because of spherical symmetry, we can write $\mathrm{d}^3x = 4\pi r^2\mathrm{d}r$ and $\mathrm{d}^3v = 4\pi v_\mathrm{t}\mathrm{d}v_\mathrm{t}\mathrm{d}v_\mathrm{r}$. We note that $F(E, J) = 0$ if $J > Jc(E)$. Let $N(E, J)\mathrm{d}E\,\mathrm{d}J$ be the number of stars with energy between E and $E + \mathrm{d}E$ and angular momentum between J and $J + \mathrm{d}J$. To convert from $F(E, J)$ to $N(E, J)$, we follow a star with energy E and angular momentum J on its orbit and integrate the volume of phase-space along the way. We use the distance from the centre r as integration variable,

$$N(E,J)\mathrm{d}E\,\mathrm{d}J = 4\pi\int_{r_\mathrm{min}(E,J)}^{r_\mathrm{max}(E,J)} r^2\mathrm{d}r\mathcal{V}_r(E,J)\mathrm{d}E\,\mathrm{d}J. \tag{4.42}$$

Here $\mathcal{V}_r(E, J)\mathrm{d}E\,\mathrm{d}J$ denotes the (infinitesimal) volume in $\boldsymbol{v}$-space with energy between E and $E + \mathrm{d}E$ and angular momentum between J and $J + \mathrm{d}J$, for a fixed r. We have

$$\mathcal{V}_r(E,J)\mathrm{d}E\,\mathrm{d}J = 4\pi v_\mathrm{t}\mathrm{d}v_\mathrm{t}\mathrm{d}v_\mathrm{r} = 4\pi v_\mathrm{t}\left\|\begin{matrix}\frac{\partial E}{\partial v_\mathrm{t}} & \frac{\partial E}{\partial v_\mathrm{r}}\\ \frac{\partial J}{\partial v_\mathrm{t}} & \frac{\partial J}{\partial v_\mathrm{r}}\end{matrix}\right\|^{-1}\mathrm{d}E\,\mathrm{d}J = 4\pi\frac{v_\mathrm{t}}{rv_\mathrm{r}}\mathrm{d}E\,\mathrm{d}J, \tag{4.43}$$

which leads to

$$N(E,J) = 8\pi P_\mathrm{orb}(E,J)J\,F(E,J). \tag{4.44}$$

In numerical applications, it is convenient to use $R \equiv (J/J_\mathrm{c}(E))^2$ as a variable instead of J. Then the density of particles per unit E and R is

$$\tilde{N}(E,R) = 4\pi J_\mathrm{c}(E)^2 P_\mathrm{orb}(E,J)\,F(E,J). \tag{4.45}$$

The FP equation for $\tilde{N}(E, R)$, in its flux-conserving form, is a direct extension of the isotropic one,

$$D_t\tilde{N}(E,R) = -\frac{\partial\mathcal{F}_E}{\partial E} - \frac{\partial\mathcal{F}_R}{\partial R}, \tag{4.46}$$

with

$$\mathcal{F}_E = m\mathcal{D}_E F - \mathcal{D}_{EE}\frac{\partial F}{\partial E} - \mathcal{D}_{ER}\frac{\partial F}{\partial R},$$
$$\mathcal{F}_R = m\mathcal{D}_R F - \mathcal{D}_{RR}\frac{\partial F}{\partial R} - \mathcal{D}_{ER}\frac{\partial F}{\partial E}. \tag{4.47}$$

The expression for the flux coefficients are significantly longer than in the isotropic case; they are given by Cohn (1979) for single-mass clusters and by Takahashi (1997) for the multi-mass case.[3] To my knowledge, in all numerical solutions of the anisotropic FP equation for stellar systems, an isotropised DF is used in the computation of the diffusion and flux coefficients. For instance, for $\mathcal{D}_{EE}$, we use

$$\mathcal{D}_{EE} = \frac{32\pi^3}{3}\lambda m_\mathrm{f}^2 \int_{r\mathrm{min}}^{r_\mathrm{max}} \frac{\mathrm{d}r}{v_\mathrm{r}}\Big[v^2\int_E^0 \mathrm{d}E'\bar{F}_\mathrm{f}(E',r)$$
$$+ v^{-1}\int_{\phi(r)}^{E} \mathrm{d}E'\bar{F}_\mathrm{f}(E',r)\left(2(\phi(r)-E')\right)^{3/2}\Big]. \tag{4.48}$$

Here, $\bar{F}_\mathrm{f}$ is the isotropised DF

$$\bar{F}_\mathrm{f}(E',r) = \frac{1}{J_\mathrm{max}}\int_0^{J_\mathrm{max}} \mathrm{d}J F_\mathrm{f}(E',J), \tag{4.49}$$

where $J_\mathrm{max}(E,r) = \sqrt{2r^2(\phi(r)-E)}$ is the maximum (scaled) angular momentum that an orbit of energy E can have if it goes through radius r and $R_\mathrm{max} = (J_\mathrm{max}/J_\mathrm{c})$.[2]

4.5 The Fokker–Planck Method in Use

To conclude this chapter, I present a quick and partial overview of the work carried out in cluster and galactic nucleus modelling using the direct resolution of the Fokker–Planck equation. My goal here is to provide pointers to the literature that will allow the reader a deeper exploration of this rich field.

4.5.1 Relaxational Evolution

The only physics included in the Fokker–Planck formalism presented here is self-gravity (through use of the Poisson equation) and two-body relaxation. This is enough to study the evolution of stellar clusters (with no or few primordial binaries) up to core collapse. The case of a single-mass cluster was

[3]Beware that in the work of these authors, E is the binding energy and has therefore the opposite sign as here, with corresponding sign changes to be tracked in the computation of the coefficients and E-derivatives.

initially computed by Cohn (1979, 1980) for a Plummer model, and revisited several times since, to explore a variety of initial cluster structures (Wiyanto et al. 1985; Quinlan 1996) or to investigate the core-collapse physics in greater detail using more sophisticated Fokker–Planck codes (Takahashi 1995; Drukier et al. 1999). Clusters with stars of different masses are much more realistic and have been considered by several authors (e.g. Merritt 1983; Inagaki & Wiyanto 1984; Inagaki & Saslaw 1985; Murphy & Cohn 1988; Chernoff & Weinberg 1990; Lee 1995; Takahashi 1997; Kim et al. 1998).

In a multi-mass cluster with a realistic mass spectrum, the evolution to core collapse is driven by mass segregation. FP simulations are the ideal tool to investigate how this process operates in the limit of a very large number of stars. They are quick and their results are not affected by any significant numerical noise, in contrast to particle-based methods such as direct N-body or Monte-Carlo codes. In Fig. 4.1, I show the evolution of the Lagrangian radii for a cluster with stellar mass spectrum, $\mathrm{d}N_*/\mathrm{d}M_* \propto M_*^{-2.35}$, covering the range 0.2–10 $\mathrm{M}_\odot$. The simulation was performed using an FP code provided by H.M. Lee (e.g. Lee et al. 1991) using 12 mass components. The initial structure is a Plummer model. In Fig. 4.2, I plot the evolution of the central "temperature" for several mass components. We see that energy equipartition is approached at the centre only amongst the most massive stars (roughly in the range 3–10 $\mathrm{M}_\odot$).

Using an energy grid of 200 elements, such an FP run requires only 1–2 min of CPU time on a laptop computer. For an anisotropic code that solves the FP equation in (E, J) space, the simulation runs for about 4 days on a desktop computer (G. Drukier 2007, personal communication). When the mass spectrum is discretised into a larger number of mass components, the computing time increases approximately linearly with the number of components. The corresponding direct N-body simulation with 256 000 particles took about 40 days using special-purpose GRAPE hardware (H. Baumgardt 2005, personal communication), and a Monte-Carlo simulation using 10^6 particles took about one week on a desktop computer (see Chap. 5).

4.5.2 Models with Additional Physics

In order to simulate more realistic and complex systems, the Fokker–Planck description of two-body relaxation has been complemented by approximate treatment of a large variety of other physical effects. Here I give a list of these effects with references to some pioneering or otherwise notable FP works where they have been considered.

- *Central massive black hole.* Assuming a quasi-stationary regime and a fixed Keplerian potential, Lightman & Shapiro (1977) and Cohn & Kulsrud (1978) used the FP formalism to determine the distribution of stars around a massive black hole (MBH) and the rate of stellar disruptions by the MBH. The treatment of the loss cone developed for these works was later

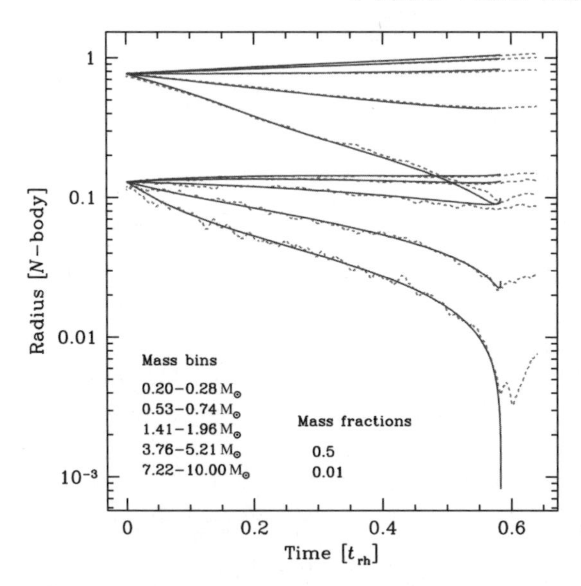

Fig. 4.1. Core collapse of a Plummer cluster model with 0.2–10 M$_\odot$ Salpeter mass function, $\mathrm{d}N_*/\mathrm{d}M_* \propto N_*^{-2.35}$. Results of an isotropic Fokker–Planck code provided by H. M. Lee, in solid lines, are compared to a direct NBODY4 simulation with 256 000 particles, in dashes (H. Baumgardt 2005, personal communication). To show mass segregation, the evolution of Lagrangian radii for mass fractions of 1 and 50 per cent is plotted for stars with masses within five different bins (corresponding to 5 of the 12 discrete mass components used for the FP simulation). The length unit is the N-body scale (see Chap. 1). The time unit is the initial half-mass relaxation time (Spitzer 1987). To convert the dynamical time units of the N-body simulation to a relaxation time, a value of $\gamma_c = 0.045$ was used for the Coulomb logarithm. Compare with Fig. 5.4

introduced in self-consistent FP codes to study the evolution of globular clusters hosting an intermediate-mass black hole or of dense galactic nuclei (Cohn 1985; David et al. 1987a, b; Murphy et al. 1991). Simplified FP codes, assuming in particular a fixed potential, have been used to investigate the segregation of stellar-mass black holes around a MBH (Hopman & Alexander 2006; Alexander 2007; O'Leary et al. 2008) and the formation of a central cusp of dark matter (Merritt et al. 2007a). Very recently, a FP code which includes the gravity of the stars self-consistently was used to study the shrinkage of a binary MBH (Merritt et al. 2007b) and the evolution of small nuclear clusters (Merritt 2008).

- *Stellar evolution.* Mass loss due to stellar evolution can be included by reducing the stellar mass represented by a mass component as a function of time (e.g. Lee 1987a; Chernoff & Weinberg 1990; Quinlan & Shapiro 1990; Murphy et al. 1991).
- *Collisions.* Some FP simulations have included the effects of collisions resulting in mergers (Lee 1987a; Quinlan & Shapiro 1989, 1990) or (partial)

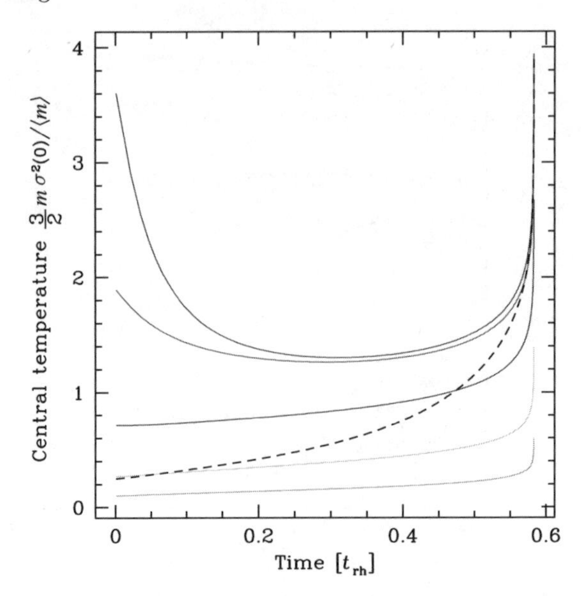

Fig. 4.2. Evolution of the central temperatures during the core collapse of a multi-mass cluster model. The temperature of component i is defined as $T_i \equiv 3/2(m_i/\langle m \rangle)\sigma_i^2(0)$, where m_i is the mass of a star of component i, $\sigma_i(0)$ the central 1D velocity dispersion of that component in N-body units and $\langle m \rangle$ the mean stellar mass. The data come from the same Fokker–Planck simulation as in Fig. 4.1. The solid lines are the temperatures for the same five mass components (highest to lowest mass from top to bottom). The dashed line represents the mass-weighted average central temperature

disruptions (David et al. 1987a, b; Murphy et al. 1991). The FP approach has also been used to follow the evolution of galaxy clusters, taking into account galaxy mergers and mass stripping due to encounters between galaxies (Merritt 1983, 1984, 1985; Takahashi et al. 2002). Collisions can only be treated in an averaged and highly approximate fashion in the FP formalism, because the mass and orbital energy of collision products of any mass have to be transferred to the predefined mass components. Furthermore, the effects of collisions on stellar evolution cannot be included in any detailed way. Finally, in the case of collisional runaway, which is the growth of one or a few stars to very high mass by successive mergers, mass components have to be introduced that contain a very small number of stars (sometimes less than one). Nevertheless, comparisons with the Monte-Carlo algorithm (Chap. 5) where collisions can be treated more accurately generally show good agreement, as far as the overall effects of collisions are concerned (Freitag & Benz 2002; Freitag et al. 2006b).

- *Binary stars.* In a cluster containing no binaries initially, some will form near the centre during core collapse when the density reaches sufficiently high values, either through dissipative two-body effects or through close three-body interactions (e.g. Aarseth 1971; Heggie & Hut 2003). Both kinds of mechanism have been included in FP codes (Statler et al. 1987; Lee et al. 1991; Takahashi & Inagaki 1991; Lee & Ostriker 1993, amongst others). In most cases the binary population is not followed explicitly. Instead, the formation, hardening and ejection of binaries are simply included as an effective central source of heating able to stop and reverse core collapse. Binary heating can result in gravothermal core oscillations in the post-collapse evolution (Cohn et al. 1989; Takahashi & Inagaki 1991; Breeden et al. 1994). A more detailed treatment of binaries would necessitate to represent them by at least one additional component (Lee 1987b; Gao et al. 1991). Only limited physical realism can be achieved because it is not practical to extend the phase space to include the internal properties of the binaries, which include mass ratio, semi-major axis and eccentricity. This limitation explains why, to the best of my knowledge, primordial binaries have only been included into the FP framework by Gao et al. (1991). Furthermore, in the case of dynamically formed binaries, only a few are expected to be present in the core at any given time (Goodman 1984; Baumgardt et al. 2002), making a description based on the distribution function inadequate.

- *Large-angle scatterings.* Goodman (1983a) included the effects of close two-body encounters in FP simulations and concluded that they do not affect appreciably the core collapse process.
- *Evaporation.* Assuming the cluster is on a circular orbit around a spherical galaxy (or in the equatorial plane of an axially symmetrical galaxy), the evaporation of stars in the steady tidal field can be approximated in a spherical FP code by an outer boundary condition. For an isotropic formulation, the condition is $F(E_{\mathrm{t}}) = 0$ with $E_{\mathrm{t}} = -GM_{\mathrm{cl}}R_{\mathrm{t}}^{-1}$ and R_{t} is the tidal truncation radius, which can be identified with the distance between the centre of the cluster and the Lagrange point L1 or L2 (e.g. Chernoff & Weinberg 1990). A more accurate condition can be used in anisotropic models by setting the DF to zero for orbits with an apocentre distance larger than R_{t} (Takahashi et al. 1997). Delayed evaporation can be simulated to account for the fact that a star can spend a significant amount of time in the cluster even when its orbital parameters would allow it to reach the Lagrange points (Lee & Ostriker 1987; Takahashi & Portegies Zwart 2000).
- *Gravitational shocking.* In general, as it orbits its host galaxy, a globular cluster can experience strongly varying external gravitational stresses. Murali & Weinberg (1997a) and Gnedin et al. (1999) have included so-called disc and bulge shocking in their FP simulations, which allowed them to study the evolution of whole globular cluster systems (Gnedin & Ostriker 1997; Murali & Weinberg 1997b, c). Thank to a new integration scheme, shocking has been studied in anisotropic FP models (Shin et al. 2008).

- *Gas dynamics.* (David et al. 1987a, b) coupled the FP algorithm with a spherical gas dynamical code to predict what amount of the gas released by stars through evolution and collisions is accreted by a central MBH in AGN models. However, gas motion is likely to be highly non-spherical and to vary on time-scales much shorter than those for evolution of the stellar cluster (e.g. Williams et al. 1999; Cuadra et al. 2005).

FP simulations, including several of the above physical processes, have been used to interpret observations of a few specific globular clusters: M 15 (Grabhorn et al. 1992; Dull et al. 1997), M 71 (Drukier et al. 1992), NGC 6397 (Drukier 1993, 1995) and NGC 6624 (Grabhorn et al. 1992). In the future, it seems likely that particle-based methods will be used to produce detailed models of observed clusters (see Giersz & Heggie 2003, 2007 and Hurley et al. 2005 for pioneering examples). These codes can deal realistically with stellar populations that are rare or otherwise problematic to simulate with FP methods, such as primordial binaries, blue stragglers or X-ray binaries. However, because they are so much faster, FP codes can be an invaluable tool to carry out extensive parameter-space exploration and determine the initial conditions and physical parameters most likely to fit the observational data. Direct N-body or Monte-Carlo simulations can then be used, using these input parameters, to obtain more detailed models.

Acknowledgement

I am indebted to Gordon Drukier and Hyung Mok Lee who provided invaluable help in the preparation of my Fokker–Planck lecture and took the time to read and comment on a draft of this chapter. I also thank Hyung Mok Lee for making available his Fokker–Planck code and helping me to use it, and Holger Baumgardt for providing unpublished N-body data. My work is supported by the STFC rolling grant to the IoA.

References

Aarseth S. J., 1971, Ap&SS, 13, 324
Alexander T., 2007, in Livio M., Koekemoer A. M., eds, 2007 STScI Spring Symposium: Black Holes. (astro-ph/0708.0688)
Amaro-Seoane P., Freitag M., Spurzem R., 2004, MNRAS, 352, 655
Baumgardt H., Hut P., Heggie D. C., 2002, MNRAS, 336, 1069
Bettwieser E., Spurzem R., 1986, A&A, 161, 102
Binney J., Tremaine S., 1987, Galactic Dynamics. Princeton Univ. Press, Princeton, NJ
Breeden J. L., Cohn H. N., Hut P., 1994, ApJ, 421, 195
Chandrasekhar S., 1943, Rev. Mod. Phys., 15, 1
Chandrasekhar S., 1960, Principles of Stellar Dynamics. Dover, enlarged edition

Chang J. S., Cooper G., 1970, J. Comp. Phys., 6, 1
Chernoff D. F., Weinberg M. D., 1990, ApJ, 351, 121
Cohn H., 1979, ApJ, 234, 1036
Cohn H., 1980, ApJ, 242, 765
Cohn H., 1985, in Goodman J., Hut P., eds, Proc. IAU Symp. 113, Dynamics of Star Clusters. Reidel, Dordrecht, p. 161
Cohn H., Hut P., Wise M., 1989, ApJ, 342, 814
Cohn H., Kulsrud R. M., 1978, ApJ, 226, 1087
Cuadra J., Nayakshin S., Springel V., Di Matteo T., 2005, MNRAS, 360, L55
David L. P., Durisen R. H., Cohn H. N., 1987a, ApJ, 313, 556
David L. P., Durisen R. H., Cohn H. N., 1987b, ApJ, 316, 505
Drukier G. A., 1993, MNRAS, 265, 773
Drukier G. A., 1995, 100, 347
Drukier G. A., Cohn H. N., Lugger P. M., Yong H., 1999, ApJ, 518, 233
Drukier G. A., Fahlman G. G., Richer H. B., 1992, ApJ, 386, 106
Dull J. D., Cohn H. N., Lugger P. M., Murphy B. W., Seitzer P. O., Callanan P. J., Rutten R. G. M., Charles P. A., 1997, ApJ, 481, 267
Einsel C., Spurzem R., 1999, MNRAS, 302, 81
Einsel M., 1996, PhD thesis, Christian-Albrechts-Universität zu Kiel
Fiestas J., 2006, PhD thesis, Heidelberg University
Fiestas J., Spurzem R., Kim E., 2006, MNRAS, 373, 677
Freitag M., Amaro-Seoane P., Kalogera V., 2006a, ApJ, 649, 91
Freitag M., Benz W., 2002, A&A, 394, 345
Freitag M., Rasio F. A., Baumgardt H., 2006b, MNRAS, 368, 121
Gao B., Goodman J., Cohn H., Murphy B., 1991, ApJ, 370, 567
Giersz M., Heggie D. C., 1994, MNRAS, 268, 257
Giersz M., Heggie D. C., 1996, MNRAS, 279, 1037
Giersz M., Heggie D. C., 2003, MNRAS, 339, 486
Giersz M., Heggie D. C., 2007, in Vesperini E,. Giersz M., Sills A., eds, Dynamical Evolution of Dense Stellar Systems. Proceedings of IAU Symposium No. 246. (astro-ph/0711.0523)
Giersz M., Spurzem R., 1994, MNRAS, 269, 241
Gnedin O. Y., Lee H. M., Ostriker J. P., 1999, ApJ, 522, 935
Gnedin O. Y., Ostriker J. P., 1997, ApJ, 474, 223
Goodman J., 1983a, ApJ, 270, 700
Goodman J., 1983b, PhD thesis, Princeton University
Goodman J., 1984, ApJ, 280, 298
Goodman J., Heggie D. C., Hut P., 1993, ApJ, 415, 715
Grabhorn R. P., Cohn H. N., Lugger P. M., Murphy B. W., 1992, ApJ, 392, 86
Heggie D., Hut P., 2003, The Gravitational Million-Body Problem Cambridge Univ. Press, Cambridge
Hemsendorf M., Merritt D., 2002, ApJ, 580, 606
Hénon M., 1960, Annales d'Astrophysique, 23, 668
Hénon M., 1961, Annales d'Astrophysique, 24, 369
Hénon M., 1973, in Martinet L., Mayor M., eds, Lectures of the 3rd Advanced Course of the Swiss Society for Astronomy and Astrophysics. Obs. de Gèneve, Gèneve, p. 183
Hénon M., 1975, in Hayli A., ed., Proc. IAU Symp. 69, Dynamics of Stellar Systems. Reidel, Dordrecht, p. 133

Hopman C., Alexander T., 2006, ApJ Lett., 645, L133
Hurley J. R., Pols O. R., Aarseth S. J., Tout C. A., 2005, MNRAS, 363, 293
Inagaki S., Saslaw W. C., 1985, ApJ, 292, 339
Inagaki S., Wiyanto P., 1984, PASJ, 36, 391
Jeans J. H., 1915, MNRAS, 76, 70
Kim E., Einsel C., Lee H. M., Spurzem R., Lee M. G., 2002, MNRAS, 334, 310
Kim E., Lee H. M., Spurzem R., 2004, MNRAS, 351, 220
Kim E., Yoon I., Lee H. M., Spurzem R., 2008, MNRAS, 383, 2
Kim S. S., Lee H. M., Goodman J., 1998, ApJ, 495, 786
Lee H. M., 1987a, ApJ, 319, 801
Lee H. M., 1987b, ApJ, 319, 772
Lee H. M., 1995, MNRAS, 272, 605
Lee H. M., Fahlman G. G., Richer H. B., 1991, ApJ, 366, 455
Lee H. M., Ostriker J. P., 1987, ApJ, 322, 123
Lee H. M., Ostriker J. P., 1993, ApJ, 409, 617
Lightman A. P., Shapiro S. L., 1977, ApJ, 211, 244
Lin D. N. C., Tremaine S., 1980, ApJ, 242, 789
Louis P. D., Spurzem R., 1991, MNRAS, 251, 408
Merritt D., 1983, ApJ, 264, 24
Merritt D., 1984, ApJ, 276, 26
Merritt D., 1985, ApJ, 289, 18
Merritt D., 1999, PASP, 111, 129
Merritt D., 2008, preprint (astro-ph/0802.3186)
Merritt D., Harfst S., Bertone G., 2007a, Phys. Rev. D, 75, 043517
Merritt D., Mikkola S., Szell A., 2007b, ApJ, 671, 53
Murali C., Weinberg M. D., 1997a, MNRAS, 288, 749
Murali C., Weinberg M. D., 1997b, MNRAS, 291, 717
Murali C., Weinberg M. D., 1997c, MNRAS, 288, 767
Murphy B. W., Cohn H. N., 1988, MNRAS, 232, 835
Murphy B. W., Cohn H. N., Durisen R. H., 1991, ApJ, 370, 60
Perets H. B., Hopman C., Alexander T., 2007, ApJ, 656, 709
O'Leary R. M., Kocsis B., Loeb A., 2008, preprint (astro-ph/0807.2638)
Quinlan G. D., 1996, New Astronomy, 1, 255
Quinlan G. D., Shapiro S. L., 1989, ApJ, 343, 725
Quinlan G. D., Shapiro S. L., 1990, ApJ, 356, 483
Rosenbluth M. N., MacDonald W. M., Judd D. L., 1957, Physical Review, 107, 1
Saslaw W. C., 1985, Gravitational Physics of Stellar and Galactic Systems. Cambridge Univ. Press, Cambridge
Shin J., Kim S. S., Takahashi K., 2008, MNRAS, 386, L67
Spitzer L., 1987, Dynamical evolution of globular clusters. Princeton Univ. Press, Princeton, NJ
Spitzer L. J., Shapiro S. L., 1972, ApJ, 173, 529
Spurzem R., Takahashi K., 1995, MNRAS, 272, 772
Statler T. S., Ostriker J. P., Cohn H. N., 1987, ApJ, 316, 626
Takahashi K., 1995, PASJ, 47, 561
Takahashi K., 1996, PASJ, 48, 691
Takahashi K., 1997, PASJ, 49, 547
Takahashi K., Inagaki S., 1991, PASJ, 43, 589

Takahashi K., Lee H. M., Inagaki S., 1997, MNRAS, 292, 331
Takahashi K., Portegies Zwart S. F., 2000, ApJ, 535, 759
Takahashi K., Sensui T., Funato Y., Makino J., 2002, PASJ, 54, 5
Williams R. J. R., Baker A. C., Perry J. J., 1999, MNRAS, 310, 913
Wiyanto P., Kato S., Inagaki S., 1985, PASJ, 37, 715

5

Monte-Carlo Models of Collisional Stellar Systems

Marc Freitag

University of Cambridge, Institute of Astronomy, Madingley Road, Cambridge CB3 0HA, UK
marc.freitag@gmail.com

5.1 Introduction

In this chapter I describe a fast, approximate, particle-based algorithm to compute the long-term evolution of stellar clusters and galactic nuclei. It relies on the assumptions of spherical symmetry of the stellar system, dynamical equilibrium and local, diffusive two-body relaxation. It allows for velocity anisotropy, an arbitrary stellar mass spectrum, stellar evolution, a central massive object, collision between stars, binary processes and two-body encounters leading to large deflection angles. Using one to ten million particles, a run extending over several relaxation times takes a few days to a few weeks to compute on a single-CPU personal computer and the CPU time scales as $t_{\rm CPU} \propto N_{\rm p} \ln N_{\rm p}$, where $N_{\rm p}$ is the number of particle used. Because each physical process is implemented with its explicit scaling, the number of stars simulated can be (much) larger than $N_{\rm p}$, making it possible to simulate galactic nuclei with (in particular) the correct rate of relaxation.

The Monte-Carlo (MC) numerical scheme is intermediate, both in terms of realism and computing time between Fokker–Planck or gas approaches and direct N-body codes. The former are very fast but based on a significantly idealised description of the stellar system, the latter treat (Newtonian) gravity in an essentially assumption-free way but are extremely demanding in terms of computing time (Binney & Tremaine 1987; Sills et al. 2003). The MC scheme was first introduced by Hénon to follow the relaxational evolution of globular clusters (Hénon 1971a,b; Hénon 1973a; Hénon 1975). To my knowledge there exist three independent codes based on Hénon's ideas in active development and use. The first is the one written by M. Giersz (Giersz 1998, 2001, 2006; Giersz et al. 2008), which implements many of the developments first introduced by Stodołkiewicz (1982, 1986). Second is the code written by K. Joshi (Joshi et al. 2000, 2001) and greatly improved and extended by A. Gürkan and J. Fregeau (see for instance Fregeau et al. 2003; Gürkan et al. 2004, 2006; Fregeau & Rasio 2007). These codes have been applied to the study of globular and young clusters. Finally, we developed a MC code specifically aimed at

Freitag, M.: *Monte-Carlo Models of Collisional Stellar Systems.* Lect. Notes Phys. **760**, 123–158 (2008)
DOI 10.1007/978-1-4020-8431-7_5

the study of galactic nuclei containing a central massive black hole (Freitag & Benz 2001c; Freitag & Benz 2002; Freitag et al. 2006a; Freitag et al. 2006b,c). The description of the method given here is based on this particular implementation.[1]

This chapter is organised as follows. In Sect. 5.2, the core principles and assumptions of the method are presented. In Sect. 5.3, I expose the inner workings of the code in detail: the basic algorithm which treats global self-gravity and two-body relaxation is the subject of Sect. 5.3.1 while Sect. 5.3.2 covers the additional physical processes (collisions, central object, binaries, stellar evolution, etc). Finally, in Sect. 5.4. I show a few applications and discuss possible avenues for future developments of the method, in the context of research on star clusters (Sect. 5.4.1) and on galactic nuclei (Sect. 5.4.2).

5.2 Basic Principles

The MC code shares most of its underlying assumptions with the Fokker–Planck (FP) approach presented in Chap. 4. Essentially, Hénon's algorithm can be seen as a particle-based method to solve the coupled FP and Poisson equations for a stellar cluster using Monte-Carlo sampling to determine the long-term effects of two-body relaxation. An advantage of the MC approach over FP integrations is that it can include a continuous stellar mass spectrum and extra physical ingredients such as stellar evolution, collisions, binaries or a central massive black hole in a much more straightforward and realistic way. On the downside, MC simulations require considerably more computing time. Furthermore, the MC results show numerical noise while those obtained with the FP codes are smooth and easier to analyse and manipulate.

The assumptions shared by both methods are the following:

1. Dynamical equilibrium
2. Spherical symmetry
3. Diffusive relaxation
4. Adequacy of representation with a one-particle distribution function.

An isolated system is likely to attain dynamical equilibrium after an initial phase of violent relaxation spanning a few dynamical times $t_{\rm dyn} = \sqrt{R_{\rm cl}^3/(GM_{\rm cl})}$, where $R_{\rm cl}$ is a characteristic length (such as the half-mass radius) and $M_{\rm cl}$ the mass of the cluster. The MC code developed by Spitzer and collaborators (Spitzer & Hart 1971a,b; Spitzer & Thuan 1972; Spitzer &

[1] This code is available at `http://www.ast.cam.ac.uk/research/repository/freitag/MC.html`. General information on the MC method and more references can be found on the web pages created for the MODEST consortium ("MOdeling DEnse STellar systems") at `http://www.manybody.org/modest/` (follow the link to the working group on stellar-dynamics methods, `WG5`).

Shull 1975; Spitzer & Mathieu 1980) allows for out-of-equilibrium situations, at the price of computing speed, but the assumption of spherical symmetry strongly limits the usefulness of this feature.

In practice, the strongest restriction is that of spherical symmetry. Violent relaxation generally leads to an equilibrium configuration with significant triaxiality (e.g. Aguilar & Merritt 1990; Theis & Spurzem 1999; Boily & Athanassoula 2006). Although it is likely that two-body relaxation makes the system more symmetrical, flattening owing to global rotation can persist over many relaxation times (Einsel & Spurzem 1999; Kim et al. 2002, 2004; Fiestas et al. 2006). In galactic nuclei, the interaction between the stars and a binary massive black hole (e.g. Merritt & Milosavljević 2005) or a massive accretion disc (e.g. Šubr et al. 2004) cannot be studied accurately when spherical symmetry is assumed (see Sect. 5.4.2).

The last two assumptions have been discussed in Chap. 4 on FP methods. They imply that correlations between particles, beyond random two-body encounters, are neglected but I stress that three- and four-body interactions in the form of binary processes can be included in the MC approach with much more realism than permitted by the direct FP formalism (see Sect. 5.3.2).

It should be noted at once that all these assumptions can only be valid if the system under consideration contains a large number of stars. In my experience, the MC approach is suitable if the number of particles N_p satisfies

$$N_\mathrm{p} \gtrsim 3000 \frac{m_\mathrm{max}}{\langle m \rangle}, \tag{5.1}$$

where m_max and $\langle m \rangle$ are the maximum and mean stellar mass, respectively.

In Hénon's scheme, the numerical realisation of the cluster is a set of spherical shells with zero thickness, each of which is given a mass M, a radius R, a specific angular momentum J and a specific kinetic energy T. These particles can be interpreted as spherical layers of synchronised stars that share the same stellar properties, orbital parameters and orbital phase and experience the same processes (relaxation, collision, etc.) at the same time.

From the radii and masses of all particles, the potential can be computed at any time or place and the orbital energies of all particles are straightforwardly deduced from their kinetic energies and positions. Hence the set of particles can be regarded as a discretised representation of the distribution function (DF) $f(\boldsymbol{x}, \boldsymbol{v}) = F(E, J)$. But, whereas a functional or tabulated expression of the DF (as implemented in direct FP methods) would require the integration of the Poisson equation, to yield the gravitational potential, the Monte-Carlo realisation of the cluster provides it directly. From this point of view, the Monte-Carlo method is closer to N-body philosophy than to direct FP methods.

The main difference between the MC code and a spherical 1D N-body simulation (e.g. Hénon 1973b) is that the former does not explicitly follow the continuous orbital motion of particles, which preserves E and J. However, these orbital constants, as well as other properties of the particles, are modified by collisional processes to be incorporated explicitly: two-body relaxation, stellar collisions, etc. So the MC simulation proceeds through millions to billions of steps, each of them consisting of the selection of particles, the modification of their properties to simulate the effects these physical processes and the selection of radial positions R on their new orbits.

5.3 Detailed Implementation

5.3.1 Core Algorithm

This subsection is divided into four parts. In the first, I present the treatment of relaxation and the overall structure of the code. In the following parts I explain in detail some important aspects of the algorithm, which are the selection of a pair of particles to evolve, the representation of the gravitational potential and the determination of a new orbital position for updated particles.

Two-Body Relaxation and General Organisation

The treatment of two-body relaxation is the backbone of Hénon-type Monte-Carlo schemes. It relies on the usual diffusive approximation developed by Chandrasekhar and presented in Chap. 4. I recall that the basic idea behind the concept of relaxation is that the gravitational potential of a stellar system containing a large number of bodies can be described as the sum of a dominating smooth contribution plus a small, granular, part that fluctuates over small scales and short times. When only the smooth part is taken into account, the DF of the cluster obeys the collisionless Boltzmann equation. However, in the long run the fluctuating part makes E and J change slowly and the DF evolve. The basic simplifying assumption underlying Chandrasekhar relaxation theory is to treat the effects of the fluctuating part as the sum of multiple uncorrelated two-body hyperbolic gravitational encounters with small deviation angles. Under these assumptions, if a test star of mass m travels through a field of stars with homogeneous number density n, which all have mass m_{f} and the same velocity, after a time span δt its velocity in the reference frame of the encounters will deviate from the initial direction by an angle θ such that

$$\begin{aligned} \langle\theta\rangle_{\delta t} &= 0 \text{ and} \\ \left\langle\theta^2\right\rangle_{\delta t} &\simeq 8\pi n \ln\Lambda \frac{G^2\left(m+m_{\mathrm{f}}\right)^2}{v_{\mathrm{rel}}^3}\,\delta t, \end{aligned} \tag{5.2}$$

where v_{rel} the relative velocity between the test star and the field stars and $\ln\Lambda \simeq \ln(\gamma_{\mathrm{c}} N_*)$ for a self-gravitating cluster with the value of γ_{c} depending on the mass spectrum (see Chap. 4).

Hénon's method avoids the computational burden and some of the necessary simplifications connected with the numerical evaluation of diffusion coefficients. The repeated application of (5.2) to a given particle implicitly amounts to a Monte-Carlo integration of the orbit-averaged diffusion coefficients, provided the orbital positions and properties of field particles are correctly sampled. Under the usual assumption that encounters are local, this latter constraint is obeyed if we take these properties to be those of the closest neighbouring particle. Furthermore, this allows us to actually modify the velocities of both particles at a time, each acting as a representative from the field for the other. Evolving particles in symmetrical pairs not only speeds up the simulations by a factor $\simeq 2$ but also, and more critically, ensures strict conservation of energy.

Therefore, at the heart of the MC treatment of relaxation are super-encounters, encounters between two neighbouring particles with a deflection angle θ_{SE} devised to reproduce statistically the cumulative effects of the numerous physical deflections taking place in the real system over a time span δt. Using the indices 1 and 2 to designate the particles in a pair, we see that, in order to reproduce the values of (5.2) for deflection angles corresponding to a time step δt, we must set

$$\theta_{\mathrm{SE}} = \frac{\pi}{2}\sqrt{\frac{\delta t}{\hat{t}_{\mathrm{rlx}\,1,2}}}, \tag{5.3}$$

where

$$\hat{t}_{\mathrm{rlx}\,1,2} \equiv \frac{\pi}{32} \frac{v_{\mathrm{rel}}^3}{\ln \Lambda\, G^2\, (m_1+m_2)^2\, n} \tag{5.4}$$

is the pair relaxation time.

With no other physical process than relaxation included, a single step in a MC simulation consists of the following operations.

1. Selection of a pair of adjacent particles to evolve. This procedure also determines the (local) value of the time step δt as explained below.
2. Modification of the orbital properties (E_i and J_i) of the particles through a super-encounter. This involves
   (a) estimation of the local density n entering $\hat{t}_{\mathrm{rlx}\,1,2}$ in (5.4),
   (b) random orientation of the velocity vectors $\boldsymbol{v}_i$ of the particles respecting their angular momenta $J_i = \|\boldsymbol{J_i}\|$ and specific kinetic energy $T_i = \frac{1}{2}\boldsymbol{v_i}^2$ (this sets the centre-of-mass [CM] and relative velocities $\boldsymbol{v}_{\mathrm{CM}}$ and $\boldsymbol{v}_{\mathrm{rel}}$; the former defines the encounter CM frame while the latter allows θ_{SE} to be determined through (5.3) and (5.4),
   (c) random orientation of the orbital plane in the CM frame around the direction of the relative velocity (the angle θ_{SE} is known, so computing the post-encounter velocities in the CM frame is trivial) and
   (d) transformation back to the cluster frame to obtain the modified E_i' and J_i'.

3. For each particle, selection of a new position on the (E_i',J_i')-orbit. As a particle is a spherical shell, its position is simply its radius R_i. This step comprises the update of the potential to take these new positions into account.

To compute the local density required in step 2a, we build and maintain a radial Lagrangian grid, the cells of which typically contain a few tens of particles each. Frequent updates (each time a particle gets a new position R) and occasional rebuilds of the mesh introduce only a very slight computational overhead.

Selection of a Pair of Particles and Determination of Time Step

For the sake of efficiency, we wish to use time steps that reflect the large variations of the relaxation time between the central and outer parts of a stellar cluster. The other constraint determining the selection procedure is that particles in an interacting pair must have the same δt, lest energy not be conserved.[2] But adjacent particles only form a pair momentarily and separate after their interaction as each is attributed a new position. This necessitates the use of *local* time steps, i.e. δt should be a function of R alone instead of being attached to particles.

For the time steps to be sufficiently short, we impose

$$\delta t(R) \leq f_{\delta t}\tilde{t}_{\rm rlx}(R), \tag{5.5}$$

where $\tilde{t}_{\rm rlx}$ is a locally averaged relaxation time,

$$\tilde{t}_{\rm rlx} \propto \frac{\langle v^2\rangle^{\frac{3}{2}}}{\ln\Lambda\, G^2\langle m\rangle^2 n} \tag{5.6}$$

and $0.005 \leq f_{\delta t} \leq 0.05$ typically. The time $\tilde{t}_{\rm rlx}$ is evaluated approximately with a sliding averaging procedure and tabulated from time to time to reflect the slow evolution of the cluster.

The members of a pair arrived at their present position at different times but have to leave it at the same time after a super-encounter. Building on the statistical nature of the scheme, instead of trying to maintain a particle at radius R during exactly $\delta t(R)$, we only require the *expectation value* for the residence time at R to be $\delta t(R)$. As explained by Hénon (1973a), this constraint can be fulfilled if the probability for a pair at R to be selected is proportional to $1/\delta t(R)$. This is realised in the following way.

- Because it would be difficult to define and use a selection probability $P_{\rm selec}$ that is a function of the continuous variable R, we define it to depend on

[2]When collisions are included, a shared δt also ensures that the probability for particle i to collide with particle j equates the symmetrical quantity.

the rank i of the pair (rank 1 designates the two particles that are closest to the centre, rank 2 the second and third particles, at increasing R and so on). For a given cluster's state, local relaxation times $\tilde{t}_{\mathrm{rlx}}$ are computed at the radial position of every pair. Rank-depending time steps are defined to obey inequality (5.5),

$$\delta t(i) \leq f_{\delta t} \tilde{t}_{\mathrm{rlx}}(R(i)). \tag{5.7}$$

- Normalised selection probabilities are computed by

$$P_{\mathrm{selec}}(i) = \frac{\overline{\delta t}}{\delta t(i)} \text{with } \overline{\delta t} = \left(\sum_{j=1}^{N_{\mathrm{p}}-1} \frac{1}{\delta t(j)} \right)^{-1}, \tag{5.8}$$

from which we derive a cumulative probability,

$$Q_{\mathrm{selec}}(i) = \sum_{j=1}^{i} P_{\mathrm{selec}}(j). \tag{5.9}$$

- At each evolution step another particle pair is randomly chosen according to P_{selec}. To do this, a random number X_{rand} is first generated with uniform probability between 0 and 1. The pair rank is then determined by inversion of Q_{selec},

$$i = Q_{\mathrm{selec}}^{-1}(X_{\mathrm{rand}}). \tag{5.10}$$

  The binary tree (see Sect. 5.3.1) is searched twice to find the id-numbers of the member particles, the (momentary) ranks of which are i and $i+1$.
- The pair is evolved through a super-encounter, as explained above, for a time step $\delta t(i)$.
- After a large number of elementary steps, $\delta t(i)$ and $P_{\mathrm{selec}}(i)$ are re-computed to reflect the slight modification of the overall cluster structure.

For the sake of efficiency, we must choose for Q_{selec}^{-1} a function that is quickly evaluated while $P_{\mathrm{selec}}(j)$ must approximate $1/\tilde{t}_{\mathrm{rlx}}(R(i))$ as closely as possible to avoid unnecessarily long time steps. A good compromise is to use a piecewise constant representation, i.e. divide the cluster into some 50 radial slices and use a constant P_{selec} in each. This is illustrated in Fig. 5.1 (with only 20 slices for clarity). Once the selection probabilities have been determined, the value $\overline{\delta t}$ relating them to the time step is set to $\overline{\delta t} = f_{\delta t} \max(T_{\mathrm{rel}}(i) P_{\mathrm{selec}}(i))$ so as to ensure that the constraint of (5.5) is satisfied everywhere.

It must be stressed that the probabilities $P_{\mathrm{selec}}(i)$ and corresponding time steps are computed in advance and are only updated (to reflect the evolution of the structure) after each particle has been treated several times on average. Once the pair of adjacent particles of rank i has been selected to be subject to a super-encounter, the time step $\delta t(i)$ is imposed and the encounter relaxation

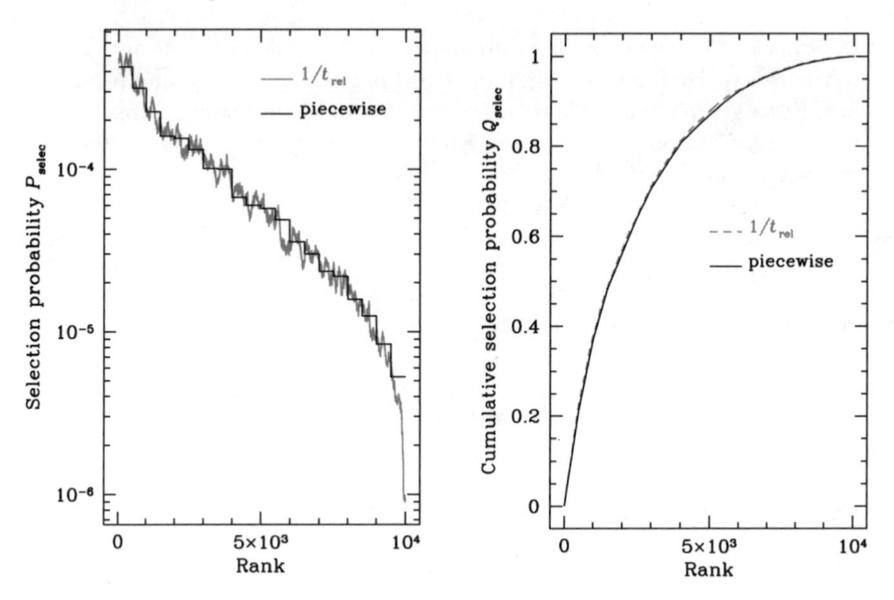

Fig. 5.1. Selection probabilities in a King $W_0 = 5$ cluster model consisting of 10 000 particles. The inverse of the locally estimated relaxation time is compared to the piecewise approximation used to set the probabilities in the MC code

time, $\hat{t}_{\mathrm{rlx}\,1,2}$, is determined by the particles' properties and the local density (5.4). This imposes the value of the deflection angle (5.3). In order to perform a proper orbit averaging and sampling over the field particles, θ_{SE} should be small so that a given particle would have experienced a large number of super-encounters by the time its orbit has changed significantly. Unfortunately, this is impossible to enforce strictly as the $\delta t(i)$ values are based on an estimate of the typical local relaxation time, while $\hat{t}_{\mathrm{rlx}\,1,2}$ can happen to be much shorter. Using a sufficiently small value of $f_{\delta t}$, we can keep the fraction of encounters leading to large values of θ_{SE} to a low level.

Representation of the Gravitational Potential

The smooth part of the potential of the cluster is simply approximated as the sum of the contributions of the N_{p} particles, each of which is a spherical infinitely thin shell. In other terms, compared to the potential in a system of N_{p} point-masses, we (implicitly) perform a complete smoothing over the angular variables. Between particles of rank i and $i+1$, the (smooth) potential felt by a particle at radius $R \in [R_i, R_{i+1}]$ is simply

$$\Phi(R) = -\frac{A_i}{R} - B_i \text{ with } A_i = \sum_{j=1}^{i-1} M_j \text{ and } B_i = \sum_{j=i}^{N_{\mathrm{p}}} \frac{M_j}{R_j}, \qquad (5.11)$$

where M_j and R_j are the mass and radius of the particle of rank j. Although we do not smooth the density distribution in the radial direction, tests show that, in practice, this spherically symmetric potential does not introduce significant unwanted relaxation for $N_\mathrm{p} \gtrsim 10^4$ in simulations extending to an average number of steps per particle of a few thousands (Hénon 1971b; Freitag & Benz 2001c). However, too small a time step parameter $f_{\delta t}$ can yield an artificially accelerated evolution owing to this numerical relaxation.

At each step in the simulation, two particles are selected, undergo a super-encounter and are given new positions on their slightly modified orbits. To enforce exact energy conservation, the A_i and B_i coefficients are updated after every such orbital displacement. Doing so saves much trouble connected with a potential that lags behind the actual distribution of particles' radii (and masses when stellar evolution or collisions are included). However, performing potential updates only after a large number of particle moves has advantages of its own, in particular, the possibility of algorithm parallelisation (Joshi et al. 2000), but requires special measures to ensure satisfactory energy conservation (Stodołkiewicz 1982; Giersz 1998; Fregeau & Rasio 2007).

The potential information is not represented by linear arrays (for the A_i and B_i) but by a *binary tree* (Sedgewick 1988). This tree also contains ranking information. It allows us to find a particle of a given rank, compute the potential at its position and update the potential data once the particle is moved to another radius in $\mathcal{O}(\log N_\mathrm{p})$ operations instead of $\mathcal{O}(N_\mathrm{p})$ as would be the case with simple arrays. At any given time, each particle is represented by a node in the tree. Each node is connected to (at most) two sub-trees. All the nodes in the left sub-tree of a given node correspond to particles with smaller radii and all the nodes in its right sub-tree to particles at larger radii. The spherical potential is represented by (floating-point) δA_k and δB_k coefficients attached to nodes. A third (integer) value, δi_k, allows the determination of the radial rank of any particle. If we define $\mathcal{LT}_k$ and $\mathcal{RT}_k$ to be the sets of nodes in the left and right sub-trees of node k, these quantities are defined by

$$\begin{aligned} \delta i_k &= 1 + \text{number of nodes in } \mathcal{LT}_k, \\ \delta A_k &= M_k + \sum_{m \in \mathcal{LT}_k} M_m \text{ and } \delta B_k = \frac{M_k}{R_k} + \sum_{m \in \mathcal{RT}_k} \frac{M_m}{R_m}. \end{aligned} \tag{5.12}$$

An example of binary tree is shown in Fig. 5.2. After a large number of specified steps, the binary tree is rebuilt from scratch to keep it well balanced.

Selection of a New Orbital Position

In a spherical potential $\Phi(R)$, a star of specific orbital energy E and angular momentum J spends, during one complete radial oscillation, a time $\mathrm{d}t = v_\mathrm{rad}^{-1}(R)\mathrm{dR}$ in an infinitesimal interval of radius $[R, R + \mathrm{d}R]$, with

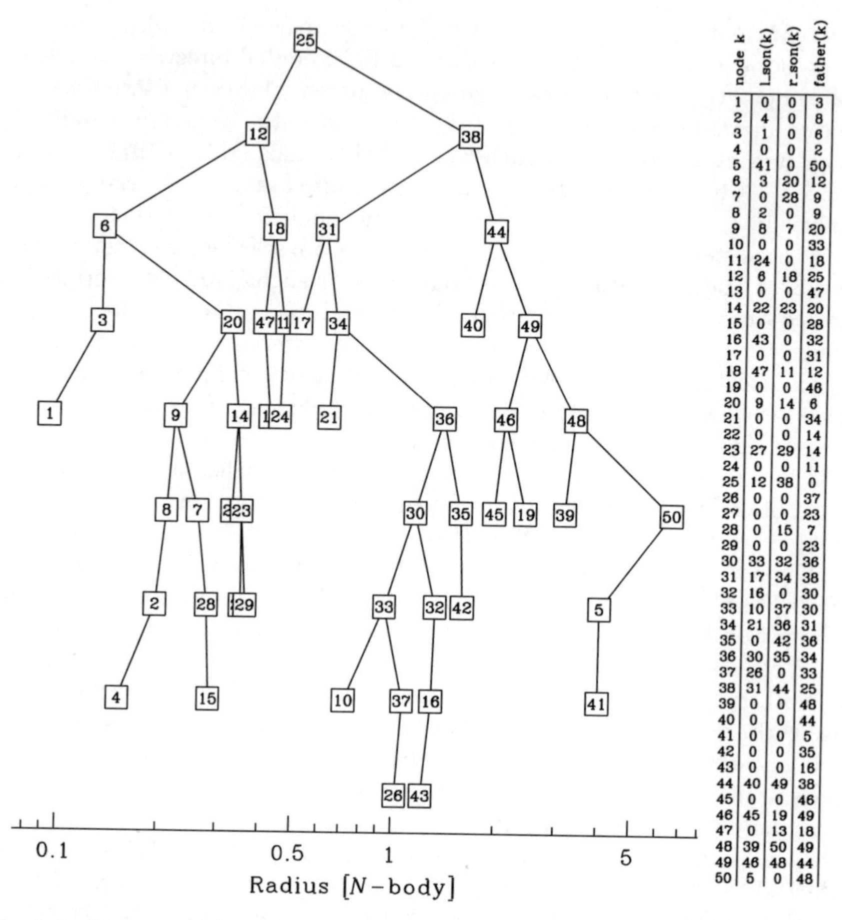

node k	l_son(k)	r_son(k)	father(k)
1	0	0	3
2	4	0	8
3	1	0	6
4	0	0	2
5	41	0	50
6	3	20	12
7	0	28	9
8	2	0	9
9	8	7	20
10	0	0	33
11	24	0	18
12	6	18	25
13	0	0	47
14	22	23	20
15	0	0	28
16	43	0	32
17	0	0	31
18	47	11	12
19	0	0	46
20	9	14	6
21	0	0	34
22	0	0	14
23	27	29	14
24	0	0	11
25	12	38	0
26	0	0	37
27	0	0	23
28	0	15	7
29	0	0	23
30	33	32	36
31	17	34	38
32	16	0	30
33	10	37	30
34	21	36	31
35	0	42	36
36	30	35	34
37	26	0	33
38	31	44	25
39	0	0	48
40	0	0	44
41	0	0	5
42	0	0	35
43	0	0	16
44	40	49	38
45	0	0	46
46	45	19	49
47	0	13	18
48	39	50	49
49	46	48	44
50	5	0	48

Fig. 5.2. Binary tree for a cluster of 50 particles. The structure of the tree is shown after many particles have been moved around since the tree was built. The lower axis shows the radius of each particle. The tree keeps the particles sorted in radius. The table on the right is the content of the three arrays used in the FORTRAN-77 code to implement the logical structure of the tree. Arrays `l_son(k)` and `r_son(k)` indicate the root nodes for the left and right sub-trees of node `k`. Array `father(k)` allows us to climb back to the root

$$v_{\rm rad}^2 = 2E - 2\Phi(R) - \frac{J^2}{R^2}. \tag{5.13}$$

Without knowledge of orbital phase, the probability density of finding the star at R is thus

$$\frac{\mathrm{d}P_{\rm orb}}{\mathrm{d}R} = \frac{2}{P_{\rm orb}}\frac{1}{v_{\rm rad}(R)}, \tag{5.14}$$

where

$$P_{\rm orb} = 2\int_{R_{\rm peri}}^{R_{\rm apo}} \frac{{\rm d}R}{v_{\rm rad}(R)} \qquad (5.15)$$

is the radial orbital period.

Since dynamical equilibrium is assumed, the knowledge of the explicit orbital motion $R(t)$ is not necessary. Instead, once a particle is updated, its position R is picked up at random, but with the requirement of correct statistical sampling. This means that the fraction of time spent at R must follow (5.14). Let the sought-for probability of placing the particle at $R \in [R_{\rm peri}, R_{\rm apo}]$ be $f_{\rm plac}(R) \equiv {\rm d}P_{\rm plac}/{\rm d}R$. We have to compensate for the fact that if the particle is placed at R, it will stay there for an average time $\overline{\delta t}/P_{\rm selec}(R)$. The average ratio of times spent at two different radii R_1 and R_2 on the orbit is

$$\left\langle \frac{t_{\rm stay}(R_1)}{t_{\rm stay}(R_2)} \right\rangle = \frac{f_{\rm plac}(R_1)P_{\rm selec}(R_2)}{f_{\rm plac}(R_2)P_{\rm selec}(R_1)} = \frac{v_{\rm rad}(R_2)}{v_{\rm rad}(R_1)}. \qquad (5.16)$$

This imposes the relation

$$f_{\rm plac}(R) \propto \frac{P_{\rm selec}(R)}{v_{\rm rad}(R)}. \qquad (5.17)$$

The numerical implementation of this probability law is complicated by the fact that $v_{\rm rad}(R)^{-1}$ is not known analytically and becomes infinite at the pericentre and apocentre. However, $v_{\rm rad}(R)^{-1}$ can always be capped by the Keplerian value with the same J, $R_{\rm peri}$ and $R_{\rm apo}$, allowing the use of an efficient rejection method (Press et al. 1992, Sect. 7.3) to pick up R according to (5.17).[3]

5.3.2 Additional Physics

Because it is based on particle representation, it is relatively easy to add a variety of physical ingredients to the MC algorithm in order to improve the realism of the simulations or the domain of applicability of the methods.

Collisions

Direct collisions are likely to occur in very dense stellar systems, from young clusters to core-collapsed globular clusters to nuclei of small galaxies (e.g. the various contributions in Shara 2002).

Let us consider a close approach between two stars with masses and radii m_1, r_1 and m_2, r_2, respectively. The relative velocity at infinity is $v_{\rm rel}$ and the

[3] This is the only significant improvement of the relaxation-only MC algorithm over the method described by Hénon. He also used a binary tree in the latest versions of his code although he did not describe it in his articles.

impact parameter b. Neglecting tidal effects, a collision requires the centres of the stars to come closer than $d_{\rm coll} = r_1 + r_2$. Although neglected in our MC code (because rare in galactic nuclei), tidal captures (Fabian et al. 1975) can be be considered using $d_{\rm capt} = \eta(r_1 + r_2)$ with $\eta > 1$ a numerical coefficient dependent on the velocity, masses and structures of the stars (e.g. Lee & Ostriker 1986; Kim & Lee 1999). Treating the approach until physical contact as a point-mass problem (assuming hyperbolic trajectories), we obtain the largest impact parameter leading to contact, $b_{\rm max}$, and the cross section,

$$S_{\rm coll\,1,2} = \pi b_{\rm max}^2 = \pi(r_1 + r_2)^2 \left[1 + \left(\frac{v_{*\,1,2}}{v_{\rm rel}}\right)^2\right], \tag{5.18}$$

where

$$v_{*\,1,2}^2 = \frac{2G(m_1 + m_2)}{r_1 + r_2} \tag{5.19}$$

is the relative velocity the stars would have at contact on a parabolic orbit. It is of the order a few 100 km s^{-1} for main-sequence (MS) objects. The second term in the bracket of (5.18) is the gravitational focusing, which highly enhances the cross section over the geometrical value $\pi(r_1+r_2)^2$ as long as $v_{\rm rel} < v_{*\,1,2}$. So, the collision rate for a star 1 travelling through a field of stars 2 with identical masses, sizes and velocities with number density n_2 is simply

$$\left.\frac{{\rm d}N_{\rm coll}}{{\rm d}t}\right|_{1,2} = n_2 v_{\rm rel} S_{\rm coll\,1,2} \equiv t_{\rm coll\,1,2}^{-1} \tag{5.20}$$

which defines the collision time $t_{\rm coll\,1,2}$. If all stars have the same mass m and size r, a number density n and their velocities follow a Maxwellian distribution with 1D dispersion σ_v^2, the average collision rate is (Binney & Tremaine 1987)

$$t_{\rm coll}^{-1} = 16\sqrt{\pi} n \sigma_v r^2 \left(1 + \frac{Gm}{2\sigma_v^2 r}\right). \tag{5.21}$$

Adding stellar collisions to the MC algorithm is relatively straightforward, thanks to the use of particles to represent the cluster (as opposed to DFs, as done in FP codes).

First, the determination of time steps (and corresponding pair-selection probabilities) has to include, in addition to (5.5), the following constraint

$$\delta t(R) \le f_{\delta t} \tilde{t}_{\rm coll}(R), \tag{5.22}$$

with

$$\tilde{t}_{\rm coll}^{-1} = 16\sqrt{\pi} n \sigma_v \langle r^2 \rangle \left(1 + \frac{G\langle mr \rangle}{2\sigma_v^2 \langle r^2 \rangle}\right), \tag{5.23}$$

where $\sigma_v^2 = 1/3\langle v^2 \rangle_m$. The notations $\langle \cdots \rangle$ and $\langle \cdots \rangle_m$ denote number- and mass-weighted averaged quantities, respectively.[4] The choice of quantities to

[4]Note that (15) of Freitag & Benz (2002) is slightly incorrect.

average is such that we retrieve the correct value for the average collision rate in the limits $\sigma_v^2 \ll G\langle m\rangle\langle r\rangle^{-1}$ and $\sigma_v^2 \gg G\langle m\rangle\langle r\rangle^{-1}$.

Next, when a pair is selected for update, and once the local density and relative velocity have been determined, the pair collision time is computed, using (5.18), (5.19) and (5.20) but with n instead of n_2. Hence, the probability of collision between the pair during the time step δt is

$$P_{\mathrm{coll}\,1,2} = n v_{\mathrm{rel}} S_{\mathrm{coll}\,1,2}\,\delta t. \tag{5.24}$$

The use of n rather than n_2 is of central importance. This way, the collision probabilities are symmetric, as they should be, $P_{\mathrm{coll}\,1,2} = P_{\mathrm{coll}\,2,1}$. Furthermore, it would be impossible to estimate the local density of each population particularly because, in MC codes as in N-body, each particle can represent a star (or stars) with properties different from any other particle. What makes this simplification possible is that for a given particle, the (local) probability that the neighbouring particle is of type x (whatever the definition of a type is) is simply n_x/n, so the process of selecting the next particle as interaction partner will, statistically, produce a rate of collisions with objects of type x proportional to n_x *because* n rather than n_x is used to compute the pair collision time. Including the estimate of the collision time in the determination of the time steps ensures that, in a vast majority of cases, $P_{\mathrm{coll}\,1,2} \lesssim f_{\delta t} \ll 1$, avoiding time steps during which more than one collision should have occurred. In the MC algorithm, a collision between two particles has a statistical weight of N_*/N_{p}. This means that every star in the first particle collides with a star of the second particle and that all these collisions are identical so that the outcome can be represented by (at most) two particles corresponding to N_*/N_{p} collision products each.

Then a random number X_{rand}, with uniform deviate between 0 and 1, is generated and a collision between the two particles has to be implemented if $X_{\mathrm{rand}} < P_{\mathrm{coll}\,1,2}$. In low-velocity environments, it is justified to assume that collisions result in mergers with negligible mass loss (Freitag et al. 2006b), but this simplification breaks down in galactic nuclei where $\sigma_v > 100\,\mathrm{km\,s^{-1}}$ (Freitag & Benz 2002). We use prescriptions for the boundary between mergers and fly-bys and for the amount of mass and energy lost based on a large set of SPH simulations of collisions between MS stars (Freitag & Benz 2005). The impact parameter is selected at random with uniform probability in b^2 between 0 and b^2_{max}. Because evolution on the MS is neglected, a collision is entirely determined by the values of m_1, m_2, v_{rel} and b and its outcome is determined using 4D interpolation and extrapolation from the SPH results (Freitag & Benz 2002; Freitag et al. 2006c). The properties of the particles are updated from the post-collision values of m_1, m_2 and $\boldsymbol{v}_{\mathrm{rel}}$.

The particles are then placed at random radii on their new orbits, according to (5.17). This concludes the step as two-body relaxation is not implemented when a collision is detected. In highly collisional systems, this can lead to an underestimate of relaxation effects and we have experimented with a modified scheme in which every second step is collisional and the others are reserved for relaxation. This makes the code approximately twice as slow but does not seem to affect the results significantly. In case of a merger or if one or both stars are completely disrupted (a rare outcome requiring velocities in excess of about $5\,v_{*\,1,2}$), the number of particles in the simulation is reduced correspondingly.

One major theoretical uncertainty still to be tackled when it comes to the effects of collisions in stellar dynamics is how they affect stellar evolution. In case of mergers, the problem is made particularly difficult by the very high rotation rate of the collision product (e.g. Sills et al. 1997, 2001; Lombardi et al. 2002). In the face of this uncertainty we adopt a simple approach in which we set the effective age of the collision product based on its mass and the amount of core helium and assume no collisional mixing at all (see Portegies Zwart et al. 1999 for another prescription).

While the hydrodynamics of collisions between two MS stars is now relatively well understood (Sills et al. 2002; Freitag & Benz 2005; Dale & Davies 2006; Trac et al. 2007, and references therein), our knowledge about encounters featuring other stellar types is still very limited, mostly because the physics involved is more challenging. Collisions between a giant and a more compact object are probably more common than MS–MS encounters, at least in galactic nuclei where gravitational focusing is weaker, but only a few authors have attempted to model such events (Davies et al. 1991; Rasio & Shapiro 1991; Bailey & Davies 1999; Lombardi et al. 2006). The main question mark concerns the evolution of the common envelope system resulting from the capture of the more compact star (see, e.g. Taam & Ricker 2006 and Chap. 11). Collisions between a compact remnant and a MS (or giant) star have been studied numerically in a larger number of papers (Regev & Shara 1987; Benz et al. 1989; Różyczka et al. 1989; Davies et al. 1992; Ruffert 1993, to mention a few), but clear and comprehensive predictions for their outcome are still missing. This is unfortunate because, in our models for galactic nuclei, collisions between a MS star and a remnant occur at a rate comparable to collisions between two MS stars (a few $10^{-6}\,\mathrm{yr}^{-1}$ in a Milky-Way-like nucleus, see Freitag et al. 2006a). Finally, in young dense clusters, where mergers may contribute to the formation of massive stars ($m > 10\,\mathrm{M}_\odot$) or lead to the build-up of very massive stars ($m > 100\,\mathrm{M}_\odot$, e.g. Bally & Zinnecker 2005 and Sect. 5.4.1), collisions involving pre-MS objects are likely, a type of event only simulated very recently (Laycock & Sills 2005; Davies et al. 2006).[5]

Central Massive Object

To study the structure and evolution of galactic nuclei with a central massive black hole (MBH, $M_{\mathrm{BH}} \gtrsim 10^4\,\mathrm{M}_\odot$) or globular clusters hosting an

[5]For more pointers to the literature on stellar collisions and tidal disruptions by a massive black hole, see the MODEST web pages at `http://www.manybody.org/modest/WG/wg4.html`.

intermediate-mass black hole (IMBH, $10^4\,\mathrm{M}_\odot \gtrsim M_{\mathrm{BH}} \gtrsim 10^2\,\mathrm{M}_\odot$) or a very massive star ($M_* \gtrsim 200\,\mathrm{M}_\odot$), the effects of a central massive object have been included in the MC code (Freitag 2000; Freitag & Benz 2002; Freitag et al. 2006a; Freitag et al. 2006b). Here I concentrate on the case of an (I)MBH (see Ferrarese & Ford 2005 for a review of the observational evidence for MBHs in centres of galaxies and Miller & Colbert 2004, van der Marel 2004 for reviews on the possible existence of IMBHs).

Recall that the MC approach is only valid for spherical systems in dynamical equilibrium and useful mostly if collisional effects such as two-body relaxation produce noticeable evolution over the period of interest. Galactic nuclei hosting MBH less massive than about $10^7\,\mathrm{M}_\odot$ are probably relaxed and therefore amenable to MC modelling. Indeed, assuming naively that the Sgr A* cluster at the centre of our Galaxy is typical as far as the total stellar mass and density are concerned (Genzel et al. 2003; Ghez et al. 2005; Schödel et al. 2007) and that we can scale to other galactic nuclei using the observed correlation between the mass of the MBH and the velocity dispersion of the host spheroid, σ, in the form $\sigma = \sigma_{\mathrm{MW}}(M_{\mathrm{BH}}/4\times10^6\,\mathrm{M}_\odot)^{1/\beta}$, with $\beta \approx 4-5$ (Ferrarese & Merritt 2000; Tremaine et al. 2002), we can estimate the relaxation time at the radius of influence (the limit of the region where the gravity of the MBH dominates) to be $t_{\mathrm{rlx}}(R_{\mathrm{infl}}) \approx 10^{10}\,\mathrm{yr}\,(M_{\mathrm{BH}}/4\times10^6\,\mathrm{M}_\odot)^{(2-3/\beta)}$.

All the key aspects of the interaction between the central MBH and its host stellar system ("cluster" in short) are included in the MC code.

Gravity of the MBH. The contribution of the MBH is treated as a central, fixed point mass. Newtonian gravity is assumed so the only modification in computing the potential ϕ is to add M_{BH} to the coefficients A_i in (5.11). The MBH is allowed to grow by accretion of material from the stars or through an ad hoc prescription to account for gas inflow. Care is taken to make the time steps significantly shorter than $\phi(\mathrm{d}\phi/\mathrm{d}t)^{-1}$ so as to ensure that the adiabatic effects of the growth of the MBH on the cluster are accounted for (Young 1980; Quinlan et al. 1995). The MBH imposes very high stellar velocities in its vicinity, causing stellar collisions to be more disruptive. The gas emitted in a collision is assumed to accrete completely and immediately onto the MBH or to accumulate in an unresolved disc around the MBH if its growth is limited by the Eddington rate.

Tidal disruptions. A star of mass M_* and radius R_*, which comes within a distance $R_{\mathrm{td}} = k\,R_*(M_{\mathrm{BH}}/M_*)^{1/3}$ of the MBH, is torn apart by the tidal forces (e.g. Fulbright 1996; Diener et al. 1997; Ayal et al. 2000; Kobayashi et al. 2004). Here k is a constant of order unity depending on the structure of the star. In the present implementation, we assume that the tidal disruption is always complete and that a fixed fraction of the mass of the disrupted star is accreted immediately, usually 50 per cent as suggested by most hydrodynamical simulations. The rest is lost from the cluster. These events are predicted to trigger month- to year-long accretion flares in the UV/X domain (Hills 1975; Rees 1988), some of which might have been detected already (see

Komossa 2005 for a review and Gezari et al. 2006; Esquej et al. 2007 for recent observations).

In a spherical galactic nucleus in dynamical equilibrium, the velocity vector $\boldsymbol{v}$ of a star, at distance R from the MBH, has to point inside the *loss cone*, in direction to or away from the centre, for its orbit to pass within $R_{\rm td}$. The aperture angle of the loss cone, $\theta_{\rm LC}$, is given by the relation

$$\begin{aligned}\sin^2(\theta_{\rm LC}) &= 2\left(\frac{R_{\rm td}}{vR}\right)^2\left[\frac{v^2}{2}+\frac{GM_{\rm BH}}{R_{\rm td}}\left(1-\frac{R_{\rm td}}{R}\right)+\Phi_*(R)-\Phi_*(R_{\rm td})\right]\\ &\simeq 2\frac{GM_{\rm BH}R_{\rm td}}{(vR)^2}\approx\frac{R_{\rm td}}{R},\end{aligned} \tag{5.25}$$

where $\Phi_*(R) = \Phi(R) + GM_{\rm BH}/R$ is the cluster contribution to the gravitational potential. The first approximation is valid as long as $R \gg R_{\rm td}$, which is nearly always the case; the second is an order-of-magnitude estimate valid within the sphere of influence of the MBH, where $v^2 \approx GM_{\rm BH}R^{-1}$.

Stars on loss-cone orbits are removed on an orbital time-scale. In a spherical potential, it is generally assumed that loss-cone orbits are replenished by two-body relaxation, but orbital perturbations by resonant relaxation (see Sect. 5.4.2) or deflections by massive objects such as molecular clouds (Perets et al. 2007) may play an important role. Barring such non-standard processes, two loss-cone regimes can be distinguished (Frank & Rees 1976; Lightman & Shapiro 1977; Cohn & Kulsrud 1978). (1) The loss cone is kept full and does not induce any significant anisotropy in the velocity distribution when relaxation is strong enough to repopulate loss-cone orbits over an orbital time, corresponding to the condition $\theta_{\rm LC}^2 t_{\rm rlx} \ll P_{\rm orb}$. For stars in this regime, which typically occurs at large distances, the average time before tidal disruption is of order $t_{\rm disr,full} \simeq \theta_{\rm LC}^{-2}P_{\rm orb}$ (when averaged over all directions of $\boldsymbol{v}$). (2) The loss cone is (nearly) empty in the opposite case, $\theta_{\rm LC}^2 t_{\rm rlx} \gg P_{\rm orb}$, and corresponds to an absorbing region of phase space into which the stars diffuse. The density of stars on orbits close to but out of the loss cone is reduced. In this regime, it takes on average $t_{\rm disr,empty} \simeq t_{\rm rlx}\ln(\theta_{\rm LC}^{-2})$ for a star to be disrupted.

Plunges through the horizon. The last stable parabolic orbit around a non-spinning massive black hole corresponds to a (Newtonian) pericentre distance $R_{\rm LSPO} = 8GM_{\rm BH}c^{-2}$. Sufficiently dense stars such as compact remnants have a tidal disruption radius $R_{\rm td}$ inside $R_{\rm LSPO}$ (or even inside the horizon), meaning that such objects will be swallowed whole rather than be tidally disrupted, and produce no accretion flare.[6] From the point of view of stellar dynamics, this situation is identical to the case of tidal disruptions, with the quantity $R_{\rm td}$ replaced by $R_{\rm LSPO}$.

[6] In fact, when $R_{\rm LSPO} > R_{\rm td} > R_{\rm hor} = 2GM_{\rm BH}c^{-2}$, the star is disrupted before it disappears through the horizon. To my knowledge, the detectability of such events has not been investigated.

Inspirals by emission of gravitational waves. Significant emission of gravitational waves (GWs) occurs during very close encounters with the MBH (Peters & Mathews 1963). For a compact, massive stellar object on a very eccentric orbit, GW emission may dominate orbital evolution over two-body relaxation, yielding to progressive circularisation and shrinking of the semi-major axis (Peters 1964) until it plunges through the horizon of the MBH (or is tidally disrupted). For a 1–10 $M_{\odot}$ object orbiting a MBH with a mass between 10^4 and $10^7\,M_{\odot}$, the final months or years of inspiral should be detectable by the future spaceborn GW observatory LISA[7] to distances of several Gpc. Such extreme mass ratio inspirals (EMRIs) yield an unprecedented view on the direct vicinity of MBHs. The promise for physics and astrophysics is as exciting as the uncertainties about their physical rates and the challenges for data analysis are high (see Amaro-Seoane et al. 2007 for an extensive review of the various aspects of EMRI research).

I now explain in some detail how the loss-cone physics is implemented in the MC code. This treatment is adequate only for the processes requiring a single passage within a well-defined critical distance of the MBH to be successful, such as tidal disruption, plunges or non-repeating GW bursts emitted by stars on quasi-parabolic orbits (Hopman et al. 2007). In contrast, an EMRI is a progressive process that will only be successful (as a potential source for LISA) if the stellar object experiences a very large number of successive dissipative close encounters with the MBHs (Alexander & Hopman 2003). The ability of the MC approach to deal with this situation is discussed in Amaro-Seoane et al. (2007).

At the end of the step in which two particles have experienced an encounter (to simulate two-body relaxation), each particle is tested for entry into the loss cone, $J < J_{\mathrm{LC}}$, where $J_{\mathrm{LC}} = RV\sin(\theta_{\mathrm{LC}}) \simeq \sqrt{2GM_{\mathrm{BH}}R_{\mathrm{td}}}$ (5.25). A complication arises because the time step δt used in the MC code is a fraction $f_{\delta t} = 10^{-3} - 10^{-2}$ of the local relaxation time $t_{\mathrm{rlx}}(R)$, which is much larger than the critical timescale $\theta_{\mathrm{LC}}^2 t_{\mathrm{rlx}}$. In other words, the super-encounter deflection angle θ_{SE} (5.3) is much larger than θ_{LC}. This keeps the loss cone effectively and artificially full. However, in contrast with direct N-body simulations, this is not due to the overall relaxation rate being too large when $N_{\mathrm{p}} < N_*$.

To treat the empty loss-cone regime in the most accurate fashion, we would need to use time steps as short as the orbital period. Unfortunately, it is not possible to give short time steps only to particles with eccentric orbits (and hence at risk of entering the loss cone), because the time step is a function of the position R and cannot be attached to a particle. Hence, at least all particles within the critical radius, defined by $\bar{t}_{\mathrm{disr,full}}(R_{\mathrm{crit}}) = \bar{t}_{\mathrm{disr,empty}}(R_{\mathrm{crit}})$, where $\bar{t}$ quantities are some local average, would need to have much shorter time steps, which would slow down the code considerably. Instead, an approximate

[7] Laser Interferometer Space Antenna, see `http://www.lisa-science.org`.

procedure is used to ensure that entry into the loss cone happens diffusively when $\theta_{LC}^2 t_{rlx} \gg P_{orb}$.

After the super-encounter deflection angle θ_{SE} has been computed (5.3) and before the particles in the pair are given their new energies, angular momenta and positions, we check each of them for entry into the loss cone in the following manner. First, the orbital period is computed by integrating (5.15) using Chebyshev quadrature (Press et al. 1992). We consider that during $P_{orb} \ll \delta t$, the direction of the velocity of the particle would have changed by an r.m.s. angle $\theta_{orb} = (P_{orb}/\delta t)^{1/2}\theta_{SE}$. We then assume that the tip of the velocity vector of the particle executes a random walk of $N_{RW} = \delta t/P_{orb}$ sub-steps of length θ_{orb} during δt. The modulus of the velocity is kept constant. Entry into the loss cone is tested at each of these sub-steps. This random walk is executed in the reference frame of the super-encounter, but independently for each particle of the pair because they have different θ_{orb} and N_{RW}. If a particle is found on a loss-cone orbit, it is immediately removed and (part of) its mass is added to the MBH. If the random walk never crosses into the loss cone, the particle is kept and, in order to ensure exact energy conservation the particle is given the velocity computed in the super-encounter, *not* that reached at the end of the random walk. The random walk is a refinement of the super-encounter from a statistical point of view but, because of its stochastic nature, it cannot produce velocity vectors anti-parallel to each other for the particles in a pair. This means that energy in the reference frame of the cluster (as opposed to that of the pair) would not be conserved. It might be possible to improve this procedure by performing the random walk in the cluster reference frame and leaving the particle with the velocity attained at the end of it. This would permit us to obtain the correct decrease of density on the orbits close to the loss cone.

In the context of loss-cone physics, I mention another type of Monte-Carlo code developed by Shapiro and collaborators at Cornell University (Shapiro 1985, for a review and references). Their approach was essentially a hybrid between that presented here, entirely based on particles and with no explicit computation of diffusion coefficients and the direct Fokker–Planck integration (Chap. 4). Instead of having particles interacting in pairs, their density in the (E, J) phase space was tabulated in order to compute diffusion coefficients used to modify their orbital parameters during the next global step. Within a global step, each particle could be evolved independently of the others (and on its own time step) until the updated phase-space density (and potential) is recomputed. This permitted to endow the particles in or close to the loss cone with time steps as short as their orbital time. Extending this scheme to a multi-mass situation seems feasible without explicit use of an augmented (and sparsely populated) (E, J, M_*) phase space. Unfortunately, to my knowledge, such a development was not attempted.

Binary Stars

The MC code presented so far in this chapter only deals with the dynamics and evolution of single stars. This is a reasonable simplification as long as the overall dynamics of galactic nuclei is concerned because in such environments most binaries are very soft, meaning that their internal orbital velocity is much smaller than velocity dispersion, at least in the vicinity of a MBH where the density and interaction probability are the highest. However, binaries play a major role in the evolution of globular clusters where the hard ones act as an efficient central source of heat by being shrunk and eventually ejected during interactions with other stars (Aarseth 1974; Spitzer & Mathieu 1980; Gao et al. 1991; Hut et al. 1992; Heggie & Hut 2003; Giersz 2006; Fregeau & Rasio 2007, amongst many others). For a given stellar density, binaries also highly increase the rate of direct collision between stars (Portegies Zwart et al. 1999; Portegies Zwart & McMillan 2002; Portegies Zwart et al. 2004; Fregeau et al. 2004). Beside their dynamical role, binary interactions in dense clusters are also of high interest as a way to create a whole zoo of "stellar exotica" and phenomena, including blue stragglers, millisecond pulsars, and mergers between compact stars as sources of supernovae, gamma-ray bursts or gravitational waves (e.g. Hurley et al. 2001; Davies 2002; Shara & Hurley 2002; Benacquista 2006; Grindlay et al. 2006; O'Leary et al. 2007). Including binaries in models of galactic nuclei is also important to explain X-ray observations at the Galactic centre (Muno et al. 2005), hyper-velocity stars (e.g. Hills 1988; Brown et al. 2005) and as a possible channel to create extreme-mass ratio sources of gravitational waves for LISA (Miller et al. 2005).

Here I put aside the very thorny question of binary evolution and how it might be affected by dynamics (see Chaps. 11 and 12) and concentrate on the dynamical aspects. Binaries have been included in MC simulations with various levels of sophistication (Spitzer & Mathieu 1980; Stodołkiewicz 1985, 1986; Giersz 1998, 2001, 2006; Giersz & Spurzem 2000, 2003; Fregeau et al. 2003; Gürkan et al. 2006; Fregeau & Rasio 2007; Spurzem et al. 2006). The approach of Fregeau & Rasio (2007) is based on our own treatment of collisions and is the most direct and accurate one, at least when each particle represents a single system (single star or binary). This treatment does not include formation of binaries through three-body interactions (see the works of Stodołkiewicz and Giersz).

To include binaries in a MC code, we first need to allow some of the particles to represent binaries instead of single stars, which requires extra data to keep track of the internal structure, masses and evolutionary phase of the member stars, semi-major axis $a_{\rm bin}$ and eccentricity $e_{\rm bin}$. In the absence of interaction with another star or binary, these parameters are updated by the use of some binary evolution prescription. Then, similar to stellar collisions, including binary dynamics amounts to (1) determining the probability of a binary interaction $P_{\rm bin}$ between two neighbouring particle if at least one of them is a binary, (2) generating a random number $X_{\rm rand}$ and, if $X_{\rm rand} < P_{\rm bin}$, (3) implementing a single–binary or binary–binary encounter.

Steps (1) and (2) are the same as in the implementation of collision between single stars. Actually, at this level, binary interactions do not need to be distinguished from stellar collisions. We only need to give to binaries a radius $\eta a_{\rm bin}$ where $\eta > 1$ is a safety factor to ensure that all interactions that can perturb the binaries significantly are taken into account. Fregeau & Rasio (2007) chose $\eta = 2$ and checked that a value $\eta = 4$ (which could cause the time steps to be about twice as short) do not lead to statistically different results, as far as the overall evolution of the cluster and binary population is concerned. More complex forms of the criterion for the most distant encounter to be included have been used by other authors (e.g. Bacon et al. 1996; Giersz & Spurzem 2003). The simple rule described here, based on proximity at the closest approach (when each binary is treated as a point mass) should yield correct results if η is made sufficiently large but, in studies of small perturbations to binaries (or planetary systems), it may be less than optimal in the sense that large η values will yield small time steps. Indeed, for binaries, we have to substitute $\eta a_{\rm bin}$ for r in (5.23). Roughly speaking, with binaries at the hard–soft boundary ($Gm_{\rm bin}a_{\rm bin}^{-1} \simeq \sigma_v^2$), the time step will be limited by binary processes rather than by two-body relaxation if $\eta > \ln \Lambda$.

Between interactions, binaries are treated as unperturbed and their properties are updated using binary evolution prescriptions. Note that this is also the case in N-body codes unless another object comes within a distance $d_{\rm pert} = \gamma_{\rm min}^{-1/3}(2m_{\rm pert}/m_{\rm bin})^{1/3}(1+e_{\rm bin})a_{\rm bin}$, where $m_{\rm pert}$ is the mass of the perturber and $\gamma_{\rm min}$ is the tidal perturbation parameter (Aarseth 2003 and Chap. 1). In most cases $\gamma_{\rm min}$ is set to 10^{-6}. Hence, in a similar-mass situation ($m_{\rm pert} \approx m_{\rm bin}$) the N-body prescription corresponds to $\eta \approx 100 - 200$ in the MC collision formalism. Whether this much more conservative condition yields significantly different results in the evolution of the binaries and their host cluster has not been investigated in depth (see Giersz & Spurzem 2003; Spurzem et al. 2006 for some discussion). Incidentally, such research may open the possibility of a more approximate but much faster treatment of binary interactions in direct N-body codes.

The most direct and accurate (but also time-consuming) way of implementing step (3), i.e. of determining the outcome of a binary encounter occurring in a MC simulation, is to switch to a direct few-body integrator (see Chap. 2 for algorithms). First, the quantities not specified by the MC particles have to be picked at random. These are the orbital phase(s) and orientation(s) and the impact parameter.[8] One difficulty arises with binary–binary encounters as they often result in the formation of a stable triple system. As

[8]In principle, we could keep track of the orbital phase of a binary between interactions. However, the MC method relies on the assumptions that strong interactions are rare and that binaries are much smaller than any length scale in the cluster. This effectively randomises the orbital phase between interactions.

mentioned by Giersz & Spurzem (2003) and Fregeau & Rasio (2007), it is in principle possible to have some particles representing triples (or higher-order stable groups) in the MC framework, with the appropriate book-keeping, but this has not been implemented so far. Instead, triple systems are forcefully broken apart into a binary and a single star just unbound to the binary. Another type of outcome that may require special treatment is the formation of a very wide, soft binary with a size not much smaller than the typical size of the cluster. Such pairs cannot be treated accurately in the MC formalism, but they are unlikely to survive the next interaction so they can be artificially broken up without affecting the results. Finally, as mentioned above, it is probably important to allow for direct collisions during binary interactions. One source of uncertainty is the size of a merged star just after a collision. It is likely to be several times the MS radius, leading to a significant probability of a triple or quadruple collision (Goodman & Hernquist 1991; Lombardi et al. 2003; Fregeau et al. 2004).

Once the outcome of a binary–single or binary–binary interaction has been determined, the products of the interaction are turned back into MC particles representing single or binary stars with the adequate internal and orbital properties and a position in the cluster is selected for each according to the procedure presented in Sect. 5.3.1.

Integrating the few-body encounters in a cluster with a large fraction of binaries can account for a significant fraction of the computing time. A much faster way to deal with binary dynamics is to use "recipes", which are fitting formulae for the cross section and outcome of interactions based on large pre-computed sets of scattering experiments (e.g. Heggie 1975; Hut 1993; Heggie et al. 1996). However, for stars of unequal masses, the parameter space is too vast to be reliably covered by such recipes. Even in the idealised case where all stars have the same mass, for which comprehensive binary-interaction cross sections are available, the use of such recipes rather than explicit few-body integrations seems to yield quantitatively inaccurate results (Fregeau et al. 2003; Fregeau & Rasio 2007).

Other Physical Ingredients

MC codes can include a few other physical processes that I describe more succinctly.

Stellar evolution – Evolution of stars (single or binaries) can be taken into account with various levels of refinement. In our MC code, a very simple prescription is used, which assumes that a star of initial mass M_* spends a time $t_{\rm MS}(M_*)$ on the MS without any evolution and abruptly turns into a compact remnant at the end of this period. Thus the giant phase is neglected. The relation $t_{\rm MS}(M_*)$ and the prescriptions for the nature and mass of the remnant are taken from stellar evolution models (Hurley et al. 2000; Belczynski et al. 2002). To ensure that stellar evolution time-scales are resolved, a supplementary constraint on the time step is introduced, $\delta t_i \leq f_{\delta t,*} t_{*,i}$, where $t_{*,i}$ is an estimate for the stellar evolution time-scale of stars at rank i and $f_{\delta t,*} = 0.025$ typically. In the present implementation, $t_{*,i}$ is simply the MS lifetime of the particle, which has rank i at the moment the time steps are computed. Because we use a piecewise constant representation of δt, the time step will generally be shorter than a fraction $f_{\delta t,*}$ of the smallest local value of $t_{\rm MS}$. Once a pair of particles is selected, it is first checked for stellar evolution and its masses and radii are updated if required, before the super-encounter (or collision) is carried out. Natal kicks can be given to newborn neutrons stars and black holes (Freitag et al. 2006a).

This simplistic treatment can be improved by the use of detailed stellar evolution packages (Portegies Zwart & Verbunt 1996; Portegies Zwart & Yungelson 1998; Hurley et al. 2000, 2001. See also Chaps. 10 and 13). A difficulty to confront, however, is that this will involve shorter time-scales t_*, e.g. to resolve the giant phase. In general, stars with short t_* can be found anywhere in a cluster, imposing (unlike relaxation or collision) uniformly short time steps. This could be prevented by using a time-stepping scheme for stellar evolution independent of the dynamical one. For instance, using a heap structure (Press et al. 1992), we could keep track of the next particle requiring update of its stellar parameters and realise this update when due, without changing the orbital parameters (except if a natal kick is imparted).

Large-angle scatterings – Gravitational encounters between stars of mass m_1 and m_2 at a relative velocity $v_{\rm rel}$ with an impact parameter smaller than a few $b_0 \equiv G(m_1+m_2)v_{\rm rel}^{-2}$ lead to deflection angles too large to be accounted for in the standard, diffusive theory of relaxation. On average, a star will experience an encounter with impact parameter smaller than $f_{\rm LA} b_0$ (with $f_{\rm LA}$ of order a few) over a time-scale

$$t_{\rm LA} \simeq \left[\pi (f_{\rm LA} b_0)^2 n \sigma\right]^{-1} \approx \frac{\ln \Lambda}{f_{\rm LA}^2} t_{\rm rlx}. \tag{5.26}$$

The effects of large-angle scatterings on the overall evolution of a cluster are negligible in comparison with diffusive relaxation (Hénon 1975; Goodman 1983). However, unlike the latter process, they can produce velocity changes strong enough to eject stars from an isolated cluster (Hénon 1960, 1969; Goodman 1983) or, more important, from the region of influence around a MBH (Lin & Tremaine 1980; Baumgardt et al. 2004; O'Leary & Loeb 2008). Large-angle scatterings are easily included in MC simulations as a special case of collision, with a cross section $\pi(f_{\rm LA} b_0)^2$ (Freitag et al. 2006a), but the time steps will be limited by this (rare) process rather than by diffusive relaxation for $f_{\rm LA} \gtrsim 4$.

Tidal evaporation – Stellar clusters are subject to the tidal influence of their host galaxy. Assuming spherical symmetry, the MC code cannot deal with the galactic field accurately but it is easy to include in an approximate way the most important effect, which is the evaporation of stars from the cluster.

A star can escape from a cluster on a circular orbit of radius $R_{\rm G}$ around a spherical host galaxy if its orbit allows it to reach the Lagrange point away from or in the direction of the galaxy. These locations are approximately at a distance $R_{\rm L} = R_{\rm G}(M_{\rm cl}/(2M_{\rm G}))^{1/3}$ from the cluster's centre, where $M_{\rm cl}$ and $M_{\rm G}$ are the masses of the cluster and a point-mass galaxy, respectively. In the spherical approximation, we assume that a star escapes when its apocentre distance is larger than $R_{\rm L}$. As the total mass of the cluster decreases, the value of $R_{\rm L}$ is adjusted. This can lead to more stars being lost if their apocentre distances happen to lie beyond the new $R_{\rm L}$ value, so we have to iterate until convergence is reached for the bound mass of the cluster. Using such treatment of tidal evaporation, combined with a prescription for the orbital decay of the cluster owing to dynamical friction, Gürkan & Rasio (2005) have simulated the internal and orbital evolution of clusters at the Galactic centre.

5.4 Some Results and Possible Future Developments

Monte-Carlo codes have been used in a variety of problems involving the collisional evolution of globular clusters and galactic nuclei. I do not attempt to review this variety of works but invite the reader to sample the references cited in Sect. 5.1. Here I limit myself to the quick presentation of a few typical results to give a flavour of the capabilities of the method.

5.4.1 Young Clusters and Globular Clusters

In Figs. 5.3 and 5.4, I show the evolution to core collapse of single-mass and multi-mass Plummer models, computed with the MC code described here with no other physics than two-body relaxation. I compare with direct NBODY4 results (H. Baumgardt 2005, personal communication). Provided the value of γ_c needed to convert N-body time units (see Chap. 1) to relaxation time is adjusted in an ad hoc fashion, very good agreement between the methods is obtained for these cases. We find $\gamma_c \simeq 0.15$ for the single-mass model and $\gamma_c \simeq 0.03$ for Salpeter mass function ($\mathrm{d}N_*/\mathrm{d}M_* \propto M_*^{-2.35}$) extending from 0.2 to 10 $\mathrm{M}_\odot$, in agreement with theoretical expectations and previous numerical determinations (Hénon 1975; Giersz & Heggie 1994, 1996; Freitag et al. 2006c). We note that in N-body simulations core collapse is always halted and reversed by the formation and hardening of binaries through close three-body interactions (e.g. Aarseth 1971; Heggie & Hut 2003), a process not included in the MC code. When the mass function is extended to 120 $\mathrm{M}_\odot$, the agreement between MC and N-body simulations is poorer but the time to core collapse is found to be approximately the same, in terms of relaxation time, namely a surprising 10–20 per cent of the initial *central* relaxation time (Spitzer 1987),

$$t_{\rm rc}(0) \equiv 0.339 \frac{\sigma_v^3}{\ln \Lambda\, G^2 \langle m \rangle^2 n}, \tag{5.27}$$

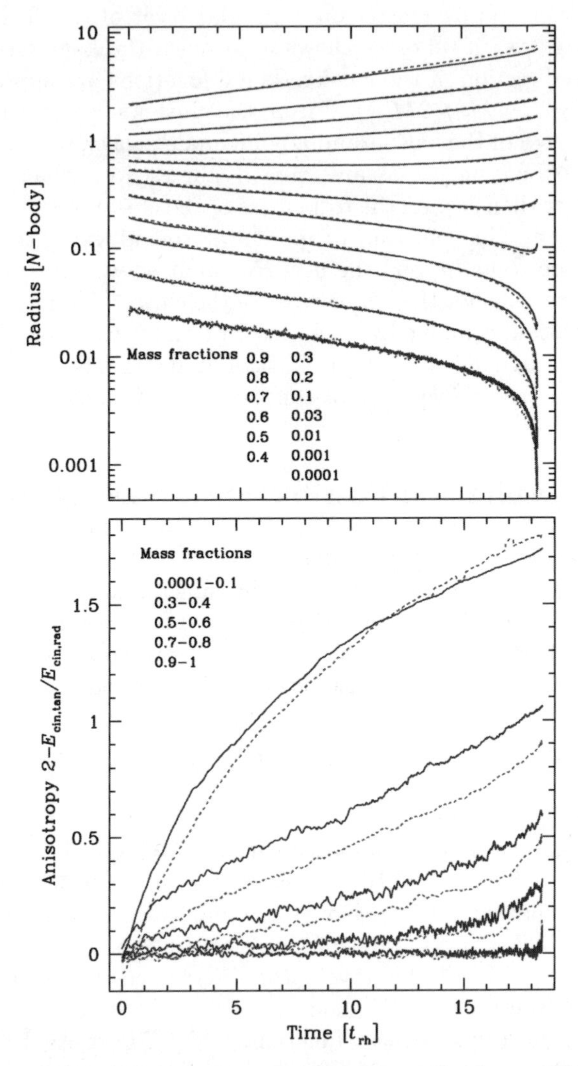

Fig. 5.3. Core collapse of a single-mass cluster initialised as a Plummer model. The results of the MC code using 250 000 particles, in solid lines, are compared to a direct NBODY4 simulation using 64 000 particles, in dashes (H. Baumgardt 2005, personal communication). *Top panel*: evolution of radii of the Lagrangian spheres containing the indicated fraction of the mass. *Bottom panel*: evolution of the anisotropy parameter, averaged over Lagrangian shells bounded by the indicated mass fractions. The length unit is the N-body scale (see Chap. 1). The time unit is the initial half-mass relaxation time (Spitzer 1987). To convert the dynamical time units of the N-body simulation to a relaxation time, a value of $\gamma_c = 0.15$ was used for the Coulomb logarithm

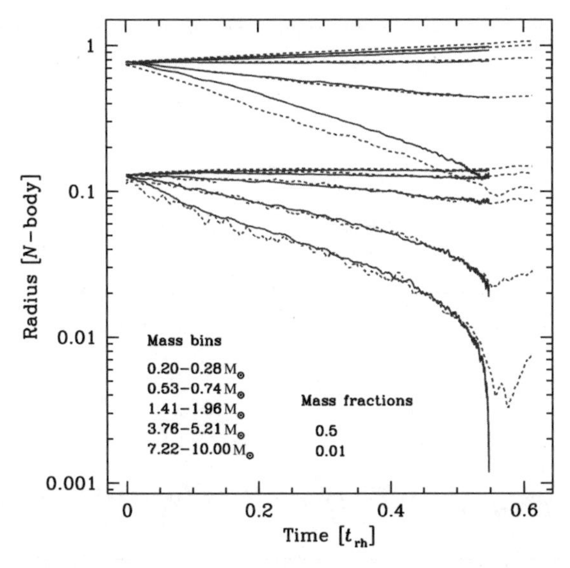

Fig. 5.4. Core collapse of a Plummer cluster with 0.2–10 M$_\odot$ Salpeter mass function. A MC code simulation with 10^6 particles, in solid lines, is compared to a direct NBODY4 simulation with 256 000 particles, in dashes (H. Baumgardt 2005, personal communication). To show mass segregation, the evolution of Lagrangian radii is plotted for mass fractions of 1 and 50 per cent for stars with masses within five different bins. To convert the dynamical time units of the N-body simulation to a relaxation time, a value of $\gamma_c = 0.03$ was used for the Coulomb logarithm. Compare with Fig. 4.1

where the quantities $\langle m \rangle$, n and σ_v are determined at the centre. This is a result of great interest as it raises the possibility of triggering a phase of runaway collisions in young dense clusters (Quinlan & Shapiro 1990; Portegies Zwart et al. 1999; Portegies Zwart & McMillan 2002; Gürkan et al. 2004; Portegies Zwart et al. 2004; Freitag et al. 2006b,c).

A domain where MC simulations are bound to play a unique role in the next few years is the evolution of large clusters with a high fraction of primordial binaries. This is one of the most challenging situations for direct N-body codes because the evolution of regularised binaries cannot be computed on special-purpose GRAPE hardware. At the time of writing, the published N-body simulations tallying the largest number of binaries are those by Hurley et al. (2005) with 12 000 binaries amongst 36 000 stars and by Portegies Zwart et al. (2007) with 13 107 binaries amongst 144 179 stars. In contrast, Fregeau & Rasio (2007) present tens of MC simulations for 10^5 particles, some with 100 per cent binaries and a few 3×10^5 particle cases with up to 1.5×10^5 binaries (see also Gürkan et al. 2006). Although single and binary stellar evolution were not included in these simulations, they can be incorporated into MC codes in the same way and with the same level of realism as in direct N-body

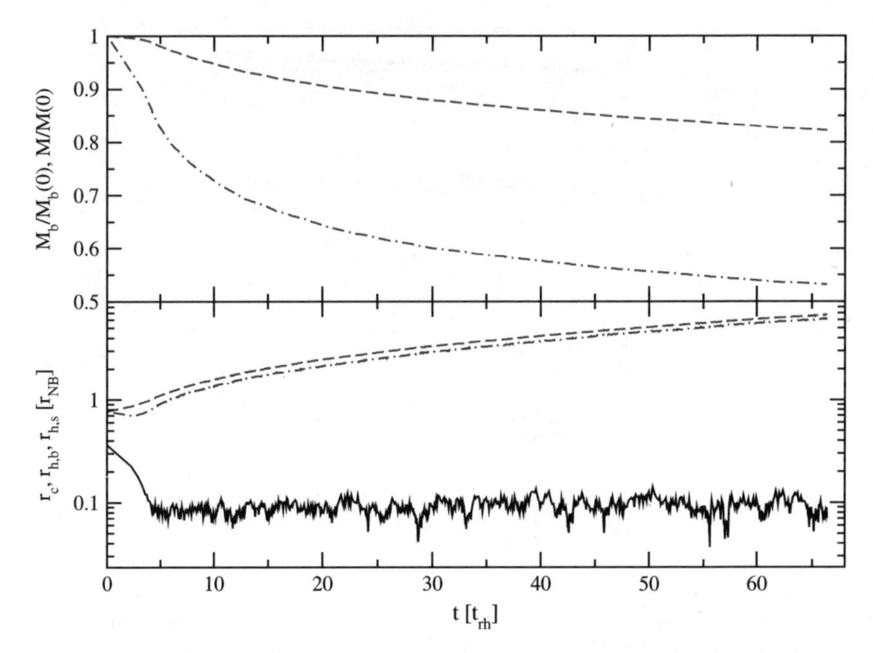

Fig. 5.5. Evolution of a cluster containing 30 per cent of (hard) primordial binaries (J. Fregeau 2007, personal communication). The cluster is set up as a Plummer model of 10^5 particles, with masses distributed according to a Salpeter IMF between 0.2 and $1.2\,\mathrm{M}_\odot$. Stellar evolution is not simulated. *Top panel*: total cluster mass (*dashed line*) and mass in binaries (*dot-dashes*), normalised to the initial values. *Bottom panel*: core radius (*solid line*), half-mass radius of single stars (*dashes*) and half-mass radius of binaries (*dot-dashes*), in N-body units. Time is in units of the initial half-mass relaxation time. For more information on this work, see Fregeau & Rasio (2007)

codes. In Fig. 5.5, I show the results from a simulation of a cluster with 30 per cent primordial binaries, i.e. $N_{\mathrm{bin}}/(N_{\mathrm{bin}} + N_{\mathrm{single}}) = 0.3$ (J. Fregeau 2007, personal communication). Binaries stabilise the core against collapse for a duration of tens of half-mass relaxation times, corresponding to more than the Hubble time when applied to real globular clusters. The quasi-equilibrium size of the core, maintained during this long phase of binary burning appears to be too small to explain the observed core size of most non-collapsed Galactic clusters. It is not yet clear whether this discrepancy is to be blamed on the neglect of stellar evolution and other well-known physical effects (collisions, non-stationary Galactic tides, etc.) or can only be resolved by assuming some more exotic physics such as the presence of IMBHs in many clusters (Baumgardt et al. 2005; Miocchi 2007; Trenti et al. 2007), but it seems that MC simulations are the ideal tool to investigate this issue.

Monte-Carlo codes that treat the dynamics and evolution of single and binary stars in great detail should be available very soon, allowing the simulation of clusters containing up to 10^7 stars on a star-by-star basis with a high level of realism as long as the assumptions of spherical symmetry and dynamical equilibrium are justified. I now mention a few strong motivations to try and extend the realm of MC cluster simulations beyond these assumptions.

- *Galactic tides.* The treatment of stellar evaporation from a cluster can be improved significantly. First, stars have to find the narrow funnels around the Lagrange points to exit the cluster (e.g. Fukushige & Heggie 2000; Ross 2004). Hence it takes a star several dynamical times to find the "exit door", even when some approximate necessary condition for the escape is reached, such as an apocentre distance (in the spherical potential) larger than the distance to the Lagrange point. Therefore, a significant fraction of the stars in a cluster can be potential escapers (Fukushige & Heggie 2000; Baumgardt 2001). Using (semi)analytical prescriptions from the cited studies, one could take this effect into account in MC simulations by giving potential escapers a finite lifetime before they are actually removed from the cluster (see Takahashi & Portegies Zwart 2000 for a similar approach applied to Fokker–Planck simulations). Other important effects of the galactic gravitational field absent from MC simulations (and most other cluster simulations) come from its non-steadiness. A cluster on an eccentric orbit experiences a stronger tidal stress at pericentre, an effect dubbed bulge shocking while compressive disc shocking happens when the cluster crosses the plane of the galactic disc (e.g. Spitzer 1987; Gnedin & Ostriker 1997; Baumgardt & Makino 2003; Dehnen et al. 2004). Such effects can be included in MC codes using the same (semi)analytical prescriptions as in some Fokker–Planck integrations (Gnedin & Ostriker 1997; Gnedin et al. 1999). Alternatively, because shocking occurs on a time-scale much shorter than the relaxation time, we could switch back and forth between a fast, non-collisional N-body algorithm (such as SUPERBOX, see Chap. 6) to compute the effects of the shocks and a MC code to evolve the cluster between shocks. Another possibility would be a hybrid non-spherical MC/N-body method, suggested in the next point.
- *Rotating clusters.* Observational evidence and theoretical models indicate that clusters may be born with significant rotation, possibly as a result of the merger of two clusters (see references in Amaro-Seoane & Freitag 2006). The MC approach exposed here is not appropriate to study non-spherical systems but, as already suggested by Hénon (1971a), it might be possible to develop a hybrid approach where a collisionless N-body code is used for fast orbit sampling in a non-spherical geometry (by actual orbital integration!) and collisional effects are included explicitly in a MC fashion, by realising super-encounters between neighbouring pairs. A combination of the Self-Consistent Field N-body method with Fokker–Planck relaxation terms was developed by S. Sigurdsson to study the evolution

of globular clusters orbiting a galaxy (Johnston et al. 1999) but, to my knowledge, no MC/N-body hybrid has ever been developed. Such a code would also be of great interest in the study of galactic nuclei, as mentioned in Sect. 5.4.2.

- *Primordial gas.* Observations show that when a cluster forms, not more than 30 per cent of the gas is eventually turned into stars (Lada 1999). In relatively small clusters, the gas is expelled by the ionising radiation and winds of OB stars within the first 1–2 Myr. In clusters with an escape velocity larger than about $\sim 10\,\mathrm{km\,s^{-1}}$, complete expulsion of the gas probably only occurs when the first SN explodes (Kroupa et al. 2001; Boily & Kroupa 2003a,b; Baumgardt & Kroupa 2007 and references therein. See also Sect. 7.4). When still present in the cluster, the gas dominates the gravitational potential. Furthermore, it can strongly affect the orbits and mass of stars as they accrete and slow down to conserve momentum, thus shaping the mass function and producing strong segregation (Bonnell et al. 2001a,b; Bonnell & Bate 2002). Such effects can be included in MC codes if the gas is treated as a smooth, parametrised component. However, to follow the reaction of the cluster to the fast gas expulsion, we would have to switch to a (collisionless) N-body code or Spitzer-type dynamical MC scheme because the Hénon algorithm can only treat adiabatic potential evolution.

5.4.2 Galactic Nuclei

In addition to the study of globular and young clusters, the MC code is also a method of choice for the study of small galactic nuclei (Freitag 2001; Freitag & Benz 2001a,b, 2002; Freitag 2003; Freitag et al. 2006a). Massive black holes (MBHs) less massive than about $10^7\,\mathrm{M_\odot}$ are probably generally surrounded by a stellar nucleus with a relaxation time shorter than 10^{10} yr at the distance where the mass in stars is equal to the mass of the MBH (e.g. Lauer et al. 1998; Genzel et al. 2003; Freitag et al. 2006a; Merritt & Szell 2006). Although direct N-body codes with GRAPE hardware can now be used to study some important aspects of the collisional evolution of galactic nuclei (Preto et al. 2004; Merritt & Szell 2006; Merritt et al. 2007b), they are still limited to $\lesssim 10^6$ particles for this kind of application, which falls short of the number of stars in galactic nuclei.

In Fig. 5.6, I show the evolution of a small galactic nucleus, computed with the MC code described in this chapter. In addition to two-body relaxation, the physics include the effects of a (growing) central MBH (tidal disruption, direct mergers for objects too compact to be disrupted) and stellar collisions. Large-angle scatterings were found to be of secondary importance for such systems and stellar evolution can be taken into account, but this raises the question of how much gas from stellar evolution will be accreted by the MBH (Freitag et al. 2006a). For the model presented, segregation of stellar-mass black holes

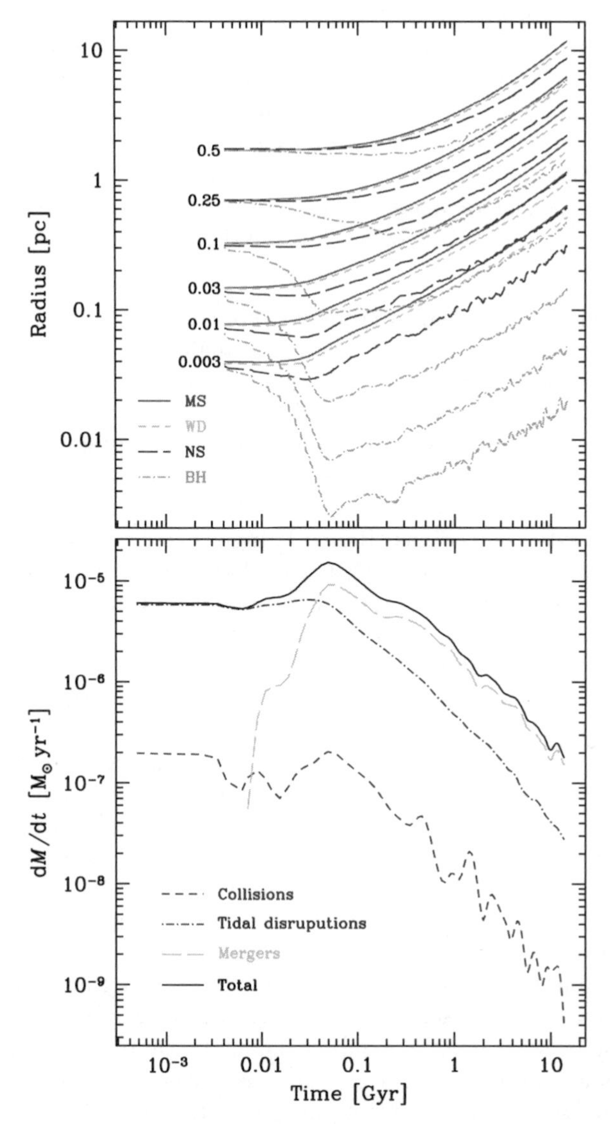

Fig. 5.6. Evolution of the model for a small galactic nucleus hosting a MBH with a mass of $3.5 \times 10^4\,M_\odot$ with 2.1×10^6 particles (model GN84 of Freitag et al. 2006a). *Top panel*: evolution of Lagrangian radii for the various stellar species (MS: main-sequence, WD: white dwarfs, NS: neutron stars, BH: stellar black holes). The stellar population has a fixed age of 10 Gyr. *Bottom panel*: accretion of stellar material by the MBH. For tidal disruptions, 50 per cent of the mass of the star is accreted. "Mergers" are events in which an object crosses the horizon whole. Collisions between MS stars are also taken into account with all the released gas being accreted by the MBH

to the centre occurs within some 50 Myr, after which their swallowing by the MBH drives the expansion of the nucleus. For models with parameters pertaining to the Milky Way nucleus, mass segregation takes about 3–5 Gyr and only little expansion occurs in a Hubble time. The segregation of stellar black holes is of key importance for the formation of EMRI sources for LISA (Hopman & Alexander 2006b; Amaro-Seoane et al. 2007, and references therein).

Simulations of galactic nuclei have not yet reached as high a level of realism as one might wish. Several aspects of the physics are still laking, including the following elements.

- *Binary stars.* Binary stars are probably not effective as a source of heat because the ambient velocity dispersion is so high in galactic nuclei. However, this population is of interest in its own right as mentioned in Sect. 5.3.2.
- *Resonant relaxation.* Close to the MBH, stars travel on approximately fixed Keplerian orbits exerting torques on each other, causing the eccentricities to fluctuate randomly on a time-scale shorter than that of standard two-body relaxation (Rauch & Tremaine 1996). This might affect moderately the rate of tidal disruptions (Rauch & Ingalls 1998) and very significantly that of EMRIs (Hopman & Alexander 2006a) but, being an intrinsically non-local effect, it can probably only be included in an approximate fashion in MC models.
- *Motion of the central MBH.* Direct N-body simulations have established the importance of MBH wandering (e.g. Merritt et al. 2007, and references therein). Because this is a dynamical, non-spherical perturbation to the idealised cluster representation used in the MC approach, it can only be included through ad hoc prescriptions determining, for example, the probability for a star to be tidally disrupted. It is not yet clear whether the wandering would affect the results appreciably and justify such modifications to the MC code.
- *Interplay between accretion disc and stars.* The orbits of stars repeatedly impacting a dense disc tend to align with it (e.g. Syer et al. 1991; Šubr et al. 2004; Miralda-Escudé & Kollmeier 2005). Stars may therefore be a major contributor to nuclear activity and the growth of SMBHs. Testing this idea is challenging since what is required is a numerical scheme coupling stellar dynamics for several millions of stars, disc physics and some prescription for the stellar and orbital evolution of the stars embedded in the disc. A non-spherical hybrid MC/N-body code, as suggested above, could form the backbone of this complex scheme.
- *Binary massive black hole.* Galaxy mergers lead to the formation of massive binaries, the evolution and fate of which is still debated. The key question is whether interactions with stars and gas are efficient at shrinking the binary to the point where it merges by the emission of gravitational waves (Begelman et al. 1980; Merritt & Milosavljević 2005; Berczik et al. 2006; Merritt 2006; Sesana et al. 2007, amongst others). If the binary instead stalls for a very long time, the next galactic merger can bring about a

highly dynamical three-body interaction involving MBHs, likely to lead to a merger and the ejection of a single MBH (Hoffman & Loeb 2007). If the parent galaxies are devoid of gas, once its separation has become smaller than about $\sim 4G\mu/\sigma^2$, where μ is the reduced mass and σ the stellar velocity dispersion, the MBH binary can only shrink by ejecting passing stars out of the nucleus. These interactions also determine the evolution of the eccentricity, which might play a key role in bringing the binary to coalescence. While only N-body methods can implement the non-symmetrical geometry of this situation (e.g. Mikkola & Aarseth 2002), they cannot include the $> 10^7$ stars present in even a moderately small nucleus. An axially symmetrical (hybrid) MC code would make it possible to simulate the interaction of a massive binary with its host nucleus employing a realistic mass ratio between the stars and the MBHs and, hence, the correct rate of relaxation into the loss cone for interaction with the massive binary.

Acknowledgement

It is a pleasure to thank M. Atakan Gürkan and John Fregeau for discussions and comments on a draft of this chapter. I also thank John Fregeau and Holger Baumgardt for providing unpublished simulation results. My work is supported by the STFC rolling grant to the IoA.

References

Aarseth S. J., 1971, Ap&SS, 13, 324
Aarseth S. J., 1974, A&A, 35, 237
Aarseth S. J., 2003, Gravitational N-body Simulations. Cambridge Univ. Press, Cambridge
Aguilar L. A., Merritt D., 1990, ApJ, 354, 33
Alexander T., Hopman C., 2003, ApJ Lett., 590, L29
Amaro-Seoane P., Freitag M., 2006, ApJ Lett., 653, L53
Amaro-Seoane P., Gair J. R., Freitag M., Miller M. C., Mandel I., Cutler C. J., Babak S., 2007, Classical and Quantum Gravity, 24, 113
Ayal S., Livio M., Piran T., 2000, ApJ, 545, 772
Bacon D., Sigurdsson S., Davies M. B., 1996, MNRAS, 281, 830
Bailey V. C., Davies M. B., 1999, MNRAS, 308, 257
Bally J., Zinnecker H., 2005, AJ, 129, 2281
Baumgardt H., 2001, MNRAS, 325, 1323
Baumgardt H., Kroupa P., 2007, MNRAS, 380, 1589
Baumgardt H., Makino J., 2003, MNRAS, 340, 227
Baumgardt H., Makino J., Ebisuzaki T., 2004, ApJ, 613, 1133
Baumgardt H., Makino J., Hut P., 2005, ApJ, 620, 238
Begelman M. C., Blandford R. D., Rees M. J., 1980, Nature, 287, 307
Belczynski K., Kalogera V., Bulik T., 2002, ApJ, 572, 407

Benacquista M. J., 2006, Living Reviews in Relativity, 9, 2
Benz W., Hills J. G., Thielemann, 1989, ApJ, 342, 986
Berczik P., Merritt D., Spurzem R., Bischof H.-P., 2006, ApJ Lett., 642, L21
Binney J., Tremaine S., 1987, Galactic Dynamics. Princeton Univ. Press, Princeton, NJ
Boily C. M., Athanassoula E., 2006, MNRAS, 369, 608
Boily C. M., Kroupa P., 2003a, MNRAS, 338, 665
Boily C. M., Kroupa P., 2003b, MNRAS, 338, 673
Bonnell I. A., Bate M. R., 2002, MNRAS, 336, 659
Bonnell I. A., Bate M. R., Clarke C. J., Pringle J. E., 2001a, MNRAS, 323, 785
Bonnell I. A., Clarke C. J., Bate M. R., Pringle J. E., 2001b, MNRAS, 324, 573
Brown W. R., Geller M. J., Kenyon S. J., Kurtz M. J., 2005, ApJ Lett., 622, L33
Cohn H., Kulsrud R. M., 1978, ApJ, 226, 1087
Dale J. E., Davies M. B., 2006, MNRAS, 366, 1424
Davies M. B., 2002, in van Leeuwen F., Hughes J. D.,Piotto G., eds, ASP Conf. Ser. Vol. 265, Omega Centauri, A Unique Window into Astrophysics. Astron. Soc. Pac., San Francisco, p. 215
Davies M. B., Benz W., Hills J. G., 1991, ApJ, 381, 449
Davies M. B., Benz W., Hills J.G., 1992, ApJ, 401, 246
Davies M. B., Bate M. R., Bonnell I. A., Bailey V. C., Tout C. A., 2006, MNRAS, 370, 2038
Dehnen W., Odenkirchen M., Grebel E. K., Rix H.-W., 2004, AJ, 127, 2753
Diener P., Frolov V. P., Khokhlov A. M., Novikov I. D., Pethick C. J., 1997, ApJ, 479, 164
Einsel C., Spurzem R., 1999, MNRAS, 302, 81
Esquej P., Saxton R. D., Freyberg M. J., Read A. M., Altieri B., Sanchez-Portal M., Hasinger G., 2007, A&A, 462, L49
Fabian A. C., Pringle J. E., Rees M. J., 1975, MNRAS, 172, 15
Ferrarese L., Ford H., 2005, Space Science Reviews, 116, 523
Ferrarese L., Merritt D., 2000, ApJ Lett., 539, L9
Fiestas J., Spurzem R., Kim E., 2006, MNRAS, 373, 677
Frank J., Rees M. J., 1976, MNRAS, 176, 633
Fregeau J. M., Cheung P., Portegies Zwart S. F., Rasio F. A., 2004, MNRAS, 352, 1
Fregeau J. M., Gürkan M. A., Joshi K. J., Rasio F. A., 2003, ApJ, 593, 772
Fregeau J. M., Rasio F. A., 2007, ApJ, 658, 1047
Freitag M., 2000, PhD thesis, Université de Genève
Freitag M., 2001, Classical and Quantum Gravity, 18, 4033
Freitag M., 2003, ApJ Lett., 583, L21
Freitag M., Amaro-Seoane P., Kalogera V., 2006a, ApJ, 649, 91
Freitag M., Benz W., 2001a, in Deiters S., Fuchs B., Just R., Spurzem, R., eds, ASP Conf. Ser. Vol. 228, Dynamics of Star Clusters and the Milky Way. Astron. Soc. Pac., San Francisco, p. 428
Freitag M., Benz W., 2001b, in Kaper L., van den Heuvel E. P. J., Woudt P. A., ESO Astrophysics Symposia, Black Holes in Binaries and Galactic Nuclei. p. 269
Freitag M., Benz W., 2001c, A&A, 375, 711
Freitag M., Benz W., 2002, A&A, 394, 345
Freitag M., Benz W., 2005, MNRAS, 358, 1133
Freitag M., Gürkan M. A., Rasio F. A., 2006b, MNRAS, 368, 141
Freitag M., Rasio F. A., Baumgardt H., 2006c, MNRAS, 368, 121

Fukushige T., Heggie D. C., 2000, MNRAS, 318, 753
Fulbright M. S., 1996, PhD thesis, University of Arizona
Gao B., Goodman J., Cohn H., Murphy B., 1991, ApJ, 370, 567
Genzel R., Schödel R., Ott T., Eisenhauer F., Hofmann R., Lehnert M., Eckart A., Alexander T., Sternberg A., Lenzen R., Clénet Y., Lacombe F., Rouan D., Renzini A., Tacconi-Garman L. E., 2003, ApJ, 594, 812
Gezari S., Martin D. C., Milliard B., Basa S., Halpern J. P., Forster K., Friedman P. G., Morrissey P., Neff S. G., Schiminovich D., Seibert M., Small T., Wyder T. K., 2006, ApJ Lett., 653, L25
Ghez A. M., Salim S., Hornstein S. D., Tanner A., Lu J. R., Morris M., Becklin E. E., Duchêne G., 2005, ApJ, 620, 744
Giersz M., 1998, MNRAS, 298, 1239
Giersz M., 2001, MNRAS, 324, 218
Giersz M., 2006, MNRAS, 371, 484
Giersz M., Heggie D. C., 1994, MNRAS, 268, 257
Giersz M., Heggie D. C., 1996, MNRAS, 279, 1037
Giersz M., Heggie D. C., Hurley J. R., 2008, MNRAS, 388, 429
Giersz M., Spurzem R., 2000, MNRAS, 317, 581
Giersz M., Spurzem R., 2003, MNRAS, 343, 781
Gnedin O. Y., Lee H. M., Ostriker J. P., 1999, ApJ, 522, 935
Gnedin O. Y., Ostriker J. P., 1997, ApJ, 474, 223
Goodman J., 1983, ApJ, 270, 700
Goodman J., Hernquist L., 1991, ApJ, 378, 637
Grindlay J., Portegies Zwart S., McMillan S., 2006, Nature Physics, 2, 116
Gürkan M. A., Fregeau J. M., Rasio F. A., 2006, ApJ Lett., 640, L39
Gürkan M. A., Freitag M., Rasio F. A., 2004, ApJ, 604
Gürkan M. A., Rasio F. A., 2005, ApJ, 628, 236
Heggie D. C., 1975, MNRAS, 173, 729
Heggie D., Hut P., 2003, The Gravitational Million-Body Problem: A Multidisciplinary Approach to Star Cluster Dynamics. Cambridge Univ. Press, Cambridge
Heggie D. C., Hut P., McMillan S. L. W., 1996, ApJ, 467, 359
Hénon M., 1960, Annales d'Astrophysique, 23, 668
Hénon M., 1969, A&A, 2, 151
Hénon M., 1971a, Ap&SS, 14, 151
Hénon M., 1971b, Ap&SS, 13, 284
Hénon M., 1973a, in Martinet L., Mayor M., eds, Lectures of the 3rd Advanced Course of the Swiss Society for Astronomy and Astrophysics. Obs. de Genève, Genève, p. 183
Hénon M., 1973b, A&A, 24, 229
Hénon M., 1975, in Hayli A., ed., Proc. IAU Symp. 69, Dynamics of Stellar Systems. Reidel, Dordrecht, p. 133
Hills J. G., 1975, Nature, 254, 295
Hills J. G., 1988, Nature, 331, 687
Hoffman L., Loeb A., 2007, MNRAS, 334
Hopman C., Alexander T., 2006a, ApJ, 645, 1152
Hopman C., Alexander T., 2006b, ApJ Lett., 645, L133
Hopman C., Freitag M., Larson S. L., 2007, MNRAS, 378, 129
Hurley J. R., Pols O. R., Aarseth S. J., Tout C. A., 2005, MNRAS, 363, 293
Hurley J. R., Pols O. R., Tout C. A., 2000, MNRAS, 315, 543

Hurley J. R., Tout C. A., Aarseth S. J., Pols O. R., 2001, MNRAS, 323, 630
Hut P., 1993, ApJ, 403, 256
Hut P., McMillan S., Goodman J., Mateo M., Phinney E. S., Pryor C., Richer H. B., Verbunt F., Weinberg M., 1992, PASP, 104, 981
Johnston K. V., Sigurdsson S., Hernquist L., 1999, MNRAS, 302, 771
Joshi K. J., Nave C. P., Rasio F. A., 2001, ApJ, 550, 691
Joshi K. J., Rasio F. A., Portegies Zwart S., 2000, ApJ, 540, 969
Kim E., Einsel C., Lee H. M., Spurzem R., Lee M. G., 2002, MNRAS, 334, 310
Kim E., Lee H. M., Spurzem R., 2004, MNRAS, 351, 220
Kim S. S., Lee H. M., 1999, A&A, 347, 123
Kobayashi S., Laguna P., Phinney E. S., Mészáros P., 2004, ApJ, 615, 855
Komossa S., 2005, in Merloni A., Nayakshin S., Sunyaev R. A., eds, Growing Black Holes: Accretion in a Cosmological Context. Springer, Berlin, p. 269
Kroupa P., Aarseth S., Hurley J., 2001, MNRAS, 321, 699
Lada E. A., 1999, in Lada C. J., Kylafis N. D., eds, NATO ASIC Proc. 540, The Origin of Stars and Planetary Systems. Kluwer Academic Publishers, p. 441
Lauer T. R., Faber S. M., Ajhar E. A., Grillmair C. J., Scowen P. A., 1998, AJ, 116, 2263
Laycock D., Sills A., 2005, ApJ, 627, 277
Lee H. M., Ostriker J. P., 1986, ApJ, 310, 176
Lightman A. P., Shapiro S. L., 1977, ApJ, 211, 244
Lin D. N. C., Tremaine S., 1980, ApJ, 242, 789
Lombardi Jr. J. C., Proulx Z. F., Dooley K. L., Theriault E. M., Ivanova N., Rasio F. A., 2006, ApJ, 640, 441
Lombardi J. C., Thrall A. P., Deneva J. S., Fleming S. W., Grabowski P. E., 2003, MNRAS, 345, 762
Lombardi J. C., Warren J. S., Rasio F. A., Sills A., Warren A. R., 2002, ApJ, 568, 939
Merritt D., 2006, ApJ, 648, 976
Merritt D., Berczik P., Laun F., 2007, AJ, 133, 553
Merritt D., Mikkola S., Szell A., 2007b, ApJ, 671, 53
Merritt D., Milosavljević M., 2005, Living Reviews in Relativity, 8, 8
Merritt D., Szell A., 2006, ApJ, 648, 890
Mikkola S., Aarseth S., 2002, Celes. Mech. Dyn. Ast., 84, 343
Miller M. C., Colbert E. J. M., 2004, International J. Modern Phys. D, 13, 1
Miller M. C., Freitag M., Hamilton D. P., Lauburg V. M., 2005, ApJ Lett., 631, L117
Miocchi P., 2007, MNRAS, 381, 103
Miralda-Escudé J., Kollmeier J. A., 2005, ApJ, 619, 30
Muno M. P., Pfahl E., Baganoff F. K., Brandt W. N., Ghez A., Lu J., Morris M. R., 2005, ApJ Lett., 622, L113
O'Leary R. M., Loeb A., 2008, MNRAS, 383, 86
O'Leary R. M., O'Shaughnessy R., Rasio F. A., 2007, Phys. Rev. D, 76, 061504
Perets H. B., Hopman C., Alexander T., 2007, ApJ, 656, 709
Peters P. C., 1964, Phys. Rev., 136, 1224
Peters P. C., Mathews J., 1963, Phys. Rev., 131, 435
Portegies Zwart S. F., Baumgardt H., Hut P., Makino J., McMillan S. L. W., 2004, Nature, 428, 724
Portegies Zwart S. F., Makino J., McMillan S. L. W., Hut P., 1999, A&A, 348, 117

Portegies Zwart S. F., McMillan S. L. W., 2002, ApJ, 576, 899
Portegies Zwart S. F., McMillan S. L. W., Makino J., 2007, MNRAS, 374, 95
Portegies Zwart S. F., Verbunt F., 1996, A&A, 309, 179
Portegies Zwart S. F., Yungelson L. R., 1998, A&A, 332, 173
Press W. H., Teukolsky S. A., Vetterling W. T., Flannery B. P., 1992, Numerical Recipes in FORTRAN. Cambridge Univ. Press, Cambridge
Preto M., Merritt D., Spurzem R., 2004, ApJ Lett., 613, L109
Quinlan G. D., Hernquist L., Sigurdsson S., 1995, ApJ, 440, 554
Quinlan G. D., Shapiro S. L., 1990, ApJ, 356, 483
Rasio F. A., Shapiro S. L., 1991, ApJ, 377, 559
Rauch K. P., Ingalls B., 1998, MNRAS, 299, 1231
Rauch K. P., Tremaine S., 1996, New Astronomy, 1, 149
Rees M. J., 1988, Nature, 333, 523
Regev O., Shara M. M., 1987, MNRAS, 227, 967
Ross S. D., 2004, PhD thesis, Calif. Inst. Technology
Różyczka M., Yorke H. W., Bodenheimer P., Müller E., Hashimoto M., 1989, A&A, 208, 69
Ruffert M., 1993, A&A, 280, 141
Schödel R., Eckart A., Alexander T., Merritt D., Genzel R., Sternberg A., Meyer L., Kul F., Moultaka J., Ott T., Straubmeier C., 2007, A&A, 469, 125
Sedgewick R., 1988, Algorithms. Second Edition. Addison-Wesley
Sesana A., Haardt F., Madau P., 2007, ApJ, 660, 546
Shapiro S. L., 1985, in Goodman J., Hut P., eds, Proc. IAU Symp. 113, Dynamics of Star Clusters. Reidel, Dordrecht, p. 373
Shara M., ed., 2002, ASP Conf. Ser. 263, Stellar Collisions & Mergers and their Consequences. Astron. Soc. Pac., San Francisco
Shara M. M., Hurley J. R., 2002, ApJ, 571, 830
Sills A., Adams T., Davies M. B., Bate M. R., 2002, MNRAS, 332, 49
Sills A., Deiters S., Eggleton P., Freitag M., Giersz M., Heggie D., Hurley J., Hut P., Ivanova N., Klessen R. S., Kroupa P., Lombardi J. C., McMillan S., Portegies Zwart S. F., Zinnecker H., 2003, New Astron., 8, 605
Sills A., Faber J. A., Lombardi J. C., Rasio F. A., Warren A. R., 2001, ApJ, 548, 323
Sills A., Lombardi J. C., Bailyn C. D., Demarque P., Rasio F. A., Shapiro S. L., 1997, ApJ, 487, 290
Spitzer L., 1987, Dynamical Evolution of Globular Clusters. Princeton Univ. Press, Princeton, NJ
Spitzer L. J., Hart M. H., 1971a, ApJ, 164, 399
Spitzer L. J., Hart M. H., 1971b, ApJ, 166, 483
Spitzer L. J., Thuan T. X., 1972, ApJ, 175, 31
Spitzer L., Mathieu R. D., 1980, ApJ, 241, 618
Spitzer L., Shull J. M., 1975, ApJ, 201, 773
Spurzem R., Giersz M., Heggie D. C., Lin D. N. C., 2006, preprint (astro-ph/0612757)
Stodołkiewicz J. S., 1982, Acta Astron., 32, 63
Stodołkiewicz J. S., 1985, in Goodman J., Hut P., eds, Proc. IAU Symp. 113, Dynamics of Star Clusters. Reidel, Dordrecht, p. 361
Stodołkiewicz J. S., 1986, Acta Astron., 36, 19
Šubr L., Karas V., Huré J.-M., 2004, MNRAS, 354, 1177

Syer D., Clarke C. J., Rees M. J., 1991, MNRAS, 250, 505
Taam R. E., Ricker P. M., 2006, preprint (astro-ph/0611043)
Takahashi K., Portegies Zwart S. F., 2000, ApJ, 535, 759
Theis C., Spurzem R., 1999, A&A, 341, 361
Trac H., Sills A., Pen U.-L., 2007, MNRAS, 337
Tremaine S., Gebhardt K., Bender R., Bower G., Dressler A., Faber S. M., Filippenko A. V., Green R., Grillmair C., Ho L. C., Kormendy J., Lauer T. R., Magorrian J., Pinkney J., Richstone D., 2002, ApJ, 574, 740
Trenti M., Ardi E., Mineshige S., Hut P., 2007, MNRAS, 374, 857
van der Marel R. P., 2004, in Ho, L., ed., Coevolution of Black Holes and Galaxies, from the Carnegie Observatories Centennial Symposia. Cambridge Univ. Press, Cambridge, p. 37
Young P. J., 1980, ApJ, 242, 1232

6

Particle-Mesh Technique and Superbox

Michael Fellhauer

University of Cambridge, Institute of Astronomy, Madingley Road, Cambridge CB3 0HA, UK
madf@ast.cam.ac.uk

6.1 Introduction

Many problems in astronomy ranging from celestial mechanics via stellar dynamics to cosmology require the solution of Newton's laws:

$$\boldsymbol{F} = \boldsymbol{a} \cdot m \;=\; m\frac{\mathrm{d}\boldsymbol{v}}{\mathrm{d}t} \tag{6.1}$$

$$\boldsymbol{v} = \frac{\mathrm{d}\boldsymbol{r}}{\mathrm{d}t}, \tag{6.2}$$

where $\boldsymbol{F}$ is the gravitational force of all other $(N-1)$ masses,

$$\boldsymbol{F}_j = \sum_{i=1, i\neq j}^{N} \frac{G m_j m_i}{r_{ij}^3} \boldsymbol{r_{ij}} \tag{6.3}$$

acting on mass j (index ij denotes the vectors connecting particle i and j).

While there is an analytical solution for the two-body system, systems involving three or more masses do not have an analytical solution. Thus computer simulations of the time-evolution of multi-body systems are very common in astronomy.

The tools used for these purposes are diverse and widely range from high-precision integrators for the dynamics of the planetary systems to programmes using up to a billion particles to investigate the structure formation in the universe. This article focuses on the particle-mesh technique and a programme to simulate galaxies called SUPERBOX.

The particle-mesh (PM) technique is explained in Sect. 6.2. Then the multi-grid structure of SUPERBOX is described in Sect. 6.3.

Fellhauer, M.: *Particle-Mesh Technique and SUPERBOX*. Lect. Notes Phys. **760**, 159–169 (2008)
DOI 10.1007/978-1-4020-8431-7_6

6.2 Particle-Mesh Technique

6.2.1 Overview

In the particle-mesh technique, the density of the particles is sampled on a grid covering the simulation area and then Poisson's equation

$$\nabla^2 \Phi = 4\pi G \varrho \tag{6.4}$$

is solved on the grid-based density using a suitable Green's function to derive the grid-based gravitational potential. Particles are integrated using the forces derived from this grid-based potential.

The first step is to locate the grid-point of each particle according to its position and derive a grid of densities. This density-grid is Fourier-transformed via the Fast Fourier Transform(FFT) algorithm. This requires that the number of grid-cells per dimension is a power of 2. The Fourier-transformed density-grid is multiplied cell-by-cell with a suitable, already Fourier-transformed Green's function. Then these values are back-transformed, which results in a grid of potential values. From these potential values the forces of each particle are derived via discrete differentiation. Finally, the particle velocities and positions are integrated forward in time.

A flow-chart of a standard PM-code is shown in Fig. 6.1.

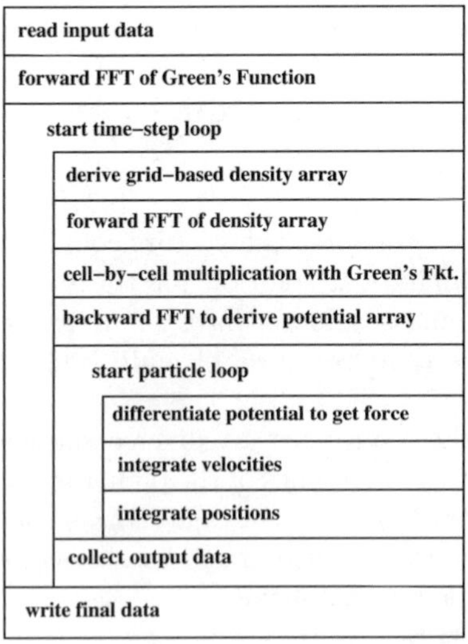

- read input data
- forward FFT of Green's Function
- start time–step loop
  - derive grid–based density array
  - forward FFT of density array
  - cell–by–cell multiplication with Green's Fkt.
  - backward FFT to derive potential array
  - start particle loop
    - differentiate potential to get force
    - integrate velocities
    - integrate positions
  - collect output data
- write final data

Fig. 6.1. Flow-chart of a standard PM-code

6.2.2 Suitable Green's Function

The usual geometry of the grid in a particle-mesh code is Cartesian and cubic. Therefore, the standard Green's function, which describes the distances between cells, looks like

$$\begin{aligned} H_{i,j,k} &= \frac{1}{\sqrt{i^2+j^2+k^2}}, \quad i,j,k = 0,\ldots,n \\ H_{0,0,0} &= \frac{1}{\xi}. \end{aligned} \tag{6.5}$$

This formula implies that the length of one grid-cell is unity, n is the number of grid-cells per dimension and has to be a power of 2.

The value for $H_{0,0,0}$ has to be chosen carefully. It describes the strength of the force between particles in the same cell, including the non-physical 'self-gravity' of the particle acting on itself. In the one-dimensional case, analytical studies by D. Pfenniger showed a value of $\xi = 3/4$ gives the best results in terms of energy conservation. Numerical experiments showed that this is also true in the three-dimensional case.

Nevertheless, in the case of very low particle numbers per cell this value could lead to spurious self-accelerations and a value that excludes the forces of particles from the same cell would be more suitable. In the SUPERBOX differentiation scheme the value to exclude self-gravity is $\xi = 1$. In a later section, we discuss why one should avoid low particle-per-cell ratios if possible.

Finally, it can be stated that the grid-array of the Green's function has to be set up and Fourier-transformed only once at the beginning of each simulation and can then be used throughout the whole simulation.

6.2.3 Deriving the Density-Grid

The actual positions and velocities of each particle (x, y, z, v_x, v_y, v_z) are stored in the particle array. From the actual positions the grid-cell in which each particle is located is derived via

$$i_x = \text{nearest integer}(\text{enh} \cdot x) + n/2. \tag{6.6}$$

i_x denotes the grid-cell number in the x-direction, *enh* is a numerical factor that stretches or compresses the physical extension of the x-direction of the simulation area to allow the grid-cell length to be unity. The grid-cell numbers in the y- and z-direction are derived accordingly.

There are two possibilities to assign the mass of the particle to the density-grid covering the simulation area. One is called nearest-grid-point scheme, and assigns the whole mass of the particle to the grid-cell that the particle is in. A second, more advanced procedure is called cloud-in-cell scheme and assigns a radius of half a cell length to each particle. The mass of the particle is now distributed to the cells, this extended particle is in, according to the actual

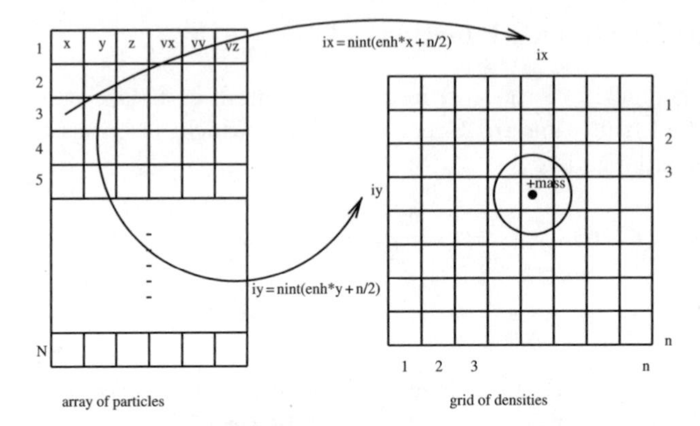

Fig. 6.2. Deriving the density-grid from the particle positions. The z-dimension is omitted for clarity. In the NGP scheme, the total mass is placed in one cell; in the CIC scheme, contributions of the mass are distributed in neighbouring cells also (denoted by the circle)

deviation of the particle position from the centre of the cell. In Fig. 6.2 this assignment is shown for two dimensions.

The CIC scheme allows for a much smoother distribution of the densities but does not allow for sub-cell-length resolution. This has to be added via direct summation of the forces of neighbouring particles within a certain sphere of influence. A code that employs direct summation in the vicinity of each particle is usually called P^{3}M-code (particle-particle particle-mesh). The CIC scheme also allows for a smooth and high accuracy derivation of the forces (this will be discussed in a sub-section below).

SUPERBOX still uses the 'old-fashioned' NGP-scheme, which results in a much faster assignment of the densities and allows for sub-cell-length resolution, if $H_{0,0,0} \neq 1$. To reach the high accuracy, we later apply a higher-order differentiation scheme to obtain the forces.

6.2.4 The FFT-Algorithm

Poisson's equation is solved for the density-grid to get the grid-based potential, Φ_{ijk}, which becomes,

$$\Phi_{ijk} = G \sum_{a,b,c=0}^{n-1} \varrho_{abc} \cdot H_{a-i,b-j,c-k}, \quad i,j,k = 0,\ldots,n-1, \tag{6.7}$$

where n denotes the number of grid-cells per dimension ($n^3 = N_{\rm gc}$ total number of grid-cells), and H_{ijk} is the Green's function. To avoid this $N_{\rm gc}^2$ procedure, the discrete Fast Fourier Transform (FFT) is used, for which $n = 2^k$, $k > 0$ being an integer. The stationary Green's function is Fourier-transformed once at the beginning of the calculation, and only the density array is transformed at each time-step:

$$\hat{\varrho}_{abc} = \sum_{i,j,k=0}^{n-1} \varrho_{ijk} \cdot \exp\left(-\sqrt{-1}\,\frac{2\pi}{n}\,(ai+bj+ck)\right)$$

$$\hat{H}_{abc} = \sum_{i,j,k=0}^{n-1} H_{ijk} \cdot \exp\left(-\sqrt{-1}\,\frac{2\pi}{n}\,(ai+bj+ck)\right). \tag{6.8}$$

The two resulting arrays are multiplied cell by cell and transformed back to get the grid-based potential,

$$\Phi_{ijk} = \frac{G}{n^3} \sum_{a,b,c=0}^{n-1} \hat{\varrho}_{abc} \cdot \hat{H}_{abc} \cdot \exp\left(\sqrt{-1}\,\frac{2\pi}{n}\,(ai+bj+ck)\right). \tag{6.9}$$

The FFT-algorithm gives the exact solution of the grid-based potential for a periodic system. For the exact solution of an isolated system, which is what simulators are interested in, the size of the density array has to be doubled ($2n$), filling all inactive grid cells with zero density and extending the Green's function in the empty regions in the following way (also shown in Fig. 6.3):

$$\begin{aligned} H_{2n-i,j,k} &= H_{2n-i,2n-j,k} = H_{2n-i,j,2n-k} = H_{2n-i,2n-j,2n-k} \\ &= H_{i,2n-j,k} = H_{i,2n-j,2n-k} = H_{i,j,2n-k} = H_{i,j,k}. \end{aligned} \tag{6.10}$$

This provides the *isolated solution* of the potential in the simulated area between $i, j, k = 0$ and $n-1$. In the inactive part the results are unphysical. To keep the data size as small as possible, only a $2n \times 2n \times n$-array is used for transforming the densities, and a $(n+1) \times (n+1) \times (n+1)$-array is used for the Green's function. For a detailed discussion see Eastwood & Brownrigg (1978) and also Hockney & Eastwood (1981).

The FFT-routine incorporated in SUPERBOX is a simple one-dimensional FFT and is taken from Werner & Schabach (1979) and Teukolsky et al. (1992). It is fast and makes the code portable and not machine-specific. The low-storage algorithm for extending the FFT to three dimensions, to obtain the 3-D potential, is taken from Hohl (1970). The performance of SUPERBOX can be increased by incorporating machine-optimised FFT routines.

A detailed description of the low-storage FFT algorithm used in SUPERBOX can be found in the manual available directly from the author (Fellhauer 2006).

6.2.5 Derivation of the Forces

After the FFT procedure has been completed, one has a grid-based potential of the simulation area. From this potential, the forces acting on each particle are derived via discrete numerical differentiation of the potential.

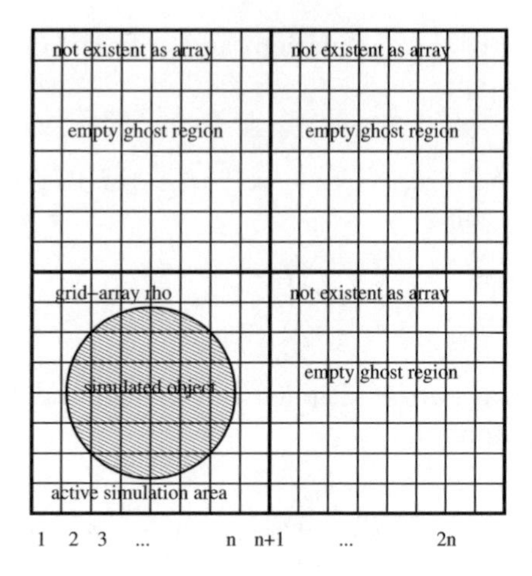

Fig. 6.3. Virtual extension of the simulation area to provide isolated solution (z-direction omitted)

As with the mass assignment of the density array, the forces are also calculated differently depending on whether a NGP or CIC scheme is used. A NGP scheme only uses the force calculated for the grid-cell the particle is in, while in a CIC scheme forces of the neighbouring cells are used with the same weights the mass was distributed to interpolate the force to the particle position.

For simplicity, the force derivation of the different schemes is given in a 1D case:

$$\text{NGP} : a(x_i + \mathrm{d}x) = \left.\frac{\partial \Phi}{\partial x}\right|_i \tag{6.11}$$

$$\textsf{SUPERBOX} : a(x_i + \mathrm{d}x) = \left.\frac{\partial \Phi}{\partial x}\right|_i + \left.\frac{\partial^2 \Phi}{\partial x^2}\right|_i \cdot \frac{\mathrm{d}x}{\Delta x} \tag{6.12}$$

$$\text{CIC} : a(x_i + \mathrm{d}x) = \left.\frac{\partial \Phi}{\partial x}\right|_i \cdot \frac{\Delta x - \mathrm{d}x}{\Delta x} + \left.\frac{\partial \Phi}{\partial x}\right|_{i+1} \cdot \frac{\mathrm{d}x}{\Delta x}, \tag{6.13}$$

where a denotes the acceleration, x_i is the position of the cell with index i the particle is located in and $\mathrm{d}x$ is the deviation of the particle from the centre of the cell. As one can see, the standard NGP scheme does not account for the deviation of the particle from the centre of the cell. The acceleration is a step function from cell to cell and is not steady at all. The CIC scheme accounts for this deviation and the acceleration of the particle is a weighted mean from the cell the particle is in and the neighbouring cell. SUPERBOX has a non-standard force calculation scheme, which is definitely NGP in nature (only the force for the cell i is used), but accounts for the deviation by using the next term of a Taylor series of the acceleration around the cell i. The steadiness of the force is not guaranteed when crossing the cell boundaries at an arbitrary angle, but anisotropies of the force are suppressed. The full 3D expression for the acceleration in SUPERBOX is

$$a_{ijk,x}(\mathrm{d}x, \mathrm{d}y, \mathrm{d}z) = \left.\frac{\partial \Phi}{\partial x}\right|_{i,j,k} + \left.\frac{\partial^2 \Phi}{\partial x^2}\right|_{i,j,k} \mathrm{d}x + \left.\frac{\partial^2 \Phi}{\partial x \partial y}\right|_{i,j,k} \mathrm{d}y + \left.\frac{\partial^2 \Phi}{\partial x \partial z}\right|_{i,j,k} \mathrm{d}z \tag{6.14}$$

The partial derivatives are replaced in the code by second-order central differentiation quotients and now the 3D expression for the acceleration in the x-direction reads

$$\begin{aligned} a_{ijk,x}(\mathrm{d}x, \mathrm{d}y, \mathrm{d}z) = & \frac{\Phi_{i+1,jk} - \Phi_{i-1,jk}}{2\Delta x} \\ & + \frac{\Phi_{i+1,jk} + \Phi_{i-1,jk} - 2 \cdot \Phi_{ijk}}{(\Delta x)^2} \cdot \mathrm{d}x \\ & + \frac{\Phi_{i+1,j+1,k} - \Phi_{i-1,j+1,k} + \Phi_{i-1,j-1,k} - \Phi_{i+1,j-1,k}}{4\Delta x \Delta y} \cdot \mathrm{d}y \\ & + \frac{\Phi_{i+1,j,k+1} - \Phi_{i-1,j,k+1} + \Phi_{i-1,j,k-1} - \Phi_{i+1,j,k-1}}{4\Delta x \Delta z} \cdot \mathrm{d}z \end{aligned} \tag{6.15}$$

Note that generally $\Delta x = \Delta y = \Delta z = 1$, i.e. the cell-length is assumed to be equal along the three axes and unity; i, j, k are the cell indices of the particle in the three Cartesian coordinates. The accelerations in y- and z-direction are calculated analogously.

6.2.6 Integrating the Particles

The orbits of the particles are integrated forward in time using the leapfrog scheme. For example, for the x-components of the velocity, v_x, and position, x, vectors of particle l,

$$\begin{aligned} v_{x,l}^{n+1/2} &= v_{x,l}^{n-1/2} + a_{x,l}^{n} \cdot \Delta t \\ x_l^{n+1} &= x_l^{n} + v_{x,l}^{n+1/2} \cdot \Delta t, \end{aligned} \tag{6.16}$$

where n denotes the nth time step and Δt is the length of the integration step.

SUPERBOX uses a fixed global time step, i.e. the time step is the same for all particles and does not vary in time.

The leapfrog integrator together with the fixed time step is very fast (no decision-making necessary) and is accurate enough for a grid-based code. It is in principle time-reversible and has very good energy conservation properties considering its simplicity.

6.3 Multi-Grid Structure of SUPERBOX

A detailed description of the code is also found in Fellhauer et al. (2000). For each galaxy, five grids with three different resolutions are used. This is made possible by invoking the additivity of the potential (Fig. 6.4).

The five grids are as follows:

- Grid 1 is the high-resolution grid that resolves the centre of the galaxy. It has a length of $2 \times R_{\rm core}$ in one dimension. In evaluating the densities, all particles of the galaxy within $r \leq R_{\rm core}$ are stored in this grid.
- Grid 2 has an intermediate resolution to resolve the galaxy as a whole. The length is $2 \times R_{\rm out}$, but only particles with $r \leq R_{\rm core}$ are stored here, i.e. the same particles as are also stored in grid 1.
- Grid 3 has the same size and resolution as grid 2, but it contains only particles with $R_{\rm core} < r \leq R_{\rm out}$.
- Grid 4 has the size of the whole simulation area (i.e. 'local universe' with $2 \times R_{\rm system}$), and has the lowest resolution. It is fixed. Only particles of the galaxy with $r \leq R_{\rm out}$ are stored in grid 4.

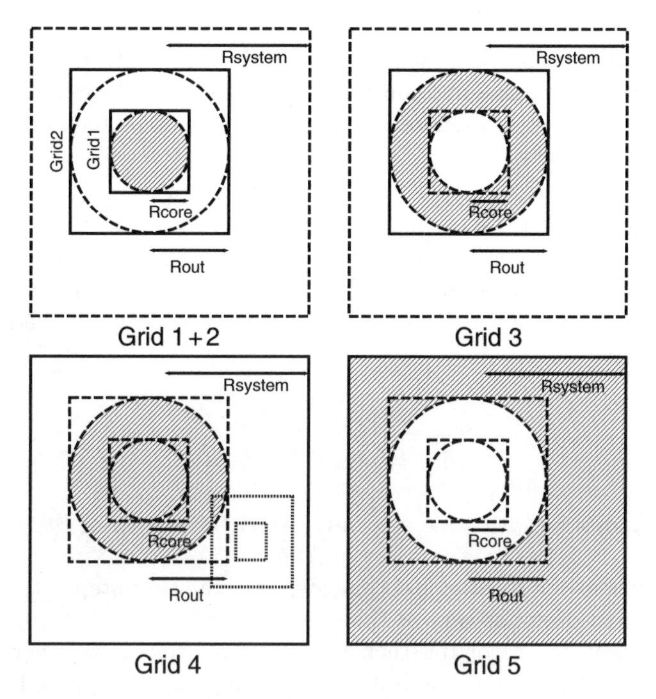

Fig. 6.4. The five grids of SUPERBOX. In each panel, solid lines highlight the relevant grid. Particles are counted in the shaded areas of the grids. The lengths of the arrows are $(N/2)-2$ grid-cells (see text). In the bottom left panel, the grids of a hypothetical second galaxy are also shown as dotted lines

- Grid 5 has the same size and resolution as grid 4. This grid treats the escaping particles of a galaxy, and contains all particles with $r > R_{\text{out}}$.

Grids 1 to 3 are focused on a common centre of the galaxy and move with it through the 'local universe', as detailed below. All grids have the same number of cells per dimension, n, for all galaxies. The boundary condition, requiring two empty cells with $\varrho = 0$ at each boundary, is open and non-periodic, thus providing an isolated system. This however means that only $n - 4$ active cells per dimension are used.

To keep the memory requirement low, all galaxies are treated consecutively in the same grid-arrays, whereby the particles belonging to different galaxies can have different masses. Each of the five grids has its associated potential Φ_{i}, $i = 1, 2, \ldots, 5$ computed by the PM technique from the particles of one galaxy located as described above. The accelerations are obtained additively from the five potentials of each galaxy in turn in the following way:

$$
\begin{aligned}
\Phi(r) &= [\theta(R_{\text{core}} - r) \cdot \Phi_1 + \theta(r - R_{\text{core}}) \cdot \Phi_2 + \Phi_3] \cdot \theta(R_{\text{out}} - r) \\
&\quad + \theta(r - R_{\text{out}}) \cdot \Phi_4 + \Phi_5 \\
\Phi(R_{\text{core}}) &= \Phi_1 + \Phi_3 + \Phi_5 \\
\Phi(R_{\text{out}}) &= \Phi_2 + \Phi_3 + \Phi_5,
\end{aligned}
\tag{6.17}
$$

where $\theta(\xi) = 1$ for $\xi > 0$ and $\theta(\xi) = 0$ otherwise. This means:

- For a particle in the range $r \leq R_{\text{core}}$, the potentials of grids 1, 3 and 5 are used to calculate the acceleration.
- For a particle with $R_{\text{core}} < r \leq R_{\text{out}}$, the potentials of grids 2, 3 and 5 are combined.
- And finally, if $r > R_{\text{out}}$, the acceleration is calculated from the potentials of grids 4 and 5.
- Any particle with $r > R_{\text{system}}$ is removed from the computation.

Due to the additivity of the potential (and hence its derivatives, the accelerations) the velocity changes originating from the potentials of each of the galaxies can be separately updated and accumulated in the first of the leapfrog formulae (6.16). The final result does not depend on the order by which the galaxies are taken into account and it could be computed even in parallel, if a final accumulation takes place. After all velocity changes have been applied to all galaxies, the positions of the particles are finally updated.

As long as the galaxies are well separated, they feel only the low-resolution potentials of the outer grids. But as the galaxies approach each other, their high-resolution grids overlap, leading to a high-resolution force calculation during the interaction.

6.3.1 Grid Tracking

Two alternative schemes to position and track the inner and middle grids can be used. The most useful scheme centres the grids on the density maximum of each galaxy at each step. The position of the density maximum is found by constructing a sphere of neighbours centred on the densest region, in which the centre of mass is computed. This is performed iteratively. The other option is to centre the grids during run-time on the position of the centre of mass of each galaxy using all its particles remaining in the computation.

6.3.2 Edge-Effects

It can be seen in Fig. 6.4 that only spherical regions of the cubic grids contain particles (except for grid 5). Particles with eccentric orbits can cross the border of two grids, thus being subject to forces resolved differently. No interpolation of the forces is done at the grid boundaries. This keeps the code fast and slim, but the grid sizes have to be chosen properly in advance to minimise the boundary discontinuities. It leads to some additional but negligible relaxation effects, because the derived total potential has insignificant discontinuities at the grid boundaries (Wassmer 1992). The best way to avoid these edge-effects is to place the grid boundaries at 'places' where the slope of the potential is not steep.

6.3.3 Choice of Parameters

Finally, we make some comments on the right choice of parameters. In principle, SUPERBOX works with all sets of parameters, but the outcome might be unphysical. The user has to check if the choice makes sense or not. There are a few rules that help to ensure that the simulation is not unrealistic. First, one should check if there are enough particles for the given resolution. As a rule-by-thumb, one can divide the number of particles by the total number of cells of one grid. If the mean number of particles per cell amounts to a few, then one is on the safe side (conservative $< N > \approx 10-15$). Second, one should check the time-step. Particles should not travel much more than one grid-cell per time step, otherwise one again loses resolution. Another rule-by-thumb is: take the shortest crossing-time of all objects and divide it by 10 (conservative: 50–70). This ensures that this object stays stable. It is also not useful to have large resolution steps between the grid levels. At least one should avoid them in all places of interest.

References

Eastwood J. W., Brownrigg D. R. K., 1978, J. Comput. Phys., 32, 24

Fellhauer M., 2006, Superbox manual, madf@ast.cam.ac.uk

Fellhauer M., Kroupa P., Baumgardt H., Bien R., Boily C. M., Spurzem R., Wassmer N., 2000, NewA, 5, 305

Hockney R. W., Eastwood J. W., 1981, Computer Simulations Using Particles. McGraw-Hill

Hohl F., 1970, NASA Technical Report, R-343
Teukolsky S. A., Vetterling W. T., Flannery B. P., 1992, Numerical Recipes in Fortran. Cambridge University Press, Cambridge
Wassmer N., 1992, Diploma thesis, University Heidelberg
Werner H., Schabach R., 1979, Praktische Mathematik II. Springer

7

Dynamical Friction

Michael Fellhauer

University of Cambridge, Institute of Astronomy, Madingley Road, Cambridge CB3 0HA, UK
madf@ast.cam.ac.uk

7.1 What is Dynamical Friction?

Dynamical friction is, as the name says, a deceleration of massive objects. It occurs whenever a massive object travels through another extended object. This behaviour makes dynamical friction one of the most important effects in stellar dynamics.

It occurs on all kinds of length-scales and objects from the sinking to the centre of massive stars inside a star cluster, leading to mass segregation, via sinking of star clusters and dwarf galaxies inside the host galaxy to collisions of massive galaxies.

Dynamical friction is a pure gravitational interaction between the massive object (M) and the multitude of lighter stars (m) of the extended object it is travelling through (see Fig. 7.1, left panel). In the rest-frame of the moving object M, the lighter stars are oncoming from the front and get deflected behind the object (see Fig. 7.1, middle panel). These many gravitational interactions sum up to an effective deceleration of the object, while some of the deflected lighter particles m build up a wake behind M (see Fig. 7.1, right panel). This wake can be measured and may induce an extra drag on the moving object, but the drag is neglected in the determination of the standard description of dynamical friction. It is dynamical friction which causes the wake and *not* the wake being responsible for the dynamical friction!

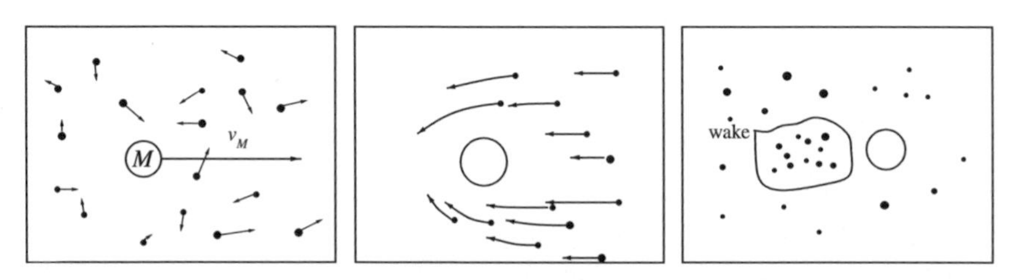

Fig. 7.1. Dynamical friction as a cartoon

Fellhauer, M.: *Dynamical Friction*. Lect. Notes Phys. **760**, 171–179 (2008)
DOI 10.1007/978-1-4020-8431-7_7

Hence dynamical friction causes a deceleration of the object M and therefore, if it was on a stable orbit before, causes a shrinking of this orbit and sinking to the centre in response to the deceleration. If the object is initially on an eccentric orbit, dynamical friction acts in a way that the orbit gets more and more circular.

7.2 How to Quantify Dynamical Friction?

Dynamical friction was first quantified by Chandrasekhar (1943). In this section, the classical way to derive the dynamical friction formula will be followed (see for example Binney & Tremaine 1987, chapter 7.1).

Before the multitude of encounters can be treated, one has to focus on a single encounter. The geometry of this encounter is shown in the left panel of Fig. 7.2. Defining $\boldsymbol{r} = \boldsymbol{x}_m - \boldsymbol{x}_M$ as the separation vector between m and M and $\boldsymbol{V} = \dot{\boldsymbol{r}}$, one gets the relative velocity change

$$\Delta \boldsymbol{V} = \Delta \boldsymbol{v}_m - \Delta \boldsymbol{v}_M. \tag{7.1}$$

Because this two-body system is conservative, one can apply momentum conservation, which leads to

$$m \Delta \boldsymbol{v}_m + M \Delta \boldsymbol{v}_M = 0. \tag{7.2}$$

Combining these two equations and eliminating $\Delta \boldsymbol{v}_m$ gives $\Delta \boldsymbol{v}_M$ as a function of $\Delta \boldsymbol{V}$:

$$\Delta \boldsymbol{v}_M = - \left(\frac{m}{m+M} \right) \Delta \boldsymbol{V}. \tag{7.3}$$

In the right panel of Fig. 7.2 we show the hyperbolic geometry of the Kepler problem in the frame of the reduced particle mass travelling in the combined potential due to both particles $(m + M)$. The conserved angular momentum

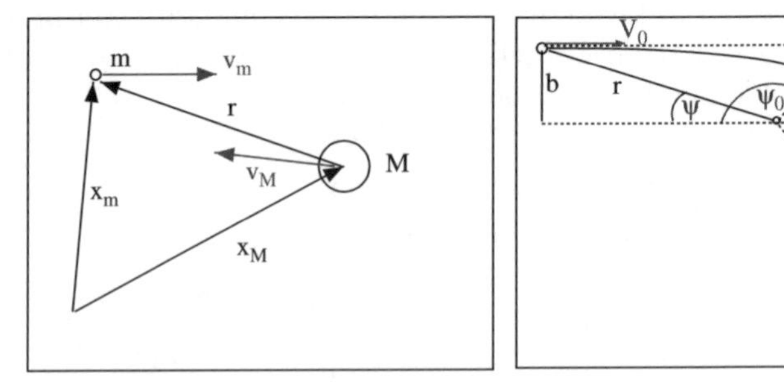

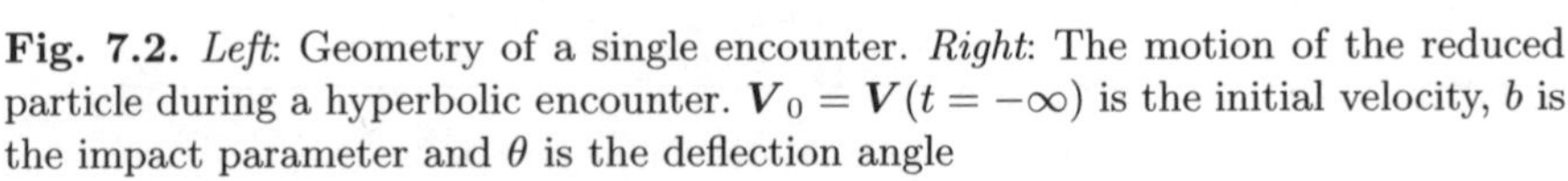
Fig. 7.2. *Left*: Geometry of a single encounter. *Right*: The motion of the reduced particle during a hyperbolic encounter. $\boldsymbol{V}_0 = \boldsymbol{V}(t = -\infty)$ is the initial velocity, b is the impact parameter and θ is the deflection angle

(per unit mass) in this system is $L = bV_0 = r^2\dot{\Psi}$. From the analytical solution of the Kepler problem, we know the equation that relates radius r and azimuthal angle Ψ:

$$\frac{1}{r} = C\cos(\Psi - \Psi_0) + \frac{G(m+M)}{b^2V_0^2}, \tag{7.4}$$

where C and Ψ_0 are constants defined by the initial conditions. If (7.4) is differentiated with respect to time, one gets

$$\frac{\mathrm{d}r}{\mathrm{d}t} = Cr^2\dot{\Psi}\sin(\Psi - \Psi_0) \;=\; CbV_0\sin(\Psi - \Psi_0). \tag{7.5}$$

Evaluating (7.4) and (7.5) at $t = -\infty$ one obtains

$$0 = C\cos(\Psi_0) + \frac{G(m+M)}{b^2V_0^2} \tag{7.6}$$

$$-V_0 = CbV_0\sin(-\Psi_0). \tag{7.7}$$

Using these two equations to eliminate C leads to

$$\tan(\Psi_0) = -\frac{bV_0^2}{G(m+M)}. \tag{7.8}$$

The point of closest approach is reached when $\Psi = \Psi_0$ and, since the orbit is symmetrical about this point, the deflection angle is $\theta = 2\Psi_0 - \pi$. By conservation of energy, the length of the relative velocity vector is the same before and after the encounter and has the value V_0. Hence the components $\Delta\boldsymbol{V}_\parallel$ and $\Delta\boldsymbol{V}_\perp$ of $\Delta\boldsymbol{V}$ are given by

$$\begin{aligned}|\Delta\boldsymbol{V}_\perp| = V_0\sin(\theta) \;&=\; V_0\,|\sin(2\Psi_0)| \;=\; \frac{2V_0\,|\tan(\Psi_0)|}{1+\tan^2(\Psi_0)} \\ &= \frac{2bV_0^3}{G(m+M)}\left[1 + \frac{b^2V_0^4}{G^2(m+M)^2}\right]^{-1}\end{aligned} \tag{7.9}$$

$$\begin{aligned}\left|\Delta\boldsymbol{V}_\parallel\right| = V_0\,[1-\cos(\theta)] \;&=\; V_0(1+\cos(2\Psi_0)) \;=\; \frac{2V_0}{1+\tan^2(\Psi_0)} \\ &= 2V_0\left[1 + \frac{b^2V_0^4}{G^2(m+M)^2}\right]^{-1}.\end{aligned} \tag{7.10}$$

$\Delta\boldsymbol{V}_\parallel$ always points in the direction opposite to $\boldsymbol{V}_0$. Using (7.3) one finally gets

$$|\Delta\boldsymbol{v}_{M\perp}| = \frac{2mbV_0^3}{G(m+M)^2}\left[1 + \frac{b^2V_0^4}{G^2(m+M)^2}\right]^{-1} \tag{7.11}$$

$$\left|\Delta\boldsymbol{v}_{M\parallel}\right| = \frac{2mV_0}{m+M}\left[1 + \frac{b^2V_0^4}{G^2(m+M)^2}\right]^{-1}. \tag{7.12}$$

Hence by (7.3), $\Delta \boldsymbol{v}_{M\parallel}$ always points in the same direction as $\boldsymbol{V}_0$.

Let us now imagine that M travels through an infinite, homogeneous "sea of particles". Then there are as many deflections from "above" as from "below" or from "right" or "left" and the changes in $\Delta \boldsymbol{v}_{M\perp}$ sum up to zero. Furthermore, one has to invoke the "Jeans swindle" to neglect the gravitational potential of the "sea of particles", so the motion of each particle is determined only by M. The changes in $\Delta \boldsymbol{v}_{M\parallel}$ are all parallel to $\boldsymbol{V}_0$ and form a non-zero resultant; i.e. the mass M suffers a steady deceleration, which is said to be dynamical friction.

To determine the deceleration, one now has to integrate over all possible impact parameters b and velocities v_m. The number density of particles m with velocity distribution $f(\boldsymbol{v})$ in the velocity-space element $\mathrm{d}^3\boldsymbol{v}_m$ at impact parameters between b and $b + \mathrm{d}b$ is

$$2\pi b\mathrm{d}b \times V_0 \times f(\boldsymbol{v}_m)\mathrm{d}^3\boldsymbol{v}_m. \tag{7.13}$$

Hence the net rate of change of $\boldsymbol{v}_M$ is

$$\left.\frac{\mathrm{d}\boldsymbol{v}_M}{\mathrm{d}t}\right|_{\boldsymbol{v}_m} = \boldsymbol{V}_0 f(\boldsymbol{v}_m)\mathrm{d}^3\boldsymbol{v}_m \int_0^{b_{\max}} \frac{2mV_0}{m+M}\left[1 + \frac{b^2V_0^4}{G^2(m+M)^2}\right]^{-1} 2\pi b\mathrm{d}b, \tag{7.14}$$

with $b_{\max}$ the largest impact parameter to be considered. Performing the integration over all b, one finds

$$\left.\frac{\mathrm{d}\boldsymbol{v}_M}{\mathrm{d}t}\right|_{\boldsymbol{v}_m} = 2\pi \ln(1+\Lambda^2)G^2m(m+M)f(\boldsymbol{v}_m)\frac{\boldsymbol{v}_m - \boldsymbol{v}_M}{|\boldsymbol{v}_m - \boldsymbol{v}_M|^3}\mathrm{d}^3\boldsymbol{v}_m \tag{7.15}$$

with

$$\Lambda = \frac{b_{\max}V_0^2}{G(m+M)} = \frac{b_{\max}}{b_{\min}}. \tag{7.16}$$

Usually Λ is very large and so one can assume that $\frac{1}{2}\ln(1+\Lambda^2) \approx \ln(\Lambda)$, which is called the Coulomb logarithm. Furthermore, one replaces V_0 by the typical speed v_{typ}. Equation (7.15) states that particles that have velocity $\boldsymbol{v}_m$ exert a force on M that acts parallel to $\boldsymbol{v}_m - \boldsymbol{v}_M$ and is inversely proportional to the square of this vector. The problem to integrate over all velocities $\boldsymbol{v}_m$ is equivalent to finding the gravitational field at the point with position vector in velocity space $\boldsymbol{v}_M$, which is generated by the "mass density" $\rho(\boldsymbol{v}_m) = 4\pi\ln(\Lambda)Gm(m+M)f(\boldsymbol{v}_m)$. If the particles move isotropically, the density distribution is spherical and, according to Newton's first and second theorem, the total acceleration of M is equal to G/v_M^2 times the total "mass" at $v_m < v_M$. Hence

$$\frac{\mathrm{d}\boldsymbol{v}_M}{\mathrm{d}t} = -16\pi^2\ln(\Lambda)G^2m(m+M)\frac{\int_0^{v_M} f(\boldsymbol{v}_m)v_m^2\mathrm{d}v_m}{v_M^3}\boldsymbol{v}_M, \tag{7.17}$$

i.e. only particles m with velocities slower than M contribute to the force that always opposes the motion of M and this equation is henceforth called the Chandrasekhar dynamical friction formula.

If $f(\boldsymbol{v}_m)$ is Maxwellian with dispersion σ, then

$$f = \frac{n_0}{(2\pi\sigma^2)^{3/2}} \exp\left(-\frac{v^2}{2\sigma^2}\right), \tag{7.18}$$

and introducing $\rho = n_0 m$ as the background density, one can perform the integration, which gives:

$$\frac{\mathrm{d}\boldsymbol{v}_M}{\mathrm{d}t} = -\frac{4\pi \ln(\Lambda) G^2 \rho M}{v_M^3} \left[\operatorname{erf}(X) - \frac{2X}{\sqrt{\pi}} \exp(-X^2)\right] \boldsymbol{v}_M \tag{7.19}$$

with $X = v_M/\sqrt{2}\sigma$. This formula holds for $M \gg m$.

With this formula one can derive some useful relations. If keeping $\ln\Lambda$ constant, we can determine the time a star cluster or dwarf galaxy needs to spiral into the centre of its host system:

$$t_{\mathrm{fric}} = \frac{1.17 D_0^2 v_{\mathrm{circ}}}{\ln(\Lambda) G M} = \frac{2.64 \times 10^{11}}{\ln(\Lambda)} \left(\frac{D_0}{2\,\mathrm{kpc}}\right)^2 \left(\frac{v_{\mathrm{circ}}}{250\,\mathrm{km\ s^{-1}}}\right) \left(\frac{10^6\,\mathrm{M}_\odot}{M}\right) \mathrm{yr}. \tag{7.20}$$

Furthermore, McMillan & Portegies Zwart (2003) derived a formula for the sinking rate if the background is a mass distribution following a power law of the form $M(D) = A \cdot D^\alpha$. Then the distance D of an object to the centre of the host system vs. time is given by

$$D(t) = D_0 \left[1 - \frac{\alpha(\alpha+3)}{\alpha+1} \sqrt{\frac{G}{A D_0^{\alpha+3}}}\, \chi M \ln(\Lambda) t\right]^{2/3+\alpha}, \tag{7.21}$$

with

$$\chi = \operatorname{erf}(X) - \frac{2X}{\sqrt{\pi}} \exp(-X^2), \tag{7.22}$$

where $X = v_M/\sqrt{2}\sigma$.

Even though one might think that the derivation of Chandrasekhar's formula has too many vague definitions and approximations in it, it has been shown that it is a really powerful tool to describe dynamical friction in all kinds of environments.

7.3 Dynamical Friction in Numerical Simulations

Especially in numerical simulations, the validity of Chandrasekhar's formula has been verified throughout the decades. Still some words of caution have to be added. In the previous section, it was shown that $\Lambda = b_{\mathrm{max}}/b_{\mathrm{min}}$, with $b_{\mathrm{min,theo}} = G(m+M)/v_M^2$ in the extreme case of a point mass being a very small quantity (e.g. for a $10^6\,M_\odot$ black hole with a velocity of 50 km s^{-1} gives

$b_{\min} \approx 2$ pc). For extended objects like a star cluster $b_{\min}$ is of the order of the size of the cluster.

However, even if one uses a point mass to determine dynamical friction, it is not easy to reach the correct result. All standard N-body codes are resolution-limited. Even if one does not introduce softening and uses a direct summation N-body code, the limitation gets introduced through the finite particle number. In a study how dynamical friction is influenced by the resolution of the simulation code (i.e. the softening length used), Spinnato et al. (2003) showed that with a given softening length ϵ (or in the case of a particle-mesh code the cell-length ℓ),

$$b_{\min,\text{eff}} \approx b_{\min,\text{theo}} + \epsilon \text{ (or } \ell). \tag{7.23}$$

This is shown as the actual sinking curve for two choices of resolution in a particle-mesh code in the left panel of Fig. 7.3, and for all choices of ϵ as the derived $\ln(\Lambda)$ in the comparison to a direct summation N-body codes, a tree code and a particle-mesh code in the right panel.

In this study, $\ln(\Lambda)$ was assumed to be constant during the whole simulation time independently of the actual distance D to the centre of the background. Fitting $b_{\max}$ of a constant $\ln(\Lambda)$ to the data resulted in $b_{\max} = kD_0$ with $k \approx 0.5$.

In another study, Fellhauer & Lin (2007) used the same approach but fitted $\ln(\Lambda)$ at many small time-slices during the sinking process and determined $b_{\min}$ as function of the resolution and $b_{\max}$ as function of the distance D as shown in Fig. 7.4:

$$\begin{aligned} \ln \Lambda &= \ln(b_{\max}) - \ln(b_{\min}) \\ &= \ln(k' \cdot D(t)) + b_{\min,\text{eff}}. \end{aligned} \tag{7.24}$$

The values for $b_{\min,\text{eff}}$ were in very good agreement with (7.23) for the different resolutions. SUPERBOX, the particle-mesh code used in this study, has three levels of grid-resolutions. While the point-mass starts inside the medium resolution, it crosses the grid-boundary to the high-resolution area when $D < 1$

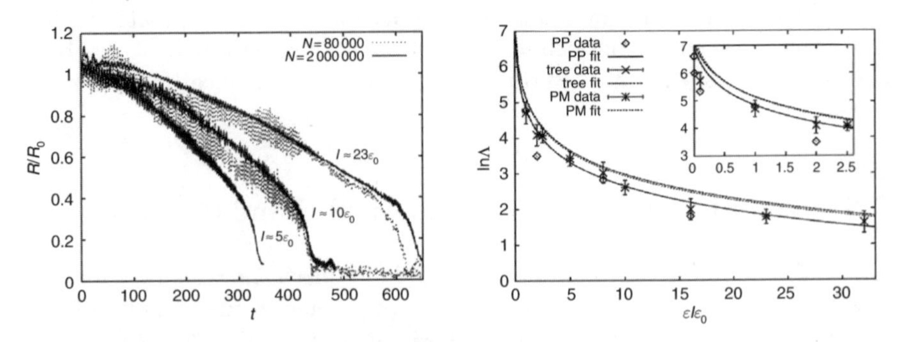

Fig. 7.3. Influence of the resolution on the dynamical friction of a point mass

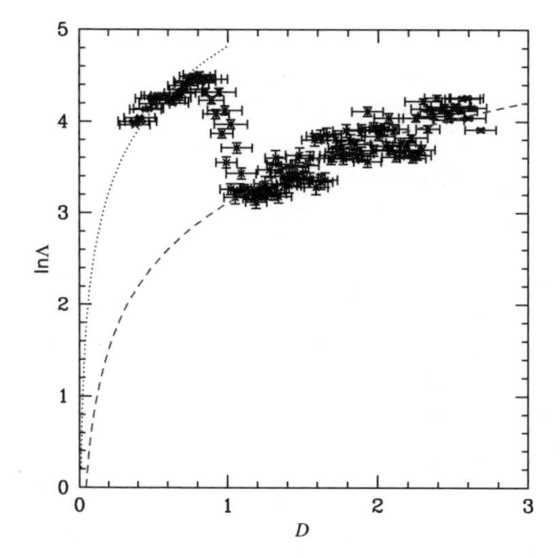

Fig. 7.4. $\ln(\Lambda)$ as a function of the distance to the centre of the background. Also visible is the change in resolution for $D < 1$ which leads to a smaller value of $b_{\min}$ and a larger value of $\ln(\Lambda)$. $\ln(\Lambda)$ is decreasing with decreasing distance. Fitting curves assume $b_{\max} \propto D$ (7.24)

in the above simulation. The values for k' differ from the value k found in the previous study and also seem to be dependent on the resolution.

7.4 Dynamical Friction of an Extended Object

In the previous section, the dependence of $\ln(\Lambda)$ on environment was investigated, which was possible because the studies involved the sinking of a point mass with constant mass. In many cases of dynamical friction the sinking object is extended, and due to tidal forces acting on it the mass is not constant. This section investigates which mass one has to insert into the dynamical friction formulae like (7.19) and (7.21).

The initial mass and orbit of the extended object (it could be a star cluster or a dwarf galaxy) is the same as the one of the point-mass of the previous section. We use again (7.21) to fit now the combined quantity $M_{\rm cl} \ln(\Lambda)$. For the left panel this quantity is converted into $\ln(\Lambda)$ in the following two ways:

$$\ln \Lambda(t)_{\rm crosses} = (M_{\rm cl} \ln \Lambda)(t)/M_{\rm bound}(t=0) \tag{7.25}$$

$$\ln \Lambda(t)_{\rm tri-pods} = (M_{\rm cl} \ln \Lambda)(t)/M_{\rm bound}(t). \tag{7.26}$$

The curves show that either way does not give the correct answer. If the mass is kept constant and the initial mass is inserted, the data points fall below the reference line of the point-mass case. This disparity is expected since an extended object should have a larger $b_{\min}$ than that of a point-mass potential. For $t < 30$ or $D > 1$, the difference between these two simulations is less than 20 per cent. However, it can also be seen that the deviation from the fitting line grows with time, especially at $t > 30$ (or equivalently as D decreases below 1). This growing difference is due to the loss of mass from the stellar cluster. This divergence shows that a constant $M_{\rm cl}$ approximation does not adequately represent the results of the simulation. If one inserts the bound mass as responsible for the dynamical friction, the measured values are systematically above the fitting line that represents the cluster with a point-mass potential. However, using the above argument that an extended object should have a larger $b_{\rm min}$ than that of a point-mass potential, the tri-pods measured from this simulation would be systematically below the fitting line if the bound stars adequately account for all the mass that contributes to the dynamical friction. This disparity is a first hint that more particles may take part in the dynamical friction than just the bound stars. In the later stages of the evolution, these values of $\ln \Lambda$ increase quite dramatically, which is a clear sign that $M_{\rm cl}$ is underestimated.

In the right panel of Fig. 7.5, the bound mass of the object as a function of time (solid line) is plotted. In the same figure, crosses and squares represent the mass of the cluster taking part in the dynamical friction process if the same $\ln \Lambda$ as that derived for a point-mass is assumed. Then one solves for $M_{\rm cl}$ with (7.21). For the crosses the actual values from the point-mass simulation is applied, while the data-points of the squares are derived using the smoothed

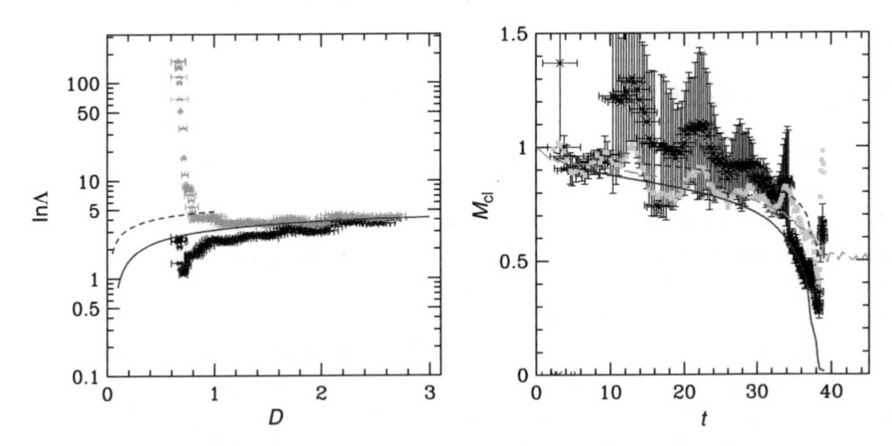

Fig. 7.5. Dynamical friction on an extended object. *Left*: Fitting $M_{\rm cl} \ln(\Lambda)$ to the sinking curve in small time-slices like in Fig. 7.4 and deriving $\ln(\Lambda)$ according to (7.25) & (7.26). *Right*: Using the values of $\ln(\Lambda)$ derived from the point-mass case to determine $M_{\rm cl}$, the mass responsible for dynamical friction (*yellow squares* using the fitting formulae; black crosses with error-bars using the actual values of the point-mass simulation). (*Red*) solid line shows the bound mass of the object, long dashed (*green*) line the bound mass plus the unbound mass in a ring around the centre of the background with size of the object. (*Red*) short dashed line is the rule-by-thumb: bound mass plus half of the unbound mass

fitting curve for $\ln \Lambda(D)$ from (7.24). (Since it has already been shown that the magnitude of $\ln \Lambda(D)$ for a cluster with a Plummer potential is smaller than that for a point mass, the actual total mass that contributes to the dynamical friction is slightly larger than both the values represented by the crosses and the squares.) Even though the uncertainties are large, the data points show that the total mass responsible which contribute to the effect of dynamical friction is systematically above the bound mass in the bound mass curve.

In addition to the bound mass the lost mass of the cluster which is located in a ring of the cluster dimension around the galaxy at the same distance is calculated, and only the particles with the same velocity signature as the cluster are counted. Adding this mass to the bound mass is shown as the short dashed line in the right panel of Fig. 7.5. This mass estimate seems to fit the data much better. This value is not easy to access and surely has to be replaced by a more elaborate formulation of dynamical friction, i.e. assigning weights to all unbound particles with respect to their position and velocity to the cluster. Thus, applying a simple rule-by-thumb by adding half of the unbound mass to the bound mass (shown as long dashed line in the right panel of Fig. 7.5) fits the data nicely, taking into account that the "actual" $\ln \Lambda$ of an extended object should be smaller than the one of a point mass, i.e. the data points have to be regarded as lower limits. Even though this simple estimate has no physical explanation and breaks down during the very final stages of the dissolution of the cluster, it gives an easy accessible estimate of the dynamical friction of an extended object suffering from mass-loss.

References

Binney J., Tremaine S., 1987, Galactic Dynamics. Princeton Univ. Press, Princeton, NJ
Chandrasekhar S., 1943, ApJ, 97, 255
Fellhauer M., Lin D. N. C., 2006, MNRAS, 375, 604
McMillan S. L., Portegies Zwart S .F., 2003, ApJ, 596, 314
Spinnato P. F., Fellhauer M., Portegies Zwart S. F., 2003, MNRAS, 344, 22

8

Initial Conditions for Star Clusters

Pavel Kroupa

Argelander-Institut für Astronomie, Auf dem Hügel 71, D-53121 Bonn, Germany
pavel@astro.uni-bonn.de

8.1 Introduction

Most stars form in dense star clusters deeply embedded in residual gas. The populations of these objects range from small groups of stars with about a dozen binaries within a volume with a typical radius of $r \approx 0.3\,\mathrm{pc}$ through to objects formed in extreme star bursts containing $N \approx 10^8$ stars within $r \approx 36\,\mathrm{pc}$. Star clusters, or more generally dense stellar systems, must therefore be seen as the fundamental building blocks of galaxies. Differentiation of the term star cluster from a spheroidal dwarf galaxy becomes blurred near $N \approx 10^6$. Both are mostly pressure-supported, that is, random stellar motions dominate any bulk streaming motions such as rotation. The physical processes that drive the formation, evolution and dissolution of star clusters have a deep impact on the appearance of galaxies. This impact has many manifestations, ranging from the properties of stellar populations, such as the binary fraction and the number of type Ia and type II supernovae, through the velocity structure in galactic discs, such as the age–velocity dispersion relation, to the existence of stellar halos around galaxies, tidal streams and the survival and properties of tidal dwarf galaxies, the existence of which challenge current cosmological perspectives. Apart from this cosmological relevance, dense stellar systems provide unique laboratories in which to test stellar evolution theory, gravitational dynamics, the interplay between stellar evolution and dynamical processes and the physics of stellar birth and stellar feedback processes during formation.

Star clusters and other pressure-supported stellar systems in the sky merely offer snap-shots from which we can glean incomplete information. Because there is no analytical solution to the equations of motion for more than two stars, these differential equations need to be integrated numerically. Thus, in order to gain an understanding of these objects in terms of the above issues, a researcher needs to resort to numerical experiments in order to test various hypotheses as to the possible physical initial conditions (to test star-formation theory) or the outcome (to quantify stellar populations in galaxies,

Kroupa, P.: *Initial Conditions for Star Clusters*. Lect. Notes Phys. **760**, 181–259 (2008)
DOI 10.1007/978-1-4020-8431-7_8

for example). The initialisation of a pressure-supported stellar system is such that the initial object is relevant for the real physical Universe, and is therefore a problem of some fundamental importance.

Here empirical constraints on the initial conditions of star clusters are discussed and some problems to which star clusters are relevant are raised. Section 8.2 contains information to set up a realistic computer model of a star cluster, including models of embedded clusters. The initial mass distribution of stars is discussed in Sect. 8.3, and Sect. 8.4 delves into the initial distribution functions of multiple stars. A brief summary is provided in Sect. 8.5.

8.1.1 Embedded Clusters

In this section an outline is given of some astrophysical aspects of dense stellar systems in order to help differentiate probable evolutionary effects from initial conditions. A simple example clarifies the meaning of this. An observer may see two young populations with comparable ages (to within 1 Myr say). They have similar observed masses but different sizes and a somewhat different stellar content and different binary fractions. Do they signify two different initial conditions derived from star-formation or can both be traced back to a $t = 0$ configuration, which is the same?

Preliminaries

Assume we observe a very young population of N stars with an age $\tau_{\rm age}$ and that we have a rough estimate of its half-mass radius, $r_{\rm h}$, and embedded stellar mass, $M_{\rm ecl}$.[1] The average mass is

$$\overline{m} = \frac{M_{\rm ecl}}{N}. \tag{8.1}$$

Also assume we can estimate the star-formation efficiency (SFE), ϵ, within a few $r_{\rm h}$. For this object,

$$\epsilon = \frac{M_{\rm ecl}}{M_{\rm ecl} + M_{\rm gas}}, \tag{8.2}$$

where $M_{\rm gas}$ is the gas left over from the star-formation process. The tidal radius of the embedded cluster can be estimated from the Jacobi limit ((Eq. (7-84) in Binney & Tremaine 1987) as determined by the host galaxy when any contributions by surrounding molecular clouds are ignored,

$$r_{\rm tid} = \left(\frac{M_{\rm ecl} + M_{\rm gas}}{3\,M_{\rm gal}}\right)^{\frac{1}{3}} D, \tag{8.3}$$

[1]Throughout all masses, m, M, etc. are in units of $M_\odot$, unless noted otherwise. "Embedded stellar mass" refers to the man in stars at the time before residual gas expulsion and when star-formation has ceased.

where $M_{\rm gal}$ is the mass of the spherically distributed galaxy within the distance D of the cluster from the centre of the galaxy. This radius is a rough estimate of that distance from the cluster at which stellar motions begin to be significantly influenced by the host galaxy.

The following quantities that allow us to judge the formal dynamical state of the system, the formal crossing time of the stars through the object, can be defined as

$$t_{\rm cr} \equiv \frac{2\,r_{\rm h}}{\sigma}, \tag{8.4}$$

where[2]

$$\sigma = \sqrt{\frac{G\,M_{\rm ecl}}{\epsilon\,r_{\rm h}}} \tag{8.5}$$

is, up to a factor of order unity, the three-dimensional velocity dispersion of the stars in the embedded cluster. Note that these equations serve to estimate the possible amount of mixing of the population. If $\tau_{\rm age} < t_{\rm cr}$, the object cannot be mixed and we are seeing it close to its initial state. It takes a few $t_{\rm cr}$ for a dynamical system out of dynamical equilibrium to return back to it. This is not to be mistaken for a relaxation process.

Once the stars orbit within the object, they exchange orbital energy through weak gravitational encounters and rare strong encounters. The system evolves towards a state of energy equipartition. The energy equipartition time-scale, $t_{\rm ms}$, between massive and average stars (Spitzer 1987, p. 74), which is an estimate of the time massive stars need to sink to the centre of the system through dynamical friction on the lighter stars, is

$$t_{\rm ms} = \frac{\overline{m}}{m_{\rm max}}\,t_{\rm relax}. \tag{8.6}$$

Here, $m_{\rm max}$ is the massive-star mass and the characteristic two-body relaxation time (e.g. Eq. (4–9) in Binney & Tremaine 1987) is

$$t_{\rm relax} = 0.1\,\frac{N}{\ln N}\,t_{\rm cr}. \tag{8.7}$$

This formula refers to a pure N-body system without embedded gas. A rough estimate of $t_{\rm relax,emb}$ for an embedded cluster can be found in Eq. (8) of Adams & Myers (2001). The above (8.7) is a measure for the time a star needs to change its orbit significantly from its initial trajectory. We often estimate it by calculating the amount of time that is required to change the velocity, v, of a star by an amount $\Delta v \approx v$.

Thus, if for example, $\tau_{\rm age} > t_{\rm cr}$ and $\tau_{\rm age} < t_{\rm relax}$, the system is probably mixed and close to dynamical equilibrium, but it is not yet relaxed. That is, it has not had sufficient time for the stars to exchange a significant amount of orbital energy. Such a cluster may have erased its sub-structures.

[2] As an aside, note that $G = 0.0045\,{\rm pc}^3/M_\odot\,{\rm Myr}^2$ and that $1\,{\rm km\,s}^{-1} = 1.02\,{\rm pc/Myr}$.

Fragmentation and Size

The very early stages of cluster evolution on a scale of a few parsecs are dominated by gravitational fragmentation of a turbulent magnetised contracting molecular cloud core (Clarke, Bonnell & Hillenbrand 2000; Mac Low & Klessen 2004; Tilley & Pudritz 2007). Gas-dynamical simulations show the formation of contracting filaments, which fragment into denser cloud cores, that form subclusters of accreting protostars. As soon as the protostars radiate or lose mass with sufficient energy and momentum to affect the cloud core, these computations become expensive because radiative transport and deposition of momentum and mechanical energy by non-isotropic outflows are difficult to handle with present computational means (Stamatellos et al. 2007; Dale, Ercolano & Clarke 2007).

Observations of the very early stages at times less than a few hundreds of thousands of years suggest that protoclusters have a hierarchical protostellar distribution: a number of subclusters with radii less than 0.2 pc and separated in velocity space are often seen embedded within a region less than a pc across (Testi et al. 2000). Many of these subclusters may merge to form a more massive embedded cluster (Scally & Clarke 2002; Fellhauer & Kroupa 2005). It is unclear though if subclusters typically merge before residual gas blow-out or if the residual gas is removed before the sub-clumps can interact significantly, nor is it clear if there is a systematic mass dependence of any such possible behaviour.

Mass Segregation

Whether or not star clusters or subclusters form mass-segregated remains an open issue. Mass segregation at birth is a natural expectation because protostars near the density maximum of the cluster have more material to accrete. For these, the ambient gas is at a higher pressure allowing protostars to accrete longer before feedback termination stops further substantial gas inflow and the coagulation of protostars is more likely there (Zinnecker & Yorke 2007; Bonnell, Larson & Zinnecker 2007). Initially mass-segregated subclusters preserve mass segregation upon merging (McMillan, Vesperini & Portegies Zwart 2007). However, for $\overline{m}/m_{\rm max} = 0.5/100$ and $N \leq 5 \times 10^3$ stars, it follows from (8.6) that

$$t_{\rm ms} \leq t_{\rm cr}. \tag{8.8}$$

That is, a $100\,M_\odot$ star sinks to the cluster centre within roughly a crossing time (see Table 8.1 below for typical values of $t_{\rm cr}$).

Currently, we cannot say conclusively if mass segregation is a birth phenomenon (e.g. Gouliermis et al. 2004), or whether the more massive stars form anywhere throughout the protocluster volume. Star clusters that have already blown out their gas at ages of one to a few million years are typically mass-segregated (e.g. R136, Orion Nebula Cluster).

Table 8.1. Notes: the Y in the O stars column indicates that the maximum stellar mass in the cluster surpasses $8\,M_\odot$ (Fig. 8.1). The average stellar mass is taken to be $\overline{m} = 0.4\,M_\odot$ in all clusters. A star-formation efficiency of $\epsilon = 0.3$ is assumed. The crossing time, $t_{\rm cr}$, is (8.4). The pre-supernova gas evacuation time-scale is $\tau_{\rm gas} = r/v_{\rm th}$, where $v_{\rm th} = 10\,{\rm km\,s^{-1}}$ is the approximate sound velocity of the ionised gas and $\tau_{\rm gas} = 0.05\,{\rm Myr}$ for $r = 0.5\,{\rm pc}$, while $\tau_{\rm gas} = 0.1\,{\rm Myr}$ for $r = 1\,{\rm pc}$

$M_{\rm ecl}/M_\odot$	N	O stars?	$t_{\rm cr}/{\rm Myr}$	$\tau_{\rm gas}/t_{\rm cr}$	$t_{\rm cr}/{\rm Myr}$	$\tau_{\rm gas}/t_{\rm cr}$
		($r_{\rm h}$ =	0.5 pc	0.5 pc	1 pc	1 pc)
40	100	N	0.9	–	2.6	–
100	250	Y/N	0.6	0.08	1.6	0.2
500	1250	Y	0.3	0.2	0.7	0.1
10^3	2.5×10^3	Y	0.2	0.25	0.5	0.2
10^4	2.5×10^4	Y	0.06	0.8	0.2	0.5
10^5	2.5×10^5	Y	0.02	2.5	0.05	2
10^6	2.5×10^6	Y	0.006	8.3	0.02	5

To affirm, natal mass segregation would impact positively on the notion that massive stars (more than about $10\,M_\odot$) only form in rich clusters and negatively on the suggestion that they can also form in isolation. For recent work on this topic see Li, Klessen & Mac Low (2003) and Parker & Goodwin (2007).

Feedback Termination

The observationally estimated SFE (8.2) is (Lada & Lada 2003)

$$0.2 \le \epsilon \le 0.4, \tag{8.9}$$

which implies that the physics dominating the star-formation process on scales less than a few parsecs is stellar feedback. Within this volume, the pre-cluster cloud core contracts under self-gravity and so forms stars ever more vigorously, until feedback energy suffices to halt the process (feedback termination).

Dynamical State at Feedback Termination

Each protostar needs about $t_{\rm ps} \approx 10^5\,{\rm yr}$ to accumulate about 95% of its mass (Wuchterl & Tscharnuter 2003). The protostars form throughout the pre-cluster volume as the protocluster cloud core contracts. The overall pre-cluster cloud-core contraction until feedback termination takes (8.4, 8.5)

$$t_{\rm cl,form} \approx {\rm few} \times \frac{2}{\sqrt{G}} \left(\frac{M_{\rm ecl}}{\epsilon}\right)^{-\frac{1}{2}} r_{\rm h}^{\frac{3}{2}}, \tag{8.10}$$

(a few times the crossing time), which is about the time over which the cluster forms. Once a protostar condenses out of the hydro-dynamical flow, it becomes a ballistic particle moving in the time-evolving cluster potential. Because many generations of protostars can form over the cluster-formation time-scale, and if the crossing time through the cluster is a few times shorter than $t_{\rm cl,form}$, the very young cluster is mostly in virial equilibrium when star-formation stops when any residual gas has been lost.[3] It is noteworthy that for $r_{\rm h} = 1\,{\rm pc}$

$$t_{\rm ps} \ge t_{\rm cl,form} \quad {\rm for} \quad \frac{M_{\rm ecl}}{\epsilon} \ge 10^{4.9}\,M_\odot \tag{8.11}$$

(the protostar-formation time formally surpasses the cluster formation time), which is near the turnover mass in the old-star cluster mass function (eg. Baumgardt 1998).

A critical parameter is thus the ratio

$$\tau = \frac{t_{\rm cl,form}}{t_{\rm cr}}. \tag{8.12}$$

If it is less than unity, protostars condense from the gas and cannot reach virial equilibrium in the potential before the residual gas is removed. Such embedded clusters may be kinematically cold if the pre-cluster cloud core was contracting, or hot if the pre-cluster cloud core was pressure confined, because the young stars do not feel the gas pressure.

In those cases where $\tau > 1$, the embedded cluster is approximately in virial equilibrium because generations of protostars that drop out of the hydrodynamic flow have time to orbit the potential. The pre-gas-expulsion stellar velocity dispersion in the embedded cluster (8.5) may reach $\sigma = 40\,{\rm pc\,Myr}^{-1}$ if $M_{\rm ecl} = 10^{5.5}\,M_\odot$, which is the case for $\epsilon\, r_{\rm h} < 1\,{\rm pc}$. This is easily achieved because the radius of one-Myr old clusters is $r_{0.5} \approx 0.8\,{\rm pc}$ with no dependence on mass. Some observationally explored cases are discussed by Kroupa (2005). Notably, using K-band number counts, Gutermuth et al. (2005) appear to find evidence for expansion after gas removal.

Interestingly, recent Spitzer results suggest a scaling of the characteristic projected radius R with mass,[4]

$$M_{\rm ecl} \propto R^2 \tag{8.13}$$

(Allen et al. 2007), so the question of how compact embedded clusters form and whether there is a mass–radius relation needs further clarification. Note though that such a scaling is obtained for a stellar population that expands freely with a velocity given by the velocity dispersion in the embedded cluster (8.5),

[3] A brief transition time $t_{\rm tr} \ll t_{\rm cl,form}$ exists during which the star-formation rate decreases in the cluster while the gas is being blown out. However, for the purpose of the present discussion this time may be neglected.

[4] Throughout this text, projected radii are denoted by R, while the 3D radius is r.

$$r(t) \approx r_o + \sigma\, t \quad \Rightarrow \quad M_{ecl} = \frac{1}{G}\left(\frac{r(t) - r_o}{t}\right)^2 , \tag{8.14}$$

where $r_o \le 1\,\mathrm{pc}$ is the birth radius of the cluster. Is the observed scaling then a result of expansion from a compact birth configuration after gas expulsion? If so, it would require a more massive system to be dynamically older, which is at least qualitatively in-line with the dynamical time-scales decreasing with mass. Note also that the observed scaling (8.13) cannot carry through to $M_{ecl} \ge 10^4\,M_\odot$ because the resulting objects would not resemble clusters.

There are two broad camps suggesting on one hand that molecular clouds and star clusters form on a free-fall time-scale (Elmegreen 2000; Hartmann 2003; Elmegreen 2007), and on the other hand that many free-fall times are needed (Krumholz & Tan 2007). The former implies $\tau \approx 1$ while the latter implies $\tau > 1$.

Thus, currently unclear issues concerning the initialisation of N-body models of embedded clusters are the ratio τ and whether a mass–radius relation exists for embedded clusters before the development of HII regions. To make progress, I assume for now that the embedded clusters are in virial equilibrium at feedback termination ($\tau > 1$) and that they form highly concentrated with $r \le 1\,\mathrm{pc}$ independently of mass.

The Mass of the Most Massive Star

Young clusters show a well-defined correlation between the mass of the most massive star, m_{max}, and the stellar mass of the embedded cluster, M_{ecl}. This appears to saturate at $m_{max*} \approx 150\,M_\odot$ (Weidner & Kroupa 2004, 2006). This is shown in Fig. 8.1. This correlation may indicate feedback termination of star-formation within the protocluster volume coupled to the most massive stars forming latest, or turning-on at the final stage of cluster formation (Elmegreen 1983).

The evidence for a universal upper mass cutoff near

$$m_{max*} \approx 150\,M_\odot \tag{8.15}$$

(Weidner & Kroupa 2004; Figer 2005; Oey & Clarke 2005; Koen 2006; Maíz Apellániz et al. 2007; Zinnecker & Yorke 2007) seems to be rather well established in populations with metallicities ranging from the LMC ($Z \approx 0.008$) to the super-solar Galactic centre ($Z \ge 0.02$) so that the stellar mass function (MF) simply stops at that mass. This mass needs to be understood theoretically (see discussion by Kroupa & Weidner 2005; Zinnecker & Yorke 2007). It is probably a result of stellar structure stability, but may be near $80\,M_\odot$ as predicted by theory if the most massive stars reside in near-equal component-mass binary systems (Kroupa & Weidner 2005). It may also be that the calculated stellar masses are significantly overestimated (Martins, Schaerer & Hillier 2005).

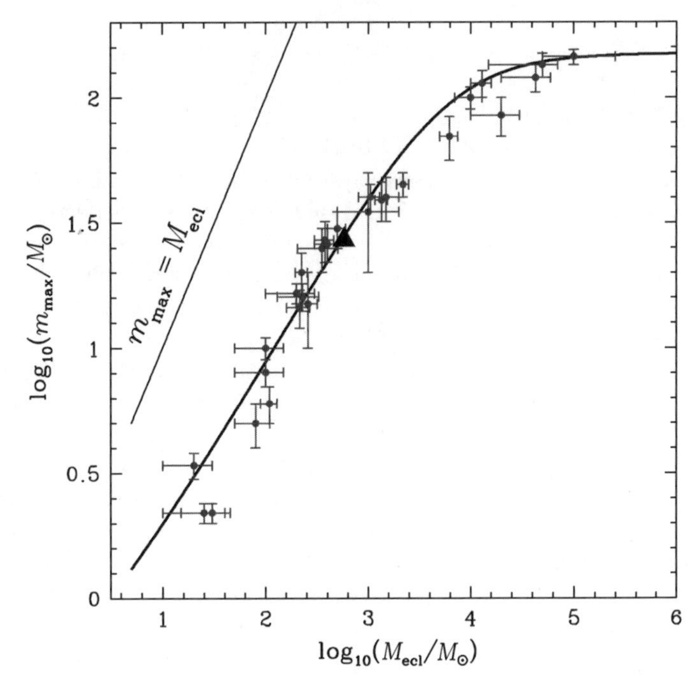

Fig. 8.1. The maximum stellar mass, m_{max}, as a function of the stellar mass of the embedded cluster, M_{ecl} (Weidner, private communication, an updated version of the data presented by Weidner & Kroupa 2006). The *solid triangle* is an SPH model of star-cluster formation by Bonnell, Bate & Vine (2003), while the *solid curve* stems from stating that there is exactly one most massive star in the cluster, $1 = \int_{m_{\mathrm{max}}}^{150} \xi(m)\,\mathrm{d}m$ with the condition $M_{\mathrm{ecl}} = \int_{0.08}^{m_{\mathrm{max}}} m\,\xi(m)\,\mathrm{d}m$, where $\xi(m)$ is the stellar IMF. The solution can only be obtained numerically, but an easy-to-use, well-fitting function has been derived by Pflamm-Altenburg, Weidner & Kroupa (2007)

The Cluster Core of Massive Stars

Irrespective of whether the massive stars (more than about $10\,M_{\odot}$) form at the cluster centre or whether they segregate there owing to energy equipartition (8.6), they ultimately form a compact sub-population that is dynamically highly unstable. Massive stars are ejected from such cores very efficiently on a core-crossing time-scale and, for example, the well-studied Orion Nebula cluster (ONC) has probably already shot out 70% of its stars more massive than $5\,M_{\odot}$ (Pflamm-Altenburg & Kroupa 2006). The properties of O and B runaway stars have been used by Clarke & Pringle (1992) to deduce the typical birth configuration of massive stars. They find them to form in binaries with similar-mass components in compact small-N groups devoid of low-mass stars. Among others, the core of the Orion Nebula Cluster (ONC) is just such a system.

The Star-Formation History in a Cluster

The detailed star-formation history in a cluster contains information about the events that build up the cluster. Intriguing is the recent evidence for some clusters that while the bulk of the stars have ages that differ by less than a few 10^5 yr, a small fraction of older stars are often encountered (Palla & Stahler 2000 for the ONC, Sacco et al. 2007 for the σ Orionis cluster). This may be interpreted to mean that clusters form over about 10 Myr with a final highly accelerated phase, in support of the notion that turbulence of a magnetised gas determines the early cloud-contraction phase (Krumholz & Tan 2007).

A different interpretation would be that as a pre-cluster cloud core contracts on a free-fall time-scale, it traps surrounding field stars which then become formal cluster members. Most clusters form in regions of a galaxy that has seen previous star-formation. The velocity dispersion of the previous stellar generation, such as an expanding OB association, is usually rather low, around a few $\mathrm{km\,s^{-1}}$ to $10\,\mathrm{km\,s^{-1}}$. The deepening potential of a newly contracting pre-cluster cloud core is able to capture some of the preceding generation of stars so that these older stars become formal cluster members although they did not form in the cluster. Pflamm-Altenburg & Kroupa (2007) study this problem for the ONC and show that the age spread reported by Palla et al. (2007) can be accounted for in this way. This suggests that the star-formation history of the ONC may in fact not have started about 10 Myr ago, supporting the argument by Elmegreen (2000), Elmegreen (2007) and Hartmann (2003) that clusters form on a time-scale comparable to the crossing time of the pre-cluster cloud core. Additionally, the sample of cluster stars may be contaminated by enhanced fore- and back-ground densities of field stars by focussing of stellar orbits during cluster formation (Pflamm-Altenburg & Kroupa 2007).

For very massive clusters such as ω Cen, Fellhauer, Kroupa & Evans (2006) show that the potential is sufficiently deep that the pre-cluster cloud core may capture the field stars of a previously existing dwarf galaxy. Up to 30% or more of the stars in ω Cen may be captured field stars. This would explain an age spread of a few Gyr in the cluster and is consistent with the notion that ω Cen formed in a dwarf galaxy that was captured by the Milky Way. The attractive aspect of this scenario is that ω Cen need not have been located at the centre of the incoming dwarf galaxy as a nucleus but within its disc, because it opens a larger range of allowed orbital parameters for the putative dwarf galaxy moving about the Milky Way. The currently preferred scenario in which ω Cen was the nucleus of the dwarf galaxy implies that the galaxy was completely stripped while falling into the Milky Way leaving only its nucleus on its current retrograde orbit (Zhao 2004). The new scenario allows the dwarf galaxy to be absorbed into the bulge of the Milky Way with ω Cen being stripped from it on its way in.

Another possibility for obtaining an age spread of a few Gyr in a massive cluster such as ω Cen is gas accretion from a co-moving inter-stellar medium (Pflamm-Altenburg & Kroupa 2008). This could only have worked for ω Cen before it became unbound from its mother galaxy, though. That is, the cluster must have spent about 2–3 Gyr in its mother galaxy before it was captured by the Milky Way.

This demonstrates beautifully how an improved understanding of dynamical processes on scales of a few pc impinges on problems related to the formation of galaxies and cosmology (through the sub-structure problem). Finally, the increasingly well-documented evidence for stellar populations in massive clusters with different metallicities and ages, and in some cases even significant He enrichment, may also suggest secondary star-formation occurring from material that has been pre-enriched from a previous generation of stars in the cluster. Different IMFs need to be invoked for the populations of different ages (see Piotto 2008 for a review).

Expulsion of Residual Gas

When the most massive stars are O stars, they destroy the protocluster nebula and quench further star-formation by first ionising most of it (feedback termination). The ionised gas, at a temperature near 10^4 K and in serious over-pressure, pushes out and escapes the confines of the cluster volume at the sound speed (near $10\,\mathrm{km\,s^{-1}}$) or faster if the winds blow off O stars with velocities of thousands of $\mathrm{km\,s^{-1}}$ and impart sufficient momentum.

There are two analytically tractable regimes of behaviour, instantaneous gas removal and slow gas expulsion over many crossing times.

- First consider instantaneous gas expulsion, $\tau_{\mathrm{gas}} = 0$. The binding energy of the object of mass M and radius r is

$$E_{\mathrm{cl,bind}} = -\frac{G\,M^2}{r} + \frac{1}{2}\,M\,\sigma^2 \;< 0. \tag{8.16}$$

Before gas expulsion, $M = M_{\mathrm{init}} = M_{\mathrm{gas}} + M_{\mathrm{ecl}} \;\rightarrow M$ and

$$\sigma^2_{\mathrm{init}} = \frac{G\,M_{\mathrm{init}}}{r_{\mathrm{init}}} \;\longrightarrow\; \sigma. \tag{8.17}$$

After instantaneous gas expulsion, $M_{\mathrm{after}} = M_{\mathrm{ecl}} \;\rightarrow M$, but $\sigma_{\mathrm{after}} = \sigma_{\mathrm{init}} \rightarrow \sigma$, and the new binding energy is

$$E_{\mathrm{cl,bind,after}} = -\frac{G\,M^2_{\mathrm{after}}}{r_{\mathrm{init}}} + \frac{1}{2}\,M_{\mathrm{after}}\,\sigma^2_{\mathrm{init}}. \tag{8.18}$$

But the cluster relaxes into a new equilibrium, so that, by the scalar virial theorem[5]

[5]The scalar virial theorem states that $2\,K + W = 0 \;\Rightarrow\; E = K + W = (1/2)\,W$, where K, W are the kinetic and potential energy and E is the total energy of the system.

$$E_{\mathrm{cl,bind,after}} = -\frac{1}{2}\frac{G\,M_{\mathrm{after}}}{r_{\mathrm{after}}}, \tag{8.19}$$

and on equating these two expressions for the final energy and using (8.17) we find that

$$\frac{r_{\mathrm{after}}}{r_{\mathrm{init}}} = \frac{M_{\mathrm{ecl}}}{M_{\mathrm{ecl}} - M_{\mathrm{gas}}}. \tag{8.20}$$

Thus, as $M_{\mathrm{gas}} \rightarrow M_{\mathrm{ecl}}$, then $\epsilon \rightarrow 0.5$ from above, $r_{\mathrm{after}} \rightarrow \infty$. This means that as the SFE approaches 50% from above, the cluster unbinds itself. But by (8.9), this result would imply either (see Kroupa, Aarseth & Hurley 2001, and references therein)

- all clusters with OB stars (and thus $\tau_{\mathrm{gas}} \ll t_{\mathrm{cr}}$) do not survive gas expulsion, or
- the clusters expel their gas slowly, $\tau_{\mathrm{gas}} \gg t_{\mathrm{cr}}$. This may be the case if surviving clusters such as the Pleiades or Hyades formed without OB stars.

- Now consider slow gas removal, $\tau_{\mathrm{gas}} \gg t_{\mathrm{cr}}$, $\tau_{\mathrm{gas}} \rightarrow \infty$. By (8.20) and the assumption that an infinitesimal mass of gas is removed instantaneously,

$$\frac{r_{\mathrm{init}} - \delta r}{r_{\mathrm{init}}} = \frac{M_{\mathrm{init}} - \delta M_{\mathrm{gas}}}{M_{\mathrm{init}} - \delta M_{\mathrm{gas}} - \delta M_{\mathrm{gas}}}. \tag{8.21}$$

For infinitesimal steps and, for convenience, $\mathrm{d}M < 0$ but $\mathrm{d}r > 0$,

$$\frac{r - \mathrm{d}r}{r} = \frac{M + \mathrm{d}M}{M + 2\,\mathrm{d}M}. \tag{8.22}$$

Re-arranging this, we find

$$\frac{\mathrm{d}r}{r} = \frac{\mathrm{d}M}{M}\left(1 - 2\frac{\mathrm{d}M}{M}\ldots\right), \tag{8.23}$$

so that

$$\frac{\mathrm{d}r}{r} = \frac{\mathrm{d}M}{M} \Rightarrow \ln\frac{r_{\mathrm{after}}}{r_{\mathrm{init}}} = \ln\frac{M_{\mathrm{init}}}{M_{\mathrm{after}}}, \tag{8.24}$$

upon integration of the differential equation. Thus,

$$\frac{r_{\mathrm{after}}}{r_{\mathrm{init}}} = \frac{M_{\mathrm{ecl}} + M_{\mathrm{gas}}}{M_{\mathrm{ecl}}} = \frac{1}{\epsilon}, \tag{8.25}$$

and for example, for a SFE of 20%, the cluster expands by a factor of 5, $r_{\mathrm{after}} = 5\,r_{\mathrm{init}}$, without dissolving.

Table 8.1 gives an overview of the type of behaviour one might expect for clusters with increasing number of stars, N, and stellar mass, M_{ecl}, for two characteristic radii of the embedded stellar distribution, r_{h}. It can be seen that the gas-evacuation time-scale becomes longer than the crossing time through the cluster for $M_{\rm ecl} \geq 10^5\,M_\odot$. Such clusters would thus undergo adiabatic expansion as a result of gas blow out. Less-massive clusters are more likely to undergo an evolution that is highly dynamic and that can be described as an explosion (the cluster pops). For clusters without O and massive B stars, nebula disruption probably occurs on the cluster-formation time-scale of about a million years and the evolution is again adiabatic. A simple calculation of the amount of energy deposited by an O star into its surrounding cluster-nebula suggests it is larger than the nebula binding energy (Kroupa 2005). This, however, only gives, at best, a rough estimate of the rapidity with which gas can be expelled. An inhomogeneous distribution of gas leads to the gas removal preferentially along channels and asymmetrically, so that the overall gas-excavation process is highly non-uniform and variable (Dale et al. 2005).

The reaction of clusters to gas expulsion is best studied numerically with N-body codes. Pioneering experiments were performed by Tutukov (1978) and then Lada, Margulis & Dearborn (1984). Goodwin (1997a,b, 1998) studied gas expulsion by supernovae from young globular clusters. Figure 8.2 shows the evolution of an ONC-type initial cluster with a stellar mass $M_{\rm ecl} \approx 4000\,M_\odot$ and a canonical IMF (8.124) and stellar evolution, a 100% initial binary population (Sect. 8.4.2) in a solar-neighbourhood tidal field, $\epsilon = 1/3$ and spherical gas blow-out on a thermal time-scale ($v_{\rm th} = 10\,{\rm km\,s^{-1}}$). The figure demonstrates that the evolution is far more complex than the simple analytical estimates above suggest, and in fact a substantial Pleiades-type cluster emerges after losing about two-thirds of its initial stellar population (see also p. 195). Subsequent theoretical work based on an iterative scheme according to which the mass of unbound stars at each radius is removed successively shows that

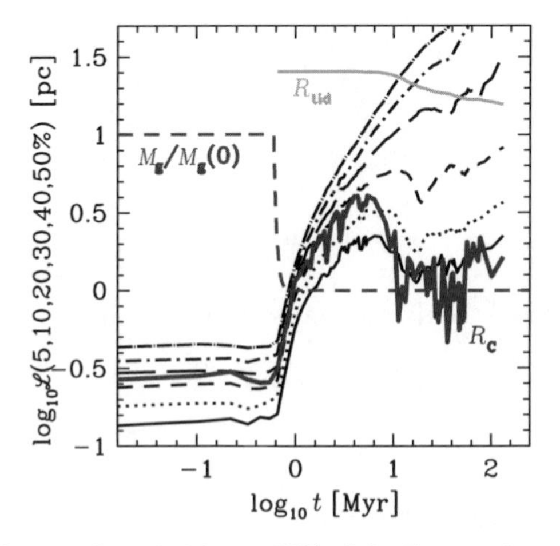

Fig. 8.2. The evolution of 5, 10, 20, ..., 50% of the Lagrangian radius and the core radius ($R_c = r_c$, *thick lower curve*) of the ONC-type cluster discussed in the text. The gas mass is shown as the *dashed line*. The cluster spends 0.6 Myr in an embedded phase before the gas is blown out on a thermal time-scale. The tidal radius (8.3) is shown by the *upper thick solid curve* (Kroupa, Aarseth & Hurley 2001)

the survival of a cluster depends not only on ϵ, τ_{gas}/t_{cr} and r_{tid} but also on the detailed shape of the stellar distribution function (Boily & Kroupa 2003). For instantaneous gas removal, $\epsilon \approx 0.3$ is a lower limit for the SFE below which clusters cannot survive rapid gas blow-out. This is significantly smaller than the critical value of $\epsilon = 0.5$ below which the stellar system becomes formally unbound (8.20). However, if clusters form as complexes of subclusters, each of which pop in this way, then overall cluster survival is enhanced to even smaller values of $\epsilon \approx 0.2$ (Fellhauer & Kroupa 2005).

Whether clusters pop and what fraction of stars remain in a post-gas expulsion cluster depend critically on the ratio between the gas-removal time-scale and the cluster crossing time. This ratio thus mostly defines which clusters succumb to infant mortality and which clusters merely suffer cluster infant weight loss. The well-studied observational cases do indicate that the removal of most of the residual gas does occur within a cluster-dynamical time, $\tau_{gas}/t_{cr} \leq 1$. Examples noted (Kroupa 2005) are the ONC and R136 in the LMC both of which have significant super-virial velocity dispersions. Other examples are the Treasure-Chest cluster and the very young star-bursting clusters in the massively interacting Antennae galaxy that appear to have HII regions expanding at velocities so that the cluster volume may be evacuated within a cluster dynamical time. However, improved empirical constraints are needed to develop further an understanding of cluster survival. Such observations would best be the velocities of stars in very young star clusters, as they should show a radially expanding stellar population.

Indeed, Bastian & Goodwin (2006) note that many young clusters have the radial-density profile signature expected if they are expanding rapidly. This supports the notion of fast gas blow out. For example, the 0.5–2 Myr old ONC, which is known to be super-virial with a virial mass about twice the observed mass (Hillenbrand & Hartmann 1998), has already expelled its residual gas and is expanding rapidly. It has therefore probably lost its outer stars (Kroupa, Aarseth & Hurley 2001). The super-virial state of young clusters makes measurements of their mass-to-light ratio a bad estimate of the stellar mass within them (Goodwin & Bastian 2006) and rapid dynamical mass-segregation likewise makes naive measurements of the M/L ratio wrong (Boily et al. 2005; Fleck et al. 2006). Goodwin & Bastian (2006) and de Grijs & Parmentier (2007) find the dynamical mass-to-light ratios of young clusters to be too large, strongly implying they are in the process of expanding after gas expulsion.

Weidner et al. (2007) attempted to measure infant weight loss with a sample of young but exposed Galactic clusters. They applied the maximal-star-mass to cluster mass relation from above to estimate the birth mass of the clusters. The uncertainties are large, but the data firmly suggest that the typical cluster loses at least about 50% of its stars.

Binary Stars

Most stars form as binaries with, as far as can be stated today, universal orbital distribution functions (Sect. 8.4). Once a binary system is born in a dense environment, it is perturbed. This changes its eccentricity and semi-major axis. Or it undergoes a relatively strong encounter that disrupts the binary or hardens it perhaps with exchanged companions. The initial binary population therefore evolves on a cluster crossing time-scale and most soft binaries are disrupted. It has been shown that the properties of the Galactic field binary population can be explained in terms of the binary properties observed for very young populations if these go through a dense cluster environment (dynamical population synthesis, Kroupa 1995d). A dense cluster environment hardens existing binaries (p. 240). This increases the SN Ia rate in a galaxy with many dense clusters (Shara & Hurley 2002).

Binaries are significant energy sources (see also Sect. 8.4). A hard binary that interacts via a resonance with a cluster field star occasionally ejects one star with a terminal velocity $v_{\rm ej} \gg \sigma$. The ejected star either leaves the cluster causing cluster expansion so that σ drops, or it shares some of its kinetic energy with the other cluster field stars through gravitational encounters causing cluster expansion. Binaries in a cluster core can thus halt and reverse core collapse (Meylan & Heggie 1997; Heggie & Hut 2003).

Mass Loss from Evolving Stars

An old globular cluster with a turn-off mass near $0.8\,M_\odot$ has lost 30% of the mass that remained in it after gas expulsion by stellar evolution (Baumgardt & Makino 2003). Because the mass loss is most rapid during the earliest times after the cluster returned to virial equilibrium once the gas was expelled, the cluster expands further during this time. This is nicely seen in the Lagrangian radii of realistic cluster-formation models (Kroupa, Aarseth & Hurley 2001).

8.1.2 Some Implications for the Astrophysics of Galaxies

In general, the above have a multitude of implications for galactic and stellar astrophysics.

1. The heaviest-star–star-cluster-mass correlation constrains feedback models of star cluster formation (Elmegreen 1983). It also implies that the sum of all IMFs in all young clusters in a galaxy, the integrated galaxy initial mass function (IGIMF), is steeper than the invariant stellar IMF observed in star clusters. This has important effects on the mass–metallicity relation of galaxies (Koeppen, Weidner & Kroupa 2007). Additionally, star-formation rates (SFRs) of dwarf galaxies can be underestimated by up to three orders of magnitude because Hα-dark star-formation becomes possible (Pflamm-Altenburg, Weidner & Kroupa 2007). This indeed constitutes an

important example of how sub-pc processes influence the physics on cosmological scales.
2. The deduction that type-II clusters probably pop (p. 190) implies that young clusters will appear to an observer to be super-virial, i.e. to have a dynamical mass larger than their luminous mass (Bastian & Goodwin 2006; de Grijs & Parmentier 2007).
3. It further implies that galactic fields can be heated and may also lead to galactic thick discs and stellar halos around dwarf galaxies (Kroupa 2002b).
4. The variation of the gas expulsion time-scale among clusters of different type implies that the star-cluster mass function (CMF) is re-shaped rapidly, on a time-scale of a few tens of Myr (Kroupa & Boily 2002).
5. Associated with this re-shaping of the CMF is the natural production of population II stellar halos during cosmologically early star-formation bursts (Kroupa & Boily 2002; Parmentier & Gilmore 2007; Baumgardt, Kroupa & Parmentier 2008).
6. The properties of the binary-star population observed in Galactic fields are shaped by dynamical encounters in star clusters before the stars leave their cluster (Sect. 8.4).

Points 2–5 are considered in more detail in the rest of Sect. 8.1.

Stellar Associations, Open Clusters and Moving Groups

As one of the important implications of point 2, a cluster in the age range 1–50 Myr has an unphysical M/L ratio because it is out of dynamical equilibrium rather than because it has an abnormal stellar IMF (Bastian & Goodwin 2006; de Grijs & Parmentier 2007).

Another implication is that a Pleiades-like open cluster would have been born in a very dense ONC-type configuration and that, as it evolves, a moving-group-I is established during the first few dozen Myr. This comprises roughly two-thirds of the initial stellar population and the cluster is expanding with a velocity dispersion that is a function of the pre-gas-expulsion configuration (Kroupa, Aarseth & Hurley 2001). These computations were among the first to demonstrate, with high-precision N-body modelling, that the re-distribution of energy within the cluster during the embedded phase and during the expansion phase leads to the formation of a substantial remnant cluster despite the inclusion of all physical effects that are disadvantageous for this to happen (explosive gas expulsion, low SFE $\epsilon = 0.33$, galactic tidal field and mass loss from stellar evolution and an initial binary-star fraction of 100%, see Fig. 8.2). Thus, expanding OB associations may be related to star-cluster birth and many OB associations ought to have remnant star clusters as nuclei (see also Clark et al. 2005).

As the cluster expands becoming part of an OB association, the radiation from its massive stars produce expanding HII regions that may trigger further star-formation in the vicinity (e.g. Gouliermis, Quanz & Henning 2007).

A moving-group-II establishes later – the classical moving group made up of stars that slowly diffuse or evaporate out of the readjusted cluster remnant with relative kinetic energy close to zero. The velocity dispersion of moving-group-I is thus comparable to the pre-gas-expulsion velocity dispersion of the cluster, while moving-group-II has a velocity dispersion close to zero.

The Velocity Dispersion of Galactic-Field Populations and Galactic Thick Discs

Thus, the moving-group-I would be populated by stars that carry the initial kinematic state of the birth configuration into the field of a galaxy. Each generation of star clusters would, according to this picture, produce overlapping moving-groups-I (and II) and the overall velocity dispersion of the new field population can be estimated by adding the squared velocities for all expanding populations. This involves an integral over the embedded-cluster mass function, $\xi_{\rm ecl}(M_{\rm ecl})$, which describes the distribution of the stellar mass content of clusters when they are born. Because the embedded cluster mass function is known to be a power-law, this integral can be calculated for a first estimate (Kroupa 2002b, 2005). The result is that, for reasonable upper cluster mass limits in the integral, $M_{\rm ecl} \leq 10^5\, M_\odot$, the observed age–velocity dispersion relation of Galactic field stars can be reproduced.

This idea can thus explain the much debated energy deficit: namely that the observed kinematic heating of field stars with age could not, until now, be explained by the diffusion of orbits in the Galactic disc as a result of scattering by molecular clouds, spiral arms and the bar (Jenkins 1992). Because the velocity-dispersion for Galactic-field stars increases with stellar age, this notion can also be used to map the star-formation history of the Milky Way disc by resorting to the observed correlation between the star-formation rate in a galaxy and the maximum star-cluster mass born in the population of young clusters (Weidner, Kroupa & Larsen 2004).

An interesting possibility emerges concerning the origin of thick discs. If the star-formation rate was sufficiently high about 11 Gyr ago, star clusters in the disc with masses up to $10^{5.5}\, M_\odot$ would have been born. If they popped a thick disc with a velocity dispersion near $40\,{\rm km\,s^{-1}}$ would result naturally (Kroupa 2002b). This notion for the origin of thick discs appears to be qualitatively supported by the observations of Elmegreen, Elmegreen & Sheets (2004) who find galactic discs at a red shift between 0.5 and 2 to show massive star-forming clumps.

Structuring the Initial Cluster Mass Function

Another potentially important implication from this picture of the evolution of young clusters is that if the ratio of the gas expulsion time to the crossing time or the SFE varies with initial (embedded) cluster mass, an initially featureless power-law mass function of embedded clusters rapidly evolves to one with peaks, dips and turnovers at cluster masses that characterise changes in the broad physics involved.

As an example, Adams (2000) and Kroupa & Boily (2002) assumed that the function

$$M_{\rm icl} = f_{\rm st}(M_{\rm ecl})M_{\rm ecl} \tag{8.26}$$

exists, where $M_{\rm ecl}$ is as above and $M_{\rm icl}$ is the classical initial cluster mass and $f_{\rm st} = f_{\rm st}(M_{\rm ecl})$. According to Kroupa & Boily (2002), the classical initial cluster mass is that mass which is inferred by standard N-body computations without gas expulsion (in effect this assumes $\epsilon = 1$, which is however, unphysical). Thus, for example, for the Pleiades, $M_{\rm cl} \approx 1000\,M_\odot$ at the present time (age about 100 Myr). A classical initial model would place the initial cluster mass near $M_{\rm icl} \approx 1500\,M_\odot$ by standard N-body calculations to quantify the secular evaporation of stars from an initially bound and relaxed cluster (Portegies Zwart et al. 2001). If, however, the SFE was 33% and the gas-expulsion time-scale were comparable to or shorter than the cluster dynamical time, the Pleiades would have been born in a compact configuration resembling the ONC and with a mass of embedded stars of $M_{\rm ecl} \approx 4000\,M_\odot$ (Kroupa, Aarseth & Hurley 2001). Thus, $f_{\rm st}(4000\,M_\odot) = 0.38\,(= 1500/4000)$.

By postulating that there exist three basic types of embedded clusters (Kroupa & Boily 2002), namely

Type I: clusters without O stars ($M_{\rm ecl} \leq 10^{2.5}\,M_\odot$, e.g. Taurus-Auriga pre-main sequence stellar groups, ρ Oph),

Type II: clusters with a few O stars ($10^{2.5} \leq M_{\rm ecl}/M_\odot \leq 10^{5.5}$, e.g. the ONC),

Type III: clusters with many O stars and with a velocity dispersion comparable to or higher than the sound velocity of ionized gas ($M_{\rm ecl} \geq 10^{5.5}\,M_\odot$),

it can be argued that $f_{\rm st} \approx 0.5$ for type I, $f_{\rm st} < 0.5$ for type II and $f_{\rm st} \approx 0.5$ for type III. The reason for the high $f_{\rm st}$ values for types I and III is that gas expulsion from these clusters may last longer than the cluster dynamical time because there is no sufficient ionizing radiation for type I clusters, or the potential well is too deep for the ionized gas to leave (type III clusters). The evolution is therefore adiabatic ((8.25) above). Type II clusters undergo a disruptive evolution and witness a high infant mortality rate (Lada & Lada 2003). They are the pre-cursors of OB associations and Galactic clusters. This broad categorisation has easy-to-understand implications for the star-cluster mass function.

Under these conditions and an assumed functional form for $f_{\rm st} = f_{\rm st}(M_{\rm ecl})$, the power-law embedded cluster mass function transforms into a cluster mass function with a turnover near $10^5\,M_\odot$ and a sharp peak near $10^3\,M_\odot$ (Kroupa & Boily 2002). This form is strongly reminiscent of the initial globular cluster mass function, which is inferred by, for example, Vesperini (1998, 2001), Parmentier & Gilmore (2005) and Baumgardt (1998) to be required for a match with the evolved cluster mass function that is seen to have a universal turnover near $10^5 M_\odot$. By the reasoning given above, this "initial" CMF is, however, unphysical, being a power-law instead.

This analytical formulation of the problem has been verified nicely with N-body simulations combined with a realistic treatment of residual gas expulsion by Baumgardt, Kroupa & Parmentier (2008), who show the Milky Way globular cluster mass function to emerge from a power-law embedded-cluster mass function. Parmentier et al. (2008) expand on this by studying the effect that different assumptions on the physics of gas removal have on shaping the star-cluster mass function within about 50 Myr. The general ansatz that residual gas expulsion plays a dominant role in early cluster evolution may thus solve the long-standing problem that the deduced initial cluster mass function needs to have this turnover, while the observed mass functions of young clusters are featureless power-law distributions.

The Origin of Population II Stellar Halos

The above view implies naturally that a major field-star component is generated whenever a population of star clusters forms. About 12 Gyr ago, the Milky Way began its assembly by an initial burst of star-formation throughout a volume spanning about 10 kpc in radius. In this volume, the star-formation rate must have reached $10\,M_\odot\,\mathrm{yr}^{-1}$ so that star clusters with masses up to $\approx 10^6\,M_\odot$ formed (Weidner, Kroupa & Larsen 2004), probably in a chaotic, turbulent early interstellar medium. The vast majority of embedded clusters suffered infant weight loss or mortality. The surviving long-lived clusters evolved to globular clusters. The so-generated field population is the spheroidal population-II halo, which has the same chemical properties as the surviving (globular) star clusters, apart from enrichment effects evident in the most massive clusters. All of these characteristics emerge naturally in the above model, as pointed out by Kroupa & Boily (2002), Parmentier & Gilmore (2007) and most recently by Baumgardt, Kroupa & Parmentier (2008).

8.1.3 Long-Term, or Classical, Cluster Evolution

The long-term evolution of star clusters that survive infant weight loss and the mass loss from evolving stars is characterised by three physical processes, the drive of the self-gravitating system towards energy equipartition, stellar evolution processes and the heating or forcing of the system through external tides. One emphasis of star-cluster work in this context is to test the theory of stellar evolution and to investigate the interrelation of stellar astrophysics with stellar dynamics. The stellar-evolution and the dynamical-evolution time-scales are comparable. The reader is directed to Meylan & Heggie (1997) and Heggie & Hut (2003) for further details.

Tidal Tails

Tidal tails contain the stars evaporating from long-lived star clusters (the moving-group-II above). The typical S-shaped structure of tidal tails close to the cluster are easily understood: stars that leave the cluster with a slightly higher galactic velocity than the cluster are on slightly outward-directed galactic orbits and therefore fall behind the cluster as the angular velocity about the galactic centre decreases with distance. The outward-directed trailing arm develops. Stars that leave the cluster with slower galactic velocities than the cluster fall towards the galaxy and overtake the cluster.

Given that energy equipartition leads to a filtering in energy space of the stars that escape at a particular time, one expects a gradient in the stellar mass function progressing along a tidal tail towards the cluster so that the mass function becomes flatter, richer in more massive stars. This effect is difficult to detect but, for example, the long tidal tails found emanating from Pal 5 (Odenkirchen et al. 2003) may show evidence for it.

As emphasised by Odenkirchen et al. (2003), tidal tails have another very interesting use: they probe the gravitational potential of the Milky Way if the differential motions along the tidal tail can be measured. They are thus important future tests of gravitational physics.

Death and Hierarchical Multiple Stellar Systems

Nothing lasts forever and star clusters that survive initial relaxation to virial equilibrium after residual gas expulsion and mass loss from stellar evolution ultimately cease to exist after all member stars evaporate to leave a binary or a long-lived hierarchical multiple system composed of near-equal mass components (de la Fuente Marcos 1997, 1998). Note that these need not be single stars. These cluster remnants are interesting, because they may account for most of the hierarchical multiple stellar systems in the Galactic field (Goodwin & Kroupa 2005), with the implication that these are not a product of star-formation but rather of star-cluster dynamics.

8.1.4 What is a Galaxy?

Star clusters, dwarf-spheroidal (dSph) and dwarf-elliptical (dE) galaxies as well as galactic bulges and giant elliptical (E) galaxies are all stellar-dynamical systems that are supported by random stellar motions, i.e. they are pressure-supported. But why is one class of these pressure-supported systems referred to as star clusters, while the others are galaxies? Is there some fundamental physical difference between these two classes of systems?

Considering the radius as a function of mass, we notice that systems with $M \le 10^6\, M_\odot$ do not show a mass–radius relation (MRR) and have $r \approx 4\,\mathrm{pc}$. More massive objects, however, show a well-defined MRR. In fact, Dabringhausen, Hilker & Kroupa (2008) find that massive compact objects (MCOs), which have $10^6 \leq M/M_\odot \leq 10^8$, lie on the MRR of giant E galaxies (about $10^{13}\, M_\odot$) down to normal E galaxies ($10^{11}\, M_\odot$), as is evident in Fig. 8.3:

$$R/\mathrm{pc} = 10^{-3.15} \left(\frac{M}{M_\odot}\right)^{0.60\pm0.02} . \tag{8.27}$$

Noteworthy is that systems with $M \geq 10^6\, M_\odot$ also exhibit complex stellar populations, while less massive systems have single-age, single-metallicity populations. Remarkably, Pflamm-Altenburg & Kroupa (2008) show that a stellar system with $M \geq 10^6\, M_\odot$ and a radius as observed for globular clusters can accrete gas from a co-moving warm inter-stellar medium and may re-start star-formation. The median two-body relaxation time is longer than a Hubble time for $M \geq 3 \times 10^6\, M_\odot$ and only for these systems is there evidence for a slight increase in the dynamical mass-to-light ratio. Intriguingly, $(M/L)_V \approx 2$ for $M < 10^6\, M_\odot$, while $(M/L)_V \approx 5$ for $M > 10^6\, M_\odot$ with a possible decrease for $M > 10^8\, M_\odot$ (Fig. 8.4). Finally, the average stellar density maximises at $M = 10^6\, M_\odot$ with about $3 \times 10^3\, M_\odot/\mathrm{pc}^3$ (Dabringhausen, Hilker & Kroupa 2008).

Thus,

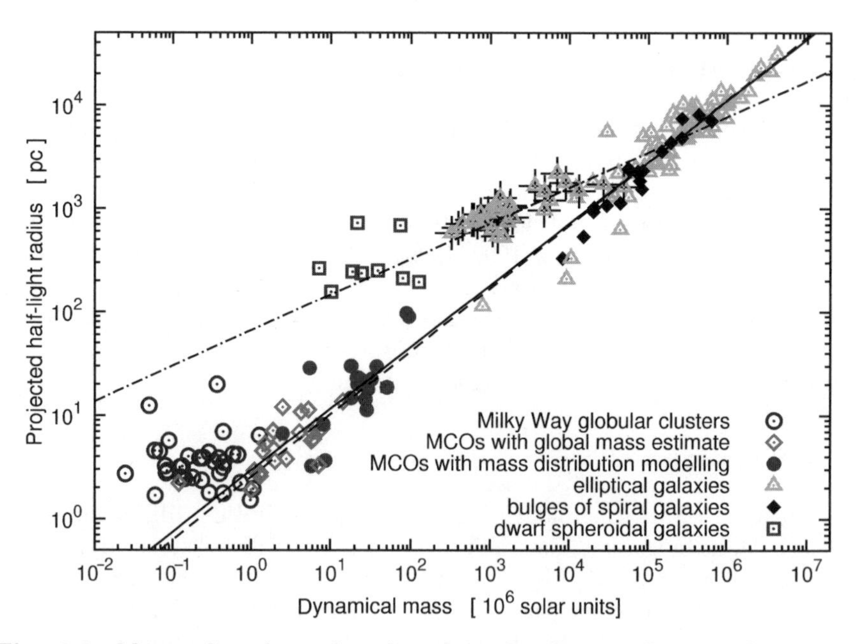

Fig. 8.3. Mass–radius data plotted against the dynamical mass of pressure-supported stellar systems (Dabringhausen, Hilker & Kroupa 2008). MCOs are massive compact objects (also referred to as ultra compact dwarf galaxies). The solid and dashed lines refer to (8.27), while the *dash-dotted line* is a fit to dSph and dE galaxies

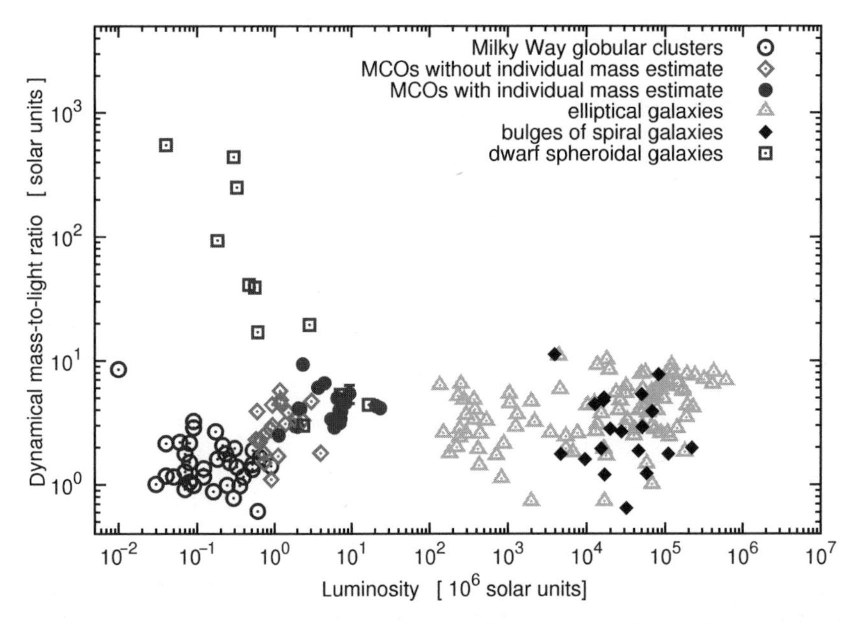

Fig. 8.4. Dynamical M/L values in dependence of the V-band luminosity of pressure-supported stellar systems (Dabringhausen, Hilker & Kroupa 2008). MCOs are massive compact objects (also referred to as ultra compact dwarf galaxies)

- the mass $10^6\,M_\odot$ appears to be special,
- stellar populations become complex above this mass,
- evidence for some dark matter only appears in systems that have a median two-body relaxation time longer than a Hubble time,
- dSph galaxies are the only stellar-dynamical systems with $10 < (M/L)_V < 1000$ and as such are total outliers and
- $10^6\,M_\odot$ is a lower accretion limit for massive star clusters immersed in a warm inter-stellar medium.

$M \approx 10^6\,M_\odot$ therefore appears to be a critical mass scale so that less-massive objects show characteristics of star clusters that are described well by Newtonian dynamics, while more massive objects show behaviour more typical of galaxies. Defining a galaxy as a stellar-dynamical object which has a median two-body relaxation time longer than a Hubble time, i.e. essentially a system with a smooth potential, may be an objective and useful way to define a galaxy (Kroupa 1998). Why only smooth systems show evidence for dark matter remains at best a striking coincidence, at worst it may be symptomatic of a problem in understanding dynamics in such systems.

8.2 Initial 6D Conditions

The previous section gave an outline of some of the issues at stake in the realm of pressure-supported stellar systems. In order to attack these and other problems, we need to know how to set up such systems in the computer. Indeed, as much as analytical solutions may be preferred, the mathematical and physical complexities of dense stellar systems leave no alternatives other than to resort to full-scale numerical integration of the $6N$ coupled first-order differential equations that describe the motion of the system through $6N$-dimensional phase space. There are three related questions to ponder. Given a well-developed cluster, how is one to set it up in order to evolve it forward in time? How does a cluster form and how does the formation process affect its later properties? How do we describe a realistic stellar population (IMF, binaries)? Each of these questions is dealt with in the following sections.

8.2.1 6D Structure of Classical Clusters

Because the state of a star cluster is never known exactly, it is necessary to perform numerical experiments with conditions that are, statistically, consistent with the cluster snap-shot. To ensure meaningful statistical results for systems with few stars, say $N < 5000$, many numerical renditions of the same object are thus necessary. For example, systems with $N = 100$ stars evolve erratically and numerical experiments are required to map out the range of possible states at a particular time: the range of half-mass radii at an age of 20 Myr in 1000 numerical experiments of a cluster, initially with $N = 100$ stars and with an initial half-mass radius $r_{0.5} = 0.5\,\mathrm{pc}$, can be compared with an actually observed object for testing consistency with the initial conditions. Excellent recent examples of this approach can be found in Hurley et al. (2005) and Portegies Zwart, McMillan & Makino (2007), with a recent review available by Hut et al. (2007) and two text books have been written dealing with computational and more general aspects of the physics of dense stellar systems (Aarseth 2003; Heggie & Hut 2003).

The six-dimensional structure of a pressure-supported stellar system at time t is conveniently described by the phase-space distribution function, $f(\boldsymbol{r}, \boldsymbol{v}; t)$, where $\boldsymbol{r}$ and $\boldsymbol{v}$ are the phase-space variables and

$$\mathrm{d}N = f(\boldsymbol{r}, \boldsymbol{v}; t)\, d^3x\, d^3v \qquad (8.28)$$

is the number of stars in 6D phase-space volume element $d^3x\, d^3v$. In the case of a steady state, the Jeans theorem (Binney & Tremaine 1987, their Sect. 4.4) allows us to express f in terms of the integrals of motion, i.e. the energy and angular momentum. The phase-space distribution function can then be written as

$$f = f(\boldsymbol{r}, \boldsymbol{v}) = f(\epsilon_e, l), \qquad (8.29)$$

where

$$\epsilon_e = \frac{1}{2} v^2 + \Phi(\boldsymbol{r}) \tag{8.30}$$

is the specific energy of a star and

$$l = |\boldsymbol{r} \times \boldsymbol{v}| \tag{8.31}$$

is the specific orbital angular momentum of a star. The Poisson equation is

$$\nabla^2 \Phi(\boldsymbol{r}) = 4\pi G \rho_m(\boldsymbol{r}) = 4\pi G \int_{\text{allspace}} m f \, d^3 v, \tag{8.32}$$

or in spherical symmetry,

$$\frac{1}{r^2} \frac{\mathrm{d}}{\mathrm{d}r} \left(r^2 \frac{\mathrm{d}\Phi}{\mathrm{d}r} \right) = 4\pi G \int_{\text{allspace}} f_m \left(\frac{1}{2} v^2 + \Phi, |\boldsymbol{r} \times \boldsymbol{v}| \right) d^3 v, \tag{8.33}$$

where f_m is the phase-space mass-density of all matter and is equal to $m\,f$ for a system with equal-mass stars. Most pressure-supported systems have a near-spherical shape and so in most numerical work it is convenient to assume spherical symmetry.

For convenience it is useful to introduce the relative potential,[6]

$$\Psi \equiv -\Phi + \Phi_0 \tag{8.34}$$

and the relative energy

$$\mathcal{E} \equiv -\epsilon_e + \Phi_0 = \Psi - \frac{1}{2} v^2, \tag{8.35}$$

where Φ_0 is a constant so that $f > 0$ for $\mathcal{E} > 0$ and $f = 0$ for $\mathcal{E} \leq 0$. The Poisson equation becomes $\nabla^2 \Psi = -4\pi G \rho_m$ subject to the boundary condition $\Psi \to \Phi_0$ as $\boldsymbol{r} \to \infty$.

One important property of stellar systems is the anisotropy of their velocity distribution function. We define the anisotropy parameter

$$\beta(r) \equiv 1 - \frac{\overline{v_\theta^2}}{\overline{v_r^2}}, \tag{8.36}$$

where $\overline{v_\theta^2}, \overline{v_r^2}$ are the mean squared tangential and radial velocities at a particular location $\boldsymbol{r}$, respectively. It follows that systems with $\beta = 0$ everywhere have an isotropic velocity distribution function.

If f only depends on the energy the mean squared radial and tangential velocities are, respectively,

$$\overline{v_r^2} = \frac{1}{\rho} \int_{\text{all vel.}} v_r^2 f \left[\Psi - \frac{1}{2} \left(v_r^2 + v_\theta^2 + v_\phi^2 \right) \right] \mathrm{d}v_r \, \mathrm{d}v_\theta \, \mathrm{d}v_\phi \tag{8.37}$$

[6]The following discussion is based on Binney & Tremaine (1987).

and

$$\overline{v_\theta^2} = \frac{1}{\rho} \int_{\text{all vel.}} v_\theta^2 \, f \left[\Psi - \frac{1}{2} \left(v_r^2 + v_\theta^2 + v_\phi^2 \right) \right] \, \mathrm{d}v_r \, \mathrm{d}v_\theta \, \mathrm{d}v_\phi . \tag{8.38}$$

If the labels θ and r are exchanged in (8.38), it can be seen that one arrives at (8.37). Equations (8.37) and (8.38) are thus identical, apart from the labelling. Thus if $f = f(\mathcal{E})$, $\beta = 0$ and the velocity distribution function is isotropic.

If f depends on the energy and the orbital angular momentum of the stars ($|\boldsymbol{l}| = |\boldsymbol{r} \times \boldsymbol{v}|$), then the mean squared radial and tangential velocities are, respectively,

$$\overline{v_r^2} = \frac{1}{\rho} \int_{\text{all vel.}} v_r^2 \, f \left[\Psi - \frac{1}{2} \left(v_r^2 + v_\theta^2 + v_\phi^2 \right), r \sqrt{v_\theta^2 + v_\phi^2} \right] \, \mathrm{d}v_r \, \mathrm{d}v_\theta \, \mathrm{d}v_\phi \tag{8.39}$$

and

$$\overline{v_\theta^2} = \frac{1}{\rho} \int_{\text{all vel.}} v_\theta^2 \, f \left[\Psi - \frac{1}{2} \left(v_r^2 + v_\theta^2 + v_\phi^2 \right), r \sqrt{v_\theta^2 + v_\phi^2} \right] \, \mathrm{d}v_r \, \mathrm{d}v_\theta \, \mathrm{d}v_\phi . \tag{8.40}$$

If the labels θ and r are exchanged in (8.40), it can be seen that this time one does not arrive at (8.39). Thus if $f = f(\mathcal{E}, l)$, then $\beta \neq 0$ and the velocity distribution function is not isotropic. This serves to demonstrate an elementary but useful property of the phase-space distribution function.

A very useful series of distribution functions can be arrived at from the simple form

$$f_m(\mathcal{E}) = \begin{cases} F \, \mathcal{E}^{n-\frac{3}{2}} & : \quad \mathcal{E} > 0, \\ 0 & : \quad \mathcal{E} \leq 0. \end{cases} \tag{8.41}$$

The mass density,

$$\rho_m(r) = 4 \pi F \int_0^{\sqrt{2\Psi}} \left(\Psi - \frac{1}{2} v^2 \right)^{n-\frac{3}{2}} v^2 \, \mathrm{d}v, \tag{8.42}$$

where the upper integration bound is given by the escape condition, $\mathcal{E} = \Psi - (1/2)v^2 = 0$. Substituting $v^2 = 2 \Psi \cos^2\theta$ for some θ leads to

$$\rho_m(r) = \begin{cases} c_n \, \Psi^n & : \quad \Psi > 0, \\ 0 & : \quad \Psi \leq 0. \end{cases} \tag{8.43}$$

For c_n to be finite, $n > 1/2$, i.e. homogeneous ($n = 0$) systems are excluded.

The Lane–Emden equation follows from the spherically symmetric Poisson equation after introducing dimensionless variables $s = r/b, \psi = \Psi/\Psi_0$, where $b = (4 \pi G \, \Psi_0^{n-1} \, c_n)^{-1/2}$ and $\Psi_0 = \Psi(0)$,

$$\frac{1}{s^2} \frac{\mathrm{d}}{\mathrm{d}s} \left(s^2 \frac{\mathrm{d}\psi}{\mathrm{d}s} \right) = \begin{cases} -\psi^n & : \quad \psi > 0, \\ 0 & : \quad \psi \leq 0. \end{cases} \tag{8.44}$$

H. Lane and R. Emden worked with this equation in the context of self-gravitating polytropic gas spheres, which have an equation of state

$$p = K\,\rho_m^{\gamma}, \tag{8.45}$$

where K is a constant and p the pressure. It can be shown that $\gamma = 1 + 1/n$. That is, the density distribution of a stellar polytrope of index n is the same as that of a polytropic gas sphere with index γ.

The natural boundary conditions to be imposed on (8.44) are at $s = 0$,

1. $\psi = 1$, because $\Psi(0) = \Psi_0$ and
2. $\mathrm{d}\psi/\mathrm{d}s = 0$ because the gravitational force must vanish at the centre.

Analytical solutions to the Lane–Emden equation are possible only for a few values of n, and we remember that a homogeneous ($n = 0$) stellar density distribution has already been excluded as a viable solutions of the general power-law phase-space distribution function.

The Plummer Model

A particularly useful case is

$$\psi = \frac{1}{\sqrt{1 + \frac{1}{3}\,s^2}}. \tag{8.46}$$

It follows immediately that this is a solution of the Lane–Emden equation for $n = 5$ and it also satisfies the two boundary conditions above and so constitutes a physically sensible potential. By integrating the Poisson equation, it can be shown that the total mass of this distribution function is finite,

$$M_\infty = \sqrt{3}\,\Psi_0\,b/G, \tag{8.47}$$

although the density distribution has no boundary. The distribution function is

$$f_m(\mathcal{E}) = \begin{cases} F\,\left(\Psi - \frac{1}{2}\,v^2\right)^{\frac{7}{2}} & , \quad v^2 < 2\Psi, \\ 0 & , \quad v^2 \geq 2\Psi, \end{cases} \tag{8.48}$$

with the relative potential

$$\Psi = \frac{\Psi_0}{\sqrt{1 + \frac{1}{3}\left(\frac{r}{b}\right)^2}} \tag{8.49}$$

and density law

$$\rho_m = \frac{\rho_{m,0}}{\left(1 + \frac{1}{3}\left(\frac{r}{b}\right)^2\right)^{\frac{5}{2}}} \tag{8.50}$$

with the above total mass. This density distribution is known as the Plummer model, named after Plummer (1911) who showed that the density distribution that results from this model provides a reasonable and, in particular, very simple analytical description of globular clusters. The Plummer model is, in fact, a work-horse for many applications in stellar dynamics because many of its properties such as the projected velocity dispersion profile can be calculated analytically. Such formulae are useful for checking numerical codes used to set up models of stellar systems.

Properties of the Plummer Model

Some useful analytical results can be derived for the Plummer density law (see also Heggie & Hut 2003, their p. 73 for another compilation). For the Plummer law of mass $M_{\rm ecl}$, the mass-density profile (8.50) can be written as

$$\rho_m(r) = \frac{3\,M_{\rm ecl}}{4\,\pi\,r_{\rm pl}^3}\,\frac{1}{\left[1+\left(\frac{r}{r_{\rm pl}}\right)^2\right]^{\frac{5}{2}}}, \tag{8.51}$$

where $r_{\rm pl}$ is the Plummer scale length. The central number density is thus

$$\rho_{\rm c} = \frac{3\,N}{4\,\pi\,r_{\rm pl}^3}. \tag{8.52}$$

The mass within radius r follows from $M(r) = 4\,\pi\,\int_0^r \rho_m(r')\,r'^2\,{\rm d}r'$,

$$M(r) = M_{\rm ecl}\,\frac{\left(\frac{r}{r_{\rm pl}}\right)^3}{\left[1+\left(\frac{r}{r_{\rm pl}}\right)^2\right]^{\frac{3}{2}}}. \tag{8.53}$$

Thus,

$r_{\rm pl}$ contains 35.4% of the mass,
$2\,r_{\rm pl}$ contain 71.6%,
$5\,r_{\rm pl}$ contain 94.3% and
$10\,r_{\rm pl}$ contain 98.5% of the total mass.

For the half-mass radius we have

$$r_{\rm h} = (2^{\frac{2}{3}} - 1)^{-\frac{1}{2}}\,r_{\rm pl} \approx 1.305\,r_{\rm pl}. \tag{8.54}$$

The projected surface mass density, $\Sigma_M(R) = 2\,\int_0^\infty \rho_m(r)\,{\rm d}z$, where R is the projected radial distance from the cluster centre and Z is the integration variable along the line-of-sight ($r^2 = R^2 + Z^2$) is

$$\Sigma_\rho(R) = \frac{M_{\rm ecl}}{\pi\,r_{\rm pl}^2}\,\frac{1}{\left[1+\left(\frac{R}{r_{\rm pl}}\right)^2\right]^2}. \tag{8.55}$$

We assume there is no mass segregation so that the mass-to-light ratio, $\Upsilon \equiv (M/L)$, measured in some photometric system is independent of radius. The integrated light within projected radius R is

$$I(R) = (1/\Upsilon) \int_0^R \Sigma_\rho(R')\, 2\,\pi\, R'\, \mathrm{d}R', \tag{8.56}$$

$$I(R) = \frac{M_{\mathrm{ecl}}\, r_{\mathrm{pl}}^2}{\Upsilon} \left[\frac{1}{r_{\mathrm{pl}}^2} - \frac{1}{R^2 + r_{\mathrm{pl}}^2} \right]. \tag{8.57}$$

Thus, r_{pl} is the half-light radius of the projected star cluster, $I(r_{\mathrm{pl}}) = 0.5\, I(\infty)$.

In the above equations $\rho(r) = \rho_m(r)/\overline{m}$, $N(r) = M(r)/\overline{m}$ and $\Sigma_n = \Sigma_\rho/\overline{m}$ are, respectively, the stellar number density, the number of stars within radius r and the projected surface number density profile if there is no mass segregation within the cluster. Thus the average stellar mass, $\overline{m}$, is constant.

The velocity dispersion can be calculated at any radius from the Jeans equation (8.120). For an isotropic velocity distribution ($\sigma_\theta^2 = \sigma_\phi^2 = \sigma_r^2$), such as the Plummer model, the Jeans equation yields

$$\sigma_r^2(r) = \frac{1}{\rho(r)} \int_r^\infty \rho(r')\, \frac{G\, M(r')}{r^2}\, \mathrm{d}r', \tag{8.58}$$

because $\mathrm{d}\phi(r)/\mathrm{d}r = GM(r)/r^2$ and the integration bounds have been chosen to make use of the vanishing $\rho_m(r)$ as $r \to \infty$. Note that the above equation is also valid if $M(r)$ consists of more than one spherical component such as a distinct core plus an extended halo. Combining (8.51), (8.53) and (8.58) we are led to

$$\sigma^2(r) = \left(\frac{G\, M_{\mathrm{ecl}}}{2\, r_{\mathrm{pl}}} \right) \frac{1}{\left[1 + \left(\frac{r}{r_{\mathrm{pl}}} \right)^2 \right]^{\frac{1}{2}}}, \tag{8.59}$$

where $\sigma(r)$ is the three-dimensional velocity dispersion of the Plummer sphere at radius r, $\sigma^2(r) = \sum_{k=r,\theta,\phi} \sigma_k^3(r)$ or $\sigma^2(r) = 3\, \sigma_{\mathrm{1D}}^2(r)$ because isotropy is assumed.

A star with mass m positioned at r and with speed $v = \left(\sum_{k=1}^3 v_k^2 \right)^{1/2}$ can escape from the cluster if it has a total energy $e_{\mathrm{bind}} = e_{\mathrm{kin}} + e_{\mathrm{pot}} = 0.5\, m\, v^2 + m\, \phi(r) \geq 0$ so that $v \geq v_{\mathrm{esc}}(r)$. So the escape speed at radius r is $v_{\mathrm{esc}}(r) = \sqrt{2\, |\phi(r)|}$. The potential at r is given by the mass within r plus the potential contributed by the surrounding matter. It is calculated by integrating the contributions from each radial mass shell,

$$\begin{aligned} \phi(r) &= - \left[G\, \frac{M(r)}{r} + \int_r^\infty G\, \frac{1}{r'}\, \rho(r')\, 4\, \pi\, r'^2\, \mathrm{d}r' \right], \\ &= - \left(\frac{G\, M_{\mathrm{ecl}}}{r_{\mathrm{pl}}} \right) \frac{1}{\left[1 + (r/r_{\mathrm{pl}})^2 \right]^{1/2}}. \end{aligned} \tag{8.60}$$

so that

$$v_{\mathrm{esc}}(r) = \left(\frac{2\, G\, M_{\mathrm{ecl}}}{r_{\mathrm{pl}}} \right)^{1/2} \frac{1}{\left[1 + (r/r_{\mathrm{pl}})^2 \right]^{1/4}}. \tag{8.61}$$

The circular speed, v_c, of a star moving on a circular orbit at a distance r from the cluster centre is obtained from centrifugal acceleration, $v_c^2/r = \mathrm{d}\phi(r)/\mathrm{d}r = G\,M(r)/r^2$,

$$v_c^2 = \left(\frac{G\,M_{\rm ecl}}{r_{\rm pl}}\right) \frac{(r/r_{\rm pl})^2}{[1+(r/r_{\rm pl})^2]^{3/2}}. \tag{8.62}$$

In many but not all instances of interest, the initial cluster model is chosen to be in the state of virial equilibrium. That is, the kinetic and potential energies of each star balance so that the whole cluster is stationary. The scalar virial theorem,

$$2\,K + W = 0, \tag{8.63}$$

where K and W are the total kinetic and potential energy of the cluster,[7]

$$K = \frac{1}{2}\int_0^\infty \rho(r)\,\sigma^2(r)\,4\pi r^2 \mathrm{d}r,$$

$$= \frac{3\pi}{64}\frac{G\,M_{\rm ecl}^2}{r_{\rm pl}}, \quad \text{for the Plummer sphere,} \tag{8.64}$$

$$W = \frac{1}{2}\int_0^\infty \phi(r)\,\rho(r)\,4\pi r^2 \mathrm{d}r,$$

$$= -\frac{3\pi}{32}\frac{G\,M_{\rm ecl}^2}{r_{\rm pl}} \quad \text{for the Plummer sphere.} \tag{8.65}$$

The total, or binding, energy of the cluster, $E_{\rm tot} = W + K$, is

$$E_{\rm tot} = -K = \frac{1}{2}\,W. \tag{8.66}$$

The characteristic three-dimensional velocity dispersion of a cluster can be defined as $\sigma_{\rm cl}^2 \equiv 2\,K/M_{\rm ecl}$ so that

$$\sigma_{\rm cl}^2 = \frac{3\,\pi}{32}\frac{G\,M_{\rm ecl}}{r_{\rm pl}}, \tag{8.67}$$

$$\equiv \frac{G\,M_{\rm ecl}}{r_{\rm grav}}, \tag{8.68}$$

$$\equiv s^2\left(\frac{G\,M_{\rm ecl}}{2\,r_{\rm h}}\right), \tag{8.69}$$

which introduces the gravitational radius of the cluster, $r_{\rm grav} \equiv G\,M_{\rm ecl}^2/|W|$. For the Plummer sphere $r_{\rm grav} = (32/3\,\pi) r_{\rm pl} = 3.4\,r_{\rm pl}$ and the structure factor

$$s = \left(\frac{6\times 1.305\,\pi}{32}\right)^{\frac{1}{2}},$$

$$\approx 0.88. \tag{8.70}$$

[7]Equation (3.251.4) on p. 295 of Gradshteyn & Ryzhik (1980) is useful to solve the integrals for the Plummer sphere.

We define the virial ratio by

$$Q = \frac{K}{|W|}, \tag{8.71}$$

so that a cluster can initially be in three possible states

$$Q \begin{cases} = \frac{1}{2} & , \text{ virial equilibrium,} \\ > \frac{1}{2} & , \text{ expanding,} \\ < \frac{1}{2} & , \text{ collapsing.} \end{cases} \tag{8.72}$$

Note that if initially $Q < 1/2$, the value $Q = 1/2$ will be reached temporarily during collapse, after which Q increases further until the cluster settles in virial equilibrium after this violent relaxation phase (Binney & Tremaine 1987, p. 271).

The characteristic crossing time through the Plummer cluster,

$$t_{\rm cr} \equiv \frac{2\, r_{\rm pl}}{\sigma_{\rm 1D,cl}}, \tag{8.73}$$

$$= \left(\frac{128}{\pi\, G}\right)^{\frac{1}{2}} M_{\rm ecl}^{-\frac{1}{2}}\, r_{\rm pl}^{\frac{3}{2}}, \tag{8.74}$$

with the characteristic one-dimensional velocity dispersion, $\sigma_{\rm 1D,cl} = \sigma_{\rm cl}/\sqrt{3}$.

Observationally, the core radius is that radius where the projected surface density falls to half its central value. For a real cluster it is much easier to determine than the other characteristic radii. For the Plummer sphere,

$$R_{\rm core} = \left(\sqrt{2} - 1\right)^{\frac{1}{2}} r_{\rm pl} = 0.64\, r_{\rm pl}, \tag{8.75}$$

from (8.55), with the assumption that the mass-to-light ratio, Υ, is independent of radius. For a King model

$$R_{\rm core}^{king} = \left(\frac{9}{4\pi\, G}\, \frac{\sigma^2}{\rho_m(0)}\right)^{\frac{1}{2}}, \tag{8.76}$$

is the King radius. From (8.59), $\sigma^2(0) = G\, M_{\rm ecl}/(2\, r_{\rm pl})$ and from (8.51), $\rho_m(0) = 3\, M_{\rm ecl}/(4\pi\, r_{\rm pl}^3)$ so that

$$r_{\rm pl} = \left(\frac{6}{4\pi\, G}\, \frac{\sigma(0)^2}{\rho_m(0)}\right)^{\frac{1}{2}} = 0.82\, R_{\rm core}^{\rm king}. \tag{8.77}$$

The Singular Isothermal Model

Another useful set of distribution functions can be arrived at by considering $n = \infty$. The Lane–Emden equation is not well defined in this limit, but for a polytropic gas sphere (8.45) implies $\gamma \to 1$ as $n \to \infty$. Thus $p = K\,\rho_m$, which is the equation of state of an isothermal ideal gas with $K = k_{\rm B}\,T/m_{\rm p}$, where $k_{\rm B}$ is Boltzmann's constant, T the temperature and $m_{\rm P}$ the mass of a gas particle. From the equation of hydrostatic support, $\mathrm{d}p/\mathrm{d}r = -\rho_m(G\,M(r)/r^2)$, where $M(r)$ is the mass within r, the following equation can be derived

$$\frac{\mathrm{d}}{\mathrm{d}r}\left(r^2\frac{\mathrm{d}\ln\rho_m}{\mathrm{d}r}\right) = -\frac{G\,m_{\rm p}}{k_{\rm B}\,T}\,4\,\pi\,r^2\,\rho_m. \tag{8.78}$$

For a distribution function (our ansatz)

$$f_m(\mathcal{E}) = \frac{\rho_{m,1}}{(2\,\pi\,\sigma^2)^{\frac{3}{2}}}\,e^{\frac{\mathcal{E}}{\sigma^2}}, \tag{8.79}$$

where σ^2 is a new quantity related to a velocity dispersion and $\mathcal{E} = \Psi - v^2/2$, one obtains, from $\rho_m = \int f_m(\mathcal{E})\,4\,\pi\,v^2\,\mathrm{d}v$,

$$\Psi(r) = \ln\left(\frac{\rho_m(r)}{\rho_{m,1}}\right)\,\sigma^2. \tag{8.80}$$

From the Poisson equation it then follows that

$$\sigma = \text{const} = \frac{k_{\rm B}\,T}{m_p} \tag{8.81}$$

for consistency with (8.78).

Therefore, the structure of an isothermal, self-gravitating sphere of ideal gas is identical to the structure of a collisionless system of stars whose phase-space mass-density distribution function is given by (8.79). Note that $f(\mathcal{E})$ is non-zero at all $\mathcal{E}$ (cf. King's models below).

The number-distribution function of velocities is $F(v) = \int_{\text{all}\,\boldsymbol{x}} f(\mathcal{E})\,d^3x$, i.e.

$$F(v) = F_0\,e^{-\frac{v^2}{2\,\sigma^2}}. \tag{8.82}$$

This is the Maxwell–Boltzmann distribution, which results from the kinetic theory of atoms in a gas at temperature T that are allowed to bounce off each other elastically. This exact correspondence between a stellar-dynamical system and a gaseous polytrope holds only for an isothermal case ($n = \infty$).

The total number of stars in the system is $N_{\rm tot} = N_{\rm tot}\int_0^\infty F(v)\,4\,\pi\,v^2\,\mathrm{d}v$ and the number of stars in the speed interval v to $v + \mathrm{d}v$ is

$$dN = F(v)\,4\,\pi\,v^2\,\mathrm{d}v = N_{\rm tot}\frac{1}{(2\,\pi\sigma^2)^{\frac{3}{2}}}e^{-\frac{v^2}{2\,\sigma^2}}\,4\,\pi\,v^2\,\mathrm{d}v, \tag{8.83}$$

which is the Maxwell–Boltzmann distribution of speeds. The mean-square speed of stars at a point in the isothermal sphere is

$$\overline{v^2} = \frac{4\pi \int_0^\infty \sigma^2 F(v)\,\mathrm{d}v}{4\pi \int_0^\infty F(v)\,\mathrm{d}v} = 3\sigma^2 \tag{8.84}$$

and the 1D velocity dispersion is $\sigma_{\mathrm{1D}} = \sigma_\alpha = \sigma$, where $\alpha = r, \theta, \phi, x, y, z, \ldots$.

To obtain the radial mass-density of this model, the ansatz $\rho_m = C\,r^{-b}$ together with the Poisson equation (8.78) implies

$$\rho_m(r) = \frac{\sigma^2}{2\pi G}\,\frac{1}{r^2}. \tag{8.85}$$

That is, a singular isothermal sphere.

The Isothermal Model

The above model has a singularity at the origin. This is unphysical. In order to remove this problem, it is possible to force the central density to be finite. To this end new dimensionless variables are introduced, $\tilde{\rho_m} \equiv \rho_m/\rho_{m,0}$, $\tilde{r} \equiv r/r_0$. The density $\tilde{\rho_m}$ is the finite central density, while $r_0 = R_{\mathrm{core}}^{\mathrm{King}}$ is the King radius (8.76) at which the projected density falls to 0.5013 (i.e. about half) its central value. The radius r_0 is also sometimes called the core radius (but see further below for King models on p. 211). The Poisson equation (8.78) then becomes

$$\frac{\mathrm{d}}{\mathrm{d}\tilde{r}}\left(\tilde{r}^2\,\frac{\mathrm{d}\ln\tilde{\rho_m}}{\mathrm{d}\tilde{r}}\right) = -9\,\tilde{\rho_m}\,\tilde{r}^2. \tag{8.86}$$

This differential equation must be solved numerically for $\tilde{\rho_m}(\tilde{r})$ subject to the boundary conditions (as before),

$$\tilde{\rho_m}(\tilde{r}=0) = 1, \qquad \left.\frac{\mathrm{d}\tilde{\rho_m}}{\mathrm{d}\tilde{r}}\right|_{\tilde{r}=0} = 0. \tag{8.87}$$

The solution is the isothermal sphere.

By imposing physical reality (central non-singularity) on our mathematical ansatz, we end up with a density profile that cannot be arrived at analytically but only numerically. The isothermal density sphere must be tabulated in the computer with entries such as

$$r/r_0, \quad \log_{10}\left(\frac{\rho}{\rho_0}\right) \quad \text{and} \quad \log_{10}\left(\frac{\Sigma}{r_0\,\rho_0}\right), \tag{8.88}$$

where Σ is the projected density (Binney & Tremaine 1987, for example see their Table 4.1 and Fig. 4.7 of). The circular velocity, $v_c(r) = G\,M(r)/r$ of the isothermal sphere is obtained by integrating Poisson's equation (8.78) from $r=0$ to $r=r'$ with $r^2(\mathrm{d}\ln\rho_m/\mathrm{d}r) = -(G/\sigma^2)\,M(r)$ and

$$v_c^2(r) = -\sigma^2\,\frac{\mathrm{d}\ln\rho_m(r)}{\mathrm{d}\ln r}. \tag{8.89}$$

Numerical solution of differential (8.86) shows that $v_c \to \sqrt{2}\,\sigma$ (constant) for large r.

The isothermal sphere is a useful model for describing elliptical galaxies, within a few core radii, and disc galaxies because of the constant rotation curve. However, combining the two equations for v_c^2 above, one finds that $M(r) \approx (2\,\sigma^2/G)\,r$ for large r, i.e. the isothermal sphere has an infinite mass as it is not bounded.

The Lowered Isothermal or King Model

We have thus seen that the class of models with $n = \infty$ contain as the simplest case the singular isothermal sphere. By forcing the central density to be finite, we are led to the isothermal sphere which, however, has an infinite mass. The final model considered here within this class is the lowered isothermal model, or the King model,[8] which forces not only a finite central density but also a cutoff in radius. These have a distribution function similar to that of the isothermal model, except for a cutoff in energy,

$$f_m(\mathcal{E}) = \begin{cases} \frac{\rho_{m,1}}{(2\,\pi\,\sigma^2)^{\frac{3}{2}}} \left(e^{\frac{\mathcal{E}}{\sigma^2}} - 1\right) & : \ \mathcal{E} > 0, \\ 0 & : \ \mathcal{E} \le 0. \end{cases} \tag{8.90}$$

The density distribution becomes

$$\rho_m = \rho_{m,1} \left[e^{\frac{\Psi}{\sigma^2}} \operatorname{erf}\left(\frac{\sqrt{\Psi}}{\sigma}\right) - \sqrt{\frac{4\,\Psi}{\pi\,\sigma^2}} \left(1 + \frac{2\,\Psi}{3\sigma^2}\right) \right] \tag{8.91}$$

with integration only to $\mathcal{E} = 0$ as before. The Poisson (8.78) becomes

$$\frac{\mathrm{d}}{\mathrm{d}\tilde{r}} \left(\tilde{r}^2 \frac{\mathrm{d} \ln \tilde{\rho_m}}{\mathrm{d}\tilde{r}}\right) = -4\,\pi\,G\,\rho_{m,1}\,r^2 \left[e^{\frac{\Psi}{\sigma^2}} \operatorname{erf}\left(\frac{\sqrt{\Psi}}{\sigma}\right) - \sqrt{\frac{4\,\Psi}{\pi\,\sigma^2}} \left(1 + \frac{2\,\Psi}{3\sigma^2}\right) \right]. \tag{8.92}$$

Again, this differential equation must be solved numerically for $\Psi(r)$ subject to the boundary conditions,

$$\Psi(0), \quad \frac{\mathrm{d}\Psi}{\mathrm{d}r}\,|_{r=0} = 0. \tag{8.93}$$

The density vanishes at $r = r_{\rm tid}$ (the tidal radius), where $\Psi(r = r_{\rm tid}) = 0$ also. A King model is thus limited in mass and has a finite central density,

[8] Note that King (1962) suggested a three-parameter (mass, core radius and cut-off/tidal radius), empirical, projected (2D) density law that fits globular clusters very well. These do not have information on the velocity structure of the clusters. The King (non-analytical) 6D models, which are solutions of the Jeans equation ((8.120) below) and discussed here, are published by King (1966).

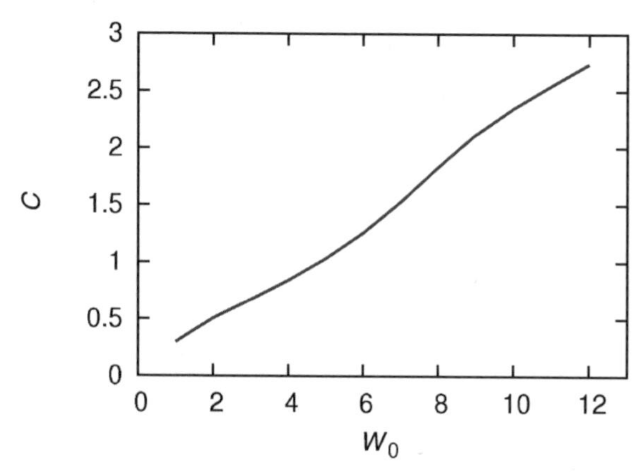

Fig. 8.5. The King concentration parameter W_0 as a function of c (cf. with Fig. 4–10 of Binney & Tremaine 1987). This figure has been produced by Andreas Küpper

but the parameter σ is not the velocity dispersion. It is rather related to the depth of the potential via the concentration parameter

$$W_o \equiv \frac{\Psi(0)}{\sigma^2}. \tag{8.94}$$

The concentration is defined as

$$c \equiv \log_{10}\left(\frac{r_{\rm tid}}{r_o}\right). \tag{8.95}$$

For globular clusters, $3 < W_o < 9$, $0.75 < c < 1.75$ and the relation between W_o and c is plotted in Fig. 8.5. Note also that the true core radius defined as $\Sigma(R_c) = (1/2)\,\Sigma(0)$, where $\Sigma(R)$ is the projected density profile and R is the projected radius, is unequal in general to the King radius, r_0 (8.76). Finally, it should be emphasised that it is not physical to use an arbitrary $r_{\rm tid}$. The tidal radius must always match the value dictated by the cluster mass and the host galaxy (e.g. (8.3)).

8.2.2 Comparison: Plummer vs King Models

The above discussion has served to show how various popular models can be followed through from a power-law distribution function (8.41) with different indices n. The Plummer model (p. 205) and the King model (p. 212) are particularly useful for describing star clusters. The Plummer model is determined by two parameters, the mass, M, and the scale radius, $r_{\rm h} \approx 1.305\, r_{\rm pl}$. The King model requires three parameters, M, a scale radius, $r_{\rm h}$, and a concentration parameter, W_0 or c. Which subset of parameters yield models that are similar in terms of the overall density profile?

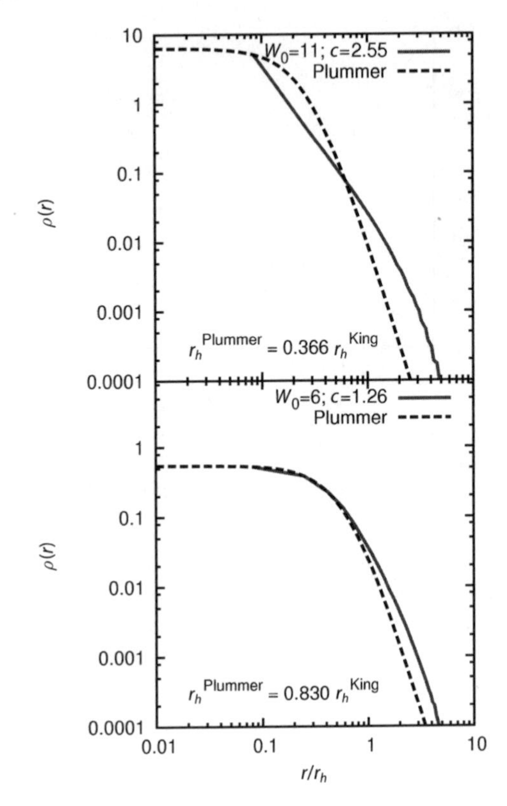

Fig. 8.6. Comparison of a King model (*solid curve*) with a Plummer model (*dashed curve*). Both have the same mass and that Plummer model is sought, which minimises the unweighted reduced chi-squared between the two models. The *upper panel* shows a high-concentration King model with $c = 2.55$ and $W_0 = 11$ and the best-fit Plummer model has $r_h^{\rm Plummer} = 0.366\, r_h^{King}$ ($r_h \equiv r_h$), as stated in the panel. The *lower panel* compares the two best matching models for the case of an intermediate-concentration King model. This figure was produced by Andreas Küpper

To answer this, the mass is set to be constant. King models with different W_0 and $r_{\rm h}$ are computed and Plummer models are sought, which minimise the reduced chi-squared value between the two density profiles. Figure 8.6 shows two examples of best-matching density profiles, and Fig. 8.7 reveals the family of Plummer profiles that best match King models with different concentrations. Note that a good match between the two is only obtained for intermediate-concentration King models ($2.5 \le W_0 \le 7.5$).

8.2.3 Discretisation

To set up a computer model of a stellar system with N particles (e.g. stars), the distribution functions need to be sampled N times. The relevant distribution

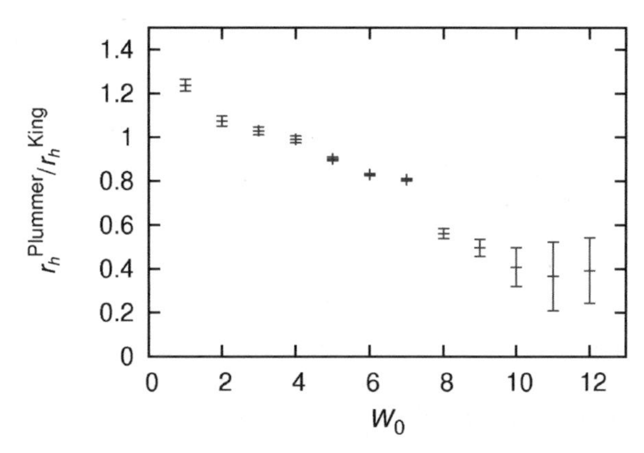

Fig. 8.7. The ratio $r_h^{\mathrm{Plummer}}/r_h^{\mathrm{King}}$ ($r_h \equiv r_h$) for the best-matching Plummer and King models (Fig. 8.6) are plotted as a function of the King concentration parameter W_0. The uncertainties are unweighted reduced chi-squared values between the two density profiles. It is evident that there are no well-matching Plummer models for low- ($c < 2.5$) and high-concentration ($c > 7.5$) King models. This figure was produced by Andreas Küpper

functions are the phase-space distribution function, the stellar initial mass function and the three distribution functions governing the properties of binary stars (periods, mass-ratios and eccentricities).

Assume the distribution function depends on the variable $\zeta_{\min} \leq \zeta \leq \zeta_{\max}$ (e.g. stellar mass, m). There are various ways of sampling from a distribution function (Press et al. 1992), but the most efficient way is to use a generating function if one exists. Consider the probability $X(\zeta)$ of encountering a value for the variable in the range $\zeta_{\min}$ to ζ,

$$X(\zeta) = \int_{\zeta_{\min}}^{\zeta} p(\zeta')\,\mathrm{d}\zeta', \tag{8.96}$$

with $X(\zeta_{\min}) = 0 \leq X(\zeta) \leq X(\zeta_{\max}) = 1$, and $p(\zeta)$ is the distribution function normalised so that the latter equal sign holds ($X = 1$). $p(\zeta)$ is the probability density. The inverse of (8.96), $\zeta(X)$, is the generating function. It is a one-to-one map of the uniform distribution $X \in [0, 1]$ to $\zeta \in [\zeta_{\min}, \zeta_{\max}]$. If an analytical inverse does not exist, it can be found numerically in a straightforward manner, for example, by constructing a table of X, ζ and then interpolating this table to obtain a ζ for a given X.

Example: The Power-Law Stellar Mass Function

As an example, consider the distribution function

$$\xi(m) = k\,m^{-\alpha}, \quad \alpha = 2.35; \quad 0.5 \leq \frac{m}{M_\odot} \leq 150. \tag{8.97}$$

The probability density is $p(m) = k_p\, m^{-\alpha}$ and $\int_{0.5}^{150} p(m)\,\mathrm{d}m = 1 \Rightarrow k_p = 0.53$. Thus

$$X(m) = \int_{0.5}^{m} p(m)\,\mathrm{d}m = k_p\,\frac{150^{1-\alpha} - 0.5^{1-\alpha}}{1-\alpha} \tag{8.98}$$

and the generating function for stellar masses becomes

$$m(X) = \left[X\frac{1-\alpha}{k_p} + 0.5^{1-\alpha}\right]^{\frac{1}{1-\alpha}}. \tag{8.99}$$

It is easy to programme this into an algorithm. Obtain a random variate X from a random number generator and use the above generating function to get a corresponding mass, m. Repeat N times.

Generating a Plummer Model

Perhaps the most useful and simplest model of a bound stellar system is the Plummer model (p. 205). It is worth introducing the discretisation of this model in some detail, because analytical formulae go a long way, which is important for testing codes. A condensed form of this material is available in Aarseth, Hénon and Wielen (1974).

The mass within radius r is ($r_{\rm pl} = b$ here)

$$M(r) = \int_0^r \rho_m(r')\,4\,\pi\,r'^2\,\mathrm{d}r' = M_{\rm cl}\,\frac{(r/r_{\rm pl})^3}{\left[1+(r/r_{\rm pl})^2\right]^{\frac{3}{2}}}. \tag{8.100}$$

A number uniformly distributed between zero and one can then be defined,

$$X_1(r) = \frac{M(r)}{M_{\rm cl}} = \frac{\zeta^3}{[1+\zeta^2]}, \tag{8.101}$$

where $\zeta \equiv r/r_{\rm pl}$ and $X_1(r=\infty) = 1$. This function can be inverted to yield the generating function for particle distances distributed according to a Plummer density law,

$$\zeta(X_1) = \left(X_1^{-\frac{2}{3}} - 1\right)^{-\frac{1}{2}}. \tag{8.102}$$

The coordinates of the particles $x, y, z, r^2 = (\zeta\,r_{\rm pl})^2 = x^2+y^2+z^2$ can be obtained as follows. For a given particle we already have r. For all possible x and y, z has a uniform distribution, $p(z) = \text{const} = 1/(2\,r)$ over the range $-r \le z \le +r$. Thus, for a second random variate between zero and one,

$$X_2(z) = \int_{-r}^{z} p(z')\,\mathrm{d}z' = \frac{1}{2\,r}\,(z+r)\,, \tag{8.103}$$

with $X_2(+r) = 1$. The generating function for z becomes

$$z(X_2) = 2\,r\,X_2 - r. \tag{8.104}$$

Having obtained r and z, x and y can be arrived at as follows, noting the equation for a circle, $r^2 - z^2 = x^2 + y^2$. Choose a random angle θ, which is uniformly distributed over the range $0 \le \theta \le 2\,\pi$. Thus $p(\theta) = 1/(2\,\pi)$ and the third random variate becomes

$$X_3(\theta) = \int_0^{\theta} \frac{1}{2\,\pi}\,\mathrm{d}\theta' = \frac{\theta}{2\,\pi}. \tag{8.105}$$

The corresponding generating function is

$$\theta(X_3) = 2\,\pi\,X_3. \tag{8.106}$$

Finally,

$$x = \left(r^2 - z^2\right)^{\frac{1}{2}}\,\cos\theta; \quad \text{and} \quad y = \left(r^2 - z^2\right)^{\frac{1}{2}}\,\sin\theta. \tag{8.107}$$

The velocity for each particle cannot be obtained as simply as the positions. In order for the initial stellar system to be in virial equilibrium, the potential and kinetic energy need to balance according to the scalar virial theorem. This is ensured by forcing the velocity distribution function to be that of the Plummer model,

$$f_m(\epsilon_e) = \begin{cases} \left(\frac{24\,\sqrt{2}}{2\,\pi^3}\,\frac{r_{\rm pl}^2}{(G\,M_{\rm cl})^5}\right)\,(-\epsilon_e)^{\frac{7}{2}} & , \quad \epsilon_e \le 0, \\ 0 & , \quad \epsilon_e > 0, \end{cases} \tag{8.108}$$

where

$$\epsilon_e(r, v) = \Phi(r) + (1/2)\,v^2 \tag{8.109}$$

is the specific energy per star and

$$\Phi(r) = -\frac{G\,M_{\rm cl}}{r_{\rm pl}}\left(1 + \left(\frac{r}{r_{\rm pl}}\right)^2\right)^{-\frac{1}{2}} \tag{8.110}$$

is the potential. Now, the Plummer distribution function can be expressed in terms of r and v,

$$f(r, v) = f_o\left(-\Phi(r) - \frac{1}{2}\,v^2\right)^{\frac{7}{2}}, \tag{8.111}$$

for a normalisation constant f_o and dropping the mass subscript, because we assume the positions and velocities do not depend on particle mass. With the escape speed at distance r from the Plummer centre, $v_{\rm esc}(r) = \sqrt{-2\,\Phi(r)} \equiv v/\zeta$, it follows that

$$f(r, v) = f_o\left(\frac{1}{2}\,v_{\rm esc}\right)^7\,(1 - \zeta^2)^{\frac{7}{2}}. \tag{8.112}$$

The number of particles with speeds in the interval v to $v + \mathrm{d}v$ is

$$\mathrm{d}N = f(r,v)\,4\,\pi\,v^2\,\mathrm{d}v \equiv g(v)\,\mathrm{d}v. \tag{8.113}$$

Thus

$$g(v) = 16\,\pi\,f_o \left(\frac{1}{2}\,v_{\mathrm{esc}}(r)\right)^9 \left(1-\zeta^2(r)\right)^{\frac{7}{2}}\,\zeta^2(r), \tag{8.114}$$

that is,

$$g(\zeta) = g_o\,\zeta^2(r)\,\left(1-\zeta^2(r)\right)^{\frac{7}{2}}, \tag{8.115}$$

for a normalisation constant g_o determined by demanding that

$$X_4(\zeta=1) = 1 = \int_0^1 g(\zeta')\,\mathrm{d}\zeta' \tag{8.116}$$

for a fourth random number variate $X_4(\zeta) = \int_0^\zeta g(\zeta')\,\mathrm{d}\zeta'$. It follows that

$$X_4(\zeta) = \frac{1}{2}\,\left(5\,\zeta^3 - 3\,\zeta^5\right). \tag{8.117}$$

This cannot be inverted to obtain an analytical generating function for $\zeta = \zeta(X_4)$. Therefore, numerical methods need to be used to solve (8.117). For example, one way to obtain ζ for a given random variate X_4 is to find the root of the equation $0 = (1/2)\,(5\,\zeta^3 - 3\,\zeta^5) - X_4$, or one can use the Neumann rejection method (Press et al. 1992).

The following procedure can be implemented to calculate the velocity vector of a particle for which r and ζ are already known from above. Compute $v_{\mathrm{esc}}(r)$ so that $v = \zeta\,v_{\mathrm{esc}}$. Each speed v is then split into its components v_x, v_y, v_z, assuming velocity isotropy using the same algorithm as above for x, y, z:

$$v_z(X_5) = (2\,X_5 - 1)\,v; \quad \theta(X_6) = 2\,\pi\,X_6; \tag{8.118}$$

$$v_x = \sqrt{v^2 - v_z^2}\,\cos\theta; \quad v_y = \sqrt{v^2 - v_z^2}\,\sin\theta. \tag{8.119}$$

Note that a rotating Plummer model can be generated by simply switching the signs of v_x and v_y so that all particles have the same direction of motion in the x, y plane.

As an aside, an efficient numerical method to set up triaxial ellipsoids with or without an embedded rotating disc is described by Boily, Kroupa & Peñarrubia-Garrido (2001).

Generating an Arbitrary Spherical, Non-Rotating Model

In most cases an analytical density distribution is not known (e.g. the King models above). Such numerical models can nevertheless be discretised straightforwardly as follows. Assume that the density distribution, $\rho(r)$, is known. Compute $M(r)$ and M_{cl}. Define $X(r) = M(r)/M_{\mathrm{cl}}$, as above. We thus have a numerical grid of numbers $r,\ M(r),\ X(r)$. For a given random variate $X \in [0,1]$, interpolate r from this grid. Compute x, y, z as above.

If the distribution function of speeds is too complex to yield an analytical generating function $X(\zeta)$ for the speeds ζ, we can resort to the following procedure. One of the Jeans equations for a spherical system is

$$\frac{\mathrm{d}}{\mathrm{d}r}\left(\rho(r)\,\sigma_r(r)^2\right) + \frac{\rho(r)}{r}\left[2\,\sigma_r^2(r) - \left(\sigma_\theta(r)^2 + \sigma_\phi(r)^2\right)\right] = -\rho(r)\,\frac{\mathrm{d}\Phi(r)}{\mathrm{d}r}. \tag{8.120}$$

For velocity isotropy, $\sigma_r^2 = \sigma_\theta^2 = \sigma_\phi^2$, this reduces to

$$\frac{\mathrm{d}\left(\rho\,\sigma_r^2\right)}{\mathrm{d}r} = -\rho\,\frac{\mathrm{d}\Phi}{\mathrm{d}r}. \tag{8.121}$$

Integrating this by making use of $\rho \to 0$ as $r \to \infty$ and remembering that $\mathrm{d}\Phi/\mathrm{d}r = -G\,M/r^2$,

$$\sigma_r^2(r) = \frac{1}{\rho(r)} \int_r^\infty \rho(r')\,\frac{G\,M(r')}{r'^2}\,\mathrm{d}r'. \tag{8.122}$$

For each particle at distance r, a one-dimensional velocity dispersion, $\sigma_r(r)$, is thus obtained. Choosing randomly from a Gaussian distribution with dispersion $\sigma_i, i = r, \theta, \phi, x, y, z$, then gives the velocity components (e.g. v_x, v_y, v_z) for this particle.

Rotating Models

Star clusters are probably born with some rotation because the pre-cluster cloud core is likely to have contracted from a cloud region with differential motions that do not cancel. How large this initial angular momentum content of an embedded cluster is remains uncertain because the dominant motions are random and chaotic owing to the turbulent velocity field of the gas. Once the star-formation process is quenched as a result of gas blow-out (Sect. 8.1.1), the cluster expands. This must imply substantial reduction in the rotational velocity. A case in point is ω Cen, which has been found to rotate with a peak velocity of about $7\,\mathrm{km\,s^{-1}}$ (Pancino et al. 2007, and references therein).

A setup for rotating cluster models is easily made, for instance, by increasing the tangential velocities of stars by a certain factor. A systematic study of relaxation-driven angular momentum re-distribution within star clusters has become available through the work of the group of Rainer Spurzem and Hyung-Mok Lee and the interested reader is directed to that body of work (Kim et al. 2008, and references therein). One important outcome of this work is that core collapse is substantially accelerated in rotating models. The primary reason for this is that increased rotational support reduces the role of support through random velocities for the same cluster dimension. Thus, the relative stellar velocities decrease and the stars exchange momentum and energy more efficiently, enhancing two-body relaxation and thence the approach towards energy equipartition.

8.2.4 Cluster Birth and Young Clusters

Some astrophysical issues related to the initial conditions of star clusters have been raised in Sect. 8.1.1. In order to address most of these issues numerical experiments are required. The very initial phase, the first 0.5 Myr, can only be treated through gas-dynamical computations that, however, lack the numerical resolution for the high-precision stellar-dynamical integrations, which are the essence of collisional dynamics during the gas-free phase of a cluster's life. This gas-free stage sets in with the blow out of residual gas at an age of about 0.5–1.5 Myr. The time 0.5–1.5 Myr is dominated by the physics of stellar feedback and radiation transport in the residual gas as well as energy and momentum transfer to it through stellar outflows. The gas-dynamical computations cannot treat all the physical details of the processes acting during this critical time, which also include early stellar-dynamical processes such as mass segregation and binary–binary encounters.

One successful procedure to investigate the dominant macroscopic physical processes of these stellar-dynamical processes, gas blow-out and the ensuing cluster expansion, through to the long-term evolution of the remnant cluster, is to approximate the residual gas component as a time-varying potential in which the young stellar population is trapped. The pioneering work using this approach has been performed by Lada, Margulis & Dearborn (1984), whereby the earlier numerical work by Tutukov (1978) on open clusters and later N-body computations by Goodwin (1997a,b, 1998) on globular clusters must also be mentioned in this context.

The physical key quantities that govern the emergence of embedded clusters from their clouds and their subsequent appearance are (Baumgardt, Kroupa & Parmentier 2008, Sect. 8.1.1):

- sub-structuring,
- initial mass segregation,
- the dynamical state at feedback termination (dynamical equilibrium?, collapsing? or expanding?),
- the star-formation efficiency, ϵ,
- the ratio of the gas-expulsion time-scale to the stellar crossing time through the embedded cluster, $\tau_{\rm gas}/t_{\rm cross}$ and
- the ratio of the embedded-cluster half-mass radius to its tidal radius, $r_{\rm h}/r_{\rm t}$.

It becomes rather apparent that the physical processes governing the emergence of star clusters from their natal clouds is terribly messy, and the research-field is clearly observationally driven. Observations have shown that star clusters suffer substantial infant weight loss and probably about 90% of all clusters disperse altogether (infant mortality). This result is consistent with the observational insight that clusters form in a compact configuration with a low star-formation efficiency ($0.2 \leq \epsilon \leq 0.4$) and that residual-gas blow-out occurs on a time-scale comparable or even faster than an embedded-cluster crossing time-scale (Kroupa 2005). Theoretical work can give a reasonable description of these empirical findings by combining some of the above parameters, such as an effective star-formation efficiency as a measure of the amount of gas removed for a cluster of a given stellar mass if this cluster were in dynamical equilibrium at feedback termination, and that the gas and stars were distributed according to the same radial density function with the same scaling radius.

Embedded Clusters: One way to parameterise an embedded cluster is to set up a Plummer model in which the stellar positions follow a density law with the parameters $M_{\rm ecl}$ and $r_{\rm pl}$ and the residual gas is a time-varying Plummer potential initially with the parameters $M_{\rm gas}$ and $r_{\rm pl}$, i.e. modelled with the same radial density law. The effective star-formation efficiency is then given by (8.2). Stellar velocities must then be calculated from a Plummer law with total mass $M_{\rm ecl} + M_{\rm gas}$ following the recipes of Sect. 8.2.3. The gas can be removed by evolving $M_{\rm gas}$ or $r_{\rm pl}$. For example, Kroupa, Aarseth & Hurley (2001) and Baumgardt, Kroupa & Parmentier (2008) assumed the gas mass decreases exponentially after an embedded phase lasting about 0.5 Myr during which the cluster is allowed to evolve in dynamical equilibrium. Bastian & Goodwin (2006), as another example, do not include a gas potential but take the initial velocities of stars to be $1/\sqrt{\epsilon}$ times larger, $v_{\rm embedded} = (1/\sqrt{\epsilon})\, v_{\rm no\ gas}$, to model the effect of instantaneous gas removal. Many variations of these assumptions are possible and Adams (2000), for example, investigated the fraction of stars left in a cluster remnant if the radial scale length of the gas is different to that of the stars, i.e. for a radially dependent star-formation efficiency, $\epsilon(r)$.

Subclustering: Initial subclustering has been barely studied. Scally & Clarke (2002) considered the degree of sub-structuring of the ONC allowed by its current morphology, while Fellhauer & Kroupa (2005) computed the evolution of massive star-cluster complexes, assuming each member cluster in the complex undergoes its own individual gas-expulsion process. McMillan, Vesperini & Portegies Zwart (2007) showed that initially mass-segregated subclusters retain mass segregation upon merging. This is an interesting mechanism for accelerating dynamical mass segregation because it occurs faster in smaller-N systems, which have a shorter relaxation time.

The simplest initial conditions for such numerical experiments are to set up the star-cluster complex (or protoONC-type cluster, for example) as a Plummer model, where each particle is a smaller subcluster. Each subcluster is also a Plummer model, embedded in a gas potential given as a Plummer model. The gas-expulsion process from each subcluster can be treated as above.

Mass Segregation and Gas Blow-Out: The problem of how initially mass-segregated clusters react to gas blow-out has not been studied at all in the past. This is due partially to the lack of convenient algorithms to set up mass-segregated clusters that are in dynamical equilibrium and which do not go into core collapse as soon as the N-body integration begins. An interesting consequence here is that gas blow-out will unbind mostly the low-mass stars, while the massive stars are retained. These, however, evolve rapidly so that the mass lost from the remnant cluster owing to the evolution of the massive stars can become destructive, enhancing infant mortality.

Ladislav Šubr has developed a numerically efficient method to set up initially mass-segregated clusters close to core-collapse based on a novel concept that uses the potentials of subsets of stars ordered by their mass (Šubr, Kroupa & Baumgardt 2008).[9] An alternative algorithm based on ordering the stars by increasing mass and increasing total energy that leads to total mass segregation, and also to a model that is not in core collapse and which therefore evolves towards core collapse, has been developed by Baumgardt, Kroupa & de Marchi (2008). An application concerning the effect on the observed stellar mass function in globular clusters shows that gas expulsion leads to bottom-light stellar mass functions in clusters with a low concentration, consistent with observational data (Marks, Kroupa & Baumgardt 2008).

8.3 The Stellar IMF

The stellar initial mass function (IMF), $\xi(m)\,\mathrm{d}m$, where m is the stellar mass, is the parent distribution function of the masses of stars formed in one event. Here, the number of stars in the mass interval m to $m + \mathrm{d}m$ is

$$\mathrm{d}N = \xi(m)\,\mathrm{d}m. \tag{8.123}$$

The IMF is, strictly speaking, an abstract theoretical construct because any observed system of N stars merely constitutes a particular representation of this universal distribution function, if such a function exists (Elmegreen 1997; Maíz Apellániz & Úbeda 2005). The probable existence of a unique $\xi(m)$ can be inferred from the observations of an ensemble of systems each consisting of N stars (e.g. Massey 2003). If, after corrections for (a) stellar evolution, (b) unknown multiple stellar systems and (c) stellar-dynamical biases, the individual distributions of stellar masses are similar within the expected statistical scatter, we (the community) deduce that the hypothesis that the stellar mass distributions are not the same can be excluded. That is, we make the case for a universal, standard or canonical stellar IMF within the physical conditions probed by the relevant physical parameters (metallicity, density, mass) of the populations at hand.

Related overviews of the IMF can be found in Kroupa (2002a); Chabrier (2003); Bonnell, Larson & Zinnecker (2007); Kroupa (2007a), and a review with an emphasis on the metal-rich problem is available in Kroupa (2007b). Zinnecker & Yorke (2007) provide an in-depth review of the formation and distribution of massive stars. Elmegreen (2007) discusses the possibility that star-formation occurs in different modes with different IMFs.

[9] The C-language software package *plumix* may be downloaded from the website `http://www.astro.uni-bonn.de/~webaiub/english/downloads.php`.

8.3.1 The Canonical or Standard Form of the Stellar IMF

The canonical stellar IMF is a two-part-power law (8.128). The only structure found with confidence so far is the change of index from the Salpeter/Massey value to a smaller one near $0.5\,M_\odot$[10]

$$\begin{array}{ll} \xi(m) \propto m^{-\alpha_i}, & i = 1, 2 \\ \alpha_1 = 1.3 \pm 0.3, \quad , & 0.08 \le m/M_\odot \le 0.5, \\ \alpha_2 = 2.3 \pm 0.5, \quad , & 0.5 \le m/M_\odot \le m_{\rm max}, \end{array} \tag{8.124}$$

where $m_{\rm max} \le m_{\rm max*} \approx 150\,M_\odot$ follows from Fig. 8.1. Brown dwarfs have been found to form a separate population with $\alpha_0 \approx 0.3 \pm 0.5$, (8.129) (Thies & Kroupa 2007).

It has been corrected for bias through unresolved multiple stellar systems in the low-mass ($m < 1\,M_\odot$) regime (Kroupa, Gilmore & Tout 1991) by a multi-dimensional optimisation technique. The general outline of this technique is as follows (Kroupa, Tout & Gilmore 1993). First, the correct form of the stellar-mass–luminosity relation is extracted using observed stellar binaries and theoretical constraints on the location, amplitude and shape of the minimum of its derivative, $\mathrm{d}m/\mathrm{d}M_V$, near $m = 0.3\,M_\odot, M_V \approx 12, M_I \approx 9$ in combination with the observed shape of the nearby and deep Galactic-field stellar luminosity function (LF)

$$\Psi(M_V) = -\left(\frac{\mathrm{d}m}{\mathrm{d}M_V}\right)^{-1} \xi(m), \tag{8.125}$$

where $\mathrm{d}N = \Psi(M_V)\,\mathrm{d}M_V$ is the number of stars in the magnitude interval M_V to $M_V + \mathrm{d}M_V$. Once the semi-empirical mass–luminosity relation of stars, which is an excellent fit to the most recent observational constraints by Delfosse et al. (2000), is established, a model of the Galactic field is calculated with the assumption that a parameterised form for the MF and different values for the scale-height of the Galactic disc and different binary fractions in it. Measurement uncertainties and age and metallicity spreads must also be considered in the theoretical stellar population. Optimisation in this multi-parameter space (MF parameters, scale-height and binary population) against observational data leads to the canonical stellar MF for $m < 1\,M_\odot$.

One important result from this work is the finding that the LF of main-sequence stars has a universal sharp peak near $M_V \approx 12, M_I \approx 9$. It results from changes in the internal constitution of stars that drive a non-linearity in the stellar mass–luminosity relation. A consistency check is then performed as follows. The above MF is used to create young populations of binary systems (Sect. 8.4.2) that are born in modest star clusters consisting of a few hundred stars. Their dissolution into the Galactic field is computed with an

[10] The uncertainties in α_i are estimated from the alpha-plot (Sect. 8.3.2), as shown in Fig. 5 of Kroupa (2002b), to be about 95% confidence limits.

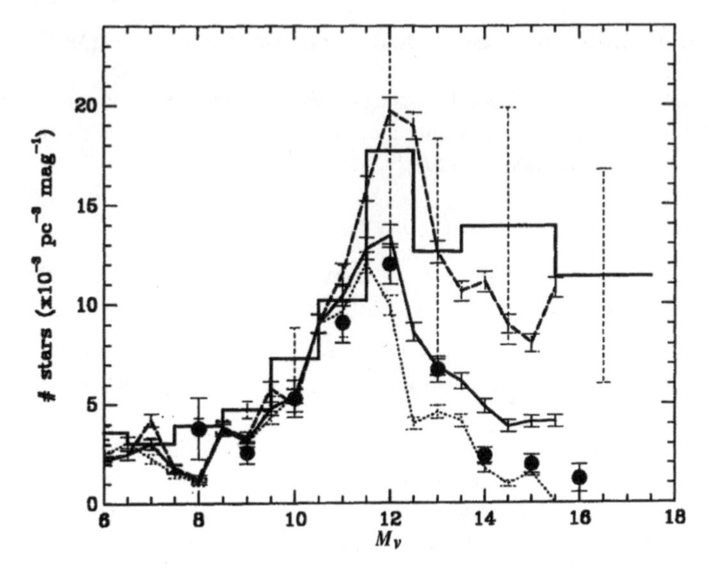

Fig. 8.8. The Galactic field population that results from disrupted star clusters, unification of both the nearby (*solid histogram*) and deep (*filled circles*) LFs with one parent MF (8.124). The theoretical nearby LF (*dashed line*) is the LF of all individual stars, while the *solid curve* is a theoretical LF with a mixture of about 50 per cent unresolved binaries and single stars from a clustered star-formation mode. According to this model, all stars are formed as binaries in modest clusters, which disperse to the field. The resulting Galactic field population has a binary fraction and a mass-ratio distribution as observed. The dotted curve is the initial system LF (100% binaries) (Kroupa 1995a,b). Note the peak in both theoretical LFs. It stems from the extremum in the derivative of the stellar-mass–luminosity relation in the mass range 0.2–0.4 $M_{\odot}$ (Kroupa 2002b)

N-body code and the resulting theoretical field is compared to the observed LFs (Fig. 8.8). Further confirmation of the form of the canonical IMF comes from independent sources, most notably by Reid, Gizis & Hawley (2002) and also Chabrier (2003).

In the high-mass regime, Massey (2003) reports the same slope or index $\alpha_3 = 2.3 \pm 0.1$ for $m \geq 10\,M_{\odot}$ in many OB associations and star clusters in the Milky Way and the Large and Small Magellanic clouds (LMC, SMC, respectively). It is therefore suggested to refer to $\alpha_2 = \alpha_3 = 2.3$ as the Salpeter/Massey slope or index, given the pioneering work of Salpeter (1955) who derived this value for stars with masses 0.4–10 $M_{\odot}$.

Multiplicity corrections await publication once we learn more about how the components are distributed in massive stars (cf. Preibisch et al. 1999; Zinnecker 2003). Weidner & Kroupa (private communication) are in the process of performing a very detailed study of the influence of unresolved binary and higher-order multiple stars on determinations of the high-mass IMF.

Contrary to the Salpeter/Massey index ($\alpha = 2.3$), Scalo (1986) found $\alpha_{\rm MWdisc} \approx 2.7$ ($m \geq 1\,M\odot$) from a very thorough analysis of OB star counts in the Milky Way disc. Similarly, the star-count analysis of Reid, Gizis & Hawley (2002) leads to $2.5 \leq \alpha_{\rm MWdisc} \leq 2.8$, and Tinsley (1980), Kennicutt (1983) (his extended Miller-Scalo IMF), Portinari, Sommer-Larsen & Tantalo (2004) and Romano et al. (2005) find $2.5 \leq \alpha_{\rm MWdisc} \leq 2.7$. That $\alpha_{\rm MWdisc} > \alpha_2$ follows naturally is shown in Sect. 8.3.4.

Below the hydrogen-burning limit (see also Sect. 8.3.3) there is substantial evidence that the IMF flattens further to $\alpha_0 \approx 0.3 \pm 0.5$ (Martín et al. 2000; Chabrier 2003; Moraux et al. 2004). Therefore, the canonical IMF most likely has a peak at $0.08\,M_\odot$. Brown dwarfs, however, comprise only a few per cent of the mass of a population and are therefore dynamically irrelevant (Table 8.2). The logarithmic form of the canonical IMF,

$$\xi_{\rm L}(m) = \log_{10} m\,\xi(m), \qquad (8.126)$$

which gives the number of stars in $\log_{10} m$-intervals, also has a peak near $0.08\,M_\odot$. However, the system IMF (of stellar single and multiple systems combined to system masses) has a maximum in the mass range 0.4–0.6 $M_\odot$ (Kroupa et al. 2003).

The above canonical or standard form has been derived from detailed considerations of Galactic field star counts and so represents an average IMF. For low-mass stars, it is a mixture of stellar populations spanning a large range of ages (0–10 Gyr) and metallicities ([Fe/H]≥ -1). For the massive stars it constitutes a mixture of different metallicities ([Fe/H]≥ -1.5) and star-forming conditions (OB associations to very dense star-burst clusters: R136 in the LMC). Therefore, it can be taken as a canonical form and the aim is to test the

IMF UNIVERSALITY HYPOTHESIS that the canonical IMF constitutes the parent distribution of all stellar populations.

Negation of this hypothesis would imply a variable IMF. Note that the work of Massey (2003) has already established the IMF to be invariable for $m \geq 10\,M_\odot$ and for densities $\rho \leq 10^5$ stars pc^{-3} and metallicity $Z \geq 0.002$.

Finally, Table 8.2 compiles some numbers that are useful for simple insights into stellar populations.

8.3.2 Universality of the IMF: Resolved Populations

The strongest test of the IMF UNIVERSALITY HYPOTHESIS (p. 225) is obtained by studying populations that can be resolved into individual stars. Because we also seek co-eval populations with stars at the same distance and with the same metallicity to minimise uncertainties, star clusters and stellar associations would seem to be the test objects of choice. But before contemplating such work, some lessons from stellar dynamics are useful.

Table 8.2. The number fraction $\eta_N = 100 \int_{m_1}^{m_2} \xi(m)\,dm / \int_{m_l}^{m_u} \xi(m)\,dm$ and the mass fraction $\eta_M = 100 \int_{m_1}^{m_2} m\,\xi(m)\,dm / M_{\rm cl}$, $M_{\rm cl} = \int_{m_l}^{m_u} m\,\xi(m)\,dm$, in per cent of BDs or main-sequence stars in the mass interval m_1 to m_2 and the stellar contribution, $\rho^{\rm st}$, to the Oort limit and to the Galactic-disc surface mass-density, $\Sigma^{\rm st} = 2\,h\rho^{\rm st}$, near to the Sun, with $m_l = 0.01\,M_\odot$, $m_u = 120\,M_\odot$ and the Galactic-disc scale-height $h = 250\,{\rm pc}$ ($m < 1\,M_\odot$ Kroupa, Tout & Gilmore 1993) and $h = 90\,{\rm pc}$ ($m > 1\,M_\odot$, Scalo 1986). Results are shown for the canonical IMF (8.124) for the high-mass-star IMF approximately corrected for unresolved companions ($\alpha_3 = 2.7, m > 1\,M_\odot$) and for the present-day mass function (PDMF, $\alpha_3 = 4.5$, Scalo 1986; Kroupa, Tout & Gilmore 1993), which describes the distribution of stellar masses now populating the Galactic disc. For gas in the disc, $\Sigma^{\rm gas} = 13 \pm 3\,M_\odot/{\rm pc}^2$ and remnants $\Sigma^{\rm rem} \approx 3\,M_\odot/{\rm pc}^2$ (Weidemann 1990). The average stellar mass is $\overline{m} = \int_{m_l}^{m_u} m\,\xi(m)\,dm / \int_{m_l}^{m_u} \xi(m)\,dm$. $N_{\rm cl}$ is the number of stars that have to form in a star cluster so that the most massive star in the population has the mass $m_{\rm max}$. The mass of this population is $M_{\rm cl}$ and the condition is $\int_{m_{\rm max}}^{\infty} \xi(m)\,dm = 1$ with $\int_{0.01}^{m_{\rm max}} \xi(m)\,dm = N_{\rm cl} - 1$. $\Delta M_{\rm cl}/M_{\rm cl}$ is the fraction of mass lost from the cluster due to stellar evolution if we assume that, for $m \geq 8\,M_\odot$, all neutron stars and black holes are kicked out by asymmetrical supernova explosions but that white dwarfs are retained (Weidemann et al. 1992) and have masses $m_{\rm WD} = 0.084\,m_{\rm ini} + 0.444\,[M_\odot]$. This is a linear fit to the data of Weidemann (2000, their Table 3) for progenitor masses $1 \leq m/M_\odot \leq 7$ and $m_{\rm WD} = 0.5\,M_\odot$ for $0.7 \leq m/M_\odot < 1$. The evolution time for a star of mass $m_{\rm to}$ to reach the turn-off age is available in Fig. 20 of Kroupa (2007a)

Mass range $[M_\odot]$	η_N [%]			η_M [%]			$\rho^{\rm st}$ $[M_\odot/{\rm pc}^3]$	$\Sigma^{\rm st}$ $[M_\odot/{\rm pc}^2]$
		α_3			α_3		α_3	α_3
	2.3	2.7	4.5	2.3	2.7	4.5	4.5	4.5
0.01–0.08	37.2	37.7	38.6	4.1	5.4	7.4	3.2×10^{-3}	1.60
0.08–0.5	47.8	48.5	49.7	26.6	35.2	48.2	2.1×10^{-2}	10.5
0.5–1	8.9	9.1	9.3	16.1	21.3	29.2	1.3×10^{-2}	6.4
1–8	5.7	4.6	2.4	32.4	30.3	15.1	6.5×10^{-3}	1.2
8–120	0.4	0.1	0.0	20.8	7.8	0.1	3.6×10^{-5}	6.5×10^{-3}
$\overline{m}/M_\odot =$	0.38	0.29	0.22				$\rho^{\rm st}_{\rm tot} = 0.043$	$\Sigma^{\rm st}_{\rm tot} = 19.6$

$m_{\rm max}$ $[M_\odot]$	$\alpha_3 = 2.3$ $N_{\rm cl}$	$M_{\rm cl}$ $[M_\odot]$	$\alpha_3 = 2.7$ $N_{\rm cl}$	$M_{\rm cl}$ $[M_\odot]$	$m_{\rm to}$ $[M_\odot]$	$\Delta M_{\rm cl}/M_{\rm cl}$ [%] $\alpha_3 = 2.3$	$\alpha_3 = 2.7$
1	16	2.9	21	3.8	80	3.2	0.7
8	245	74	725	195	60	4.9	1.1
20	806	269	3442	967	40	7.5	1.8
40	1984	703	1.1×10^4	2302	20	13	4.7
60	3361	1225	2.2×10^4	6428	8	22	8.0
80	4885	1812	3.6×10^4	1.1×10^4	3	32	15
100	6528	2451	5.3×10^4	1.5×10^4	1	44	29
120	8274	3136	7.2×10^4	2.1×10^4	0.7	47	33

Star Clusters and Associations

To access a pristine population one would consider observing star-clusters that are younger than a few Myr. However, such objects carry rather serious disadvantages. The pre-mainsequence stellar evolution tracks are unreliable (Baraffe et al. 2002; Wuchterl & Tscharnuter 2003) so that the derived masses are uncertain by at least a factor of about two. Remaining gas and dust lead to patchy obscuration. Very young clusters evolve rapidly. The dynamical crossing time is given by (8.4) where the cluster radii are typically $r_{\rm h} < 1\,{\rm pc}$ and for pre-cluster cloud-core masses $M_{\rm gas+stars} > 10^3\,M_\odot$ the velocity dispersion $\sigma_{\rm cl} > 2\,{\rm km\,s^{-1}}$ so that $t_{\rm cr} < 1\,{\rm Myr}$.

The inner regions of populous clusters have $t_{\rm cr} \approx 0.1\,{\rm Myr}$ and thus significant mixing and relaxation occurs there by the time the residual gas has been expelled by any winds and photo-ionising radiation from massive stars. This is the case in clusters with $N \geq$ few $\times$ 100 stars (Table 8.1).

Massive stars ($m > 8\,M_\odot$) are either formed at the cluster centre or get there through dynamical mass segregation, i.e. energy equipartition (Bonnell et al. 2007). The latter process is very rapid ((8.6), p. 184) and can occur within 1 Myr. A cluster core of massive stars is therefore either primordial or forms rapidly because of energy equipartition in the cluster and it is dynamically highly unstable decaying within a few $t_{\rm cr,\,core}$. The ONC, for example, should not be hosting a Trapezium because it is extremely unstable. The implication for the IMF is that the ONC and other similar clusters and the OB associations which stem from them must be very depleted in their massive star content (Pflamm-Altenburg & Kroupa 2006).

Important for measuring the IMF are corrections for the typically high multiplicity fraction of the very young population. However, these are very uncertain because the binary population is in a state of change (Fig. 8.14 below). The determination of an IMF relies on the assumption that all stars in a very young cluster formed together. However, trapping and focussing of older field or OB association stars by the forming cluster has been found to be possible (Sect. 8.1.1).

Thus, be it at the low-mass end or the high-mass end, the stellar mass function seen in very young clusters cannot be the true IMF. Statistical corrections for the above effects need to be applied and comprehensive N-body modelling is required.

Old open clusters, in which most stars are on or near the main sequence, are no better stellar samples. They are dynamically highly evolved, because they have left their previous concentrated and gas-rich state and so they contain only a small fraction of the stars originally born in the cluster (Kroupa & Boily 2002; Weidner et al. 2007; Baumgardt & Kroupa 2007). The binary fraction is typically high and comparable to the Galactic field, but does depend on the initial density and the age of the cluster as does the mass-ratio distribution of companions. So, simple corrections cannot be applied equally for all old clusters. The massive stars have died and secular evolution begins to affect the remaining stellar population (after gas expulsion) through energy equipartition. Baumgardt & Makino (2003) have quantified the changes of the MF for clusters of various masses and on different Galactic orbits. Near the half-mass radius, the local MF resembles the global MF in the cluster but the global MF is already significantly depleted of its lower-mass stars by about 20% of the cluster disruption time.

Given that we are never likely to learn the exact dynamical history of a particular cluster, it follows that we can never ascertain the IMF for any individual cluster. This can be summarised concisely with the following conjecture.

CLUSTER IMF CONJECTURE: The IMF cannot be extracted for any individual star cluster.

JUSTIFICATION: For clusters younger than about 0.5 Myr, star-formation has not ceased and the IMF is therefore not yet assembled and the cluster cores consisting of massive stars have already dynamically ejected members (Pflamm-Altenburg & Kroupa 2006). For clusters with an age between 0.5 and a few Myr, the expulsion of residual gas has lead to loss of stars (Kroupa, Aarseth & Hurley 2001). Older clusters are either still losing stars owing to residual gas expulsion or are evolving secularly through evaporation driven by energy equipartition (Baumgardt & Makino 2003). Furthermore, the birth sample is likely to be contaminated by captured stars (Fellhauer, Kroupa & Evans 2006; Pflamm-Altenburg & Kroupa 2007). There exists no time when all stars are assembled in an observationally accessible volume (i.e. a star cluster).

Note that the CLUSTER IMF CONJECTURE implies that individual clusters cannot be used to make deductions on the similarity or not of their IMFs, unless a complete dynamical history of each cluster is available. Notwithstanding this pessimistic conjecture, it is nevertheless necessary to observe and study star clusters of any age. Combined with thorough and realistic N-body modelling the data, do lead to essential statistical constraints on the IMF UNIVERSALITY HYPOTHESIS. Such an approach is discussed in the next section.

The Alpha Plot

Scalo (1998) conveniently summarised a large part of the available observational constraints on the IMF of resolved stellar populations with the alpha plot, as used by Kroupa (2001, 2002b) for explicit tests of the IMF UNIVERSALITY HYPOTHESIS given the CLUSTER IMF CONJECTURE. One example is presented in Fig. 8.9, which demonstrates that the observed scatter in $\alpha(m)$ can be readily understood as being due to Poisson uncertainties (see also Elmegreen 1997, 1999) and dynamical effects, as well as arising from biases through unresolved multiple stars. Furthermore, there is no evident systematic change of α at a given m with metallicity or density of the star-forming cloud.

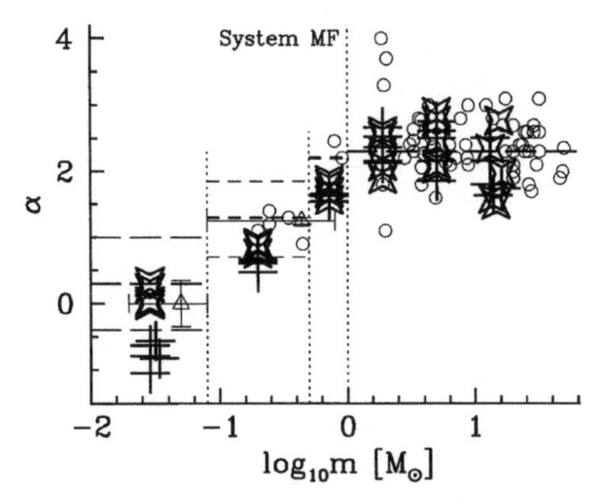

Fig. 8.9. The alpha plot. The power-law index, α, is measured over stellar mass-ranges and plotted at the mid-point of the respective mass range. The power-law indices are measured on the mass function of system masses, where stars not in binaries are counted individually. *Open circles* are the observations from open clusters and associations of the Milky Way and the Large and Small Magellanic clouds collated mostly by Scalo (1998). The open stars (*crosses*) are theoretical star clusters observed in the computer at an age of 3 (0) Myr and within a radius of 3.2 pc from the cluster centre. The 5 clusters have 3000 stars in 1500 binaries initially and the assumed IMF is the canonical one. The theoretical data nicely show a similar spread to the observational ones; note the binary-star-induced depression of α_1 in the mass range 0.1–0.5 $M_{\odot}$. The IMF UNIVERSALITY HYPOTHESIS can therefore not be discarded given the observed data. Models are from Kroupa (2001)

More exotic populations such as the Galactic bulge have also been found to have a low-mass MF indistinguishable from the canonical form (e.g. Zoccali et al. 2000). Thus the IMF UNIVERSALITY HYPOTHESIS cannot be falsified for known resolved stellar populations.

Very Ancient and/or Metal-Poor Resolved Populations

Witnesses of the early formation phase of the Milky Way are its globular clusters. Such 10^4–$10^6\,M_{\odot}$ clusters formed with individual star-formation rates of 0.1–1 $M_{\odot}\,\mathrm{yr}^{-1}$ and densities of about 5×10^3–$10^5\,M_{\odot}\,\mathrm{pc}^{-3}$. These are relatively high values, when compared with the current star-formation activity in the Milky Way disc. For example, a $5\times10^3\,M_{\odot}$ Galactic cluster forming in 1 Myr corresponds to a star-formation rate of $0.005\,M_{\odot}\,\mathrm{yr}^{-1}$. The alpha plot, however, does not support any significant systematic difference between the IMF of stars formed in globular clusters and present-day low-mass star-formation. For massive stars, it can be argued that the mass in stars more massive than $8\,M_{\odot}$ cannot have been larger than about half the cluster mass, because otherwise the globular clusters would not be as compact as they are today. This constrains the IMF to have been close to the canonical IMF (Kroupa 2001).

A particularly exotic star-formation mode is thought to have occurred in dwarf-spheroidal (dSph) satellite galaxies. The Milky Way has about 19 such satellites at distances from 50 to 250 kpc (Metz & Kroupa 2007). These objects have stellar masses and ages comparable to those of globular clusters, but are 10–100 times larger and are thought to have 10–1000 times more mass in dark matter than in stars. They also show evidence for complex star-formation activity and metal-enrichment histories and must therefore have formed under rather exotic conditions. Nevertheless, the MFs in two of these satellites are found to be indistinguishable from those of globular clusters in the mass range 0.5–0.9 $M_\odot$. So again there is consistency with the canonical IMF (Grillmair et al. 1998; Feltzing, Gilmore & Wyse 1999).

The work of Yasui et al. (2006) and Yasui et al. (2008) have been pushing studies of the IMF in young star clusters to the outer, metal-poor regions of the Galactic disc. They find the IMF to be indistinguishable, within the uncertainties, from the canonical IMF.

The Galactic Bulge and Centre

For low-mass stars the Galactic bulge has been shown to have a MF indistinguishable from the canonical form (Zoccali et al. 2000). However, abundance patterns of bulge stars suggest the IMF was top-heavy (Ballero, Kroupa & Matteucci 2007). This may be a result of extreme star-burst conditions prevailing in the formation of the bulge (Zoccali et al. 2006).

Even closer to the Galactic centre, models of the Hertzsprung–Russell diagram of the stellar population within 1 pc of Sgr A* suggest the IMF was always top-heavy there (Maness et al. 2007). Perhaps, this is the long-sought after evidence for a variation of the IMF under very extreme conditions, in this case a strong tidal field and higher temperatures (but note Fig. 8.10 below).

Extreme Star Bursts

As noted on p. 199, objects with a mass $M \geq 10^6\,M_\odot$ have an increased M/L ratio. If such objects form in 1 Myr, their star-formation rates SFR $\geq 1\,M_\odot$/yr and they probably contain more than 10^4 O stars packed within a region spanning at most a few parsecs, given their observed present-day mass–radius relation. Such a star-formation environment is presently outside the reach of theoretical investigation. However, it is conceivable that the higher M/L ratios of such objects may be due to a non-canonical IMF. One possibility is that the IMF is bottom-heavy as a result of intense photo-destruction of accretion envelopes of intermediate to low-mass stars (Mieske & Kroupa 2008). Another possibility is that the IMF becomes top-heavy leaving many stellar remnants that inflate the M/L ratio (Dabringhausen & Kroupa 2008). Work is in progress to achieve observational constraints on these two possibilities.

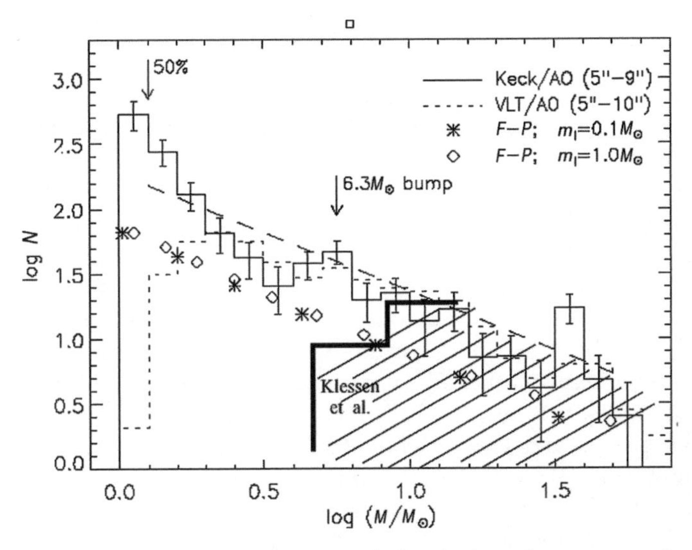

Fig. 8.10. The observed mass function of the Arches cluster near the Galactic centre by Kim et al. (2006) shown as the thin histogram is confronted with the theoretical MF for this object calculated with the SPH technique by Klessen, Spaans & Jappsen (2007), marked as the hatched histogram. The latter has a down-turn (bold steps near $10^{0.7}$) incompatible with the observations. This rules out a theoretical understanding of the stellar mass spectrum because one counter-example suffices to bring-down a theory. One possible reason for the theoretical failure may be the assumed turbulence driving. For details of the figure see Kim et al. (2006)

Population III: The Primordial IMF

Most theoretical workers agree that the primordial IMF ought to be top-heavy because the ambient temperatures were much higher and the lack of metals did not allow gas clouds to cool and to fragment into sufficiently small cores (Larson 1998). The existence of extremely metal-poor, low-mass stars with chemical peculiarities is interpreted to mean that low-mass stars could form under extremely metal-poor conditions but that their formation was suppressed in comparison to later star-formation (Tumlinson 2007). Models of the formation of stellar populations during cosmological structure formation suggest that low-mass population-III stars should be found within the Galactic halo if they formed. Their absence to-date would imply a primordial IMF depleted in low-mass stars (Brook et al. 2007).

Thus, the last three sub-sections hint at physical environments in which the IMF universality hypothesis may be violated.

8.3.3 Very Low-Mass Stars (VLMSs) and Brown Dwarfs (BDs)

The origin of BDs and some VLMSs is being debated fiercely. One camp believes these objects to form as stars, because the star-formation process does not know where the hydrogen burning mass limit is (e.g. Eislöffel & Steinacker 2008). The other camp believes that BDs cannot form exactly like stars through continued accretion because the conditions required for this to occur in molecular clouds are far too rare (e.g. Reipurth & Clarke 2001; Goodwin & Whitworth 2007).

If BDs and VLMSs form like stars, they should follow the same pairing rules. In particular, BDs and G dwarfs would pair in the same manner, i.e. according to the same mathematical rules, as M dwarfs and G dwarfs. Kroupa et al. (2003) tested this hypothesis by constructing N-body models of Taurus-Auriga-like groups and Orion-Nebula-like clusters, finding that it leads to far too many star–BD and BD–BD binaries with the wrong semi-major axis distribution. Instead, star–BD binaries are very rare (Grether & Lineweaver 2006), while BD–BD binaries are rarer than stellar binaries (BDs have a 15% binary fraction as opposed to 50% for stars) and BDs have a semi-major axis distribution significantly narrower than that of star–star binaries. The hypothesis of a star-like origin of BDs must therefore be discarded. BDs and some VLMSs form a separate population, which is however linked to that of the stars.

Thies & Kroupa (2007) re-addressed this problem with a detailed analysis of the underlying MF of stars and BDs given observed MFs of four populations: Taurus, Trapezium, IC348 and the Pleiades. By correcting for unresolved binaries in all four populations and taking into account the different pairing rules of stellar and VLMS and BD binaries, they discovered a significant discontinuity of the MF. BDs and VLMSs therefore form a truly separate population from that of the stars. It can be described by a single power-law MF (8.129), which implies that about one BD forms per five stars in all four populations.

This strong correlation between the number of stars and BDs and the similarity of the BD MF in the four populations implies that the formation of BDs is closely related to the formation of stars. Indeed, the truncation of the binary binding energy distribution of BDs at a high energy suggests that energetic processes must be operating in the production of BDs, as discussed by Thies & Kroupa (2007). Two such possible mechanisms are embryo ejection (Reipurth & Clarke 2001) and disc fragmentation (Goodwin & Whitworth 2007).

8.3.4 Composite Populations: The IGIMF

The vast majority of all stars form in embedded clusters and so the correct way to proceed to calculate a galaxy-wide stellar IMF is to add up all the IMFs of all star clusters born in one star-formation epoch. Such epochs may be identified with the Zoccali et al. (2006) star-burst events that create the Galactic bulge. In disc galaxies they may be related to the time-scale of transforming the interstellar matter to star clusters along spiral arms. Addition of the clusters born in one epoch gives the integrated galactic initial mass function, the IGIMF (Kroupa & Weidner 2003).

> IGIMF DEFINITION: The IGIMF is the IMF of a composite population, which is the integral over a complete ensemble of simple stellar populations.

Note that a simple population has a mono-metallicity and a mono-age distribution and is therefore a star cluster. Age and metallicity distributions emerge for massive populations with $M_{\rm cl} \geq 10^6\, M_\odot$ (e.g. ω Cen). This indicates that such objects, which also have relaxation times comparable to or longer than a Hubble time, are not simple (Sect. 8.1.4). A complete ensemble is a statistically complete representation of the initial cluster mass function (ICMF) in the sense that the actual mass function of $N_{\rm cl}$ clusters lies within the expected statistical variation of the ICMF.

> IGIMF CONJECTURE: The IGIMF is steeper than the canonical IMF if the IMF UNIVERSALITY HYPOTHESIS holds.

JUSTIFICATION: Weidner & Kroupa (2006) calculate that the IGIMF is steeper than the canonical IMF for $m \geq 1\, M_\odot$ if the IMF UNIVERSALITY HYPOTHESIS holds. The steepening becomes negligible if the power-law mass function of embedded star clusters,

$$\xi_{\rm ecl}(M_{\rm ecl}) \propto M_{\rm ecl}^{-\beta}, \tag{8.127}$$

is flatter than $\beta = 1.8$.

It may be argued that IGIMF = IMF (e.g. Elmegreen 2006) because, when a star cluster is born, its stars are randomly sampled from the IMF up to the most massive star possible. On the other hand, the physically motivated ansatz of Weidner & Kroupa (2005, 2006) to take the mass of a cluster as the constraint and to include the observed correlation between the maximal star mass and the cluster mass (Fig. 8.1) yields an IGIMF which is equal to the canonical IMF for $m \leq 1.5\, M_\odot$ but which is systematically steeper above this mass. By incorporating the observed maximal-cluster-mass vs star-formation rate of galaxies, $M_{\rm ecl,max} = M_{\rm ecl,max}(SFR)$, for the youngest clusters (Weidner, Kroupa & Larsen 2004) it follows for $m \geq 1.5\, M_\odot$ that low-surface-brightness (LSB) galaxies ought to have very steep IGIMFs, while normal or L_* galaxies have Scalo-type IGIMFs, i.e. $\alpha_{\rm IGIMF} = \alpha_{\rm MWdisc} > \alpha_2$ (Sect. 8.3.1) follows naturally. This systematic shift of $\alpha_{\rm IGIMF}$ $(m \geq 1.5\, M_\odot)$ with galaxy type implies that less massive galaxies have a significantly suppressed supernova II rate per low-mass star. They also show a slower chemical enrichment so that the observed metallicity–galaxy-mass relation can be nicely accounted for (Koeppen, Weidner & Kroupa 2007). Another very important implication is that the SFR–Hα-luminosity relation for galaxies flattens so that the SFR becomes greater by up to three orders of magnitude for dwarf galaxies than the value calculated from the standard (linear) Kennicutt relation (Pflamm-Altenburg, Weidner & Kroupa 2007).

Strikingly, the IGIMF variation has now been directly measured by Hoversten & Glazebrook (2008) using galaxies in the Sloan Digital Sky Survey. Lee et al. (2004) have indeed found LSBs to have bottom-heavy IMFs, while Portinari, Sommer-Larsen & Tantalo (2004) and Romano et al. (2005) find the Milky Way disc to have a an IMF steeper than Salpeter's for massive stars which is, in comparison with Lee et al. (2004), much flatter than the IMF of LSBs, as required by the IGIMF CONJECTURE.

8.3.5 Origin of the IMF: Theory vs Observations

General physical concepts such as coalescence of protostellar cores, mass-dependent focussing of gas accretion on to protostars, stellar feedback and fragmentation of molecular clouds lead to predictions of systematic variations of the IMF with changes of the physical conditions of star-formation (Murray & Lin 1996; Elmegreen 2004). (But see Casuso & Beckman 2007 for a simple cloud coagulation/dispersal model that leads to an invariant mass distribution.) Thus, the thermal Jeans mass of a molecular cloud decreases with temperature and increasing density. This implies that for higher metallicity (stronger cooling) and density the IMF should shift on average to smaller stellar masses (e.g. Larson 1998; Bonnell et al. 2007). The entirely different notion that stars regulate their own masses through a balance between feedback and accretion also implies smaller stellar masses for higher metallicity due, in part, to more dust and thus more efficient radiation pressure on the gas through the dust grains. Also, a higher metallicity allows more efficient cooling and thus a lower gas temperature, a lower sound speed and therefore a lower accretion rate (Adams & Fatuzzo 1996; Adams & Laughlin 1996). As discussed above, a systematic IMF variation with physical conditions has not been detected. Thus, theoretical reasoning, even at its most elementary level, fails to account for the observations.

A dramatic case in point has emerged recently: Klessen, Spaans & Jappsen (2007) report state-of-the art calculations of star-formation under physical conditions as found in molecular clouds near the Sun and they are able to reproduce the canonical IMF. Applying the same computational technology to the conditions near the Galactic centre, they obtain a theoretical IMF in agreement with the previously reported apparent decline of the stellar MF in the Arches cluster below about $6\,M_\odot$. Kim et al. (2006) published their observations of the Arches cluster on the astrophysics preprint archive shortly after Klessen, Spaans & Jappsen (2007) and performed N-body calculations of the dynamical evolution of this young cluster, revising our knowledge significantly. In contradiction to the theoretical prediction, they find that the MF continues to increase down to their 50% completeness limit ($1.3\,M_\odot$) with a power-law exponent only slightly shallower than the canonical Massey/Salpeter value once mass-segregation has been corrected for. This situation is demonstrated in Fig. 8.10. It therefore emerges that there does not seem to exist any solid theoretical understanding of the IMF.

Observations of cloud cores appear to suggest that the canonical IMF is already frozen in at the pre-stellar cloud-core level (Motte, Andre & Neri 1998; Motte et al. 2001). Nutter & Ward-Thompson (2007) and Alves, Lombardi & Lada (2007) find, however, the pre-stellar cloud cores are distributed according to the same shape as the canonical IMF but shifted to larger masses by a factor of about three or more. This is taken to perhaps mean a star-formation efficiency per star of 30% or less independently of stellar mass. The interpretation of such observations in view of multiple star-formation in each cloud-core is being studied by Goodwin et al. (2008), while Krumholz (2008) outlines current theoretical understanding of how massive stars form out of massive pre-stellar cores.

8.3.6 Conclusions: IMF

The IMF UNIVERSALITY HYPOTHESIS, the CLUSTER IMF CONJECTURE and the IGIMF CONJECTURE have been stated. In addition, we may make the following assertions.

1. The stellar luminosity function has a pronounced maximum at $M_V \approx 12$, $M_I \approx 9$, which is universal and well understood as a result of stellar physics. Thus by counting stars in the sky we can look into their interiors.
2. Unresolved multiple systems must be accounted for when the MFs of different stellar populations are compared.
3. BDs and some VLMSs form a separate population that correlates with the stellar content. There is a discontinuity in the MF near the star/BD mass transition.
4. The canonical IMF (8.124) fits the star counts in the solar neighbourhood and all resolved stellar populations available to-date. Recent data at the Galactic centre suggest a top-heavy IMF, perhaps hinting at a possible variation with conditions (tidal shear, temperature).
5. Simple stellar populations are found in individual star clusters with $M_{\rm cl} \leq 10^6\ M_\odot$. These have the canonical IMF.
6. Composite populations describe entire galaxies. They are a result of many epochs of star-cluster formation and are described by the IGIMF CONJECTURE.
7. The IGIMF above about $1\,M_\odot$ is steep for LSB galaxies and flattens to the Scalo slope ($\alpha_{\rm IGIMF} \approx 2.7$) for L_* disc galaxies. This is nicely consistent with the IMF UNIVERSALITY HYPOTHESIS in the context of the IGIMF CONJECTURE.

8. Therefore, the IMF UNIVERSALITY HYPOTHESIS cannot be excluded despite the CLUSTER IMF CONJECTURE for conditions $\rho \leq 10^5\,\mathrm{stars\,pc^{-3}}$, $Z \geq 0.002$ and non-extreme tidal fields.
9. Modern star-formation computations and elementary theory give wrong results concerning the variation and shape of the stellar IMF, as well as the stellar multiplicity (Goodwin & Kroupa 2005).
10. The stellar IMF appears to be frozen-in at the pre-stellar cloud-core stage. So it is probably a result of the processes that lead to the formation of self-gravitating molecular clouds.

8.3.7 Discretisation

As discussed above, a theoretically motivated form of the IMF that passes observational tests does not exist. Star-formation theory gets the rough shape of the IMF right. There are fewer massive stars than low-mass stars. However, other than this, it fails to make any reliable predictions whatsoever as to how the IMF should look in detail under different physical conditions. In particular, the overall change of the IMF with metallicity, density or temperature predicted by theory is not evident. An empirical multi-power-law form description of the IMF is therefore perfectly adequate and has important advantages over other formulations. A general formulation of the stellar IMF in terms of multiple power-law segments is

$$\xi(m) = k \begin{cases} \left(\frac{m}{m_{\rm H}}\right)^{-\alpha_0} & , m_{\rm low} \leq m \leq m_{\rm H} \\ \left(\frac{m}{m_{\rm H}}\right)^{-\alpha_1} & , m_{\rm H} \leq m \leq m_0 \\ \left(\frac{m_0}{m_{\rm H}}\right)^{-\alpha_1} \left(\frac{m}{m_0}\right)^{-\alpha_2} & , m_0 \leq m \leq m_1 \\ \left(\frac{m_0}{m_{\rm H}}\right)^{-\alpha_1} \left(\frac{m_1}{m_0}\right)^{-\alpha_2} \left(\frac{m}{m_1}\right)^{-\alpha_3} & , m_1 \leq m \leq m_{\rm max} \end{cases}, \quad (8.128)$$

where $m_{\rm max} \leq m_{\rm max*} \approx 150\,M_\odot$ depends on the stellar mass of the embedded cluster (Fig. 8.1). The empirically determined stellar IMF is a two-part-form (8.124), with a third power-law for BDs, whereby BDs and VLMSs form a separate population from that of the stars (p. 232),

$$\xi_{\rm BD} \propto m^{-\alpha_0}, \quad \alpha_0 \approx 0.3, \quad (8.129)$$

(Martín et al. 2000; Chabrier 2003; Moraux, Bouvier & Clarke 2004) and

$$\xi_{\rm BD}(0.075\,M_\odot) \approx 0.25 \pm 0.05\,\xi(0.075\,M_\odot),$$

(Thies & Kroupa 2007) where ξ is the canonical stellar IMF (8.124). This implies that about one BD forms per five stars.

One advantage of the power-law formulation is that analytical generating functions and other quantities can be readily derived. Another important advantage is that with a multi-power-law form, different parts of the IMF can be varied in numerical experiments without affecting the other parts. A practical numerical formulation of the IMF is prescribed in Pflamm-Altenburg & Kroupa (2006). Thus, for example, the canonical two-part power-law IMF can be changed by adding a third power-law above $1\,M_\odot$ and making the IMF top-heavy ($\alpha_{m>1\,M_\odot} < \alpha_2$), without affecting the shape of the late-type stellar luminosity function as evident in Fig. 8.8. The KTG93 (Kroupa, Tout & Gilmore 1993) IMF is such a three-part power-law form relevant to the overall young population in the Milky Way disc. This is top-light ($\alpha_{m>1\,M_\odot} > \alpha_2$, Kroupa & Weidner 2003).

A log-normal formulation does not offer these advantages and requires power-law tails above about $1\,M_\odot$ and for brown dwarfs, for consistency with the observations discussed above. However, while not as mathematically convenient, the popular Chabrier log-normal plus power-law IMF (Table 1 of Chabrier 2003) formulation leads to an indistinguishable stellar mass distribution to the two-part power-law IMF (Fig. 8.11). Various analytical forms for the IMF are compiled in Table 3 of Kroupa (2007a).

A generating function for the two-part power-law form of the canonical IMF (8.124) can be written down by following the steps taken in Sect. 8.2.3. The corresponding probability density is

$$\begin{array}{ll} p_1 = k_{p,1}\, m^{-\alpha_1}, & 0.08 \le m \le 0.5\,M_\odot \\ p_2 = k_{p,2}\, m^{-\alpha_2}, & 0.5 < m \le m_{\max}, \end{array} \tag{8.130}$$

where $k_{p,i}$ are normalisation constants ensuring continuity at $0.5\,M_\odot$ and

$$\int_{0.08}^{0.5} p_1\, \mathrm{d}m + \int_{0.5}^{m_{\max}} p_2\, \mathrm{d}m = 1, \tag{8.131}$$

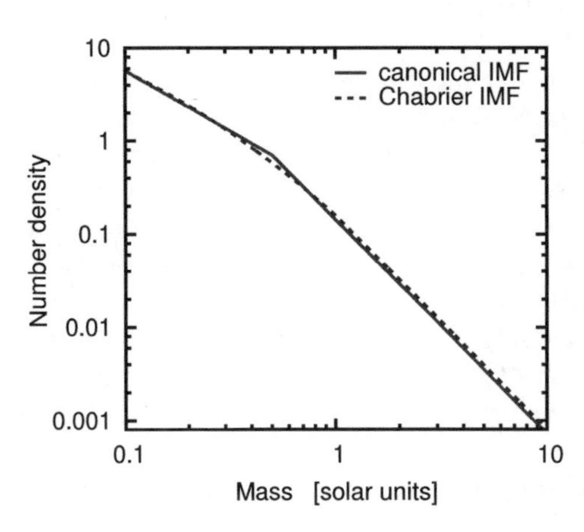

Fig. 8.11. Comparison between the popular Chabrier IMF (log-normal plus power-law extension above $1\,M_\odot$, *dashed curve*, Table 1 in Chabrier 2003) with the canonical two-part power-law IMF (*solid line*, (8.124)). The figure is from Dabringhausen, Hilker & Kroupa (2008)

whereby $m_{\rm max}$ follows from Fig. 8.1. Defining

$$X_1' = \int_{0.08}^{0.5} p_1(m)\, {\rm d}m, \tag{8.132}$$

it follows that

$$X_1(m) = \int_{0.08}^{m} p_1(m)\, {\rm d}m, \quad \text{if } m \le 0.5\, M_\odot, \tag{8.133}$$

or

$$X_2(m) = X_1' + \int_{0.5}^{m} p_2(m)\, {\rm d}m, \quad \text{if } m > 0.5\, M_\odot. \tag{8.134}$$

The generating function for stellar masses follows from inversion of the above two equations $X_i(m)$. The procedure is then to choose a random variate $X \in [0,1]$ and to select the generating function $m(X_1 = X)$ if $0 \le X \le X_1$, or $m(X_2 = X)$ if $X_1 < X \le 1$.

This algorithm is readily generalised to any number of power-law segments (8.128), such as including a third segment for brown dwarfs and allowing the IMF to be discontinuous near $0.08\, M_\odot$ (Thies & Kroupa 2007). Such a form has been incorporated into the NBODY4/6/7 programmes, but hitherto without the discontinuity. However, Jan Pflamm-Altenburg has developed a more powerful and general method of generating stellar masses (or any other quantities) given an arbitrary distribution function (Pflamm-Altenburg & Kroupa 2006).[11]

8.4 The Initial Binary Population

It has already been demonstrated that corrections for unresolved multiple stars are of much importance to derive correctly the shape of the stellar MF given an observed LF (Fig. 8.8). Binary stars are also of significant importance for the dynamics of star clusters because a binary has intrinsic dynamical degrees of freedom that a single star does not possess. A binary can therefore exchange energy and angular momentum with the cluster. Indeed, binaries are very significant energy sources, as for example, a binary composed of two $1\, M_\odot$ main-sequence stars and with a semi-major axis of 0.1 AU has a binding energy comparable to that of a $1000\, M_\odot$ cluster of size 1 pc. Such a binary can interact with cluster-field star accelerating them to higher velocities and thereby heating the cluster.

The dynamical properties describing a multiple system are

- the period P (in days throughout this text) or semi-major axis a (in AU),
- the system mass $m_{\rm sys} = m_1 + m_2$,

[11] The C-language software package, libimf, can be downloaded from the website `http://www.astro.uni-bonn.de/~webaiub/english/downloads.php`.

- the mass ratio $q \equiv \frac{m_2}{m_1} \leq 1$, where m_1, m_2 are, respectively, the primary and secondary-star masses and
- the eccentricity $e = (r_{\rm apo} - r_{\rm peri})/(r_{\rm apo} + r_{\rm peri})$, where $r_{\rm apo}, r_{\rm peri}$ are, respectively, the apocentric and pericentric distances.

Given a snapshot of a binary, the above quantities can be computed from the relative position, $\boldsymbol{r}_{\rm rel}$ and velocity, $\boldsymbol{v}_{\rm rel}$, vectors and the masses of the two companion stars by first calculating the binding energy,

$$E_{\rm b} = \frac{1}{2}\,\mu\, v_{\rm rel}^2 - \frac{G\,m_1\,m_2}{r_{\rm rel}} = -\frac{G\,m_1\,m_2}{2\,a} \Rightarrow a, \tag{8.135}$$

where $\mu = m_1\,m_2\,/(m_1 + m_2)$ is the reduced mass. From Kepler's third law we have

$$m_{\rm sys} = \frac{a_{AU}^3}{P_{\rm yr}^2} \Rightarrow P = P_{\rm yr} \times 365.25 \text{ days}, \tag{8.136}$$

where $P_{\rm yr}$ is the period in years and a_{AU} is in AU. Finally, the instantaneous eccentricity can be calculated using

$$e = \left[\left(1 - \frac{r_{\rm rel}}{a}\right)^2 + \frac{\left(\boldsymbol{r}_{\rm rel}\cdot\boldsymbol{v}_{\rm rel}\right)^2}{a\,G\,m_{\rm sys}}\right]^{\frac{1}{2}}, \tag{8.137}$$

which can be derived from the orbital angular momentum too,

$$\boldsymbol{L} = \mu\,\boldsymbol{v}_{\rm rel} \times \boldsymbol{r}_{\rm rel}, \tag{8.138}$$

with

$$L = \left[\frac{G}{m_{\rm sys}}\,a\,(1 - e^2)\right]^{\frac{1}{2}}\,m_1\,m_2. \tag{8.139}$$

The relative equation of motion is

$$\frac{{\rm d}^2\boldsymbol{r}_{\rm rel}}{{\rm d}t^2} = -G\frac{m_{\rm sys}}{r_{\rm rel}^3}\boldsymbol{r}_{\rm rel} + \boldsymbol{a}_{\rm pert}(t), \tag{8.140}$$

where $\boldsymbol{a}_{\rm pert}(t)$ is the time-dependent perturbation from other cluster members. It follows that the orbital elements of a binary in a cluster are functions of time, $P = P(t)$ and $e = e(t)$. Also, $q = q(t)$ during strong encounters when partners are exchanged. Because most stars form in embedded clusters, the binary-star properties of a given population cannot be taken to represent the initial or primordial values.

The following conjecture can be proposed.

DYNAMICAL POPULATION SYNTHESIS CONJECTURE: if initial binary populations are invariant, a dynamical birth configuration of a stellar population can be inferred from its observed binary population. This birth configuration is not unique, however, but defines a class of dynamically equivalent solutions.

The proof is simple. Set up initially identical binary populations in clusters with different radii and masses and calculate the dynamical evolution with an N-body programme. For a given snapshot of a population, there is a scalable starting configuration in terms of size and mass (Kroupa 1995c,d).

Binaries can absorb energy and thus cool a cluster. They can also heat a cluster. There are two extreme regimes that can be understood with a Gedanken experiment. Define

$$E_{\rm bin} \equiv -E_{\rm b} > 0, \\ E_{\rm k} \equiv (1/2)\,\overline{m}\,\sigma^2 \approx (1/N) \times \text{ kinetic energy of cluster.} \tag{8.141}$$

Soft binaries have $E_{\rm bin} \ll E_{\rm k}$, while hard binaries have $E_{\rm bin} \gg E_{\rm k}$. A useful equation in this context is the relation between the orbital period and circular velocity of the reduced particle,

$$\log_{10} P[\text{days}] = 6.986 + \log_{10} m_{\rm sys}[M_\odot] - 3\,\log_{10} v_{\rm orb}[\text{km s}^{-1}]. \tag{8.142}$$

Consider now the case of a soft binary, a reduced-mass particle with $v_{\rm orb} \ll \sigma$. By the principle of energy equipartition, $v_{\rm orb} \to \sigma$ (8.5) as time progresses. This implies $a \uparrow, P \uparrow$. A hard binary has $v_{\rm orb} \gg \sigma$. Invoking energy equipartition, we see that $v_{\rm orb} \downarrow$, $a \downarrow, P \downarrow$. Furthermore, the amount of energy needed to ionise a soft binary is negligible compared to the amount of energy required to ionise a hard binary. And the cross section for suffering an encounter scales with the semi-major axis. This implies that a soft binary becomes ever more likely to suffer an additional encounter as its semi-major axis increases. Therefore, it is much more probable for soft binaries to be disrupted rapidly, than for hard binaries to do so. Thus follows (Heggie 1975; Hills 1975) a law.

HEGGIE–HILLS LAW: soft binaries soften and cool a cluster while hard binaries harden and heat a cluster.

Numerical scattering experiments by Hills (1975) have shown that hardening of binaries often involves partner exchanges. Heggie (1975) derived the above law analytically. Binaries in the energy range $10^{-2}\,E_{\rm k} \le E_{\rm bin} \le 10^{2}\,E_{\rm k}$, $33^{-1}\,\sigma \le v_{\rm orb} \le 33\,\sigma$ cannot be treated analytically owing to the complex resonances that are created between the binary and the incoming star or binary. It is these binaries that may be important for the early cluster evolution, depending on its velocity dispersion, $\sigma = \sigma(M_{\rm ecl})$. Cooling of a cluster is energetically not significant but has been seen for the first time by Kroupa, Petr & McCaughrean (1999).

Figure 8.12 shows the broad evolution of the initial period distribution in a star cluster. At any time, binaries near the hard/soft boundary, with energies $E_{\rm bin} \approx E_{\rm k}$ and periods $P \approx P_{\rm th}$ ($v_{\rm orb} = \sigma$) (8.5), the thermal period, are most active in the energy exchange between the cluster field and the binary population. The cluster expands as a result of binary heating and

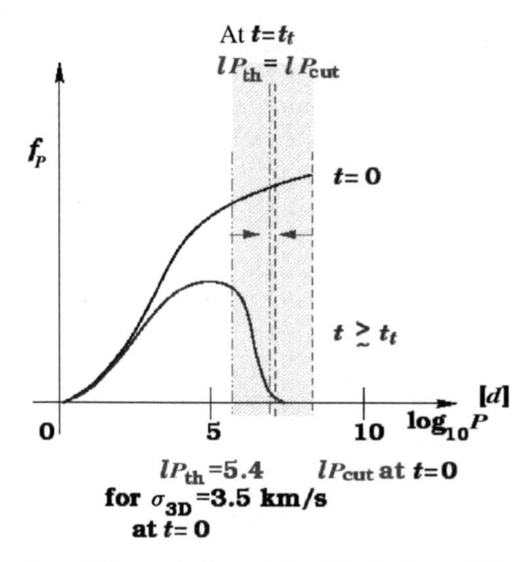

Fig. 8.12. Illustration of the evolution of the distribution of binary star periods in a cluster ($l\mathrm{P} = \log_{10} P$). A binary has orbital period P_{th} when $\sigma_{3\mathrm{D}}$ ($= \sigma$) equals its circular orbital velocity (8.142). The initial or birth distribution (8.164) evolves to the form seen at time $t > t_t$

mass segregation and the hard/soft boundary, P_{th}, shifts to longer periods. Meanwhile, binaries with $P > P_{\mathrm{th}}$ continue to be disrupted while P_{th} keeps shifting to longer periods. This process ends when

$$P_{\mathrm{th}} \geq P_{\mathrm{cut}}, \tag{8.143}$$

which is the cutoff or maximum period in the surviving period distribution. At this critical time, t_t, further cluster expansion is slowed because the population of heating sources, the binaries with $P \approx P_{\mathrm{th}}$, is significantly reduced. The details strongly depend on the initial value of P_{th}, which determines the amount of binding energy in soft binaries which can cool the cluster if significant enough.

After the critical time, t_{t}, the expanded cluster reaches a temporary state of thermal equilibrium with the remaining binary population. Further evolution of the binary population occurs with a significantly reduced rate determined by the velocity dispersion in the cluster, the cross section given by the semi-major axis of the binaries and their number density and that of single stars in the cluster. The evolution of the binary star population during this slow phase usually involves partner exchanges and unstable but also long-lived hierarchical systems. The IMF is critically important for this stage, as the initial number of massive stars determines the cluster density at $t \geq 5\,\mathrm{Myr}$ owing to mass loss from evolving stars. Further binary depletion occurs once the cluster goes into core-collapse and the kinetic energy in the core rises.

8.4.1 Frequency of Binaries and Higher-Order Multiples

The emphasis here is on late-type binary stars because higher-order multiples are rare as observed. The information on the multiplicity of massive stars is very limited (Goodwin et al. 2007). We define, respectively, the number of single stars, binaries, triples, quadruples, etc., by the numbers

$$(N_{\rm sing} : N_{\rm bin} : N_{\rm trip} : N_{\rm quad} : \ldots) = (\mathcal{S} : \mathcal{B} : \mathcal{T} : \mathcal{Q} : \ldots) \qquad (8.144)$$

and the multiplicity fraction by

$$f_{\rm mult} = \frac{N_{\rm mult}}{N_{\rm sys}} = \frac{\mathcal{B} + \mathcal{T} + \mathcal{Q} + \ldots}{\mathcal{S} + \mathcal{B} + \mathcal{T} + \mathcal{Q} + \ldots} \qquad (8.145)$$

and the binary fraction is

$$f_{\rm bin} = \frac{\mathcal{B}}{N_{\rm sys}}. \qquad (8.146)$$

In the Galactic field, Duquennoy & Mayor (1991) derive from a decade-long survey for G-dwarf primary stars, ${}^{\rm G}N_{\rm mult} = (57{:}38{:}4{:}1)$ and for M-dwarfs Fischer & Marcy (1992) find ${}^{\rm M}N_{\rm mult} = (58{:}33{:}7{:}1)$. Thus,

$$^{\rm G}f_{\rm mult} = 0.43; \quad {}^{\rm G}f_{\rm bin} = 0.38 \qquad (8.147)$$

$$^{\rm M}f_{\rm mult} = 0.41; \quad {}^{\rm M}f_{\rm bin} = 0.33. \qquad (8.148)$$

It follows that most stars are indeed binary.

After correcting for incompleteness,

$$^{\rm G}f_{\rm bin} = 0.53 \pm 0.08, \qquad (8.149)$$

$$^{\rm K}f_{\rm bin} = 0.45 \pm 0.07, \qquad (8.150)$$

$$^{\rm M}f_{\rm bin} = 0.42 \pm 0.09, \qquad (8.151)$$

where the K-dwarf data have been published by Mayor et al. (1992). It follows that

$$^{\rm G}f_{\rm bin} \approx^{\rm K} f_{\rm bin} \approx^{\rm M} f_{\rm bin} \approx 0.5 \approx f_{\rm tot} \qquad (8.152)$$

in the Galactic field, perhaps with a slight decrease towards lower masses. In contrast, for brown dwarfs, ${}^{\rm BD}f_{\rm bin} \approx 0.15 \ll {}^{\rm stars}f_{\rm bin}$ (Thies & Kroupa 2007 and references therein).

An interesting problem arises because 1 Myr old stars have $f_{\rm TTauri} \approx 1$ (e.g. Duchêne 1999). Given the above information, the following conjecture can be stated:

Binary-star conjecture: nearly all stars form in binary systems.

Justification: if a substantial fraction of stars were to form in higher-order multiple systems, or as small-N systems, the typical properties of these at birth imply their decay within typically 10^4 to 10^5 yr, leaving a predominantly single-stellar population. However, the majority of 10^6 yr old stars are observed to be in binary systems (Goodwin & Kroupa 2005).

Higher-order multiple systems do exist and can only be hierarchical to guarantee stability. Such systems are multiple stars, which are stable over many orbital times and are usually tight binaries orbited by an outer tertiary companion, or two tight binaries in orbit about each other. Stability issues are discussed in detail in Chap. 3, based on a theoretical development from first principles. In particular, a new stability criterion for the general three-body problem is derived in terms of all the orbital parameters. For comparable masses, long-term stability is typically ensured for a ratio of the outer pericentre to the inner semi-major axis of about 4. If the stability condition is not fulfilled, higher-order multiple systems usually decay on a time-scale relating to the orbital parameters. Star cluster remnants (or dead star clusters) may be the origin of most hierarchical, higher-order multiple stellar systems in the field (p. 199).

8.4.2 The Initial Binary Population – Late-Type Stars

The initial binary population is described by distribution functions that are as fundamental for a stellar population as the IMF. There are four distribution functions that define the initial dynamical state of a population,

1. the IMF, $\xi(m)$,
2. the distribution of periods (or semi-major axis), $\mathrm{d}f = f_P(\log P)\,\mathrm{d}\log P$,
3. the distribution of mass-ratios, $\mathrm{d}f = f_q(q)\,\mathrm{d}q$ and
4. the distribution of eccentricities, $\mathrm{d}f = f_e(e)\,\mathrm{d}e$,

where $\mathrm{d}f$ is the fraction of systems between f and $f+\mathrm{d}f$. Thus, for example, ${}^{\mathrm{G}}f_{\log P}(\log_{10} P = 4.5) = 0.11$, i.e. of all G-dwarfs in the sky, 11% have a companion with a period in the range 4–5 d (Fig. 8.16).

These distribution functions have been measured for late-type stars in the Galactic field and in star-forming regions (Fig. 8.13). According to Duquennoy & Mayor (1991) and Fischer & Marcy (1992), both G-dwarf and M-dwarf binary systems in the Galactic field have period distribution functions that are well described by log-normal functions,

$$f_P(\log_{10} P) = f_{\mathrm{tot}} \left(\frac{1}{\sigma_{\log_{10} P}\sqrt{2\,\pi}} \right) e^{\left[-\frac{1}{2}\frac{(\log_{10} P - \overline{\log_{10} P})^2}{\sigma^2_{\log_{10} P}}\right]}, \tag{8.153}$$

with $\overline{\log_{10} P} \approx 4.8$ and $\sigma_{\log_{10} P} \approx 2.3$ and $\int_{\mathrm{all}\,P} f_{\log_{10} P}(\log_{10} P)\,\mathrm{d}\log_{10} P = f_{\mathrm{tot}} \approx 0.5$. K-dwarfs appear to have an indistinguishable period distribution.

From Fig. 8.13 it follows that the pre-mainsequence binary fraction is larger than that of main-sequence stars (see also Duchêne 1999). Is this an evolutionary effect?

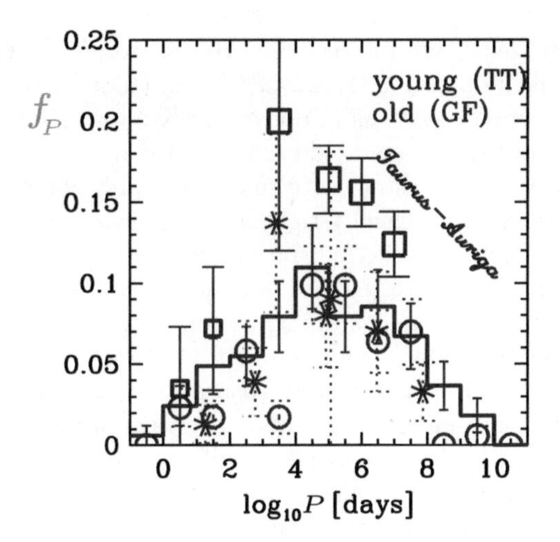

Fig. 8.13. Measured period-distribution functions for G-dwarfs in the Galactic field (*histogram*, Duquennoy & Mayor 1991), K-dwarfs (*open circles*, Mayor et al. 1992) and M-dwarfs (*asterisks*, Fischer & Marcy 1992). About 1-Myr-old T Tauri binary data (*open squares*, partially from the Taurus–Auriga stellar groups) are a compilation from various sources (see Fig. 10 in Kroupa, Aarseth & Hurley 2001). In all cases, the area under the distribution is $f_{\rm tot}$

Further, Duquennoy & Mayor (1991) derived the mass-ratio and eccentricity distributions for G-dwarfs in the Galactic field. The mass-ratio distribution of G-dwarf primaries is not consistent with random sampling from the canonical IMF (8.124), as the number of observed low-mass companions is underrepresented (Kroupa 1995c). In contrast, the pre-mainsequence mass-ratio distribution is consistent, within the uncertainties, with random sampling from the canonical IMF for $q \geq 0.2$ (Woitas, Leinert & Koehler 2001). The eccentricity distribution of Galactic-field G-dwarfs is found to be thermal for $\log_{10} P \geq 3$, while it is bell shaped with a maximum near $e = 0.25$ for $\log_{10} P \leq 3$. Not much is known about the eccentricity distribution of pre-mainsequence binaries, but numerical experiments show that f_e does not evolve much in dense clusters, i.e. the thermal distribution must be initial (Kroupa 1995d).

The thermal eccentricity distribution,

$$f_e(e) = 2\,e, \tag{8.154}$$

follows from a uniform binding-energy distribution (all energies are equally populated) as follows. The orbital angular momentum of a binary is

$$L^2 = \frac{G}{m_{\rm sys}}\,\frac{G\,m_1\,m_2}{2E_{\rm bin}}\,\left(1 - e^2\right)\,\left(m_1\,m_2\right)^2, \tag{8.155}$$

from which follows

$$e = \left(1 - 2\,E_{\rm bin}\,L^2\,\frac{m_{\rm sys}}{G^2\,(m_1\,m_2)^2}\right)^{\frac{1}{2}}. \tag{8.156}$$

Differentiation leads to

$$\frac{{\rm d}e}{{\rm d}E_{\rm bin}} = \left[-L^2\frac{m_{\rm sys}}{G^2\,(m_1\,m_2)^2}\right]\,e^{-1} \propto e^{-1}. \tag{8.157}$$

The number of binaries with eccentricities in the range $e, e + {\rm d}e$ is the same number of binaries with binding energy in the range $E_{\rm bin}, E_{\rm bin} + {\rm d}E_{\rm bin}$ (the same sample of binaries),

$$f(e)\,{\rm d}e = f(E_{\rm bin})\,{\rm d}E_{\rm bin} \propto f(E_{\rm bin})\,e\,{\rm d}e. \tag{8.158}$$

But

$$\int_0^1 f(e)\,{\rm d}e = 1. \tag{8.159}$$

That is,

$$f(E_{\rm bin})\int_0^1 e\,{\rm d}e \propto f(E_{\rm bin})\,\frac{1}{2}\,e^2|_0^1 = {\rm const}. \tag{8.160}$$

So

$$f(E_{\rm bin}) = {\rm const} \Rightarrow f(e)\,{\rm d}e = 2\,e\,{\rm d}e. \tag{8.161}$$

Thus, $f(e) = 2\,e$ is a thermalized distribution. All energies are equally occupied so $f(E_{\rm bin}) = {\rm const}$. N-body experiments have demonstrated that the period distribution function must span the observed range of periods at birth, because dynamical encounters in dense clusters cannot widen an initially narrow distribution (Kroupa & Burkert 2001). There are thus three discrepancies between main-sequence and pre-mainsequence late-type stellar binaries,

1. the binary fraction is higher for the latter,
2. the period distribution function is different and
3. the mass-ratio distribution is consistent with random paring for the latter, while it is deficient in low-mass companions in the former, for G-dwarf primaries.

Can these be unified? That is, are there unique initial $f_{\log P}, f_q$ and f_e consistent with the pre-mainsequence data that can be evolved to the observed main-sequence distributions?

This question can be solved by framing the following ansatz. Assume the orbital-parameter distribution function for binaries with primaries of mass m_1 can be separated,

$$\mathcal{D}(\log P, e, q\,:\,m_1) = f_{\log P}\,f_e\,f_q. \tag{8.162}$$

The stellar-dynamical operator, $\Omega^{N,r_{\rm h}}$, can now be introduced so that the initial distribution function is transformed to the final (Galactic-field) one,

$$D_{\rm fin}(\log P, e, q\,:\,m_1) = \Omega^{N,r_{\rm h}}\,[\mathcal{D}_{\rm in}(\log P, e, q\,:\,m_1)]\,. \tag{8.163}$$

This operator provides a dynamical environment equivalent to that of a star cluster with N stars and a half-mass radius $r_{\rm h}$ (see also the Dynamical Population Synthesis Conjecture, p. 239). Kroupa (1995c) and Kroupa (1995d) indeed show this to be the case for a cluster with $N = 200$ binaries and $r_{\rm h} = 0.77\,{\rm pc}$ and derive the initial distribution function, $\mathcal{D}_{\rm in}$, for late-type binary systems that fulfils the above requirement and also has a simple generating function (see below). It is noteworthy that such a cluster is very similar to the typical cluster from which most field stars probably originate. The full solution for Ω, so that the Galactic field is reproduced from forming and dissolving star clusters, requires full-scale inverse dynamical population synthesis for the Galactic field.

Thus, by the DYNAMICAL POPULATION SYNTHESIS CONJECTURE (p. 239), the above ansatz with $\Omega^{N,r_{\rm h}}$ leads to one solution of the inverse dynamical population synthesis problem (the 200 binary, $r_{\rm h} = 0.8\,{\rm pc}$ cluster, Fig. 8.14 i.e. most stars in the Galactic field stem from clusters dynamically similar to this one), provided the birth (or primordial) distribution functions for $\log P, e, q$ are

$$f_{\log P,{\rm birth}} = \eta \, \frac{\log P - \log P_{\rm min}}{\delta + (\log P - \log P_{\rm min})^2}. \tag{8.164}$$

This distribution function has a generating function (Sect. 8.2.3)

$$\log P(X) = \left[\delta\left(e^{\frac{2\,X}{\eta}} - 1\right)\right]^{\frac{1}{2}} + \log P_{\rm min}. \tag{8.165}$$

The solution obtained by Kroupa (1995d) has

$$\eta = 2.5, \quad \delta = 45, \quad \log P_{\rm min} = 1, \tag{8.166}$$

so that $\log P_{\rm max} = 8.43$ since $\int_{\log P_{\rm min}}^{\log P_{\rm max}} f_{\log P}\, {\rm d}\log P = f_{\rm tot} = 1$ is a requirement for stars at birth. Intriguingly, similar distributions can be arrived at semi-empirically if we assume isolated formation of binary stars in a turbulent molecular cloud (Fisher 2004).

The birth-eccentricity distribution is thermal (8.154) while the birth mass-ratio distribution is generated from random pairing from the canonical IMF. However, in order to reproduce (1) the observed data in the eccentricity–period diagram, (2) the observed eccentricity distribution and (3) the observed mass-ratio distribution for short-period ($\log P \leq 3$) systems, a correlation of the parameters needs to be introduced through eigenevolution. Eigenevolution is the sum of all dissipative physical processes that transfer mass, energy and angular momentum between the companions when they are still very young and accreting.

A formulation that is quite successful in reproducing the overall observed correlations between $\log P, e, q$ for short-period systems has been derived from tidal circularisation theory (Kroupa 1995d). The most effective orbital dissipation occurs when the binary is at periastron,

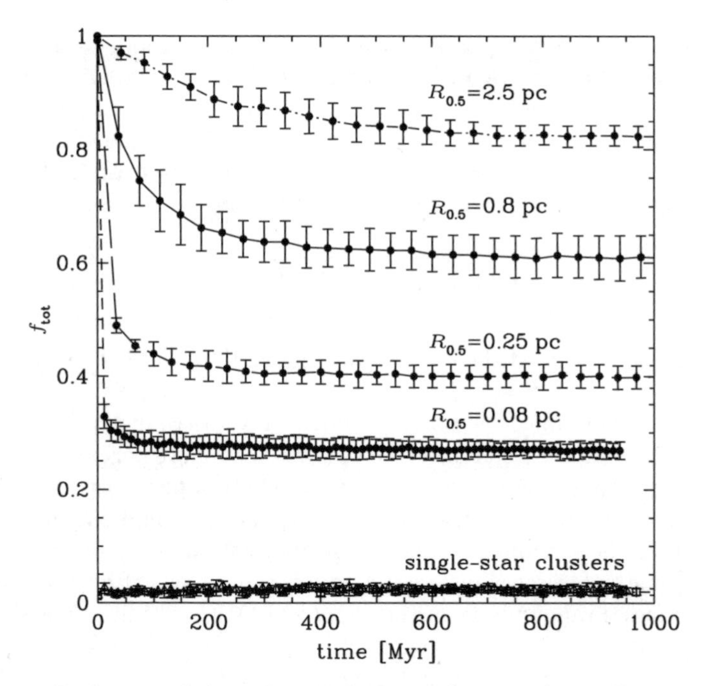

Fig. 8.14. Evolution of $f_{\rm tot}$, the total binary fraction for stellar mass $0.1 \leq m_i/M_\odot \leq 1.1$, $i = 1,2$ with time for the four star-cluster models initially with $N = 200$ binaries computed by Kroupa (1995c) in the search for the existence of an $\Omega^{r_{\rm h},N}$. The initial half-mass radius of the clusters is denoted in this text as $r_{\rm h}$. Note that the $r_{\rm h} = 0.8\,$pc cluster yields the correct $f_{\rm tot} \approx 0.5$ for the Galactic field. The period-distribution function and the mass-ratio distribution function that emerge from this cluster also fit the observed Galactic-field distribution. Some binary stars form by three-body encounters in clusters that initially consist only of single stars and the proportion of such binaries is shown for the single-star clusters (with initially $N = 400$ stars). Such dynamically formed binaries are very rare and so $f_{\rm tot}$ remains negligible

$$r_{\rm peri} = (1-e)\, P_{\rm yr}^{\frac{2}{3}}\, (m_1 + m_2)^{\frac{1}{3}}\,, \tag{8.167}$$

where $P_{\rm yr} = P/365.25$ is the period in years. Let the binary be born with eccentricity $e_{\rm birth}$, then the system evolves approximately, according to (Goldman & Mazeh 1994), as

$$\frac{1}{e}\,\frac{{\rm d}e}{{\rm d}t} = -\rho' \Rightarrow \log_{10}e_{\rm in} = -\rho + \log_{10}e_{\rm birth}, \tag{8.168}$$

where $1/\rho'$ is the tidal circularisation time-scale, $e_{\rm in}$ is the initial eccentricity and

$$\rho = \int_0^{\Delta t} \rho'\,{\rm d}t = \left(\frac{\lambda\, R_\odot}{r_{\rm peri}}\right)^{\chi}, \tag{8.169}$$

where $R_\odot$ is the Solar radius in AU, λ, χ are tidal circularisation parameters and $r_{\rm peri}$ (in AU) is assumed to be constant because the dissipational force only acts tangentially at periastron. Note that a large λ implies that tidal dissipation is effective for large separations of the companions (e.g. they are puffed-up pre-mainsequence structures) and a small χ implies the dissipation is soft, i.e. weakly varying with the separation of the companions. In this integral, $\Delta t \le 10^5\,\mathrm{yr}$ is the time-scale within which pre-mainsequence eigenevolution completes. The initial period becomes, from (8.167),

$$P_{\rm in} = P_{\rm birth} \left(\frac{m_{\rm tot,birth}}{m_{\rm tot,in}}\right)^{\frac{1}{2}} \left(\frac{1-e_{\rm birth}}{1-e_{\rm in}}\right)^{\frac{3}{2}} . \tag{8.170}$$

Kroupa (1995d) assumed the companions merge if $a_{\rm in} \le 10\,R_\odot$ in which case $m_1 + m_2 \rightarrow m$.

In order to reproduce the observed mass-ratio distribution, given random pairing at birth and to also reproduce the fact that short-period binaries tend to have similar-mass companions, Kroupa (1995d) implemented a feeding algorithm, according to which the secondary star accretes high angular momentum gas from the circumbinary accretion disc or material, so that its mass increases while the primary mass remains constant. Thus, after generating the two birth masses randomly from the canonical IMF, the initial mass-ratio is

$$q_{\rm in} = q_{\rm birth} + (1 - q_{\rm birth})\,\rho^* , \tag{8.171}$$

where

$$\rho^* = \begin{cases} \rho & : \quad \rho \le 1, \\ 1 & : \quad \rho > 1. \end{cases} \tag{8.172}$$

The above is a very simple algorithm which nevertheless reproduces the essence of orbital dissipation so that the correlations between the orbital parameters for short-period systems are well accounted for. The best parameters for the evolution

$$\text{birth} \;\rightarrow\; \text{initial} : \lambda = 28\,,\; \chi = 0.75. \tag{8.173}$$

Figure 8.15 shows an example of the overall model in terms of the eccentricity–period diagram. Figures 8.16 and 8.17 demonstrate that it nicely accounts for the period and mass-ratio distribution data, respectively.

Note that initial distributions are derived from birth distributions. This is to be understood in terms of these initial distributions being the initialisation of N-body experiments, while the birth distributions are more related to the theoretical distribution of orbital parameters before dissipational and accretion processes have had a major effect on them. The birth distributions are, however, mostly an algorithmic concept. Once the N-body integration is finished, e.g. when the cluster is dissolved, the remaining binaries can be evolved to the main-sequence distributions by applying the same eigenevolution algorithm above, but with parameters

$$\text{after Nbody integration} \;\rightarrow\; \text{mainsequence} : \lambda_{\rm ms} = 24.7\,,\; \chi_{\rm ms} = 8. \tag{8.174}$$

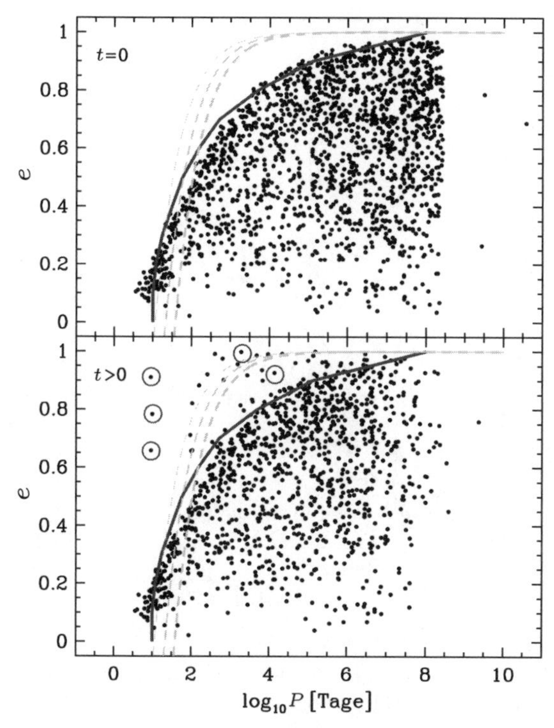

Fig. 8.15. Eccentricity–period after pre-mainsequence eigenevolution ($\lambda = 28, \chi = 0.75$) at $t = 0$ (*upper panel*) for masses $0.1 \leq m_i/M_\odot \leq 1.1$ and after cluster disintegration (*bottom panel*; note: *Tage* means days). Systems with semi-major axes, $a \leq 10\,R_\odot$ have been merged. Binaries are only observed to have $e, \log P$ below the envelope described by Duquennoy & Mayor (1991). The region above is forbidden because pre-mainsequence dissipation depopulates it within 10^5 yr. However, dynamical encounters can repopulate the eigenevolution region so that systems with forbidden parameters can be found but are short-lived. Some of these are indicated as *open circles*. Eigenevolution (tidal circularisation) on the main sequence with $\lambda_{\rm ms} = 24.7$ and $\chi_{\rm ms} = 8$, applied to the data in the *lower panel*, depopulates the eigenevolution region and circularises all orbits with periods less than about 12 d. The *dashed lines* are constant periastron distances (8.167) for $r_{\rm peri} = \lambda\,R_\odot$ and $m_{\rm sys} = 2.2, 0.64$ and $0.2\,M_\odot$ (in increasing thickness). Horizontal and vertical cuts through this diagram produce eccentricity and period distribution functions, and mass-ratio distributions that fit the observations (Kroupa 1995d)

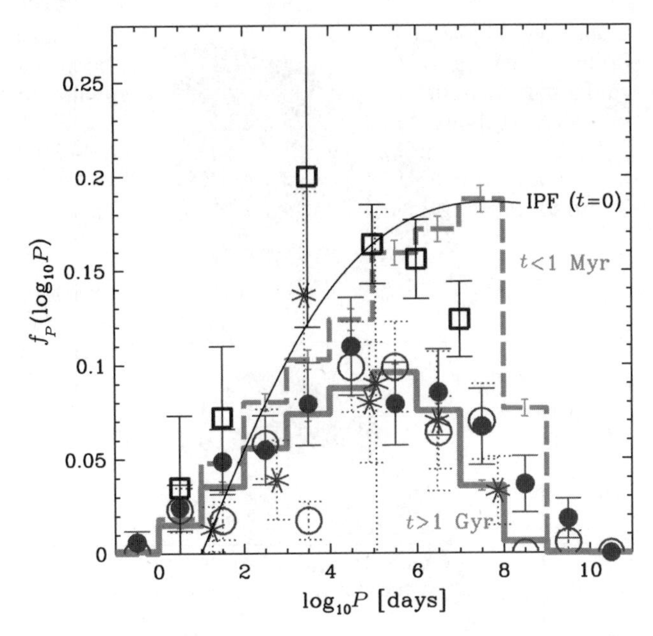

Fig. 8.16. The period distribution functions (IPF: (8.164) with (8.166) and for stellar masses $0.1 \leq m_i/M_\odot \leq 1.1$). The *dashed histogram* is derived from IPF with the eigenevolution and feeding algorithms and represents the binary population at an age of about 10^5 yr. The solid histogram follows from the dashed one after evolving a cluster with initially $N = 200$ binaries and $r_\mathrm{h} = 0.8\,$pc. The agreement of the dashed histogram with the observational pre-mainsequence data (as in Fig. 8.13) and of the *solid histogram* with the observed main sequence (Galactic field) data (also as in Fig. 8.13) is good. A full model of the Galactic field late-type binary population has been arrived at which unifies all available, but apparently discordant, observational data (see also Figs. 8.14, 8.15, and 8.17), nothing that the longest-period TTauri binary population is expected to show some disruption

The need for $\lambda_\mathrm{ms} < \lambda$ and $\chi_\mathrm{ms} > \chi$ to ensure, for example, the tidal circularisation period of 12 days for G dwarfs (Duquennoy & Mayor 1991) is nicely qualitatively consistent with the shrinking of pre-mainsequence stars and the emergence of radiative cores that essentially reduce the coupling between the stellar surface, where the dissipational forces are most effective, and the centre of the star. The reader is also directed to Mardling & Aarseth (2001) who introduce a model of tidal circularisation to the N-body code. Finally, the above work and the application to the ONC and Pleiades (Kroupa, Aarseth & Hurley 2001) suggests the following hypothesis:

> INITIAL BINARY UNIVERSALITY HYPOTHESIS: the initial period (8.166), eccentricity (8.154) and mass-ratio (random pairing from canonical IMF) distributions constitute the parent distribution of all late-type stellar populations.

Can this hypothesis be rejected?

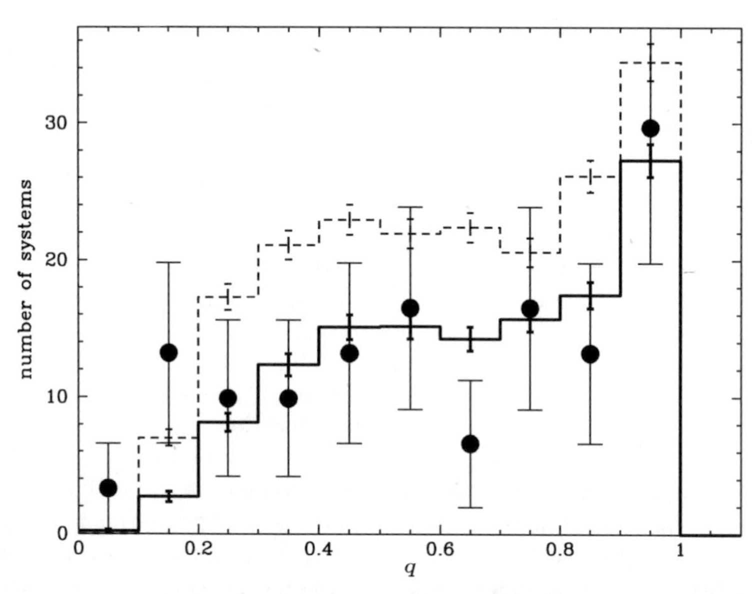

Fig. 8.17. The mass-ratio distribution for stars with $0.1 \leq m/M_{\odot} \leq 1.1$ is the solid histogram, whereas the initial mass-ratio distribution (random pairing from the canonical IMF, after eigenevolution and feeding, at an age of about 10^5 yr) is shown as the *dashed histogram*. The solid histogram follows from the dashed one after evolving a cluster with initially $N = 200$ binaries and $r_{\rm h} = 0.8$ pc. The observational data (*solid dots*, Reid & Gizis 1997) have been obtained after removing WD companions and scaling to the model. This solar neighbourhood 8 pc sample is not complete and may be biased towards $q = 1$ systems (Henry et al. 1997). Nevertheless, the agreement between model (*solid histogram*) and the data is striking. A full model of the Galactic field binary population has been arrived at which unifies all available, but apparently discordant, observational data (see also Figs. 8.14, 8.15, 8.16)

8.4.3 The Initial Binary Population – Massive Stars

The above semi-empirical distribution functions have been formulated for late-type stars (primary mass $m_1 \leq 1\,M_{\odot}$). It is for these that we have the best observations. It is not clear yet if they are also applicable to massive binaries.

An approach taken by Clarke & Pringle (1992) is to constrain the binary properties of OB stars by assuming that runaway OB stars are ejected from star-forming regions. About 10–25% of all O stars are runaway stars, while about 2% of B stars are runaways. This approach leads to the result that massive stars must form in small-N groups of binaries that are biased towards unit mass ratio. This is a potentially powerful approach but it can only constrain the properties of OB binaries when they are ejected. This occurs after many dynamical encounters in the cluster core, which typically lead to the mass-ratio evolving towards unity as the binaries harden. The true birth properties of massive binaries therefore remain obscure and we need to resort to N-body experiments to test various hypotheses given the observations. One such hypothesis could be, for example, to assume massive stars form in binaries with birth pairing properties as for low-mass stars (Sect. 8.4.2), i.e. most massive primaries would have a low-mass companion and to investigate if this hypothesis leads to the observed number of runaway massive stars through dynamical mass segregation to the cluster core and partner exchanges through dynamical encounters there between the massive stars.

Apart from the fraction of runaway stars, direct surveys have lead to some insights into the binary properties of the observed massive stars. Thus, for example, Baines et al. (2006) report a very high ($f \approx 0.7 \pm 0.1$) binary fraction among Herbig Ae/Be stars with a binary fraction that increases with increasing primary mass. Furthermore, they find that the circumbinary discs and the binary orbits appear to be coplanar. This supports a fragmentation origin rather than collisions or capture as the origin of massive binaries. Most O stars are believed to exist as short-period binaries with $q \approx 1$ (García & Mermilliod 2001), at least in rich clusters. On the other hand, small-q appear to be favoured in smaller clusters such as the ONC, consistent with random pairing (Preibisch et al. 1999). Kouwenhoven et al. (2005) report that the A and late-type B binaries in the Scorpius OB2 association have a mass-ratio distribution inconsistent with random pairing. The lower limit on the binary fraction is 0.52, while Kouwenhoven et al. (2007) update this to a binary fraction of 72%. They also find that the semi-major axis distribution contains too many close pairs compared to a Duquennoy & Mayor (1991) log-normal distribution. These are important constraints but, again, they are derived for binaries in an OB association, which is an expanded version of a dense star cluster (Sect. 8.1.2) and therefore hosts a dynamically evolved population.

Given the above results, perhaps the massive binaries in the ONC represent the primordial population, whereas in rich clusters and in OB associations the population has already evolved dynamically through hardening and companion exchanges (f_q rising towards $q = 1$). This possibility needs to be investigated with high-precision N-body computations of young star clusters. The first simplest hypothesis to test would be to extend the pairing rules of Sect. 8.4.2 to all stellar masses, perform many (because of the small number of massive stars) N-body renditions of the same basic pre-gas expulsion cluster and to quantify the properties of the emerging stellar population at various dynamical times (Kroupa 2001).

Another approach would be to constrain a and m_2 for a given $m_1 \geq 5\, M_\odot$ so that

$$E_{\rm bin} \approx E_{\rm k} \tag{8.175}$$

(8.141). Or we can test the initial massive-star population given by

$$a < \frac{r_c}{N_{\rm OB}^{\frac{1}{3}}}, \tag{8.176}$$

which follows from stating that the density of a massive binary, $2 \times 3/(a^3\, 4\,\pi)$, be larger than the cluster-core density, $N_{\rm OB}\, 3/(r_c^3\, 4\,\pi)$. So far, none of these possibilities have been tested, apart from the INITIAL BINARY UNIVERSALITY HYPOTHESIS (p. 250) extension to massive stars (Kroupa 2001).

8.5 Summary

The above material gives an outline of how to set up an initial, birth or primordial stellar population so that it resembles observed stellar populations. In Sect. 8.4.2 a subtle differentiation was made between initial and birth populations, in the sense that an initial population is derived from a birth population through processes that act too rapidly to be treated by an N-body integration.

An N-body stellar system is generated for numerical experiments by specifying its 3D structure and velocity field (Sect. 8.2), the mass distribution of its population (Sect. 8.3) and the properties of its binary population (Sect. 8.4). Given the distribution functions discussed here and the existing numerical results based on these, it is surprising how universal the stellar and binary population turns out to be at birth. A dependence of the IMF or the birth binary properties on the physical properties of star-forming clouds cannot be detected conclusively. In fact, the theoretical proposition that there should be a dependency can be rejected, except possibly (i) in the extreme tidal field environment at the Galactic centre, or (ii) in the extreme protostellar density environment of ultra-compact dwarf galaxies, or (iii) for extreme physical environments (pp. 230–231).

The unified picture that has emerged concerning the origin of stellar populations is that stars form according to a universal IMF and mostly in binary systems. They form in very dense clusters, which expel their residual gas and rapidly evolve to T- or OB-associations. If the latter are massive enough, the dense embedded clusters evolve to populous OB associations that may be expanding rapidly and contain cluster remnants, which may reach globular cluster masses and beyond, in intense star-bursts. This unified picture explains naturally the high infant weight loss and infant mortality of clusters, the binary properties of field stars, possibly thick discs of galaxies and the existence of population II stellar halos around galaxies that have old globular cluster systems.

Many open questions remain. Why is the star-formation product so universal within current constraints? How are massive stars distributed in binaries? Do they form at the centres of their clusters? Why is the cluster mass of about $10^6\, M_\odot$ special? And which star cluster population is a full solution to the inverse dynamical population synthesis problem? (p. 246). Many more observations are required. These must not only be of topical high red-shift star-burst systems but also of the more mundane low red-shift, and preferably local, star-forming objects, globular and open star clusters.

Acknowledgement

It is a pleasure to thank Sverre Aarseth for organising a splendid and much to be remembered Cambridge N-body school in the Summer of 2006, and also Christopher Tout for editing and proof-reading this chapter. I am indebted to Jan Pflamm-Altenburg who read parts of this manuscript carefully, to Andreas Küpper for producing the Plummer vs King model comparisons and for carefully reading the whole text, and to Joerg Dabringhausen, who supplied figures from his work.

References

Aarseth S. J. 2003, Gravitational N-Body Simulations. Cambridge Univ. Press, Cambridge
Aarseth S. J., Hénon M., Wielen R., 1974, A&A, 37, 183
Adams F. C., 2000, ApJ, 542, 964
Adams F. C., Fatuzzo M., 1996, ApJ, 464, 256
Adams F. C., Laughlin G., 1996, ApJ, 468, 586
Adams F. C., Myers P. C., 2001, ApJ, 553, 744
Allen L., Megeath S. T., Gutermuth R., Myers P. C., et al., 2007, in Reipurth B., Jewitt D., Keil K., eds, Protostars and Planets V. University Arizona Press, Tucson, p. 361
Alves J., Lombardi M., Lada C. J., 2007, A&A, 462, L17
Baines D., Oudmaijer R. D., Porter J. M., Pozzo M., 2006, MNRAS, 367, 737
Ballero S., Kroupa P., Matteucci F., 2007, MNRAS, 467, 117
Baraffe I., Chabrier G., Allard F., Hauschildt P. H., 2002, A&A, 382, 563
Bastian N., Goodwin S. P., 2006, MNRAS, 369, L9
Baumgardt H., 1998, A&A, 330, 480
Baumgardt H., Kroupa P., 2007, MNRAS, 380, 1589
Baumgardt H., Kroupa P., de Marchi G., 2008, MNRAS, in press (astroph/0806.0622)
Baumgardt H., Kroupa P., Parmentier G., 2008, MNRAS, 384, 1231
Baumgardt H., Makino J., 2003, MNRAS, 340, 227
Binney J., Tremaine S., 1987, Galactic Dynamics. Princeton Univ. Press, Princeton, NJ
Boily C. M., Kroupa P., 2003, MNRAS, 338, 673
Boily C. M., Kroupa P., Peñarrubia-Garrido J., 2001, New Astron., 6, 27
Boily C. M., Lançon A., Deiters S., Heggie D. C., 2005, ApJ, 620, L27
Bonnell I. A., Bate M. R., Vine S. G., 2003, MNRAS, 343, 413
Bonnell I. A., Larson R. B., Zinnecker H., 2007, in Reipurth B., Jewitt D., Keil K., eds, Protostars and Planets V. University Arizona Press, Tucson, p. 149
Brook C. B., Kawata D., Scannapieco E., Martel H., Gibson B. K., 2007, ApJ, 661, 10
Casuso E., Beckman J. E., 2007, ApJ, 656, 897
Chabrier G., 2003, PASP, 115, 763
Clark P. C., Bonnell I. A., Zinnecker H., Bate M. R., 2005, MNRAS, 359, 809

Clarke C. J., Bonnell I. A., Hillenbrand L. A., 2000, in Mannings V., Boss A. P., Russell S. S., eds, Protostars and Planets IV. University Arizona Press, Tucson, p. 151
Clarke C. J., Pringle J. E., 1992, MNRAS, 255, 423
Dabringhausen J., Hilker M., Kroupa P., 2008, MNRAS, 386, 864
Dabringhausen J., Kroupa P., Baumgardt, H., 2008, MNRAS, submitted
Dale J. E., Bonnell I. A., Clarke C. J., Bate M. R., 2005, MNRAS, 358, 291
Dale J. E., Ercolano B., Clarke C. J., 2007, MNRAS, 1056
de Grijs R., Parmentier G., 2007, Chinese J. Astron. Astrophys., 7, 155
de la Fuente Marcos R., 1997, A&A, 322, 764
de la Fuente Marcos R., 1998, A&A, 333, L27
Delfosse X., Forveille T., Ségransan D., Beuzit J.-L., Udry S., Perrier C., Mayor M., 2000, A&A, 364, 217
Duchêne G., 1999, A&A, 341, 547
Duquennoy A., Mayor M., 1991, A&A, 248, 485
Eislöffel J., Steinacker J., 2008, in The Formation of Low-Mass-Protostars and Proto-Brown Dwarfs. (in press, astro-ph/0701525), ASP conf. series Vol. 384, p. 359, ed: Gerard von Belle
Elmegreen B. G., 1983, MNRAS, 203, 1011
Elmegreen B. G., 1997, ApJ, 486, 944
Elmegreen B. G., 1999, ApJ, 515, 323
Elmegreen B. G., 2000, ApJ, 530, 277
Elmegreen B. G., 2004, MNRAS, 354, 367
Elmegreen B. G., 2006, ApJ, 648, 572
Elmegreen B. G., 2007, ApJ, 668, 1064
Elmegreen D. M., Elmegreen B. G., Sheets C. M., 2004, ApJ, 603, 74
Fellhauer M., Kroupa P., 2005, ApJ, 630, 879
Fellhauer M., Kroupa P., Evans N. W., 2006, MNRAS, 372, 338
Feltzing S., Gilmore G., Wyse R. F. G., 1999, ApJ, 516, L17
Figer D. F., 2005, Nature, 434, 192
Fischer D. A., Marcy, G. W., 1992, ApJ, 396, 178
Fisher R. T., 2004, ApJ, 600, 769
Fleck J.-J., Boily C. M., Lançon A., Deiters S., 2006, MNRAS, 369, 1392
García B., Mermilliod J. C., 2001, A&A, 368, 122
Goldman I., Mazeh T., 1994, ApJ, 429, 362
Goodwin S. P., 1997a, MNRAS, 284, 785
Goodwin S. P., 1997b, MNRAS, 286, 669
Goodwin S. P., 1998, MNRAS, 294, 47
Goodwin S. P., Bastian N., 2006, MNRAS, 373, 752
Goodwin S. P., Kroupa, P., 2005, A&A, 439, 565
Goodwin S. P., Kroupa P., Goodman, A., Burkert, A., 2007, in Reipurth B., Jewitt D., Keil K., eds, Protostars and Planets V. University Arizona Press, Tucson, p. 133
Goodwin S. P., Nutter D., Kroupa P., Ward-Thompson D., Whitworth A. P., 2008, A&A, 477, 823
Goodwin S. P., Whitworth A., 2007, A&A, 466, 943
Gouliermis D., Keller S. C., Kontizas M., Kontizas E., Bellas-Velidis I., 2004, A&A, 416, 137
Gouliermis D. A., Quanz S. P., Henning T., 2007, ApJ, 665, 306

Gradshteyn I. S., Ryzhik I. M., 1980, Table of Integrals, Series, and Products. Academic Press, New York
Grether D., Lineweaver C. H., 2006, ApJ, 640, 1051
Grillmair C. J., et al., 1998, AJ, 115, 144
Gutermuth R. A., Megeath S. T., Pipher J. L., Williams J. P., Allen L. E., Myers P. C., Raines S. N., 2005, ApJ, 632, 397
Hartmann L., 2003, ApJ, 585, 398
Heggie D. C., 1975, MNRAS, 173, 729
Heggie D., Hut P., 2003, The Gravitational Million-Body Problem. Cambridge Univ. Press, Cambridge
Henry T. J., Ianna P. A., Kirkpatrick J. D., Jahreiss H., 1997, AJ, 114, 388
Hillenbrand L. A., Hartmann L. W., 1998, ApJ, 492, 540
Hills J. G., 1975, AJ, 80, 809
Hoversten E. A., Glazebrook K., 2008 ApJ, 675, 163
Hurley J. R., Pols O. R., Aarseth S. J., Tout C. A., 2005, MNRAS, 363, 293
Hut P., Mineshige S., Heggie D. C., Makino J., 2007, Progress Theor. Phys., 118, 187
Jenkins A., 1992, MNRAS, 257, 620
Kennicutt R. C., 1983, ApJ, 272, 54
Kim S. S., Figer D. F., Kudritzki R. P., Najarro F., 2006, ApJ, 653, L113
Kim E., Yoon I., Lee H. M., Spurzem R., 2008, MNRAS, 383, 2
King I. R., 1962, AJ, 67, 471
King I. R., 1966, AJ, 71, 64
Klessen R. S., Spaans M., Jappsen A.-K., 2007, MNRAS, 374, L29
Koen C., 2006, MNRAS, 365, 590
Koeppen J., Weidner C., Kroupa P., 2007, MNRAS, 375, 673
Kouwenhoven M. B. N., Brown A. G. A., Portegies Zwart S. F., Kaper L., 2007, A&A, 474, 77
Kouwenhoven M. B. N., Brown A. G. A., Zinnecker H., Kaper L., Portegies Zwart S. F., 2005, A&A, 430, 137
Kroupa P., 1995a, ApJ, 453, 350
Kroupa P., 1995b, ApJ, 453, 358
Kroupa P., 1995c, MNRAS, 277, 1491
Kroupa P., 1995d, MNRAS, 277, 1507
Kroupa P., 1998, MNRAS, 300, 200
Kroupa P., 2000, New Astron., 4, 615
Kroupa P., 2001, MNRAS, 322, 231
Kroupa P., 2002a, Science, 295, 82
Kroupa P., 2002b, MNRAS, 330, 707
Kroupa P., 2005, in Turon C., O'Flaherty K. S., Perryman M. A. C., eds, Proc. Gaia Symp. Vol. 576, The Three-Dimensional Universe with Gaia. ESA Publications Division, Noordwijk, p. 629 (astro-ph/0412069)
Kroupa P., 2007a, in Valls-Gabaud D., Chavez M., eds, Resolved Stellar Populations. (in press, astro-ph/0703124)
Kroupa P., 2007b, in Israelian G., Meynet G., eds, The Metal Rich Universe. Cambridge Univ. Press, Cambridge (astro-ph/0703282)
Kroupa P., Aarseth S. J., Hurley J., 2001, MNRAS, 321, 699
Kroupa P., Boily C. M., 2002, MNRAS, 336, 1188
Kroupa P., Bouvier J., Duchêne G., Moraux E., 2003, MNRAS, 346, 354

Kroupa P., Burkert A., 2001, ApJ, 555, 945
Kroupa P., Gilmore G., Tout C. A., 1991, MNRAS, 251, 293
Kroupa P., Petr M. G., McCaughrean M. J., 1999, New Astron., 4, 495
Kroupa P., Tout C. A., Gilmore G., 1993, MNRAS, 262, 545
Kroupa P., Weidner C., 2003, ApJ, 598, 1076
Kroupa P., Weidner C., 2005, in Cesaroni, R., Felli M., Churchwell E., Walmsley M., eds, Proc. IAU Symp. 227, Massive Star Birth: A Crossroads of Astrophysics. Cambridge Univ. Press, Cambridge, p. 423
Krumholz M. R., 2008, in Knapen J., Mahoney T., Vazdekis A., eds, Pathways Through an Eclectic Universe. (astro-ph/0706.3702) ASP conference series, vol. 390
Krumholz M. R., Tan J. C., 2007 ApJ, 654, 304
Lada C. J., Lada E. A., 2003, ARA&A, 41, 57
Lada C. J., Margulis M., Dearborn D., 1984, ApJ, 285, 141
Larson R. B., 1998, MNRAS, 301, 569
Lee H.-C., Gibson B. K., Flynn C., Kawata D., Beasley M. A., 2004, MNRAS, 353, 113
Li Y., Klessen R. S., Mac Low M.-M., 2003, ApJ, 592, 975
Mac Low M.-M., Klessen R. S., 2004, Rev. Mod. Phys., 76, 125
Maíz Apellániz J., Úbeda L., 2005, ApJ, 629, 873
Maíz Apellániz J., Walborn N. R., Morrell N. I., Niemela V. S., Nelan E. P., 2007, ApJ, 660, 1480
Maness H., et al., 2007, ApJ, 669, 1024
Mardling R. A., Aarseth S. J., 2001, MNRAS, 321, 398
Marks M., Kroupa P., Baumgardt H., 2008, MNRAS, 386, 2047
Martín E. L., Brandner W., Bouvier J., Luhman K. L., Stauffer J., Basri G., Zapatero Osorio M. R., Barrado y Navascués, D., 2000, ApJ, 543, 299
Martins F., Schaerer D., Hillier D. J., 2005, A&A, 436, 1049
Massey P., 2003, ARA&A, 41, 15
Mayor M., Duquennoy A., Halbwachs J.-L., Mermilliod J.-C., 1992, in McAlister H. A., Hartkopf W. I., eds, ASP Conf. Ser. Vol. 32, Complementary Approaches to Double and Multiple Star Research. Astron. Soc. Pacific, San Francisco, p. 73
McMillan S. L. W., Vesperini E., Portegies Zwart S. F., 2007, ApJ655, L45
Metz M., Kroupa P., 2007, MNRAS, 376, 387
Meylan G., Heggie D. C., 1997, A&AR, 8, 1
Mieske S., Kroupa P., 2008, ApJ, 677, 276
Moraux E., Bouvier J., Clarke C., 2004, in Combes F., Barret D., Contini T., Meynadier F., Pagani L., eds, SF2A-2004: Semaine de l'Astrophysique Francaise. EdP-Sciences, Conference Series, p. 251
Motte F., Andre P., Neri R., 1998, A&A, 336, 150
Motte F., André P., Ward-Thompson D., Bontemps S., 2001, A&A, 372, L41
Murray S. D., Lin D. N. C., 1996, ApJ, 467, 728
Nutter D., Ward-Thompson D., 2007, MNRAS, 374, 1413
Odenkirchen M., et al., 2003, AJ, 126, 2385
Oey M. S., Clarke C. J., 2005, ApJ, 620, L43
Palla F., Randich S., Pavlenko Y. V., Flaccomio E., Pallavicini, R., 2007, ApJ, 659, L41
Palla F., Stahler S. W, 2000, ApJ, 540, 255
Pancino E., Galfo A., Ferraro F. R., Bellazzini M., 2007, ApJ, 661, L155

Parker R. J., Goodwin S. P., 2007, MNRAS, 380, 1271
Parmentier G., Gilmore G., 2005, MNRAS, 363, 326
Parmentier G., Gilmore G., 2007, MNRAS, 377, 352
Parmentier G., Goodwin S., Kroupa P., Baumgardt H., 2008, ApJ, 678, 347
Pflamm-Altenburg J., Kroupa P., 2006, MNRAS, 373, 295
Pflamm-Altenburg J., Kroupa P., 2007, MNRAS, 375, 855
Pflamm-Altenburg J., Kroupa P., 2008, MNRAS, submitted
Pflamm-Altenburg J., Weidner C., Kroupa P., 2007, ApJ, 671, 1550
Piotto G., 2008, in Cassisi S., Salaris M., XXI Century Challenges for Stellar Evolution. Mem. d. Soc. Astron. It., Vol. 79/2 (arXiv:0801.3175)
Plummer H. C., 1911, MNRAS, 71, 460
Portegies Zwart S. F., McMillan S. L. W., Hut P., Makino J., 2001, MNRAS, 321, 199
Portegies Zwart S. F., McMillan S. L. W., Makino J., 2007, MNRAS, 374, 95
Portinari L., Sommer-Larsen J., Tantalo, R., 2004, MNRAS, 347, 691
Preibisch T., Balega Y., Hofmann K., Weigelt G., Zinnecker H., 1999, New Astron., 4, 531
Press W. H., Teukolsky S. A., Vetterling W. T., Flannery B. P., 1992, Numerical Recipes. Cambridge Univ. Press, Cambridge, 2nd ed
Reid I. N., Gizis J. E., 1997, AJ, 113, 2246
Reid I. N., Gizis J. E., Hawley S. L., 2002, AJ, 124, 2721
Reipurth B., Clarke C., 2001, AJ, 122, 432
Romano D., Chiappini C., Matteucci F., Tosi M., 2005, A&A, 430, 491
Sacco G. G., Randich S., Franciosini E., Pallavicini R., Palla F., 2007, A&A, 462, L23
Salpeter E. E., 1955, ApJ, 121, 161
Scally A., Clarke C., 2002, MNRAS, 334, 156
Scalo J. M., 1986, Fundamentals Cosmic. Phys., 11, 1
Scalo J., 1998, in Gilmore G., Howell D., eds, ASP Conf. Ser. Vol. 142, The Stellar Initial Mass Function (38th Herstmonceux Conference). Astron. Soc. Pac., San Francisco, p. 201
Shara M. M., Hurley J. R., 2002, ApJ, 571, 830
Spitzer L., 1987, Dynamical Evolution of Globular Clusters. Princeton Univ. Press, Princeton, NJ
Stamatellos D., Whitworth A. P., Bisbas T., Goodwin S., 2007, A&A, 475, 37
Šubr L., Kroupa P., Baumgardt H., 2008, MNRAS, 385, 1673
Testi L., Sargent A. I., Olmi L., Onello J. S., 2000, ApJ, 540, L53
Thies I., Kroupa P., 2007, ApJ, 671, 767
Tilley D. A., Pudritz R. E., 2007, MNRAS, 382, 73
Tinsley B. M., 1980, Fundamentals Cosmic. Phys., 5, 287
Tumlinson J., 2007, ApJ, 665, 1361
Tutukov A. V., 1978, A&A, 70, 57
Vesperini E., 1998, MNRAS, 299, 1019
Vesperini E., 2001, MNRAS, 322, 247
Weidemann V., 1990, ARA&A, 28, 103
Weidemann V., 2000, A&A, 363, 647
Weidemann V., Jordan S., Iben I. J., Casertano S., 1992, AJ, 104, 1876
Weidner C., Kroupa, P., 2004, MNRAS, 348, 187
Weidner C., Kroupa P., 2005, ApJ, 625, 754

Weidner C., Kroupa P., 2006, MNRAS, 365, 1333
Weidner C., Kroupa P., Larsen S. S., 2004, MNRAS, 350, 1503
Weidner C., Kroupa P., Nürnberger D. E. A., Sterzik M. F., 2007, MNRAS, 376, 1879
Woitas J., Leinert C., Koehler R., 2001, A&A, 376, 982
Wuchterl G., Tscharnuter W. M., 2003, A&A, 398, 1081
Yasui C., Kobayashi N., Tokunaga A. T., Saito M., Tokoku C., 2008 (astro-ph/0801.0204)
Yasui C., Kobayashi N., Tokunaga A. T., Terada H., Saito M., 2006, ApJ, 649, 753 in Formation and Evolution of Galaxy Disks, ASP Conf. series, in press eds: J. G. Funes, E.M. Cossini (astro-ph/0801.0204)
Zhao H., 2004, MNRAS, 351, 891
Zinnecker H., 2003, in van der Hucht, K., Herrero A., Esteban C., eds, Proc. IAU Symp. 212, A Massive Star Odyssey: From Main Sequence to Supernova. Astron. Soc. Pacific, San Francisco, p. 80
Zinnecker H., Yorke H. W., 2007, ARA&A, 45, 481
Zoccali M., et al., 2006, A&A, 457, L1
Zoccali M., Cassisi S., Frogel J. A., Gould A., Ortolani S., Renzini A., Rich R. M., Stephens A. W., 2000, ApJ, 530, 418

9

Stellar Evolution

Christopher A. Tout

University of Cambridge, Institute of Astronomy, Madingley Road, Cambridge
CB3 0HA, England
cat@ast.cam.ac.uk

The bodies in any N-body system can change. The most changeable bodies are stars. In order to fully model the evolution of a cluster of stars, we need to know how they interact with their environment, particularly how much mass they lose, and how they interact with each other. Is their evolution affected by a companion or close encounter? In this chapter we describe the physics and the mathematical formulation that we use to describe it. If we could we would evolve each star in every detail (Church, Tout & Aarseth 2007) but up to now, in practice, we have had to approximate the detailed evolution by empirical models (Hurley, Tout & Pols 2000). As the number of bodies we can model increases with increasing computing power, it becomes more reasonable to include the full evolution (Chap. 13). So let us examine the physics of stars.

9.1 Observable Quantities

When we look at stars in the night sky they have two immediately discernible properties: they vary in brightness and colour. The brightness is assessed in terms of magnitudes. Historically, and we are going back to the ancient Greeks here, stars fall into six magnitude classes. The brightest stars are of first magnitude and the faintest stars visible to the naked eye are sixth magnitude, though these are rarely visible amongst today's city lights. The eye measures brightness logarithmically so that a star of magnitude 5.0 turns out to be one hundred times fainter than a star of magnitude 1.0. Modern photometry can measure the magnitude of stars extremely accurately and in different wavelength ranges. But these magnitudes are only apparent. A star can vary in brightness for two reasons. First, it may be brighter because it is intrinsically more luminous. Alternatively, it might just be brighter because it is close to us. Indeed Herschel (1783) hoped that all stars were of similar intrinsic luminosity so that he might map the Galaxy by taking variations in brightness to indicate variations in distance. Today the distances to nearby stars can be determined by accurate trigonometric parallaxes. The motion of

Tout, C. A.: *Stellar Evolution*. Lect. Notes Phys. **760**, 261–282 (2008)
DOI 10.1007/978-1-4020-8431-7_9

the star is measured against the background of distant, apparently immovable, stars and galaxies as the Earth moves around its orbit. Once the distance is known, an absolute magnitude can be calculated from the observed apparent magnitude and from this we get an estimate of the luminosity of the star.

The second observable quantity is a star's colour. Some stars appear redder while others are distinctly blue. The colour of a star is related to its surface temperature. The apparent surface or photosphere of a star represents the locus of points at which the majority of photons were last emitted or scattered before they began their journey through space to the Earth. Typically, the spectrum of radiation emitted by a star is close to that of a black body. The hotter the black body the bluer is the peak in its spectrum. Thus blue stars are hot while red stars are relatively cool. Another way of determining the surface temperature of a star is to look at the dark lines in its spectrum. These generally occur at wavelengths where an atomic transition of an electron makes the absorption of a photon particularly favourable. Historically spectra where classified by the strength of their hydrogen lines. Those with the strongest hydrogen lines are of type A while those with the weakest are of type M. Hydrogen ionizes at about $10,000\,\mathrm{K}$ and it is stars of this temperature that have the most prominent hydrogen lines. As the temperature rises fewer and fewer atoms have bound electrons and the lines disappear from the spectra. As the temperature falls the electrons around the hydrogen nuclei become more and more energetically confined to the ground-state orbits. This in turn leads to fewer hydrogen lines in the spectra. However, lines from more weakly bound electrons and bands owing to molecular rotations and vibrations become more prominent. So it is easy to distinguish the very hot O stars from the relatively very cool M stars. The sequence of spectral types from the hottest to the coolest normal stars follows

$$\mathrm{O} \quad \mathrm{B} \quad \mathrm{A} \quad \mathrm{F} \quad \mathrm{G} \quad \mathrm{K} \quad \mathrm{M}.$$

Once we know the temperature and nature of a star's atmosphere, we can relate its absolute magnitude to a bolometric luminosity. This bolometric luminosity L is the total energy radiated by the star per unit time.

In the early years of the twentieth century, Russell (1913), who worked partly in Cambridge at the time, and the Danish astronomer and chemist Hertzsprung (1905) examined the correlations of these two quantities with each other. The resulting Hertzsprung–Russell (HR) diagram has become the major tool for describing the evolution of stars over their lifetimes. Rather than populating the whole of such a diagram, we find that most of the stars lie on a band running from hot bright stars to cool faint stars (Fig. 9.1). This is the main sequence. Because the radiation from stars is very close to a black body, the temperature of the photosphere is close to the effective temperature given by

$$L = 4\pi\sigma R^2 T_{\mathrm{eff}}^4, \tag{9.1}$$

where σ is the Stefan-Boltzmann constant and R is the radius of the photosphere. This means that the loci of stars of constant radius are straight lines

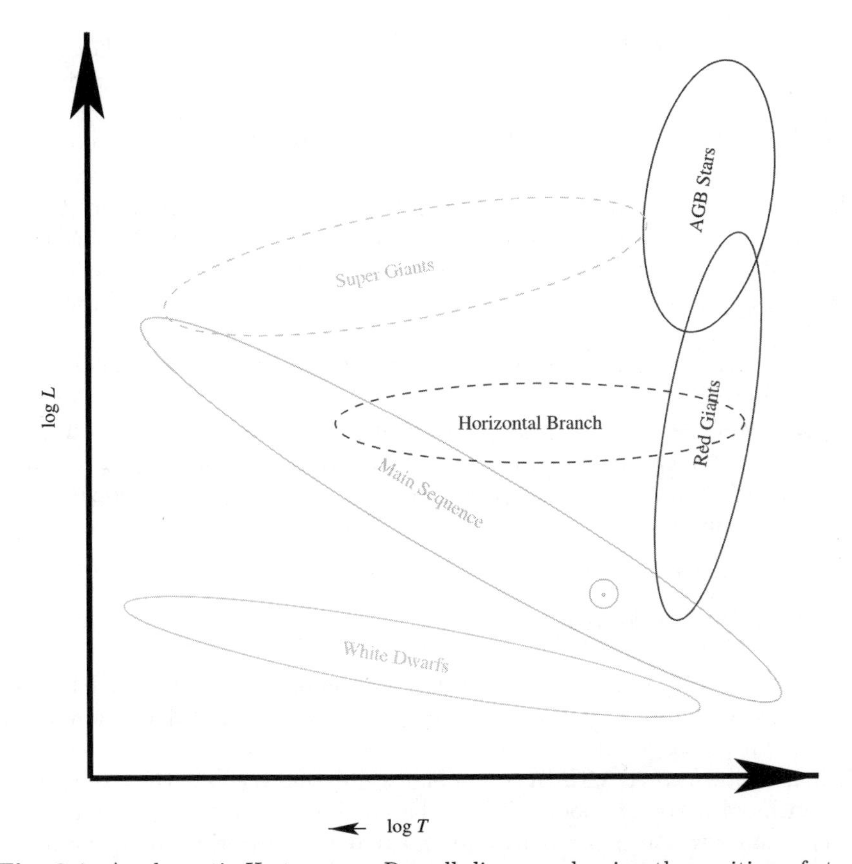

Fig. 9.1. A schematic Hertzsprung–Russell diagram showing the position of stars in a surface temperature – luminosity, or colour magnitude diagram. Temperature increases from right to left along the horizontal axis. Colour changes from *blue* to *red* from left to right. Most stars, like the Sun, lie along the main sequence but other distinct groups of stars are visible, particularly in such diagrams of clusters

of slope −4 in the HR diagram so that stars at the top left of the main sequence are blue supergiants while those at the bottom right are red dwarfs. In a diagram of the brightest stars, another region to the right and nearly vertically upwards from the main sequences is prominent. These are the red giants. In diagrams of globular clusters this giant branch splits into two distinct parts, the normal red giants and the asymptotic giant branch (AGB). We shall see later how these are populated by stars in quite distinct evolutionary phases. In HR diagrams of nearby stars the fainter but relatively common white dwarfs appear in a band below the main sequence. Also discernible as separate though not so distinct regions are the supergiants from blue to red across the very top of the diagram and the subgiants between the main sequence and the true red giants. Globular clusters have the advantage that the stars all lie at approximately the same distance so that, relatively though not absolutely, the errors associated with distance measurements are significantly reduced. Today, some very beautiful HR diagrams of globular clusters can be plotted with the data obtained with large telescopes (Pancino et al. 2000) and these reveal all sorts of detail. Of particular interest is the horizontal branch at relatively constant luminosity, extending from red to blue across from the red giants. The structure and population of this feature vary considerably from cluster to cluster and contain clues to the age and initial chemical composition of the constituent stars. The Sun itself lies right in the middle of the most populated part of the main sequence so that we can deduce that it is typical of the majority of stars. In the next sections we shall investigate the physics and the mathematical models that have allowed us to unravel the life of a star as it moves about the HR diagram from the main sequence to the red giant branch, perhaps to the horizontal branch or back to the subgiant area, then on to the AGB and finally to a white dwarf if the star has lost enough mass to avoid a supernova explosion.

9.2 Structural Equations

The structure of a star can be described in essence by four differential equations. Two of these, that describe the variation of mass and pressure with radius, can be called the structural equations. They are the subject of this section. Supplemented with an equation of state, these two are the basic building blocks of a stellar model. When the equation of state depends on two physical state variables, we must add an equation to describe the variation of temperature through the star and another to incorporate energy-generating processes to complete the set.

The first equation is easily derived by considering a thin shell of mass δm and thickness δr at radius r in the star (Fig. 9.2). The mass in the shell is just its volume multiplied by the local density $\rho(r)$ and, when we take the limit as δr tends to zero we obtain

$$\frac{\mathrm{d}m}{\mathrm{d}r} = 4\pi r^2 \rho, \tag{9.2}$$

the mass equation.

The mass interior to this shell exerts on it an attractive radial force of magnitude $\delta m\, g = 4\pi r^2 \rho g \delta r$, where $g(r) = Gm/r^2$ is the local gravitational acceleration and $m(r)$ is the mass inside radius r. This must be balanced by the differences in the pressure force on either side of the shell $4\pi r^2 [P(r+\delta r) - P(r)]$. Again taking the limit as δr tends to zero we obtain

$$\frac{\mathrm{d}P}{\mathrm{d}r} = -\frac{Gm\rho}{r^2}. \tag{9.3}$$

This is the equation of hydrostatic equilibrium. Equations (9.2) and (9.3) are special spherically symmetric cases of the more general equations of mass

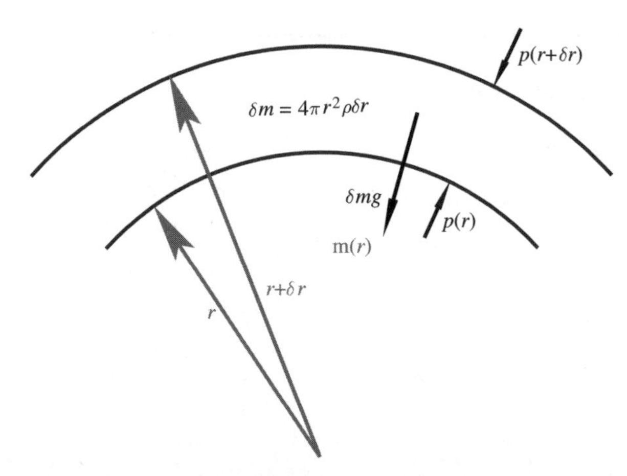

Fig. 9.2. The structure of a fluid sphere. The mass enclosed by a spherical surface of radius r is $m(r)$. A shell of mass δm of thickness δr at this radius is supported against gravity by a pressure gradient

conservation and the Euler momentum equation of fluid dynamics when the velocity in the fluid is everywhere zero.

If we can write P explicitly as a function of ρ only we can obtain a full solution to the structure of the star. The simplest boundary conditions to apply are, at $r = 0$,

$$m(0) = 0 \quad \Rightarrow \quad \frac{\mathrm{d}P}{\mathrm{d}r} = 0 \tag{9.4}$$

and, at $r = R$,

$$m(R) = M \qquad \rho(R) = 0, \tag{9.5}$$

where M is the total mass of the star. It turns out that the equation of state of very degenerate matter takes just such a form and white dwarfs can be modelled immediately (Chandrasekhar 1939).

9.3 Equation of State

In practice pressure does not depend only on density. Figure 9.3 illustrates the various contributions to the pressure as temperature and density vary. Typically, the state of stellar material depends on its composition plus any two state variables. In general, there are many contributions to the equation of state but for most normal stars the fluid behaves very similarly to an ideal gas, for which the pressure may be written as a function of density, temperature T and mean molecular weight μ,

$$P = \rho\frac{\mathcal{R}T}{\mu}, \tag{9.6}$$

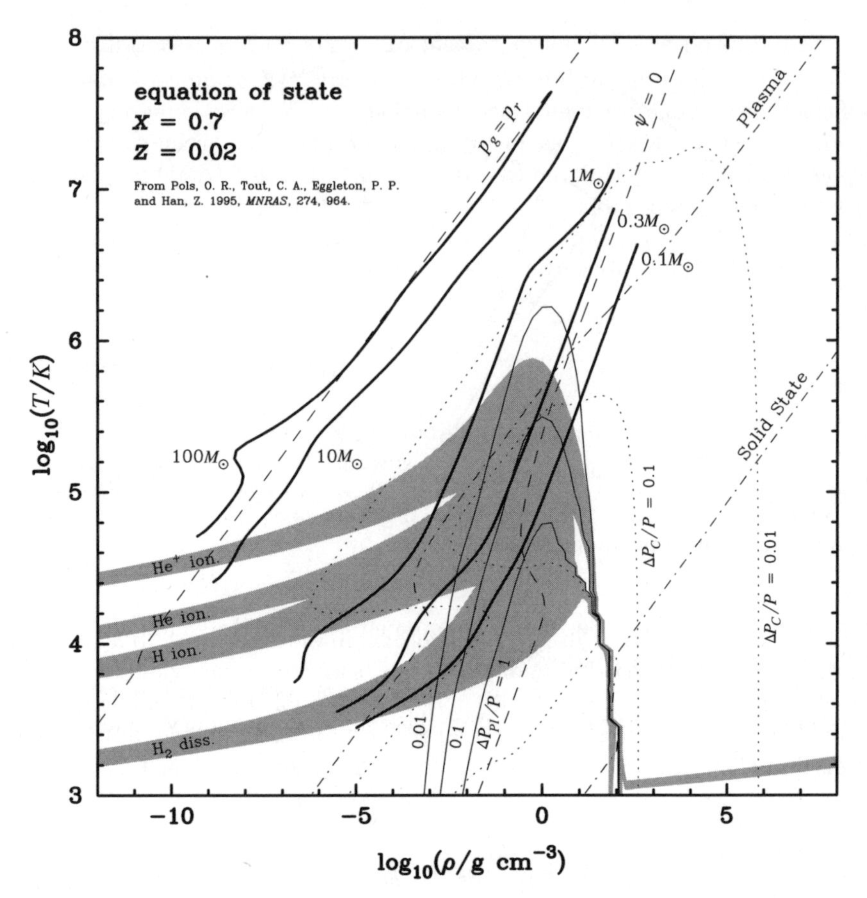

Fig. 9.3. Contributions to the equation of state as a function of temperature and density. The *thick solid lines* are the run of temperature and density through zero-age main-sequence stars of masses 0.1, 0.3, 1, 10 and 100 $M_\odot$. Their centres are towards the *top right* of the figure. A *dashed line* marks where gas and radiation pressure are equal with increasing P_r/P_g to the *left*. A *second dashed line* indicates where the electron chemical potential $\psi = 0$. To the *right* of this line material becomes more and more degenerate. The *shaded regions* represent the range over which ionisation of H, He and He^+ and dissociation of molecular hydrogen take place. *Thin solid lines* indicate the effects of pressure ionisation and *dotted lines* corrections to account for plasma effects. *Dot-dashed lines* indicate when the fluid can be considered a plasma and when it begins to crystallize into the solid state

where $\mathcal{R}$ is the gas constant per unit mass. The mean molecular weight is the reciprocal of the number of particles, each of which contributes to the pressure equally at a given temperature, per atomic mass unit. Thus neutral hydrogen contributes one particle for each mass unit and has $\mu = 1$, while fully ionized hydrogen contributes two particles, an electron and a proton, for each mass unit and so has $\mu = 1/2$. Fully ionised helium contributes two electrons and and a helium nucleus made up of two protons and two neutrons for its four mass units and so has $\mu = 4/3$. Anything heavier than hydrogen and helium is designated a metal and when fully ionized contributes approximately half as many particles as its atomic mass, because the nucleus typically consists of equal numbers of protons and neutrons and each positively charged proton is balanced by an electron. Thus metals have $\mu \approx 2$. For a fully ionized mixture, adding the numbers and masses, we find

$$\frac{1}{\mu} = 2X + \frac{3}{4}Y + \frac{1}{2}Z, \tag{9.7}$$

where X is the mass fraction of hydrogen, Y is that of helium and Z that of all metals and $X + Y + Z = 1$. In the deep interiors of stars temperatures are such that all atoms are ionized but as the temperature falls electrons recombine with their nuclei to form atoms in various ionization states. The most strongly bound electrons recombine at the highest temperatures. Thus in the Sun hydrogen recombines between about $10,000$ and $20,000\,\mathrm{K}$ while iron is still not completely ionized at $100,000\,\mathrm{K}$.

An important consequence of (9.7) is that the equation of state changes as nuclear reactions convert one element to another. This is one of the driving forces behind stellar evolution and is responsible for the Sun gradually expanding and brightening with time.

At high temperatures the pressure exerted by energetic photons becomes comparable with that exerted by the particles and we must include a term

$$P_{\mathrm{r}} = \frac{1}{3}aT^4, \tag{9.8}$$

where a is the radiation constant.

At high densities electrons contribute a degeneracy pressure. This arises because free electrons must occupy a discrete set of momentum states and, as the volume to which an electron is confined is reduced, the energies of its available states increase. Thus squeezing an electron gas increases the momenta of the electrons and this requires energy. So work must be done and the gas exerts a force against compression. The contribution to this degeneracy pressure P_{e} becomes important when the electron chemical potential ψ becomes positive. It is already becoming important in the core of the Sun and lower-mass main-sequence stars but comes into its own in the white dwarfs where it provides sufficient support against gravity even when the gas is cold. Although we might expect a cold gas to consist of neutral atoms, this is not the case at very high densities because the nuclei are so close to one another, much nearer than the radius of an atom, that the electrons are not bound to a particular nucleus but behave as a free gas similar to those in metallic elements at room temperature. This effect of pressure ionization is also important to some extent in the Sun.

There are various other corrections to the pressure P_{c} that must be included such as plasma effects at high densities and eventually liquefaction and crystallization to the solid state as density increases and temperature falls.

9.4 Radiation Transport

When temperature is important for the equation of state, we require two further equations to describe the star. The first is for the temperature gradient. This depends on the rate at which energy can be transported from where it is generated, usually at the hot centre, through the star. One of the three processes dominates energy transport under different conditions. Radiation, or the diffusion of photons, dominates in the central parts of the Sun. Conduction, or the diffusion of particles, is prevalent in degenerate material. Convection, or energy transport by bulk fluid motion, operates when the temperature gradient becomes too large for stable radiative transfer. This is the case in the outer layers of the Sun.

In radiative regions we can estimate the temperature gradient by considering two surfaces of different temperatures separated by a distance λ, the distance that a photon moves between interactions with the matter and over which it maintains memory of the conditions when it last interacted (Fig. 9.4). Deep in the star everything is in local thermodynamic equilibrium so that a surface at temperature T emits energy as a blackbody providing a flux of energy per unit area of $F = \sigma T^4$, where the Stefan Boltzmann constant

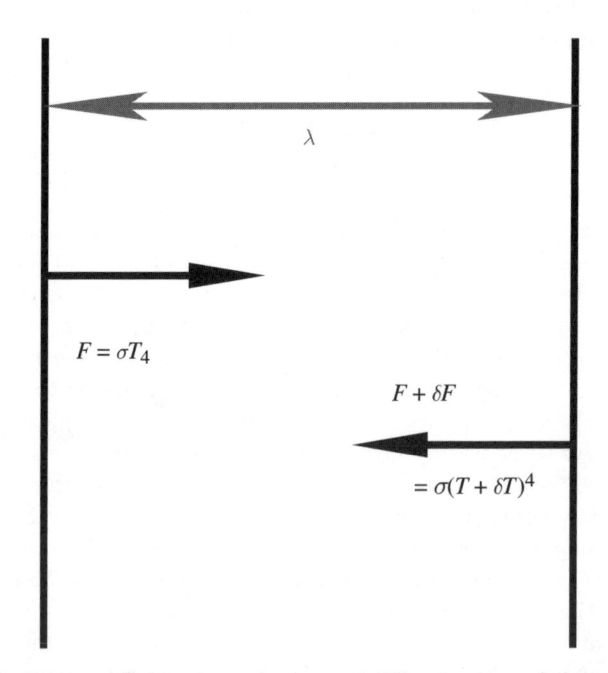

Fig. 9.4. Radiation diffuses through the star. The interior of the star is locally in thermodynamic equilibrium so that the radiation flux emitted by any surface depends on the temperature of that surface. Photons travel until they are absorbed or scattered, typically a mean free path length from where they were emitted or last scattered. In this way heat diffuses from hotter to cooler regions

$\sigma = ac/4$. Consider two such surfaces, one at temperature T and one at $T+\delta T$. In our spherically symmetric star, the surfaces are spheres of area $4\pi r^2$ and T usually decreases as r increases. We call the net energy flow through a sphere of radius r the local luminosity L_r, and we have

$$L_r = 4\pi r^2 \delta F, \tag{9.9}$$

where

$$\delta F \approx -4\sigma T^3 \delta T \tag{9.10}$$

is the difference between the inward flux from the surface at temperature $T+\delta T$ and the outward from the surface at T. The difference in temperature is just the temperature gradient multiplied by the distance between the surfaces

$$\delta T = \lambda \frac{\mathrm{d}T}{\mathrm{d}r}. \tag{9.11}$$

So we have

$$L_r \approx 16\pi\sigma r^2 \lambda T^3 \frac{\mathrm{d}T}{\mathrm{d}r}. \tag{9.12}$$

The typical distance travelled by a photon between interactions, its mean free path, depends on the opacity of the material. Opacity is defined as the effective cross-section per unit mass seen by a photon. The probability of interaction of a photon passing along a cylinder (Fig. 9.5) of cross-section equal to κ times the mass in the cylinder and length λ is unity. Thus for material of density ρ,

$$\rho\kappa\lambda = 1. \tag{9.13}$$

Combining this with (9.12) we find

$$\frac{\mathrm{d}T}{\mathrm{d}r} = \frac{-\kappa\rho L_r}{4\pi a c r^2 T^3}. \tag{9.14}$$

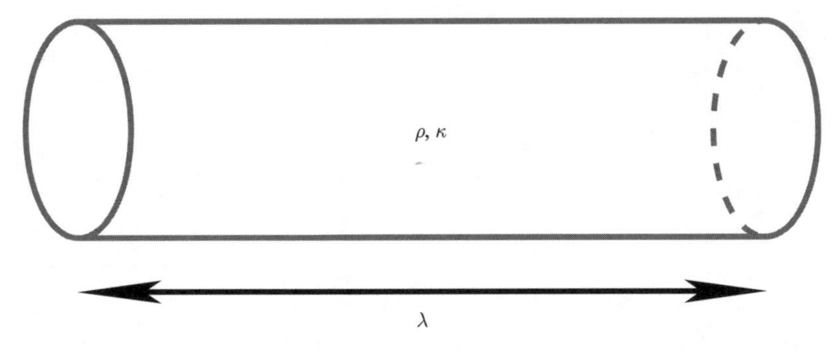

Fig. 9.5. The relation between mean free path and opacity. A photon is likely to be absorbed or scattered once within a cylinder of height λ and cross-sectional area κm, aligned with its motion, which contains one target of mass m

This is not quite correct because we have not taken proper account of the fact that the radiation field from a point on a surface is isotropic and not directed towards the other surface. With somewhat more effort we should obtain

$$\frac{\mathrm{d}T}{\mathrm{d}r} = \frac{-3\kappa\rho L_r}{16\pi acr^2T^3}, \tag{9.15}$$

which is the equation of radiative transfer.

The detailed calculation of opacity is a long and complex procedure. Figure 9.6 illustrates how it varies with temperature and density in stellar material. At high temperatures all material is ionized and the only source of opacity is scattering by electrons. This is independent of temperature and density until at very high temperatures when relativistic effects become important. At intermediate temperatures atomic processes, where electrons are moved from one state to another by absorption of a photon, dominate. The states may be either bound or free and a dependence

$$\kappa \propto \rho T^{-3.5} \tag{9.16}$$

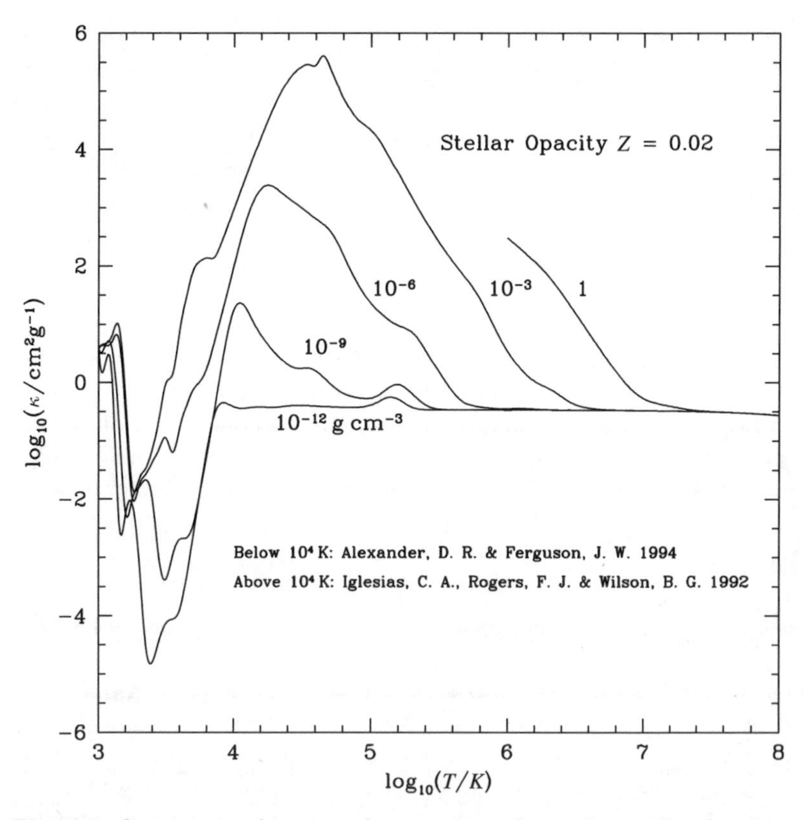

Fig. 9.6. Opacity as a function of temperature for various stellar densities

emerges. Just above 10,000 K the opacity drops rapidly with decreasing temperature, as hydrogen recombines and fewer and fewer photons have sufficient energy to change the electronic states. At lower temperatures it begins to rise again as H^- ions and various molecules become important sources but the calculation becomes even more complex.

Conductivity can be described in a similar way with electrons replacing the photons as the energy carriers. Usually, the mean free path of electrons is much shorter than that of photons so that their effective opacity is much larger and radiation transport dominates. However, in degenerate material electrons are not easily scattered because they must scatter into an empty momentum state but all neighbouring momentum states are already occupied. The mean free path becomes very large and the fluid is effectively superconducting. In practice, this means that degenerate regions of stars are close to isothermal.

9.5 Convection

The process of convection is sufficiently important to warrant a separate discussion. Fluid is convectively unstable when the temperature gradient is such that a packet of material displaced vertically, parallel to the direction of gravity, continues to rise or fall. Suppose we displace a blob of material by a small distance δz upwards in the star (Fig. 9.7), the density of the material outside the blob changes according to the ambient gradient. Let the new density within the blob be ρ'. Then the blob continues to rise if it is now less dense than its surroundings,

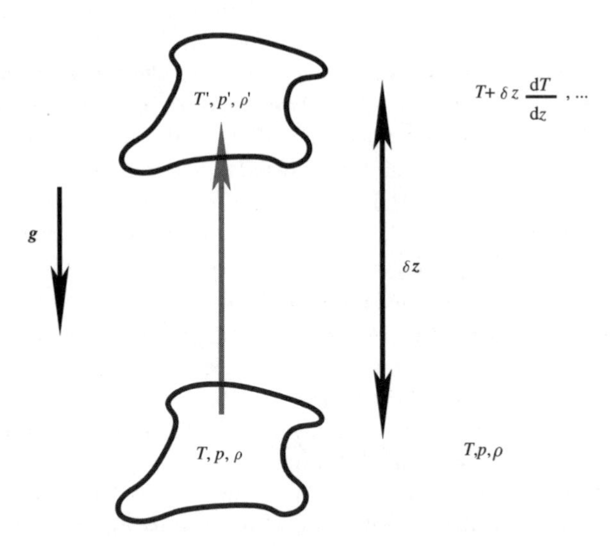

Fig. 9.7. The convective instability. A blob of fluid displaced upwards continues to rise if its density is less than that of its surroundings when it has reached pressure equilibrium adiabatically

$$\rho' < \rho + \delta z \frac{\mathrm{d}\rho}{\mathrm{d}z}, \tag{9.17}$$

and is convectively unstable.

The sound speed in the fluid is generally short so that the blob quickly reaches pressure equilibrium with its surroundings and

$$P' = P + \delta z \frac{\mathrm{d}P}{\mathrm{d}z}. \tag{9.18}$$

Initially, the displaced blob has had no time to exchange heat with its surroundings so that its density changes adiabatically, at constant entropy s. We can then write

$$\rho' - \rho = \delta \rho_s = \left(\frac{\partial \log \rho}{\partial \log P} \right)_s \frac{\rho}{P} \frac{\mathrm{d}P}{\mathrm{d}z} \delta z. \tag{9.19}$$

The adiabatic change in density with pressure can be found from the equation of state and is written as

$$\frac{1}{\Gamma_1} = \left(\frac{\partial \log \rho}{\partial \log P} \right)_s. \tag{9.20}$$

From the structure of the star we also have

$$\frac{\mathrm{d}\rho}{\mathrm{d}z} \delta z = \left(\frac{\mathrm{d} \log \rho}{\mathrm{d} \log P} \right) \frac{\rho}{P} \frac{\mathrm{d}P}{\mathrm{d}z} \delta z. \tag{9.21}$$

and we define Γ by

$$\frac{1}{\Gamma} = \left(\frac{\mathrm{d} \log \rho}{\mathrm{d} \log P} \right) \frac{\rho}{P}, \tag{9.22}$$

the density exponent with respect to pressure in the surrounding material. Because P must always fall as z increases, in order to maintain hydrostatic equilibrium, $\mathrm{d}P/\mathrm{d}z < 0$ always and so the fluid is unstable to convection if

$$\frac{1}{\Gamma} < \frac{1}{\Gamma_1}, \tag{9.23}$$

the Schwarzschild criterion.

By considering the ideal gas equation of state, we can see that Γ is large when the temperature gradient in the star is large. Thus, just as in a boiling kettle, convection is driven when there is a strong heat source that would drive a very large temperature gradient. Convection is also induced by a small value of Γ_1. This occurs in ionization regions where the number of particles, and so the pressure, increases over a small temperature range.

In unstable regions efficient turbulent mixing of the fluid takes place and this leads to an adiabatically stratified region of constant entropy,

$$\Gamma \approx \Gamma_1. \tag{9.24}$$

So, in convective regions, we write the temperature gradient as

$$\frac{\mathrm{d}T}{\mathrm{d}r} = \nabla_{\mathrm{a}} \frac{T}{P} \frac{\mathrm{d}P}{\mathrm{d}r} + \Delta\nabla T, \tag{9.25}$$

where $\Delta\nabla T$ is the superadiabatic gradient. It is one of the least certain features of stellar evolution but is usually calculated by mixing length theory (Böhm-Vitense 1958). Throughout most of a convective region it is small and not important but at the outer edge of the solar convection zone it becomes relatively large and determines the adiabat on which the whole convective zone lies. It can be calibrated by ensuring that the radius of a model of the Sun fits the measured radius but there is no guarantee that the same calibration or even the same theory can be applied to other stars.

There are further complications that have yet to be fully satisfactorily addressed. Convective overshooting might occur at Schwarzschild boundaries because, although the acceleration of a blob goes to zero at the edge of a convective region, its velocity does not. However, the deceleration of a blob that crosses a boundary is generally extremely fast and any overshooting quite negligible. Even so the concept is still popular because there is much evidence for composition mixing in radiative regions that does not have an established cause. Semiconvection occurs when there is a composition gradient. Convection may be stable according to the Schwarzschild criterion if no material is mixed across the boundary but unstable if it is. There is an equilibrium when just enough material mixes to maintain stability. What is uncertain is the timescale on which this equilibrium is attained. Varying it significantly changes some evolutionary phases and in particular the size of the burnt core at the end of helium burning (Dewi, Stancliffe & Tout, private communication).

9.6 Energy Generation

The luminosity of a star is created by various sources of energy. The change in luminosity from radius r to $r + \delta r$ is the total energy generated by material in the shell of mass δm between the two radii (Fig. 9.8). Thus for an energy generation rate per unit mass of ϵ,

$$\frac{\mathrm{d}L_r}{\mathrm{d}r} = 4\pi r^2 \rho\epsilon. \tag{9.26}$$

This is a simple equation but a great deal of complexity is hidden within the rate ϵ, which depends on the state of the fluid, particularly its temperature, and its composition.

There are three major contributions. First as a star contracts the fluid releases gravitational energy. This is the dominant source of luminosity during star formation when a gas cloud collapses to form the star and before its core is hot enough to ignite hydrogen fusion. It is occasionally important later in

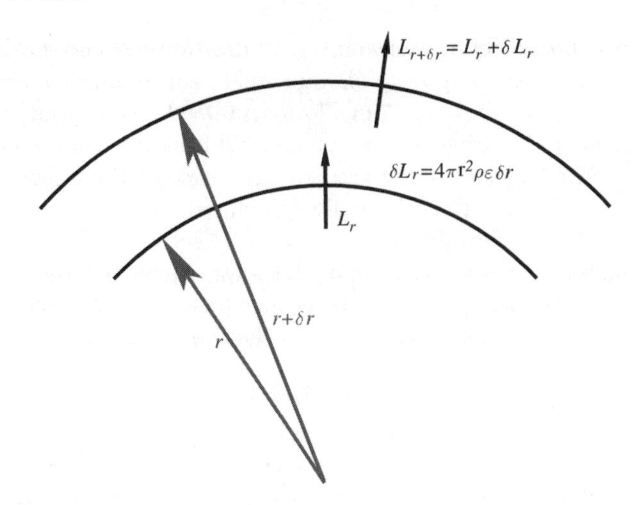

Fig. 9.8. Luminosity variation. The local luminosity L_r of a star is the energy flux outwards through the sphere of radius r within the star. Luminosity increases between r and $r+\delta r$ when there is energy generation in the shell of mass δm between these spheres

the evolution too when contraction can release energy at a comparable rate to nuclear burning. For an ideal gas the contribution is

$$\epsilon_{\text{grav}} = -C_{\text{V}} T \frac{\partial}{\partial t} \left(\log_{\text{e}} \frac{P}{\rho^{\gamma}} \right), \tag{9.27}$$

where $\gamma = C_{\text{P}}/C_{\text{V}}$ is the ratio of the specific heat at constant pressure C_{P} to the specific heat at constant volume C_{V}. This term is negative when the star is expanding but it generally does not dominate nuclear energy sources. It also introduces stellar evolution via the time derivative.

Secondly, energy is generated by nuclear reactions and the discussion of these will compose the major part of this section. Thirdly, at very high temperatures and densities, neutrino loss processes become important. Reversible weak reactions release two energetic neutrinos, both of which escape from the star because the matter cross-section to neutrinos is very small. Their mean free path is much greater than the radius of the star. The contribution ϵ_{ν} is always negative.

9.6.1 Nuclear Burning

One ^{4}He nucleus is less massive than four protons and two electrons. This is because the magnitude of the binding energy per nucleon is larger in helium-4. It is more stable. In general, the binding energy of a nucleus

$$E_{\text{B}} = (Zm_{\text{p}} + [A - Z]m_{\text{n}} - m_{\text{nuc}})c^2 \tag{9.28}$$

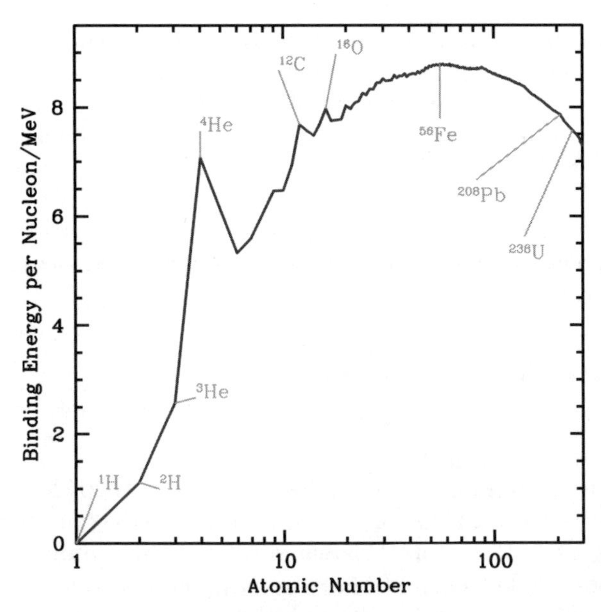

Fig. 9.9. Binding energy per nucleon for nuclides of atomic mass A. The most stable isotope is plotted for each atomic number. Up to the iron group elements around ^{56}Fe the binding energy per nucleon increases and energy is usually released in nuclear reactions that create heavier stable nuclei. For higher mass nuclei the energy per nucleon decreases with A. Energy is required to create these nuclei from less massive ones

for a nucleus of mass $m_{\rm nuc}$ containing Z protons of mass $m_{\rm p}$ and $A - Z$ neutrons of mass $m_{\rm n}$. This is zero for a hydrogen nucleus, which is just a single proton $Z = A = 1$. Figure 9.9 shows the binding energy per nucleon $E_{\rm B}/A$ as a function of atomic number A. This average binding energy tends to rise up to iron-56 and then falls again. There are notable peaks of stability at helium-4, carbon-12 and oxygen-16. When any of these are formed from less stable nuclei, the binding energy is released. As A increases beyond 56, the binding energy per nucleon falls again so that it is not energetically favourable to fuse lower-mass isotopes to form higher-mass ones.

9.6.2 Hydrogen Burning

The energy released when converting four protons to one helium-4 nucleus is 26.73 MeV. However, the actual energy available to the star depends on the reaction pathway. Energy is released in three forms, high-energy gamma rays, kinetic energy of the reacting particles and neutrinos. The first two forms are thermalized locally but once again the neutrinos can escape from the star and carry off their energy. At relatively low temperatures, as in the Sun, the reaction proceeds via the proton–proton chain. The first and slowest reaction is the combination of two protons to form a deuterium nucleus

$$^{1}\mathrm{H} + {}^{1}\mathrm{H} \rightarrow {}^{2}\mathrm{H} + \mathrm{e}^{+} + \nu. \tag{9.29}$$

The neutrino escapes with an energy of 0.26 MeV while the positron annihilates with an electron

$$\mathrm{e}^{+} + \mathrm{e}^{-} \rightarrow \gamma \tag{9.30}$$

to leave an energetic gamma ray. Another proton can then react with the deuterium nucleus,

$$^{1}\mathrm{H} + {}^{2}\mathrm{H} \rightarrow {}^{3}\mathrm{He} + \gamma, \tag{9.31}$$

and two of these $^{3}\mathrm{He}$ nuclei can then combine,

$$^{3}\mathrm{He} + {}^{3}\mathrm{He} \rightarrow {}^{4}\mathrm{He} + 2\,{}^{1}\mathrm{H} + \gamma. \tag{9.32}$$

The actual energy released to the stellar material is 26.20 MeV because two neutrinos are lost for each $^{4}\mathrm{He}$ nucleus created. This is the ppI chain. At higher temperatures the ppII and ppIII chains, which involve lithium, beryllium and boron, also operate but each of these loses more energy in neutrinos.

Above a temperature of 2×10^{7} K hydrogen burns faster via a catalytic cycle, the CNO cycle,

$$^{12}\mathrm{C}(\mathrm{p},\gamma)^{13}\mathrm{N}(,\mathrm{e}^{+}\nu)^{13}\mathrm{C}(\mathrm{p},\gamma)^{14}\mathrm{N}(\mathrm{p},\gamma) \tag{9.33}$$

$$^{15}\mathrm{O}(,\mathrm{e}^{+}\nu)^{15}\mathrm{N}(\mathrm{p},\alpha)^{12}\mathrm{C} \tag{9.34}$$

with a rare branch when $^{15}\mathrm{N}$ captures a proton before it decays

$$^{15}\mathrm{N}(\mathrm{p},\gamma)^{16}\mathrm{O}(\mathrm{p},\gamma)^{17}\mathrm{F}(,\mathrm{e}^{+}\nu)^{17}\mathrm{O}(\mathrm{p},\alpha)^{14}\mathrm{N}. \tag{9.35}$$

The component of the cycle $^{12}\mathrm{C}(\mathrm{p},\gamma)^{13}\mathrm{N}$ represents

$$^{12}C + {}^{1}H \rightarrow {}^{13}N + \gamma \tag{9.36}$$

etc. The neutrino losses are greater than those in the ppI chain so that the total energy available per $^{4}\mathrm{He}$ nucleus created is reduced to 23.8 MeV. The core temperature of main-sequence stars increases with their mass and the CNO cycle begins to dominate at about $1.5\,M_{\odot}$. Hydrogen burns faster but less efficiently because of the greater neutrino losses.

9.6.3 Reaction Rates

Quite a complicated mixture of theory and experiment is required to estimate reaction rates and details may be found in Clayton (1968). Charged-particle reactions can only occur at all because the most energetic nuclei in the tail of the Maxwellian distribution are able to quantum-mechanically tunnel through the Coulomb barrier. Once they reach the nucleus the bound states tend to be of much lower energy and they face being reflected unless they can enter a similar energy resonant state. All these lead to very strong temperature dependences for nuclear reactions. The energy generation rate of the pp chain at 10^7 K

$$\epsilon_{\mathrm{pp}} \propto \rho T^{4.6}, \tag{9.37}$$

and for the CNO cycle at 2×10^7 K

$$\epsilon_{\mathrm{CNO}} \propto \rho T^{14}. \tag{9.38}$$

In most cases these temperature dependences lead to thermostatic control of the reactions. If energy production were to rise, the star would expand in response and the temperature would fall. As a result hydrogen burning takes place at a temperature much too low for helium burning which in turn takes place at a temperature much too low for carbon burning so that stars use up one fuel, at a particular radius, at a time before igniting the next.

As mentioned before, as nuclear reactions change the composition of the material, the star evolves because the equation of state is changed. The opacities and the energy generation rates, which depend on the state, also change. Once a star has begun nuclear burning it is these composition changes that drive evolution.

9.6.4 Helium Burning

Above 10^8 K, with hydrogen long gone, helium can fuse to carbon. First, two ^{4}He nuclei react and form the unstable ^{8}Be*.

$$^4\mathrm{He} + {}^4\mathrm{He} \rightleftharpoons {}^8\mathrm{Be}^*, \tag{9.39}$$

This is a resonant state but, unlike the deuterium nucleus formed in the pp chain, there is no stable state of ^{8}Be to which it can decay. Indeed, there is no stable nucleus of atomic mass 8 at all. The ^{8}Be* nucleus has no choice but to split up into two ^{4}He nuclei again with a half life of 3×10^{-16} s. Though short, this is long enough for a third α-particle to collide if the temperature is high enough. Interestingly, there is a resonant state of ^{12}C not very different from that of the colliding nuclei. This reaction too is reversible but now there is a stable state into which the ^{12}C* nucleus can decay, by the emission of two photons to conserve spin, and complete the process,

$$^8\mathrm{Be}^* + {}^4\mathrm{He} \rightleftharpoons {}^{12}\mathrm{C}^* \rightarrow {}^{12}\mathrm{C} + \gamma + \gamma. \tag{9.40}$$

The first two reactions are endothermic. Formation of an ^{8}Be* nucleus requires 0.092 MeV and formation of the ^{12}C* requires a further 0.285 MeV. But when this decays to the stable ^{12}C the photons extract 7.65 MeV. The total energy liberated by the whole process is therefore 7.27 MeV, 0.606 MeV per nucleon or about one tenth of that released during hydrogen burning. The energy generation rate

$$\epsilon_{3\alpha} \propto \rho^2 T^{40}. \tag{9.41}$$

This is perhaps the most extreme sensitivity to temperature found in nature and in the Sun it will lead to a thermonuclear runaway when it ignites in the degenerate helium ash in the core.

At temperatures required to run this triple-α reaction it is easy to add another helium nucleus

$$^{12}\mathrm{C} + {}^4\mathrm{He} \rightarrow {}^{16}\mathrm{O} + \gamma, \tag{9.42}$$

and in many cases helium burning produces more oxygen than carbon.

9.6.5 Later Burning Stages

Hydrogen and helium burning account for most of the energy production in a star's life but stars more massive than about $8\,M_\odot$ can go on to ignite carbon at $T \approx 5 \times 10^8\,\mathrm{K}$,

$$^{12}\mathrm{C} + {}^{12}\mathrm{C} \rightarrow \begin{cases} {}^{20}\mathrm{Ne} + {}^4\mathrm{He} \\ {}^{23}\mathrm{Na} + {}^1\mathrm{H} \\ {}^{23}\mathrm{Mg} + \mathrm{n} \qquad \textit{rare}. \end{cases} \tag{9.43}$$

The next major phase is neon burning by photodisintegration. Temperatures of about 10^9 K are sufficient to provide energetic photons capable of ejecting an α-particle from a neon nucleus. At these temperatures the α-particle can readily combine with another neon nucleus and produce more stable magnesium,

$$\gamma + {}^{20}\mathrm{Ne} \rightleftharpoons {}^{16}\mathrm{O} + {}^4\mathrm{He}. \tag{9.44}$$

$$^{20}\mathrm{Ne} + {}^4\mathrm{He} \rightarrow {}^{24}\mathrm{Mg} + \gamma. \tag{9.45}$$

At 2×10^9 K oxygen can burn to produce a variety of products including silicon,

$$^{16}\mathrm{O} + {}^{16}\mathrm{O} \rightarrow {}^{28}\mathrm{Si} + {}^4\mathrm{He} + \gamma, \tag{9.46}$$

then at 3×10^9 K photons are energetic enough to break up the silicon,

$$\gamma + {}^{28}\mathrm{Si} \rightleftharpoons {}^{24}\mathrm{Mg} + {}^4\mathrm{He}. \tag{9.47}$$

This is followed by a series of α captures and photodisintegrations that culminate in the iron group elements. The actual combination of isotopes depends on the nuclear statistical equilibrium, which is controlled by the number of protons and neutrons present. When numbers are about equal the dominant product is ^{56}Ni, which is the power source of most supernovae as it decays to ^{56}Fe via ^{56}Co.

9.7 Boundary Conditions

We now have the set of four equations of stellar structure together with the time dependence that drives stellar evolution. We discussed boundary conditions in Sect. 9.2. We want the surface of a star to be what we see when we look at it. This is the surface from which the photons that reach us are emitted. Photons escape freely when the optical depth

$$\tau = \int_r^\infty \kappa\rho \, \mathrm{d}r \approx 1. \tag{9.48}$$

More carefully, we can use a thin grey atmosphere with the Eddington Closure approximation (Woolley & Stibbs 1953). Then, at $\tau = 2/3$,

$$L_r = 4\pi R^2 \sigma T^4 \tag{9.49}$$

and with hydrostatic equilibrium

$$P \approx \frac{2}{3}\frac{g}{\kappa}. \tag{9.50}$$

With yet more sophistication, we can make a full model of the radiative transfer in the atmosphere and fit it to the stellar interior. Unfortunately, the solution to this is sufficiently complex to consume as much time as a full stellar evolution sequence and so tends not to be used unless absolutely necessary.

9.8 Evolutionary Tracks

Figure 9.10 shows the path followed in the HR diagram for stars of 1, 5 and 32 $M_\odot$ as they evolve from the zero-age main sequence when no hydrogen has yet been converted to helium. They have been evolved with the Cambridge STARS code that is described in more detail in Chap. 13. There details of how to obtain and run the program can be found so that the reader can reproduce this and similar diagrams. On the ZAMS our 5 $M_\odot$ star has a radius of 2.65 $R_\odot$ and a luminosity of 540 $L_\odot$. It is burning hydrogen to helium via the CNO cycle in its core. Because of the relatively strong temperature dependence of the CNO reactions, the burning mostly occurs right at the centre but the temperature gradient drives convection out to 1.2 $M_\odot$ and the whole of this core is burnt. The core shrinks in both mass and radius as burning proceeds so that only the inner 0.53 $M_\odot$ is completely converted to helium. Just before this, after 8.24×10^7 yr, when the star's luminosity has reached 900 $L_\odot$ and its radius grown to 5.35 $R_\odot$, the fraction of hydrogen at the centre by mass has dropped to 0.05. At this point it is more energetically favourable for the whole star to contract. This is the hook in the HR diagram at the end of the main sequence. Shortly afterwards (2.3×10^6 yr later), central hydrogen is exhausted completely and burning moves to a shell surrounding the core.

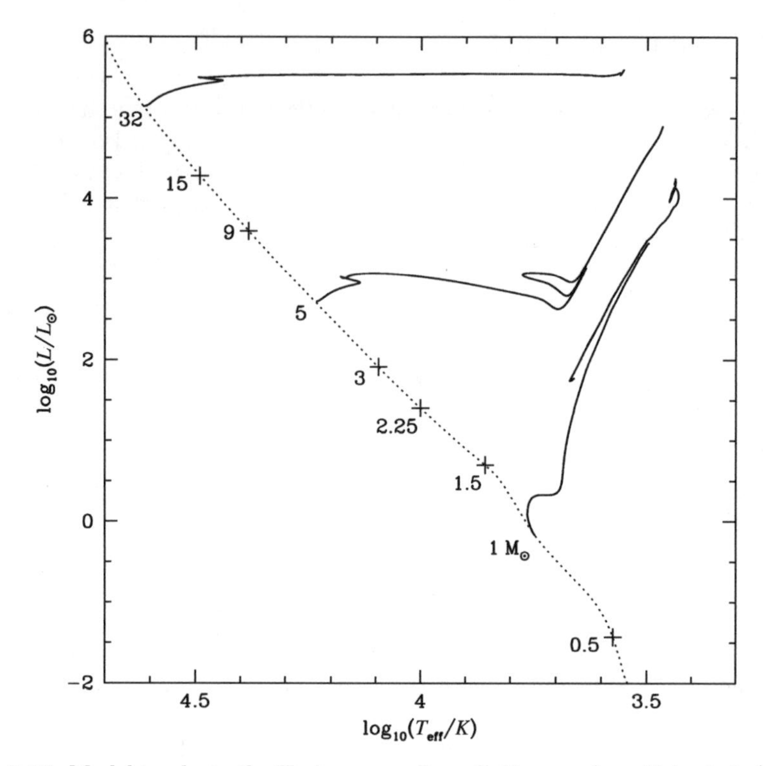

Fig. 9.10. Model tracks in the Hertzsprung–Russell diagram from Pols et al. (1995)

After another 3.9×10^6 yr this core has grown so large (to about $0.6\,M_\odot$) that it can no longer support itself with gas pressure. It starts to contract, gradually forcing the nuclei and electrons together but the core does not get very degenerate at this stage. It does, however, rapidly contract and the star moves over to the giant branch in the relatively short time of 8.4×10^5 yr. As the core contracts the envelope expands. Though no one has yet explained simply why it expands, we do appear to include all the relevant physics because our models expand. A star is complex and behaves in very non-linear ways so it is often not easy to predict what will happen or even to explain why it has! An important result of the expansion is that the surface temperature descends and convection sets in reaching right down to parts of the stellar core that have previously been processed. Once established on the giant branch, the helium core grows as hydrogen burns outwards. It contracts in radius as it does so and heats up. This raises the temperature at the burning shell so that the reactions run faster and the luminosity rises. The star makes its first ascent of the giant branch.

The core, growing in mass but contracting in radius, continues to heat up until at 1.2×10^8 K it is hot enough for helium to ignite. Once again the helium burning drives convection in the core which this time grows as the burning proceeds. Eventually, helium fuel is exhausted in the core too and helium burning moves to a shell that starts to follow the hydrogen-burning shell out through the envelope. During core helium burning, our star had settled back to a lower luminosity, shrunk and lost its deep convective envelope. It now moves back over to the giant branch but only slowly resumes the same rising track so we call this AGB.

At this point we should note that the production of elements in stars is not on its own enough to ensure their availability when a new generation of stars and planets condense. The processed material must actually be somehow driven off into the interstellar medium at a velocity that exceeds the escape velocity of the star. Indeed, stars leave behind remnants that might be white dwarfs, neutron stars or black holes depending on mass and these remnants swallow a substantial part of the processed core in the most common stars.

In comparison, two significant differences characterise the evolution of a $1\,M_{\odot}$ star. First, the central temperature on the main sequence is lower so that hydrogen burning proceeds via the pp chain rather than the CNO cycle. Then the lower core temperature on the giant branch means that the core becomes very degenerate before it reaches the temperature at which helium can ignite. Because the degenerate equation of state does not respond to the rising temperature as the reaction generates energy, it is not thermostatically controlled in the normal way. This is coupled with the incredible temperature sensitivity so that a thermonuclear runaway ensues during which the energy production reaches the luminosity of a small galaxy! But it lasts only a few seconds before the degeneracy is raised and the star drops rapidly down the giant branch to begin stable core helium burning. The energy produced is absorbed by the star's envelope and is hardly noticed at its surface. From then it evolves much like the $5\,M_{\odot}$. Once high on the AGB, it is mass loss that controls the evolution of these stars. A very strong dusty wind eventually removes all the hydrogen envelope and exposes the burning shells. These cool and extinguish leaving a white dwarf that rapidly falls in luminosity to below the main sequence and then cools from left to right across the diagram.

The $32\,M_{\odot}$ star on the other hand goes on to ignite carbon in its core, which is processed all the way to iron. When the iron core reaches the Chandrasekhar mass of $1.44\,M_{\odot}$, the maximum that can be supported by electron degeneracy, it collapses to a tiny neutron star. The energy released blows the entire envelope off in a spectacularly bright supernova.

9.9 Stellar Evolution of Many Bodies

In Chap. 10 Jarrod Hurley describes how single-star stellar evolution can be incorporated in N-body calculations. It is important to know how the masses of the stars change both by mass loss in stellar winds and any sudden mass loss in a supernova because this affects the dynamics of the cluster. As the bodies interact, dynamics can also influence the stellar evolution. This is most apparent when stars are in or form close binary systems. These form the topic of Chap.11.

References

Alexander D. R., Ferguson J. W., 1994, ApJ, 437, 879
Böhm-Vitense E., 1958, Z. Astrophys., 46, 108
Chandrasekhar S., 1939, An Introduction to the Study of Stellar Structure. Chicago Univ. Press, Chicago
Church R. P., Tout C. A., Aarseth S. J., 2007, private communication
Clayton D. D., 1968, Principles of Stellar Evolution and Nucleosynthesis. Chicago Univ. Press, Chicago
Herschel W., 1783, Phil. Trans. R. Soc., 73, 247
Hertzsprung E., 1905, Z. Wissenschaftliche Photographie, 3, 422
Hurley J. R., Tout C. A., Pols O. R., 2000, MNRAS, 315, 543
Iglesias C. A., Rogers F. J., Wilson B. G., 1992, ApJ, 397, 717
Pancino E., Ferraro F. R., Bellazzini M., Piotto G., Zoccali M., 2000, ApJ, 534, L83
Pols O. R., Tout C. A., Eggleton P. P., Han Z., 1995, MNRAS, 274, 964
Russell H. N., 1913, Obs., 36, 324
Woolley R. v. d. R., Stibbs D. W. N., 1953, The Outer Layers of a Star. Clarendon Press, Oxford

10

N-Body Stellar Evolution

Jarrod R. Hurley

Centre for Astrophysics and Supercomputing, Swinburne University of Technology, P.O. Box 218, VIC 3122, Australia
jhurley@swin.edu.au

10.1 Motivation

The advent of the Hubble Space Telescope (HST), with its ability to peer deep inside the globular clusters (GCs) of our Galaxy and resolve individual stars (Paresce et al. 1991), provided reason enough to include stellar evolution in cluster models. We only have to look at the beautiful images of stars in the core of, say, Omega Centauri[1] (Carson, Cool & Grindlay 2000) to be motivated to produce colour-magnitude diagrams (CMDs) from simulations to match those emanating from HST. There are also a number of questions relating to stellar populations in star clusters that require a combination of stellar evolution and stellar dynamics for investigation. For example, population gradients are observed, which indicate a central concentration of blue stragglers (BSs) as well as a central depletion of red giants (Yanny et al. 1994). A possible explanation is that close encounters between stars in the dense core of a GC leads to enhanced production of BSs in collisions (or mergers) of main-sequence stars. Encounters are also then expected to enhance the stripping of the envelopes of giant stars to produce blue horizontal branch stars or white dwarfs (WDs). The situation is not straightforward though, as evidenced by the classic second-parameter pair of GCs, M3 and M13 (Ferraro et al. 1997). Here we have two clusters of the same mass, density, metallicity and (apparently) age, but with dramatic differences in their blue straggler and blue horizontal branch star populations. Also, HST is not alone in exposing the cores of star clusters – the Chandra X-ray Telescope has provided a wealth of complementary information on objects such as millisecond pulsars and cataclysmic variables (Grindlay et al. 2001a,b).

Aside from a desire to produce models to match observations of stellar populations in star clusters, there is a more basic need for stellar evolution in N-body models. Here we are talking specifically about mass loss from stars as they evolve. This can have a dramatic effect on the lifetime and structure

[1] http://hubblesite.org/newscenter/archive/releases/2001/33/image/a

Hurley, J.R.: *N-Body Stellar Evolution.* Lect. Notes Phys. **760**, 283–296 (2008)
DOI 10.1007/978-1-4020-8431-7_10

of a star cluster. Put simply, mass lost from stars in stellar winds is expected to escape from a cluster and therefore weakens its potential. The cluster then expands, which leads to a temporary increase in the loss of stars across the tidal boundary. This weakening of the potential leaves the cluster more exposed to the possibility of disruption if, for example, the cluster encounters a giant molecular cloud or orbits through the Galactic disc. In the long-term stellar evolution mass loss affects the timescale for two-body relaxation and core-collapse (the reader is directed to Meylan & Heggie 1997 for an overview of the processes involved in cluster evolution). Thus stellar and cluster evolution are intertwined and an accurate description of the former in concert with the dynamics is required.

10.2 Method and Early Approaches

To meet the needs described above, there are a minimum set of variables that a stellar evolution algorithm must be able to provide within the N-body code. In order to detect and enact collisions between stars, the stellar radius is required for each star in the model. To produce CMDs requires the luminosity and effective temperature (or radius) of each star. The mass of each star is required and information on the mass and radius of the core is important for determining the nature of stellar remnants as well as the outcomes of collisions (and the inclusion of binary evolution). Therefore, the algorithm must be able to account for the mass, size and appearance of the N-body stars as the cluster evolves. Ideally, it should be able to do this with metallicity as a free parameter. The open clusters of the Galaxy typically contain stars of close to solar metallicity while the GCs are metal-poor (see Meylan & Heggie 1997), and in comparison the star clusters of the Large Magellanic Cloud exhibit a wide range of metallicity (Mackey & Gilmore 2003). This is an important distinction to make because the evolution timescale and appearance of a star depends critically on its composition as well as its mass.

When deciding on an appropriate stellar evolution method, there are three approaches from which to choose, (i) a detailed evolution code, (ii) look-up tables and (iii) fitted functions. An overriding concern is that the stellar evolution algorithm should not impede the progress of the N-body calculations: the algorithm must be robust and provide rapid updating of the necessary variables for all possible stages of evolution. The robustness requirement has always been a stumbling block for using a detailed evolution code to provide stellar evolution because the codes are liable to break down at critical stages in the evolution. However, steps have recently been taken to overcome this shortcoming and *live* stellar evolution in N-body simulations is now an exciting possibility. Computational constraints make this method more relevant to the large-N regime. At present, look-up tables constructed from the output of a series of detailed evolution calculations represent a more reliable approach.

These require interpolation and the associated data files can be very large if a fine grid in mass is used to ensure accuracy, especially when a range of metallicities is also considered. This would not be of much concern today but it was in the early to mid-1990s when including stellar evolution in dynamics codes was under serious consideration and computing memory was at a premium. As a result, the third approach – a set of functions approximating the detailed dataset – has proven to be the most popular to date. This is the most time-consuming approach to set up, but the reward is a relatively compact algorithm that lends itself well to the requirements of an N-body code.

One drawback of the fitted function approach is that much of the information provided by a detailed stellar evolution code is discarded and not available to the dynamics code. This could be important, for example, in the case of stellar collisions where the outcome of the collision and nature of the collision product depends on the internal density profiles of the colliding stars. This is circumvented somewhat by also predetermining the collision outcomes based on prior calculations (see Hurley, Tout & Pols 2002). Another potential problem with this approach to stellar evolution is that if the detailed models on which the functions are based become outdated for any reason it is non-trivial to generate a new set of functions. Nevertheless, the fitted function approach is the method of choice in the codes NBODY4 and NBODY6 and has proven successful to date.

An early approach to combining stellar and dynamical evolution was provided by the Fokker-Planck models of Chernoff & Weinberg (1990). This was a two-step method based on an expression for the main-sequence lifetime of a star as a function of stellar mass and a WD initial-final mass relation, i.e. at the end of the main sequence a star would lose mass instantaneously and become a white dwarf. Even earlier attempts had employed simple schemes to describe mass loss in supernovae events (e.g. Wielen 1968), see Aarseth (2003) for an overview. In their population synthesis work, Eggleton, Fitchett & Tout (1989) provided a more sophisticated algorithm that described the luminosity, radius and core mass of the stars for a range of evolution phases. This treatment was included in NBODY4 in 1994 and is still adopted by other dynamics codes. Improvements were made to this algorithm by Tout et al. (1997), specifically for use in NBODY4. The next major development in the fitted function approach came with the creation of the Single Star Evolution (SSE) package by Hurley, Pols & Tout (2000). This was based on an updated set of detailed stellar models that included convective overshooting and for the first time metallicity was a free parameter – all previous algorithms were solar metallicity only. It also included an expanded range of evolution phases, a more detailed description of the evolution within each phase and an updated mass-loss algorithm. SSE currently provides stellar evolution in NBODY4 and NBODY6 and is outlined below. A general introduction to stellar evolution theory has been presented in Chap. 9.

10.3 The SSE Package

The goal here is to provide an overview of the method used to construct the SSE package and to discuss some aspects relevant to inclusion in an N-body code. A full description of the SSE package is given in Hurley, Pols & Tout (2000).

The basic idea of the algorithm is to break the evolution of a star into a series of evolution phases. These are listed in Table 10.1. Each phase has an associated index, `kstar`, which identifies the stellar type.[2] The phases fall into three groupings, normal nuclear burning evolution, `kstar` $\in [1,6]$, naked helium star evolution, `kstar` $\in [7,9]$, and remnant evolution, `kstar` $\in [10,14]$.

All stars are assumed to be born on the zero-age main sequence (ZAMS) where core hydrogen burning is initiated. Stars then move through a series of phases as they evolve, although a particular star may not experience all phases. For example, a $1\,M_\odot$ star stays on the main sequence (`kstar`$=1$) for about 11 Gyr before quickly passing through the Hertzsprung gap phase (`kstar`$=2$) as hydrogen burning commences in a shell surrounding the helium core. It then ascends the giant branch (`kstar`$=3$) until helium is ignited degenerately in the core and the core helium flash brings the star to the core-helium burning, or horizontal branch, phase (`kstar`$=4$). This is as far as a $1\,M_\odot$ star would get within the age of the Galaxy. If for some reason the star was stripped of its envelope while on the giant branch, as a result of a collision or close binary evolution, it would become a helium white dwarf (`kstar`$=10$). Otherwise, given enough time it would eventually evolve to become a WD comprised primarily of carbon and oxygen (`kstar`$=11$). A $5\,M_\odot$

Table 10.1. Evolution phases identified in SSE and the assigned `kstar` index

`kstar`	Evolution phase	`kstar`	Evolution phase
1	main sequence	10	helium white dwarf
2	Hertzsprung gap	11	carbon oxygen white dwarf
3	first giant branch	12	oxygen neon white dwarf
4	core helium burning	13	neutron star
5	early asymptotic giant branch	14	black hole
6	thermally pulsing AGB		
7	helium main sequence		
8	helium Hertzsprung gap		
9	helium giant branch		

[2]There is an additional phase (`kstar`$=0$) not listed, which is used to denote low-mass main-sequence stars with mass less than $0.7\,M_\odot$. This is carried over from Tout et al. (1997) and distinguishes stars with deeply or fully convective envelopes, which respond differently to mass changes during binary evolution (see Chaps. 11 and 12).

star evolves through phases $1 \to 6$ before becoming a $1\,M_{\odot}$ carbon oxygen white dwarf (kstar = 11). This takes about 100 Myr. The asymptotic giant branch (AGB) is divided into two separate phases by the the onset of second dredge-up or, more generally, the time at which the growing carbon oxygen core reaches the helium core in mass. On the early AGB (kstar = 5), luminosity is dominated by a helium-burning shell. At the onset of the thermally pulsing AGB (kstar = 6), a hydrogen shell source is ignited and subsequently provides the bulk of the luminosity. Thermal pulses that reduce the growth of the core mass are modelled during this phase. Stars of approximately $8\,M_{\odot}$ ignite carbon on the AGB and evolve to become oxygen neon white dwarfs (kstar = 12). More massive stars (10–25 $M_{\odot}$) evolve to become neutron stars (kstar = 13) and even more massive stars become black holes (kstar = 14). A $20\,M_{\odot}$ star, for example, evolves through phases $1 \to 2 \to 4 \to 5 \to 13$ in approximately 10 Myr. In this case, central helium burning is ignited during phase 2 so that phase 3 is skipped. Furthermore, a $25\,M_{\odot}$ star sheds its envelope during phase 4 and thus becomes a naked helium main-sequence star (kstar = 7) rather than reach the AGB. It then evolves onto the helium Hertzsprung gap (kstar = 8) and giant branch (kstar = 9) before becoming a black hole. Transitions from $12 \to 13$ and $13 \to 14$ are also possible through mass accretion in a close binary (see Chap. 12 and Hurley, Tout & Pols 2002, for details). Note that the quoted evolution times and landmark masses are for solar metallicity and vary for different composition.

The SSE package comprises a set of analytical evolution functions that provide quantities such as the luminosity, radius and core mass for a star which evolves through the phases mentioned above. Input variables are the mass, M, metallicity, Z, and age of the star. The method used in constructing SSE was to first find functions to fit the end-points of the various evolutionary phases as well as the timescales. Then the behaviour within each phase was fitted. A starting point was the set of formulae provided by Tout et al. (1996) to describe the ZAMS luminosity and radius as a function of M and Z. This was then extended to fit aspects of the evolution, such as the luminosity and radius at the end of the main sequence, with rational polynomials that are continuous and differentiable where possible. For example, the formula to describe the time taken for a star to evolve from the ZAMS to the base of the giant branch is

$$t_{\rm BGB} = \frac{a_1 + a_2 M^4 + a_3 M^{5.5} + M^7}{a_4 M^2 + a_5 M^7}, \tag{10.1}$$

where the coefficients a_n are functions of Z. Data to create the functions for the standard nuclear burning phases was taken from the detailed models of Pols et al. (1998). The models cover a range in mass from 0.1 to $50\,M_{\odot}$ and a range of metallicity from 0.0001 to 0.03 with $Z \simeq 0.02$ being solar. The resulting functions are accurate to within 5% of the detailed stellar models over all phases of the evolution. The errors introduced by this approach are less than the intrinsic errors of the detailed models themselves, owing to uncertainties in the input physics. Note that the functions can be safely extrapolated up to $100\,M_\odot$, but for greater mass `SSE` evolves the star using timescales and quantities for a $100\,M_\odot$ star. Extrapolation outside of the Z range of the models is not recommended.

The functions for the naked helium star phases were fitted to models produced by Onno Pols (see Dewi et al. 2002 for some details). The luminosity evolution of white dwarfs in `SSE` was initially modelled according to standard cooling theory, but has subsequently been expanded to reflect better current white dwarf models (see Hurley & Shara 2003 for details). Radii for white dwarfs come from Eq. (17) of Tout et al. (1997) and mass-dependent luminosities and radii are also assigned to neutron stars and black holes (see Hurley et al. 2000). Another change to `SSE` subsequent to Hurley, Pols & Tout (2000) is the adoption of the prescription suggested by Belczynski, Kalogera & Bulik (2002) for calculating the masses of neutron stars and black holes. Related to this, the default maximum mass for a neutron star is now assumed to be $3.0\,M_\odot$ rather than $1.8\,M_\odot$ as suggested in Hurley, Pols & Tout (2000) – this is an adjustable input parameter.

The models of Pols et al. (1998) neglect mass loss from the surface of a star owing to a stellar wind. However, the `SSE` package supplements these models by including a prescription for mass loss in a simple subroutine form that can easily be altered or added to. This prescription is drawn from a range of current mass-loss theories available in the literature. It is applicable to all nuclear burning evolution phases (`kstar` $\in [1, 9]$) and includes standard Reimers' mass loss (Kudritzki & Reimers 1978) for giants, pulsation-driven winds for AGB stars and a Wolf-Rayet like mass loss for helium stars. The reader is referred to Sect. 7 of Hurley, Pols & Tout (2000) for full details. To achieve a smooth transition from the Pols et al. (1998) models (without mass loss) to the beginning of remnant evolution `SSE` employs perturbation functions that alter the radius and luminosity of a star as the envelope becomes small in mass. `SSE` also follows the spin evolution of a star and includes magnetic braking.

The `SSE` package can be obtained by contacting the author or from `http://astronomy.swin.edu.au/jhurley/bsedload.html` (where the associated binary evolution package is also available). It provides a rapid and reliable method for evolving stars and is therefore well suited for use in population synthesis and dynamics codes. The bulk of the `SSE` functions are contained in a subroutine called `zfuncs.f`, and before any of these are used the subroutine `zcnsts.f` must be called to set all the Z-dependent coefficients (this in turn requires the `zdata.h` data file). The routine `hrdiag.f` determines which evolution stage a star is currently at and calculates the appropriate properties such as luminosity, radius and core mass. It must be preceded by a call to `star.f`, which sets the timescales for the evolution phases (as a function of M and Z) as well as various landmark luminosities. Other associated routines are: `mlwind.f`, which calculates the current mass-loss rate, `mrenv.f`, which

Table 10.2. Subroutines in NBODY4 and NBODY6 associated with stellar evolution

SSE routines	Related routines	
hrdiag.f	fcorr.f	(← mdot)
magbrk.f	hrplot.f*	(← output)
mlwind.f	instar.f*	(← start)
mrenv.f	kick.f	(← fcorr)
star.f	mdot.f*	(← intgrt)
zcnsts.f	mix.f*	(← cmbody)
zdata.h	trdot.f*	(← instar/mdot)
zfuncs.f		
corerd.f	cmbody.f	
gntage.f	data.f	
mturn.f		

routines marked with * call **hrdiag** directly

calculates the mass and radius of the convective envelope (if one exists), and **magbrk.f**, which determines the rate of angular momentum change owing to magnetic braking. These are the main SSE routines. They are listed in the left-hand column of Table 10.2 along with some further routines that are mentioned in the next section.

10.4 *N*-Body Implementation

The core SSE routines, as described in the previous section, are included in the *N*-body codes in their entirety. That is to say, they operate independently of the structure of the *N*-body codes – if any of these routines are updated in the SSE package, they can simply be copied into NBODY4 and NBODY6 without any further concern. This also means that a routine such as **hrdiag.f** could be swapped for any other routine that sets the stellar parameters provided that the current interface, or subroutine arguments, are the same. The SSE subroutines that are involved in the *N*-body codes are shown in Table 10.2. Also shown are all NBODY4/6 subroutines that either interact with these routines directly or are associated with the stellar evolution procedure in some way. Note that the subroutines that call **hrdiag.f** have been highlighted and it was also considered instructive to identify from where in NBODY4/6 these routines were called (as shown in the parentheses on the far right).

Within Table 10.2 there exist some grey areas. For example, **trdot.f** is actually a SSE routine that calculates the appropriate stellar evolution timestep for a star based on its type and the restriction that the radius should not change by more than 10% in a single timestep. This is listed in the right-hand column of Table 10.2 as an N-body routine because it contains additional lines of code specific to NBODY4/6. The same goes for `kick.f`, which is a `SSE` routine that sets the velocity kick for newly born neutron stars and black holes. Some subsidiary `SSE` routines are utilised by NBODY4/6 and these are also listed in Table 10.2 (on the left-hand side below the dividing line). The routine `corerd.f` contains a function to calculate the core radius of a star and is rendered somewhat obsolete by the combination of `hrdiag.f` and `zfuncs.f`. However, it is still used in NBODY4/6 for convenience. The routine `mturn.f` provides an estimate of the turn-off mass of a star cluster, the most massive star that currently resides on the main sequence, based on the current time and the `SSE` function that calculates the main-sequence lifetimes of the stars. It is not a routine that is essential to the evolution algorithm. On the other hand, the `SSE` routine `gntage.f` is an essential component of a stellar evolution/dynamics interface, but its use is more relevant in a discussion of binary evolution. Given a stellar type, current mass and core mass of a star, this routine calculates an appropriate age and initial mass. Thus it is essentially an inverse of `hrdiag.f` and is used to set the parameters of stars produced in mergers and collisions.

Before proceeding to give an overview of the NBODY4/6 stellar evolution algorithm, it is first pertinent to describe the associated stellar variables. Each star has an initial mass, `body0`, a current mass, `body`, a radius, `radius`, a luminosity, `zlmsty`, spin angular momentum, `spin`, and a stellar type, `kstar`. These are all common arrays of size `NMAX` where `NMAX` is set in `params.h` and must be greater than N to accommodate binaries. A star of index `i` has quantities saved at the `i`th position of these arrays, e.g. `body0(i)`. Other quantities such as the core mass are not stored and are obtained from `hrdiag.f` as required. The need to keep track of both the current and initial masses is driven by the stellar evolution algorithm. In both `SSE` and its predecessor (Tout et al. 1997), it was recognised that the evolution timescales and landmark luminosities depend on the initial mass whereas the stellar radius is more correctly a function of the current mass. Note that both `body0` and `body` are in dimensionless N-body units, and the scale-factor `ZMBAR` (or equivalently `SMU`) is used to convert to solar masses. Similarly, `radius` is converted to solar radii using `SU` and `spin` is converted to units of $M_\odot\, R_\odot^2\, \mathrm{yr}^{-1}$ using `SPNFAC`.

To allow stars to have different update frequencies, each star has an associated stellar evolution update time specified by the `tev` array. This recognises that massive stars, and the stars in advanced evolution stages such as on the AGB, require more frequent updates than, say, low-mass main-sequence stars or white dwarfs. Thus it would not be computationally efficient to have the update frequency of all stars dictated by the most rapidly evolving star at the time. A second update variable, `tev0(i)`, is also utilised. This denotes the time at which star `i` was last updated, as opposed to `tev(i)`, which represents the next required update time, and the two are used to compute the amount of mass lost between updates. Also associated with the time-keeping for each star is a quantity called `epoch`. This is a product of the `SSE` package and is used to calculate the effective stellar evolution age of a star, i.e. if `tphys` is the current physical time in Myr, the stellar evolution age of star `i` is `tphys` – `epoch(i)`. To illustrate the need for such a variable, consider a star that has just lost its envelope on the AGB and evolved to become a white dwarf. The luminosity evolution of a white dwarf is calculated from a cooling law that is a function of the time elapsed since the birth of the white dwarf. So the evolution algorithm needs to know when the white dwarf was born. This is communicated by setting `epoch(i)` = `tphys` when the star leaves the AGB. The `epoch` variable is also used to reset the stellar evolution *clock* of stars that lose (or gain) mass during certain phases of evolution (see Hurley, Pols & Tout 2000; Hurley, Tout & Pols 2002 for more details on the use of `epoch`). Note that the units of `epoch` are Myr whereas `tev` and `tev0` are in N-body units and the scale-factor `TSTAR` is required to convert to N-body times to physical units of Myr.

The next step is to be aware of N-body input variables that are relevant to stellar evolution. These are read by the routine `data.f` and are the maximum stellar mass, `body1`, the minimum stellar mass, `bodyn`, the metallicity, `zmet`, an offset parameter for the stellar evolution time, `epoch0` and the time interval between writing stellar evolution–related output, `dtplot`. Also related are the input options `kz(19)` and `kz(20)` (actually read in `input.f`). Setting `kz(19)` = 3 is necessary to activate stellar evolution according to `SSE`. If this is indicated, `data.f` calls `zcnsts.f` with `zmet` to set the metallicity dependent coefficients. This only needs to be done once as it is assumed that all stars are of the same composition. However, if a restart is required, then `zcnsts.f` is called once more but from the main routine (`nbody4.f` or `nbody6.f`). The value of `kz(20)` affects the choice of initial mass function. Options include the distribution of masses derived by Kroupa, Tout & Gilmore (1993) from stars in the solar neighborhood (`kz(20)` = 5) and a power-law mass function (`kz(20)` = 0). If the latter is indicated, the exponent `alpha` is also required from the input file. The stellar masses, i.e. `body(i)` for `i` = $1, N$, are required to lie between the bounds of `bodyn` and `body1` and are set in `data.f` according to `kz(20)` – it is also possible to read these from a file using `kz(22)`.

After reading the input file and generating the stellar masses, the N-body stellar evolution algorithm starts by initialising the stellar variables for each of the N stars. The routine `instar.f` is responsible for this process. For each star `i` it sets `body0(i)` = `body(i)`, `kstar(i)` = 1 or 0, and `epoch(i)` = 0.0 before calling the `star.f` and `hrdiag.f` combination to set `radius(i)` and `zlmsty(i)`. The spin angular momentum, `spin(i)`, is also set using the `SSE` package (see Hurley, Pols & Tout 2000). For the stellar evolution update times, the routine sets `tev0(i)` = 0.0 and `tev(i)` are initialised by a call to `trdot.f` for each star. Note that it is possible to start the stars at an advanced evolution stage by setting the input parameter `epoch0` to some negative value (see the usage of `epoch` above). In this case `epoch(i)` = `epoch0`.

Subsequent to initialisation stellar evolution is controlled by the `mdot.f` subroutine. Frequent updates are performed in step with the dynamical integration by means of a variable `TMDOT`, the minimum of `tev(i)` for all $\mathtt{i} = 1, N$. At the end of each integration step (in `intgrt.f`), a check is made to determine if the new time exceeds `TMDOT`. If it does, then `mdot.f` is called in order to update each star that has `tev(i)` less than the current time (more than one star may be due). Within `mdot.f` the stellar variables for star `i` are updated to an age of $\mathtt{tev(i)} * \mathtt{TSTAR} - \mathtt{epoch(i)}$ by calling the `star.f` and `hrdiag.f` combination. The mass-loss rate, $\dot{m}$, for the star is obtained by a call to `mlwind.f`, which gives $\dot{m}$ and the actual mass lost in the interval `tev0(i)` $\rightarrow$ `tev(i)` is

$$\Delta m = \dot{m}\,(\mathtt{tev(i)} - \mathtt{tev0(i)}) * \mathtt{TSTAR} * 1 \times 10^6/\mathtt{ZMBAR} \qquad (10.2)$$

in N-body units. If non-zero, this correction is applied to `body(i)` to update the stellar mass. If $\mathtt{kstar(i)} \leq 2$ or $\mathtt{kstar(i)} = 7$, then `body0(i)` is reset to be equal to `body(i)` and `epoch(i)` is updated to reflect the change in mass. Note that `epoch(i)` is also updated when the stellar type changes. Also, if mass loss occurs, the spin angular momentum of the star is adjusted accordingly – a call to `magbrk.f` makes any further adjustments resulting from magnetic braking. In the case of $\Delta m > 0$ the routine `fcorr.f` is called to perform force and energy corrections for the mass loss. If a new neutron star or black hole is detected, this routine calls `kick.f` to generate the velocity kick arising from the supernova event and deals with the ramifications of the velocity change. If the mass loss is substantial ($\Delta m * \mathtt{ZMBAR} > 0.1$), or a velocity kick has occurred, it is also necessary to initialise new force polynomials for the star and its neighbours. This is performed in `mdot.f` (with calls to the appropriate subroutines). The update procedure in `mdot.f` for star `i` is then completed by setting $\mathtt{tev0(i)} = \mathtt{tev(i)}$ and calling `trdot.f` to set a new `tev(i)`. Before leaving `mdot.f`, and after dealing with each star that is due, `TMDOT` is updated to the new minimum in the `tev` array.

Output of the stellar evolution variables is performed by the routine `hrplot.f`, which is called from `output.f` at intervals of `dtplot`. Note that `dtplot` must be greater than or equal to `deltat` – the time interval in N-body units for major output – and ideally the two input variables should commensurate. A call to `hrplot.f` creates a snapshot of the model stars at the current time. This involves two output files: `fort.83` contains a line for each single star and `fort.82` contains a line for each binary. These files provide the necessary information for generating descriptions of the model in the form of colour-magnitude diagrams, radial profiles and mass functions, for example.

The possibility of stellar collisions has been mentioned and the N-body codes allow for such events. Direct hyperbolic collisions between stars are rare in the cluster simulations for which NBODY4 and NBODY6 have typically been used. Rather, two stars in a close gravitational encounter more likely form a binary and this may be followed by a merging of the two stars. As such, a discussion of how these events are dealt with falls more naturally under the banner of binary evolution and will be described in Chap. 12. Here it suffices to say that collisions of all types (eccentric, parabolic or hyperbolic) are processed by the routine `cmbody.f`, which calls `mix.f` if two stars are to merge. The routine `mix.f` determines the nature of the merger product and initialises its stellar variables through calls to `gntage.f` and `hrdiag.f` (see also Hurley, Tout & Pols 2002 for more details of this procedure).

The interested reader may find Hurley et al. (2001) and Aarseth (2003, p. 279) useful for additional discussions regarding the implementation of stellar evolution in N-body codes. To complement these discussions, the material in this section is rounded off by making the user aware of SSE parameters that are hardwired, so to speak, into various N-body routines. For example, the parameter η appears in the Reimers mass-loss formula in `mlwind.f` – in the stand-alone SSE package this is an input parameter, but in NBODY4/6 it is set in the header of the subroutine. The same goes for the maximum neutron star mass, which is set in the header of `hrdiag.f` rather than appearing as an input variable. There may be occasions when the user would wish to vary these parameters, and this requires an edit of the relevant file and recompiling the code.

10.5 Some Results

The stellar evolution capability in NBODY4 and NBODY6 has been used to good effect to produce realistic models of star clusters (Baumgardt & Makino 2003, for example). The results of such endeavours are presented in Chap. 14. Given that the option to use metallicity as a free parameter is a unique feature that SSE has added to the N-body codes, this section briefly highlights some results relating to the models of varying metallicity.

In Hurley et al. (2004) a series of NBODY4 simulations was presented in order to investigate the effect of metallicity on the evolution of open clusters. Each simulation started with 30 000 single stars. Figure 10.1 shows CMD snapshots at four times for one of these simulations at solar metallicity. This was constructed using the `fort.83` output file. Note that stellar evolution not only affects the distribution of stars in the nuclear burning phases as the cluster evolves but also affects the locus of the white dwarf stars. To illustrate how metallicity affects the CMD appearance, Fig. 10.2 shows the snapshots of four models at the same age but with different metallicities.

The models of Hurley et al. (2004) showed that clusters with low-Z stars experienced more mass loss from stellar evolution over the first 5 000 Myr of evolution compared to clusters of solar metallicity. This lead to increased expansion of the cluster and a decreased stellar mass range with a knock-on

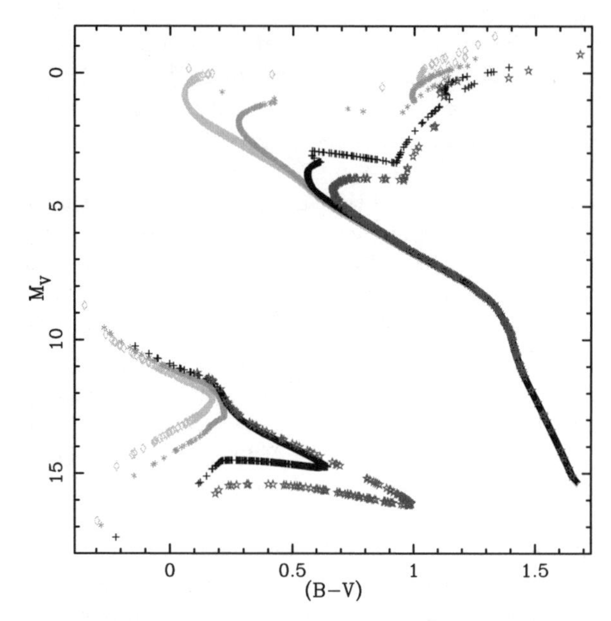

Fig. 10.1. Colour-magnitude diagram showing four N-body *isochrones*. Data are taken from a $Z = 0.02$ **NBODY4** simulation that started with 30 000 single stars. Shown are stars in the simulation at 500 Myr (*diamond symbols*), 1 000 Myr (* *symbols*), 4 000 Myr (+ *symbols*) and 9 000 Myr (*star symbols*). Stars in the upper-right of the diagram are in normal nuclear burning phases of evolution (**kstar** ≤ 6) and stars in the lower-left are white dwarfs. There are no naked helium stars present. Any neutron stars or black holes are not shown. The luminosity and effective temperature provided for each star by **SSE** have been converted to magnitude and colour with the bolometric corrections given by the models of Kurucz (1992) and, in the case of white dwarfs, Bergeron, Wesemael & Beauchamp (1995)

effect of a delay in the onset of core-collapse and binary formation. Overall, this means that low-Z clusters have extended lifetimes. Models with low-Z also produced many more double-WD binaries. This is a result of shorter main-sequence lifetimes and greater AGB core-masses producing more WDs, and more massive WDs, in comparison to high-Z models of the same age. This is a direct illustration of the interaction between stellar and dynamical evolution within the star cluster environment (see Hurley et al. 2004 for more details).

The focus so far has been on models of single stars – in Chap. 11 we shall begin to discuss the intricacies of binary evolution. This will be followed in Chap. 12 by details of the binary evolution algorithm used in NBODY4 and NBODY6.

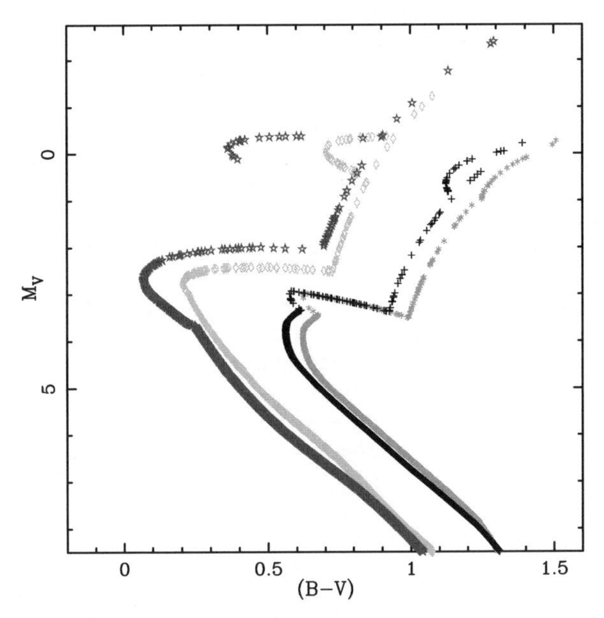

Fig. 10.2. Colour-magnitude diagram showing N-body *isochrones* at 4 000 Myr for simulations of different metallicity. Shown are stars with $Z = 0.03$ (* *symbols*), $Z = 0.02$ (+ *symbols*), $Z = 0.001$ (*diamond symbols*) and $Z = 0.0001$ (*star symbols*). Data are from `NBODY4` simulations begun with 30 000 single stars. The simulations are described in Hurley et al. (2004). Only stars with `kstar` ≤ 6 are shown

References

Aarseth S. J., 2003, Gravitational N-Body Simulations. Cambridge University Press, Cambridge
Baumgardt H., Makino J., 2003, MNRAS, 340, 227
Belczynski K., Kalogera V., Bulik T., 2002, ApJ, 572, 407
Bergeron P., Wesemael F., Beauchamp A., 1995, PASP, 107, 1047
Carson J. E., Cool A. M., Grindlay J. E., 2000, ApJ, 532, 461
Chernoff D. F., Weinberg M. D., 1990, ApJ, 351, 121
Dewi J. D. M., Pols O. R., Savonije G. J., van den Heuvel E. P. J., 2002, MNRAS, 331, 1027
Eggleton P. P., Fitchett M., Tout C. A., 1989, ApJ, 347, 998
Ferraro F. R., Paltrinieri B., Fusi Pecci F., Cacciari C., Dorman B., Rood R. T., 1997, ApJ, 484, L145
Grindlay J. E., Heinke C., Edmonds P. D., Murray S. S., 2001a, Science, 292, 2290
Grindlay J. E., Heinke C. O., Edmonds P. D., Murray S. S., Cool A. M., 2001b, ApJ, 563, L53
Hurley J. R., Pols O. R., Tout C. A., 2000, MNRAS, 315, 543
Hurley J. R., Shara M. M., 2003, ApJ, 589, 179
Hurley J. R., Tout C. A., Aarseth S. J., Pols O. R., 2001, MNRAS, 323, 630
Hurley J. R., Tout C. A., Aarseth S. J., Pols O. R., 2004, MNRAS, 355, 1207

Hurley J. R., Tout C. A., Pols O. R., 2002, MNRAS, 329, 897
Kroupa P., Tout C. A., Gilmore G., 1993, MNRAS, 262, 545
Kudritzki R. P., Reimers D., 1978, A&A, 70, 227
Kurucz R. L., 1992, in Barbuy B., Renzini A., eds, Proc. IAU Symp. 149, The Stellar Populations of Galaxies. Kluwer, Dordrecht, p. 225
Mackey A. D., Gilmore G. F., 2003, MNRAS, 338, 85
Meylan G., Heggie D. C., 1997, A&ARv, 8, 1
Paresce F., Meylan G., Shara M., Baxter D., Greenfield P., 1991, Nature, 352, 297
Pols O. R., Schröder K. -P., Hurley J. R., Tout C. A., Eggleton P. P., 1998, MNRAS, 298, 525
Tout C. A., Aarseth S. J., Pols O. R., Eggleton P. P., 1997, MNRAS, 291, 732
Tout C. A., Pols O. R., Eggleton P. P., Han Z., 1996, MNRAS, 281, 257
Wielen R., 1968, Bull. Astron., 3, 127
Yanny B., Guhathakurta P., Schneider D. P., Bahcall J. N., 1994, AJ, 435, L59

11

Binary Stars

Christopher A. Tout

University of Cambridge, Institute of Astronomy, Madingley Road, Cambridge
CB3 0HA, England
cat@ast.cam.ac.uk

In clusters there are both primordial binary stars and binaries created by dynamical interactions. Occasionally, a new binary system can be formed (Fabian, Pringle & Rees 1975) but more often new systems are the result of exchanges. In Chap. 12 Hurley describes an algorithm for including the interaction of the components of a binary star in N-body simulations. In this chapter we investigate the underlying physics and note that, though we have a good qualitative idea of what goes on, there is still much to be determined fully quantitatively.

Double stars have been known since ancient times and were referred to in written records as early as Ptolemy. But the concept of a binary star as a gravitationally bound entity did not exist before the late eighteenth century. The Revd. John Michell (Michell 1767) showed statistically that not all double stars could be chance superpositions on the sky. He concluded that "[Double Stars] were brought together by their mutual gravitation, or some other law or by the appointment of the Creator." This statement sums up well our understanding of the formation of binary stars, the physics of which still eludes us. At the time Herschel disagreed with Michell's deductions because he wanted to use stars as standard candles to map the structure of the Milky Way. If the two very different components of a double star were actually at the same distance, such a mapping would be impossible. He eventually acquiesced and himself introduced the term "binary star" in 1803 (Herschel 1803).

Binary stars are common and consequently perhaps represent the normal formation mode. The ratio of single (or unresolved) systems to binary to triple or higher multiple systems is two to five to two to one. As a requirement for the dynamical stability of a system, higher multiples must be hierarchical and can be considered as a sequence of binary stars within binary stars. For instance, a quadruple system (Fig. 11.1) can take essentially two forms. Either they are two pairs of stars both orbiting one another or a very close binary in a wider orbit with a third star and then this triple system in a yet wider orbit with the fourth star. Typically, the separations of nested pairs must be a factor of four or more smaller for long-term stability. Though no multiple

Tout, C. A.: *Binary Stars*. Lect. Notes Phys. **760**, 297–319 (2008)
DOI 10.1007/978-1-4020-8431-7_11

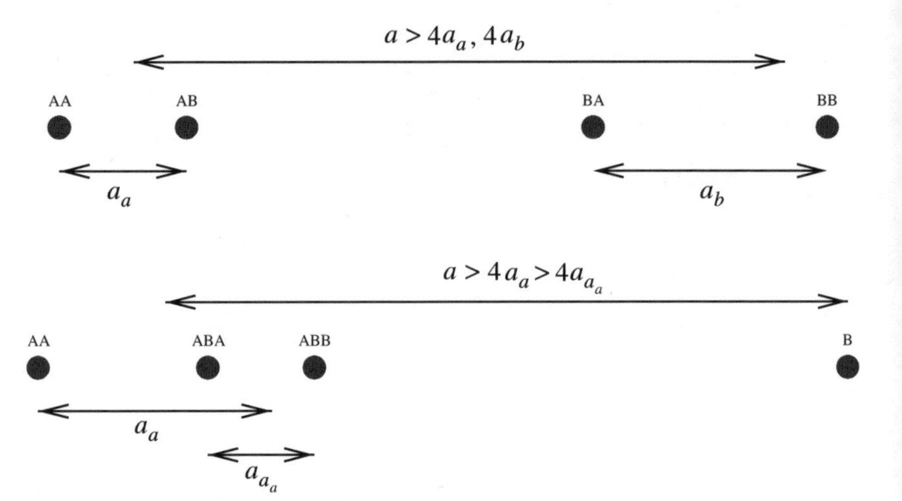

Fig. 11.1. Two possible configurations of quadruple systems with long lifetimes. The convention of labelling the stars in a binary as A and B is extended through the hierarchy. Separations of binary systems within the hierarchy must be typically a factor four or more larger when moving from one level up to the next for long-term survival

system is indefinitely stable, many can be expected to survive the current age of the Universe (Chap. 3).

11.1 Orbits

The orbits of binary stars (Fig. 11.2) obey a form of Kepler's laws generalised to the case where both stars have similar masses. First, the orbits are conic sections and bound orbits are ellipses. The diagram shows the semi-major axis a, the semi-minor axis b and the semi-latus rectum l. These are related to the eccentricity e by

$$l = a(1 - e^2) \tag{11.1}$$

and

$$e^2 = 1 - \frac{b^2}{a^2}. \tag{11.2}$$

A general point on the ellipse is given parametrically by

$$r = \frac{l}{1 + e\cos\theta}, \tag{11.3}$$

where r is the distance from the primary focus F and θ is the angle from the semi-major axis to the line joining the F to P. Secondly the line connecting the two bodies sweeps out equal areas in equal times. If one body is considered fixed at F while the other orbits at P, this is equivalent to

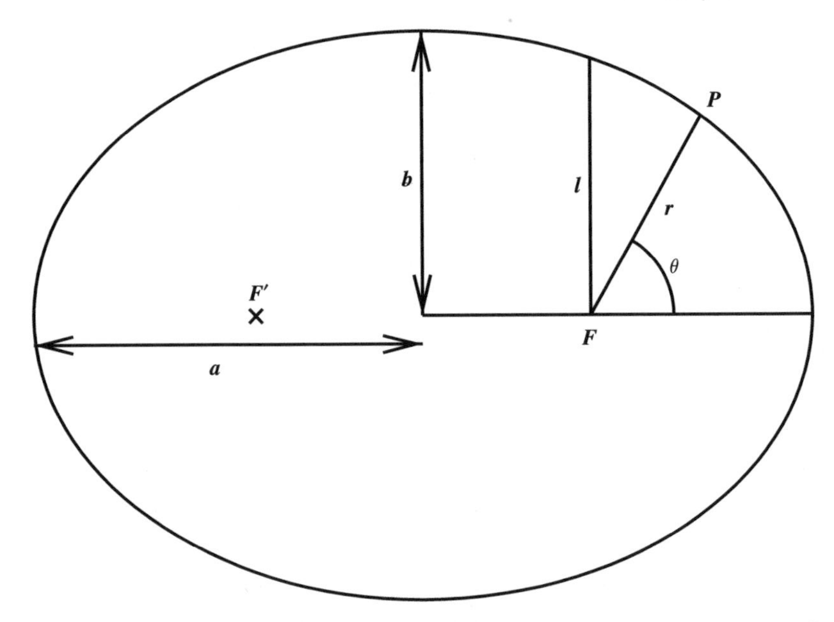

Fig. 11.2. Stars in a bound binary follow elliptical orbits. One star is at the focus F and the other orbits at P around the ellipse

$$\frac{1}{2}r^2\dot{\theta} = \frac{\pi a^2(1-e^2)^{1/2}}{P}, \tag{11.4}$$

where the numerator is the area of the ellipse and the denominator P is the period of the binary, the time taken for a complete orbit. This follows from the conservation of angular momentum. Third the period and separation are related by

$$\left(\frac{P}{2\pi}\right)^2 = \frac{a^3}{G(M_1+M_2)}, \tag{11.5}$$

where G is Newton's gravitational constant and M_1 and M_2 are the masses of the two stars.

Each of these laws is a consequence of Newton's laws of motion and his law of gravity. Both stars orbit the centre of mass in ellipses and both feel a centrally directed force so angular momentum is conserved. Again with r the instantaneous separation, we have

$$r^2\dot{\theta} = h = \frac{MJ}{M_1M_2} = \text{const}, \tag{11.6}$$

where $M = M_1 + M_2$ is the total mass, J is the total angular momentum of the system and h is the specific angular momentum per unit reduced mass. Solving the equations of motion we find that

$$l = \frac{h^2}{GM} \tag{11.7}$$

so that conservation of angular momentum fixes the semi-latus rectum of the orbit. Similarly we find the total energy, kinetic plus potential, to be

$$E = -\frac{GM_1M_2}{2a} \tag{11.8}$$

so that the energy determines the semi-major axis and thence the period of the system.

11.2 Tides

Though angular momentum can be lost in stellar winds and gravitational radiation, let us first consider the case when the total orbital angular momentum is conserved. Because the stars are luminous they can radiate orbital energy if it is converted to heat by tides or any other process. We may write the energy in terms of the angular momentum and eccentricity as

$$E = -\frac{GM_1M_2}{2h^2}GM(1-e^2) \tag{11.9}$$

from which we can see that

$$\left(\frac{\partial E}{\partial e}\right)_J \propto 2e \quad \text{and} \quad \left(\frac{\partial^2 E}{\partial e^2}\right)_J > 0 \quad \text{at} \quad e = 0. \tag{11.10}$$

Thus a circular orbit is the most stable configuration for a given angular momentum.

11.2.1 Tidal Forces

So far we have considered both stars as point masses. This is a good approximation when they are well separated but when they are closer the finite size of the stars becomes important and tidal interactions and eventually mass transfer occur between the two. Let us assume that star 2 is sufficiently small to still be considered a point mass and let star 1 have a radius R (Fig. 11.3). The potential at a point P, a distance r from the centre of star 1 along a line at an angle θ to the line joining the centres of the two stars and a distance r' from star 2, owing to star 2, can be expanded as

$$\Phi_2 = -\frac{GM_2}{r'} = \frac{-GM_2}{\sqrt{a^2+r^2-2ar\cos\theta}} = -\frac{GM_2}{a}\sum_{n=0}^{\infty}\left(\frac{r}{a}\right)^n P_n(\cos\theta), \tag{11.11}$$

where P_n is the nth Legendre polynomial. The force on material in star 1 is $-\nabla\Phi_2$. The $n=1$ term balances the overall orbital motion. Of most interest for the evolution of the system is the $n=2$ term because it is the largest that leads to both transfer of angular momentum between star 1 and the

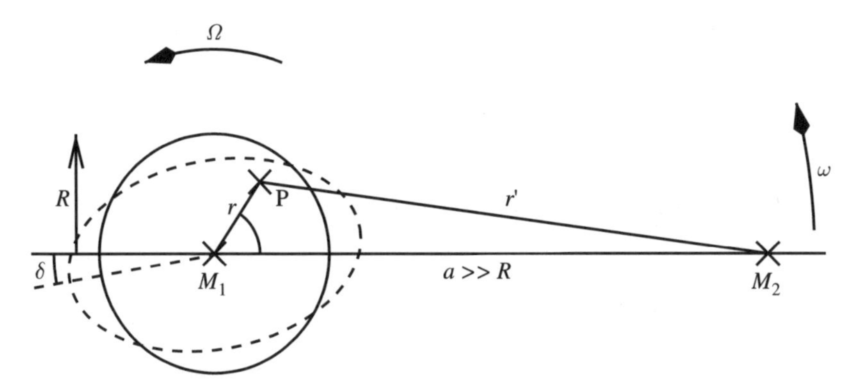

Fig. 11.3. The tidal potential of star 2 distorts star 1. If, as here, the star is spinning faster than the orbit ($\Omega > \omega$), viscosity drags the tidal bulges ahead of the orbit and dissipates energy. The force between star 2 and the two bulges provides a torque that transfers angular momentum from star 1 to the orbit

orbit and dissipation of energy. Star 1 is distorted as illustrated by the dashed curve in Fig. 11.3. If the star is not rotating synchronously with the orbit, the distortion is dragged around it. If the star is spinning more slowly, viscosity leads to a lag, of angle $-\delta$, so that the tidal bulges lag behind the line joining the stars. If the star is spinning faster than the orbit, the companion lags behind the bulges. The gravitational force between the two bulges and star 2 provides a torque that tends to synchronize the stellar spin and the orbit. At the same time, energy dissipation circularises the orbit. Tides also align the spin axes with the orbital axis (Hut 1981).

The synchronous state is not always stable (Hut 1980). Transfer of angular momentum from the orbit to a star increases both the spin of the star and the orbital angular velocity because the orbital angular momentum

$$J_{\mathrm{orb}} \propto a^2\omega \propto \omega^{-1/3}, \tag{11.12}$$

in a circular orbit with angular velocity ω. If there is insufficient total angular momentum in the system, the stars end up spiralling together. This is the expected fate of contact binary stars and some planetary systems though the process can take a very long time (Rasio, Tout & Livio 1996).

For a typical system in which the extended star, star 1, is convective, with mass ratio $q = M_1/M_2$, separation a and radius of the largest star (star 1 here) R, the circularisation time

$$\tau_{\mathrm{circ}} \approx \frac{2q^2}{1+q}\left(\frac{a}{R}\right)^8 \ \mathrm{yr}. \tag{11.13}$$

We shall see in the next section that much more drastic interaction begins when $R = R_{\mathrm{L}} \approx \frac{1}{3}a$. At this point $\tau_{\mathrm{circ}} \approx 2{,}000\,\mathrm{yr}$. Even when $R = \frac{1}{2}R_{\mathrm{L}}$, $\tau_{\mathrm{circ}} \approx 6 \times 10^5\,\mathrm{yr}$ which is still much less than the nuclear timescale for stellar evolution that ranges from 10^{10} yr for a $1\,M_\odot$ star on the main sequence to 10^6 yr for a massive giant. Synchronization times are even shorter,

$$\tau_{\rm sync} \approx q^2 \left(\frac{a}{R}\right)^6 \,\text{yr} \tag{11.14}$$

or 300 yr for $R \approx R_{\rm L}$ and 2×10^4 yr for $R \approx \frac{1}{2} R_{\rm L}$.

11.3 Mass Transfer

When the two stars are very close and $R \approx a$, we can no longer ignore the higher terms in the expansion of the tidal potential. We shall begin the analysis again and make use of the fact that by the time the radius of either star gets large enough, tides will have already circularised the orbit and synchronized the spin of the star. We can therefore work in a frame rotating at $\boldsymbol{\Omega}$ as illustrated in Fig. 11.4. Let all the material be stationary except for a test particle at P. Then in an inertial frame the velocity of P is

$$\boldsymbol{v} = \dot{\boldsymbol{r}} + \boldsymbol{\Omega} \times \boldsymbol{r} \tag{11.15}$$

and its acceleration is

$$\boldsymbol{a} = \ddot{\boldsymbol{r}} + 2\boldsymbol{\Omega} \times \dot{\boldsymbol{r}} + \boldsymbol{\Omega} \times (\boldsymbol{\Omega} \times \boldsymbol{r}), \tag{11.16}$$

where the first term may be familiar as the Coriolis force and the second as the centrifugal. We can then apply the Euler momentum equation in the inertial frame

$$\rho \boldsymbol{a} = -\nabla P - \rho \nabla \phi_{\rm G}, \tag{11.17}$$

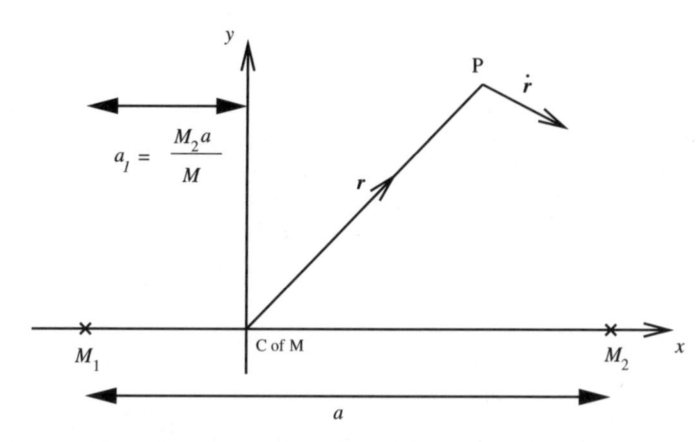

Fig. 11.4. Coordinates rotating with the binary system centred on its centre of mass with the z-axis perpendicular to the orbital plane

where ρ is the density, P is the pressure,

$$\nabla^2 \phi_{\rm G} = 4\pi G \rho \tag{11.18}$$

and $\phi_{\rm G}$ is the gravitational potential. In corotation $\dot{\boldsymbol{r}}$ and $\ddot{\boldsymbol{r}}$ vanish and aligning the z-axis with $\boldsymbol{\Omega}$ we may write

$$\boldsymbol{\Omega} \times (\boldsymbol{\Omega} \times \boldsymbol{r}) = -\nabla \phi_\Omega, \tag{11.19}$$

with

$$\phi_\Omega = -\frac{1}{2}\Omega^2 s^2, \tag{11.20}$$

where s is the distance from the z-axis. Thus the Euler equation reduces to

$$\frac{1}{\rho}\nabla P + \nabla \Phi = \boldsymbol{0}, \tag{11.21}$$

with $\Phi = \phi_{\rm G} + \phi_\Omega$. So surfaces of constant pressure are surfaces of constant Φ. In particular the surface of the star, if defined as $P = 0$, is a surface of constant Φ. Stars are centrally condensed, so to a good approximation $\phi_{\rm G}$ is just the gravitational potential of two point masses at the centres of the stars and, in Cartesian coordinates with star 1 at the origin and star 2 at $(a, 0, 0)$, we find

$$\Phi = \frac{-GM_1}{\sqrt{x^2 + y^2 + z^2}} + \frac{-GM_2}{\sqrt{(x-a)^2 + y^2 + z^2}} - \frac{1}{2}\frac{GM}{a^3}\left[\left(x - \frac{a}{1+q}\right)^2 + y^2\right], \tag{11.22}$$

which is just a function of the mass ratio $q = M_1/M_2$, GM and a. Moreover, if we scale all lengths by the separation $x \to x/a$, the shape of the equipotential surfaces is a function of q only. We plot them for $q = 2$ in Fig. 11.5. Corotating material in hydrostatic equilibrium fills up to an equipotential surface. Thus when the radii are small compared to a, the surface equipotentials are spheres. Far from the binary surfaces are again spheres. Of interest to us are the two innermost critical surfaces on which the lines meet at stationary Lagrangian points. Moving outwards from the centres of the stars the first, meeting at the inner Lagrangian point L_1, determines when material is more attracted to its companion than to the star itself. The second opens to the right at the L_2 point and determines the maximum size of a joint star, or contact binary, around the two orbiting masses. The three other stationary points are also shown but are not of interest to us now because, beyond the surface through the L_2 point, there is nothing to keep the material corotating and (3.1) is no longer valid.

Figure 11.6 shows the value of the potential along the x-axis and illustrates how stars fill their equipotential surfaces to form three different classes of binary star. In a wide binary system both stars have radii small compared to the separation and the system is said to be detached. As either star grows,

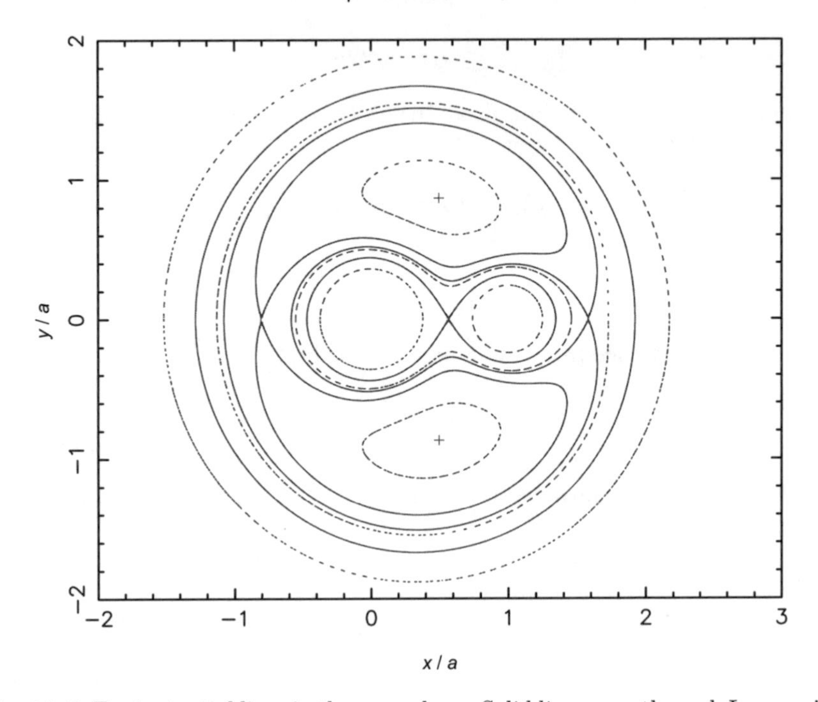

Fig. 11.5. Equipotential lines in the $x-y$ plane. *Solid lines* pass through Lagrangian points where $\nabla\Phi = 0$

it is gradually distorted until it fills the critical potential surface that crosses at the inner Lagrangian L_1 point between the two stars. This equipotential around the star is its Roche lobe. If the star grows any larger, material at L_1 is more attracted to its companion than to itself and the material can flow from it to the other star. This is known as Roche lobe overflow and the system is said to be semi-detached. Algols (Sect. 11.5.1) and cataclysmic variable stars (Sect. 11.5.3) are in this state. If the second star expands so that it too would overfill its Roche lobe, the two stars can exist in equilibrium in contact. Such systems appear to be common but do not last long. Material and heat are transferred between the two until the mass ratio becomes large and tidal instability shrinks the orbit and merges the two stars.

Even the surface through the L_1 point is almost spherical. When the mass ratio $q = 1$, the difference in extent between the x and z directions is only 5% of the diameter and this rises to only 10% when $q = 10$. We define the Roche lobe radius R_L to be the radius of a sphere with the same volume as the Roche lobe

$$V_L = \frac{4}{3} R_L^3. \tag{11.23}$$

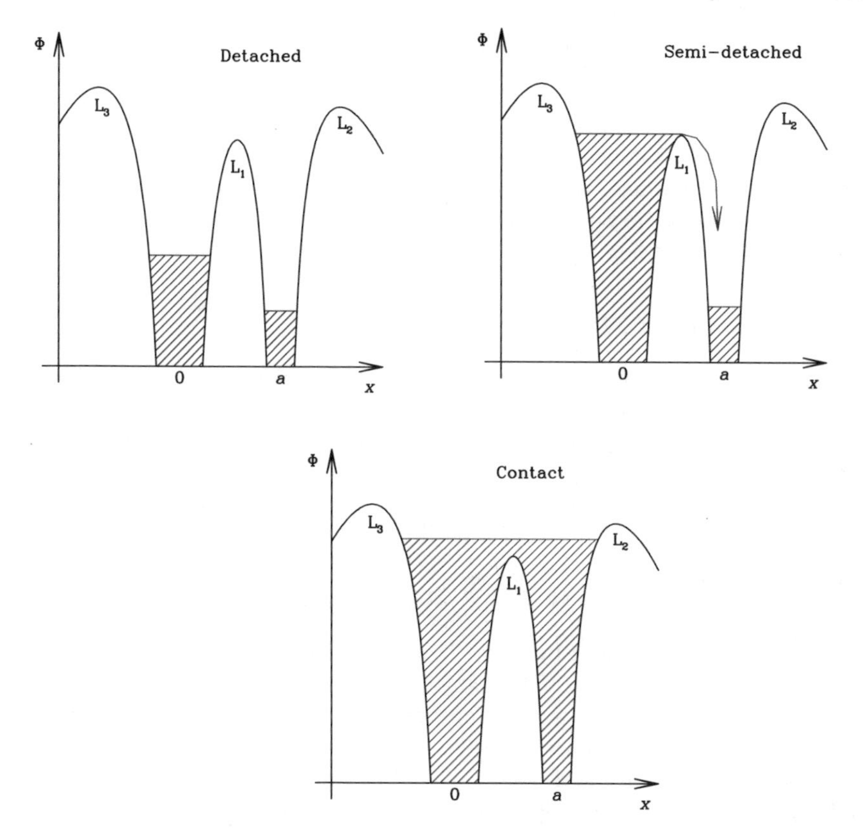

Fig. 11.6. The potential along the x-axis in Fig. 11.5. Three binary star configurations are shown

The volume can be evaluated numerically and various simple fits to $R_{\rm L}$ have been deduced. Eggleton (1983) fitted the Roche lobe radius of star 1 by

$$\frac{R_{\rm L}}{a} = \frac{0.49q^{2/3}}{0.6q^{2/3} + \log_e(1 + q^{1/3})}. \tag{11.24}$$

This is accurate to better than 1% over the whole range $0 < q < \infty$. It is the preferred form for numerical work but for analytic work a formula deduced by Paczyński (1971),

$$\frac{R_{\rm L}}{a} = 0.462 \left(\frac{M_1}{M}\right)^{\frac{1}{3}}, \tag{11.25}$$

which is accurate to better than 3% for $0 < q < 0.8$, is much more useful.

The rate of flow through the L_1 point is a rapidly rising function of the amount by which the star overfills its Roche lobe $\Delta R = R - R_{\rm L}$. So, as long as the rate at which the star expands, or the Roche lobe shrinks, is long compared with the dynamical timescale, on which hydrostatic equilibrium is regained, we can expect the mass transfer rate to adjust to maintain

$$R \approx R_{\rm L} \quad \text{and} \quad \dot{R} \approx \dot{R}_{\rm L}. \tag{11.26}$$

If this timescale is much less, we can expect ΔR and consequently $\dot{M}$ to increase on a dynamical timescale. We consider the consequences of such unstable mass transfer in Sect. 11.5.4 but first we examine under what conditions mass transfer is stable.

11.3.1 Stability of Mass Transfer

To examine the stability of mass transfer we follow Webbink (1985) and define three derivatives of radii with respect to the mass of the lobe-filling star. The first is the rate of change of the Roche lobe radius $R_{\rm L}$ for conservative mass transfer in which the angular momentum of the system J and the total mass M are conserved. Any material lost by star 1 is accreted by star 2 so that

$$\zeta_{\rm L} = \left(\frac{\partial \log R_{\rm L_1}}{\partial \log M_1}\right)_{M,J}. \tag{11.27}$$

This can be approximated by $\zeta_{\rm L} = 2.13q - 1.67$ (Eggleton 2006) and we see that it is positive for $M_1 > 0.78M_2$ so that, in this case, the Roche lobe shrinks in response to mass transfer from star 1 to star 2 and otherwise it expands. The initial response of the star to mass loss is adiabatic as it regains hydrostatic equilibrium and loses thermal equilibrium in the process. So we define a second derivate at constant entropy s and composition of each isotope X_i throughout the star

$$\zeta_{\rm ad} = \left(\frac{\partial \log R_1}{\partial \log M_1}\right)_{s,X_i}. \tag{11.28}$$

For stars with radiative envelopes $\zeta_{\rm ad} > 0$ so they shrink on mass loss, while for stars with convective envelopes $\zeta_{\rm ad} < 0$ and they expand on mass loss. On a thermal timescale the star regains full equilibrium at its new mass but still with constant composition. A third derivative,

$$\zeta_{\rm eq} = \left(\frac{\partial \log R_1}{\partial \log M_1}\right)_{X_i}, \tag{11.29}$$

describes the rate of change of radius with mass in equilibrium. For main-sequence stars $\zeta_{\rm eq} > 0$ typically, while for red giants and stars crossing the Hertzsprung gap $\zeta_{\rm eq} < 0$.

The rate at which mass transfer proceeds depends on the relative values of these derivatives. If $\zeta_{\rm L} > \zeta_{\rm ad}$, then the Roche lobe shrinks faster than the radius of the star in direct response to mass transfer. So ΔR increases and consequently $\dot{M}$ increases rapidly. There is positive feedback and the mass transfer is unstable,

$$\left|\frac{M_1}{\dot{M}_1}\right| \rightarrow \tau_{\rm dyn} \approx 10 - 100\,{\rm yr} \tag{11.30}$$

and mass transfer proceeds on a dynamical timescale. Star 2 often cannot accrete the material at such a high rate. Instead, it expands itself and the transferred material ends up in a common envelope around the two stars. We shall discuss this in detail in Sect. 11.5.4. This is typically the outcome when a giant fills its Roche lobe when in orbit with a less massive companion because the giant expands while its Roche lobe is shrinking. Positive feedback drives the mass transfer up to the dynamical rate.

If $\zeta_{\rm L} < \zeta_{\rm ad}$ but $\zeta_{\rm L} > \zeta_{\rm eq}$ then the star shrinks in its immediate response to mass transfer but then expands on its thermal timescale $\tau_{\rm th}$ and

$$\left|\frac{M_1}{\dot{M}_1}\right| \rightarrow \tau_{\rm th} \approx 10^5 - 10^6\,{\rm yr}. \tag{11.31}$$

Mass transfer proceeds on a thermal timescale. This is the case when a subgiant in the Hertzsprung gap with a radiative or thin convective envelope fills its Roche lobe.

If both $\zeta_{\rm ad} > \zeta_{\rm L}$ and $\zeta_{\rm eq} > \zeta_{\rm L}$, the star shrinks in response to mass transfer and does not expand again to fill its Roche lobe until driven to, either by its own nuclear evolution or until some angular momentum loss mechanism causes the orbit to shrink sufficiently. Either

$$\left|\frac{M_1}{\dot{M}_1}\right| \rightarrow \tau_{\rm nuc} \approx 10^7 - 10^9\,{\rm yr}, \tag{11.32}$$

the case for main-sequence stars or red giants in present-day Algols (see Sect. 11.5.1), or

$$\left|\frac{M_1}{\dot{M}_1}\right| \rightarrow \tau_{\rm J}, \tag{11.33}$$

the timescale on which angular momentum is lost from the system. This is the case for cataclysmic variables that form the subject of Sect. 11.5.3.

11.4 Period Evolution

When the angular momentum of the component stars is negligible compared to that of their orbit, we can derive simple formulae for how the orbit evolves with mass loss and mass transfer. We allow a wind from star 1 that escapes from the system and mass transfer from star 1 to star 2 so that $-\dot{M}_1$ is the mass loss rate from star 1, $\dot{M}_2$ is the rate of accretion by star 2, the mass transfer rate, and $-\dot{M}$ is the rate of mass loss from the system, the wind from star 1. Then

$$-\dot{M}_1 = -\dot{M} + \dot{M}_2, \tag{11.34}$$

with $\dot{M}$ and $\dot{M}_1 \leq 0$ and $\dot{M}_2 \geq 0$. The wind from star 1 carries off angular momentum intrinsic to the orbit of the star so that the rate of change of angular momentum of the orbit is

$$\dot{J} = \dot{M} a_1^2 \Omega. \tag{11.35}$$

We recall that

$$J = \frac{M_1 M_2}{M} a^2 \Omega \tag{11.36}$$

so that we can differentiate $\log J$ to find

$$\frac{\dot{J}}{J} = \frac{\dot{M}_1}{M_1} + \frac{\dot{M}_2}{M_2} - \frac{\dot{M}}{M} + 2\frac{\dot{a}}{a} + \frac{\dot{\Omega}}{\Omega} = \frac{M}{M_1 M_2}\left(\frac{M_2}{M}\right)^2 \dot{M} = \frac{M_2}{M_1}\frac{\dot{M}}{M}, \tag{11.37}$$

from (11.35). Differentiating Kepler's third law we find

$$2\frac{\dot{P}}{P} = -2\frac{\dot{\Omega}}{\Omega} = 3\frac{\dot{a}}{a} - \frac{\dot{M}}{M} \tag{11.38}$$

and combining these gives us

$$\frac{M_2}{M_1}\frac{\dot{M}}{M} = \frac{\dot{M}_1}{M_1} + \frac{\dot{M}_2}{M_2} - \frac{1}{3}\frac{\dot{M}}{M} + \frac{1}{3}\frac{\dot{P}}{P}. \tag{11.39}$$

When there is no mass transfer but mass loss in a wind, $\dot{M}_2 = 0$ and $\dot{M}_1 = \dot{M}$ so that

$$\frac{\dot{P}}{P} = -2\frac{\dot{M}}{M}. \tag{11.40}$$

We can integrate this to give $P^2 M = \text{const}$ or, with (11.38), $aM = \text{const}$. The period and separation increase as mass is lost. Indeed, as the Sun loses mass, so the planets of the solar system will drift further away from it.

When there is mass transfer but no mass lost from the system $\dot{M} = 0$ and $\dot{J} = 0$ so that

$$\frac{\dot{P}}{P} = -3\frac{\dot{M}_1}{M_1} - 3\frac{\dot{M}_2}{M_2}. \tag{11.41}$$

This can be integrated to give $P(M_1 M_2)^3 = \text{const}$ or $a(M_1 M_2)^2 = \text{const}$. The period and separation decrease while mass is transferred from the more massive to the less massive component, reach minima when the masses are equal and then increase as mass is transferred from the less massive to the more massive component.

11.5 Actual Types

We have described the basic physics of binary stars and their interactions. Coupling this with stellar evolution leads to a veritable zoo of different types of binary star, as described by Eggleton (1985). Observations do overlap with what we expect but often require the introduction of new physical processes such as common envelope evolution (Sect. 11.5.4) that are not fully understood. We shall illustrate with just three examples. The Algols as the prototypes, the cataclysmic variables as those studied in most detail and the type Ia supernovae that have recently been used as standard candles to measure the structure and evolution of the Universe.

11.5.1 Algols

As one of the brightest stars in the northern hemisphere, Algol or β Persei has been known for a long time. It is an eclipsing SB2 and so yields a great deal of information about its current state. Its variability was first definitely recorded by Montanari (1671) in Bologna but the name Algol suggests that it may have been recognised much earlier. Algol is derived from the Arabian *Al Ghūl*, which has been variously translated as demon or changing spirit (Kopal 1959). However, Allen (1899) felt it is more likely that the name is derived from Ptolemy who referred to it as the brightest star in the Gorgon's head, a constellation recognised by the Greeks at the time and indeed generally until quite recently (Goodricke 1783). The Hebrews called it *Rōsh-ha-Satan* or Satan's head and the Chinese *Tseih She* or the piled up corpses. Whether these names reflect the variability or not must be left to our imaginations because no actual record has been found.

Its eclipses were not noted for over a century until John Goodricke (1783) sent a short letter to the Royal Society describing how he had spotted a periodicity in the light variations of Algol. He and his friend, Edward Pigott, had by then already obtained a fairly accurate estimate of the period of 2 days and 21 h. Goodricke in a short paragraph at the end of his letter went on to suggest that the cause of the variation might be either a dark object orbiting and eclipsing the star or a dark spot on its surface. Confirmation of his first hypothesis did not come for yet another century when Vogel (1890) observed radial velocity shifts in the spectrum of Algol and found the positions of minimum light to correspond to the conjunctions of the eclipse model.

Observations improved with time giving better photometric and spectroscopic measurements of Algol and a number of similar systems. It seems that it had been apparent that something was not quite right with Algol for some time before Hoyle (1955) recorded what he described as the Algol Paradox. From the shapes of the eclipses it was clear that the fainter star was larger. Such a situation was thought not to be possible according to the theory of stellar evolution. If both stars were on the main sequence then the brighter would be larger. In fact, the fainter could only be larger if it had evolved off the main sequence and indeed Parenago (1950) had already claimed that the fainter components of Algols were in many cases sub-giants. Hoyle argued that, although it would be possible to pick the two stars from the H-R diagram, one on the main sequence and the other a much older sub-giant, all reasonable theories of the formation of binary stars suggested that the two components would have formed at the same time and would be of the same age now. Thus he had identified the paradox without the need to introduce the masses of the stars directly and went on to explain it successfully in terms of the initially brighter star evolving to such a size that its fainter companion gobbled up matter from its surface. This companion could then move up the main sequence and become the brighter of the two. In clusters such stars could later appear as blue stragglers (Sandage 1953).

At the same time, Crawford (1955) was also solving the same paradox though more specifically in terms of the limitations placed on the mass ratios by the spectroscopically determined mass functions and the assumption that the brighter component does in fact lie on the main sequence. Struve (1948) had already pointed out that these mass functions are low. Crawford also introduced the concept of the giant filling its Roche lobe. In fact, Walter (1931) had pointed out that the cool stars in Algols are close to the limit of dynamical stability but this had gone largely unnoticed.

This semi-detached nature of Algols provided mutual support for the hypothesis formulated by Struve (1949) that the existence of gaseous streams between the two stars in Algols could account for an asymmetry in the radial velocity curve. Although the photometric light curve of U Cephei showed symmetric eclipses, the radial velocity curve is asymmetric. Struve explained this in terms of the spectrum of a gaseous stream, moving faster than the two stars, superimposed on the symmetric curve of the star. Evidence had also been provided by Wood (1950) who had found that binaries with period fluctuations almost always have one star filling its Roche lobe.

With a fairly definite theory and the dawn of numerical stellar evolution, the stage was set for the construction of theoretical models of these semi-detached systems. The first step was taken by Morton (1960) who, concentrating on the initially more massive star, examined the process of mass transfer. He pointed out that since all observed Algols have the sub-giant component already less massive, the initial rate of mass transfer must have been much faster than that taking place now. It must have been sufficiently fast to make it unusual to observe a system in a state where the primary is still the more massive.

11.5.2 Critical Mass Ratio

A simple calculation reveals why. Let the mass-losing giant be star 1. Its radius

$$R_1 \approx f(L) M_1^{-0.27}, \tag{11.42}$$

where f is a function of its luminosty L, which does not vary much with mass loss. The fully convective giant envelope is isentropic so that $\zeta_{\rm ad} \approx \zeta_{\rm eq}$ and, for timescales short compared with the nuclear evolution timescale on which L varies,

$$\frac{\dot{R}_1}{R_1} = -0.27 \frac{\dot{M}_1}{M_1}. \tag{11.43}$$

For stable mass transfer we must have negative feedback

$$\dot{R} < \dot{R}_{\mathrm{L}} \qquad \text{when} \qquad R_1 = R_{\mathrm{L}}, \tag{11.44}$$

because otherwise the process of mass transfer would mean that the star overfills its Roche lobe even more and the rate of overflow would increase.

We can differentiate formula (11.25) which, recall, is valid for $q < 0.8$, to find

$$\frac{\dot{R}_{\mathrm{L}}}{R_{\mathrm{L}}} = \frac{1}{3}\frac{\dot{M}_1}{M_1} - \frac{1}{3}\frac{\dot{M}}{M} + \frac{\dot{a}}{a}. \tag{11.45}$$

Then, assuming conservative mass transfer ($\dot{M} = 0$ and $\dot{J} = 0$), we require

$$-0.27\frac{\dot{M}_1}{M_1} < \left(\frac{1}{3M_1} - \frac{2(M_2 - M_1)}{M_1 M_2}\right)\dot{M}_1. \tag{11.46}$$

But $\dot{M}_1 < 0$ so

$$M_1 < 0.7 M_2 \qquad \text{or} \qquad q < q_{\mathrm{crit}} = 0.7. \tag{11.47}$$

Over the decade following Morton's work, detailed models were made by many independent workers: Paczyński (1966), Kippenhahn & Wiegert (1967) and Plavec et al. (1968) all confirmed Morton's results. Kippenhahn and Weigert introduced the nomenclature of case A to indicate mass transfer before the exhaustion of central hydrogen burning and case B for mass transfer afterwards, when the star has evolved off the main sequence. In all of these models conservative mass transfer (all the matter lost by the primary being accreted by the secondary) was assumed but Paczyński & Ziółkowski (1967) showed that the resulting Algol systems are more realistic if half the mass lost by the primary is actually lost from the system. In order to avoid dynamical mass transfer, all Algols must have begun mass transfer before the most massive star has evolved on to the giant branch unless it has suffered sufficient mass loss that $q < q_{\mathrm{crit}} \approx 0.7$ and the Roche lobe expands faster than the star (Tout & Eggleton 1988).

11.5.3 Cataclysmic Variables

Cataclysmic variables are very close binary stars in which the primary component is a white dwarf, which is accreting material transferred from its Roche-lobe-filling companion. Figure 11.7 illustrates the basic components. The companion to the white dwarf is always less massive, often substantially, and is typically a low-mass main-sequence star for which the Roche-filling state dictates an orbital period of a few hours and a separation of about a solar radius. In a very few systems the secondary star can be slightly evolved. For example GK Per, the widest system classified as a cataclysmic variable, has an orbital period of 47 h and its white dwarf has a subgiant companion.

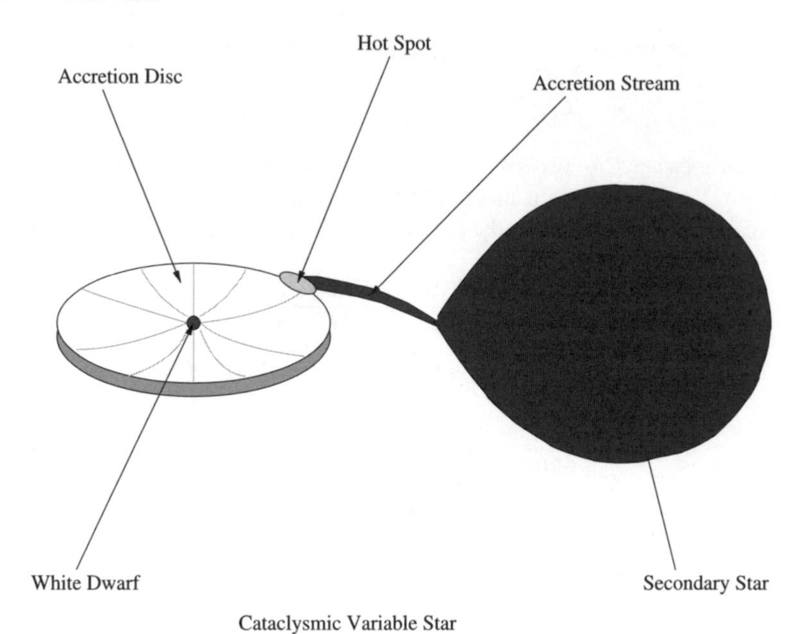

Fig. 11.7. A schematic diagram of a cataclysmic variable with the major observable components marked. According to general practice, the accreting white dwarf is star 1 and the Roche lobe filling companion is star 2

The nuclear, or in some cases mass-loss, timescales of evolved companions can be relatively short and their nature is therefore fundamentally different from those systems with unevolved low-mass secondaries. Most importantly the mass transfer rates are higher. These systems, particularly those with very large red or supergiant secondaries, are classified as symbiotic stars. At the other extreme, the companion can be another white dwarf of lower mass than the primary. AM CVn is the prototype of this class of cataclysmic variables and has a period of 89 min.

In addition to the two stars a third component, an accretion disc, is important and often dominates the light from the cataclysmic variable. It is formed because the material overflowing from the companion at the inner Lagrangian point L_1 has too much angular momentum to fall directly on to the white dwarf. Viscous dissipation allows the slow infall of the majority of the matter through the disc while angular momentum is carried outwards until it can be tidally returned to the orbit. Many cataclysmic variables are observationally very clean systems in which the light variations and spectra of each of the three main components can be separated out. Often the signature of the high-velocity accretion stream and the hot spot where it impacts the edge of the disc can also be identified. An excellent, detailed and very readable review of the observations from early times forms a substantial part of the book by Warner (1995) to which the interested reader is encouraged to turn.

Two instabilities gave cataclysmic variables their name and were responsible for their early observation. The first is the classical nova. Hydrogen-rich material transferred to the white dwarf from its companion builds up in a degenerate layer on the surface. When the base of this layer becomes dense enough and hot enough, the hydrogen ignites in a thermal nuclear runaway that leads to a large increase in brightness and probably the ejection of most of the accreted material. The second is an instability in the accretion disc. Under some conditions material can accumulate in the disc and fall through in bursts. The quasiperiodic increase in brightness of the disc makes these visible as dwarf novae. There are yet other systems that have never displayed either of these phenomena and others that are dominated by magnetic fields.

Typically, the nuclear timescale on which the donor star evolves, $\tau_{\rm N} > 10^{11}$ yr so that evolution cannot be the driving force behind the mass transfer. Rather this is direct angular momentum loss. In the closest systems, typically those with $P < 3\,$h, it is achieved in gravitational radiation (Peters & Mathews 1963) at a fractional rate

$$\frac{\dot{J}_{\rm GR}}{J} = -\frac{32G^3}{5c^5}\frac{M_1 M_2(M_1+M_2)}{a^4}. \tag{11.48}$$

In longer period systems this is too weak and the most likely mechanism is a process of magnetic braking (Fig. 11.8). A very mild wind carrying off mass at $|\dot{M}| < |\dot{M}_1|$, the mass transfer rate, can be dragged round by the star out to large distances beyond the Alfvén radius $R_{\rm A}$ at which the magnetic energy density equals the specific kinetic energy in the wind,

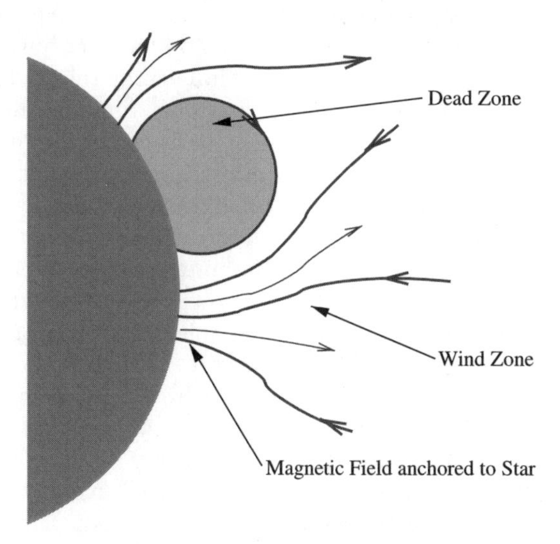

Fig. 11.8. A very weak wind can be dragged around by a magnetic field linked to a star. In dead zones the wind cannot escape but where it can open the field lines it carries of substantial angular momentum because it effectively corotates with the star to the Alfvén radius $R_{\rm A}$

$$\frac{1}{2}v_{\rm w}^2 = \frac{B^2}{2\mu_0}, \tag{11.49}$$

where $v_{\rm w}$ is the wind velocity, B is the magnetic field strength and μ_0 is the vacuum permeability. The combined angular momentum loss rate, in the wind and owing to magnetic torques, is

$$\dot{J} = \dot{M} R_{\rm A}^2 \Omega, \tag{11.50}$$

where Ω is the spin angular velocity of the star, effectively as if the wind were corotating to $R_{\rm A}$ (Mestel & Spruit 1987). This can be very effective when $R_{\rm A} \gg R$, which is usually the case when $|\dot{M}|$ is small. It is most probably magnetic braking that is responsible for bringing cataclysmic variables into the semidetached state in the first place.

11.5.4 Common Envelope Evolution

The white dwarfs in cataclysmic variables must have originally formed as the cores of giants, which must have had room to grow to 100 or even 1,000 $R_\odot$ before interaction. However, their orbital separation is now only a few solar radii. The generally accepted route by which a binary reduces its period is common-envelope evolution (Paczyński 1976). Following dynamical mass transfer from the giant, the pair becomes a common-envelope system (Fig. 11.9) in which the degenerate core of the original giant and the relatively dense red dwarf are orbiting within the low-density envelope of the giant that now engulfs both stars. From here on what happens is as much fantasy as fact. By some frictional process the two cores are supposed to spiral together towards the centre of the envelope. During this process the orbital energy released is transferred to the envelope which it drives away in a strong wind. Because the orbital energy of the cores and the binding energy of the envelope are of the same order, it can be envisaged that in some cases the balance is just such that the entire envelope is blown away when the cores reach a separation of a few solar radii. If more energy is transferred, the envelope is lost while the orbit is still quite wide. If less energy is transferred, the cores coalesce before the envelope is lost. In practice, coalescence most likely occurs when the red dwarf reaches a depth in the envelope where it has comparable density with the envelope or when it is tidally disrupted by the white dwarf.

Webbink (1984) defined a parameter $\alpha_{\rm CE}$ to be the fraction of the orbital energy released, during the spiralling-in, which goes into driving away the envelope. Knowing $\alpha_{\rm CE}$ and the binding energy of the envelope, we can calculate the final orbital separation from the initial. Note that the binding energy of the envelope is calculated differently by different authors. The most significant discrepancy is whether we use the binding energy of the single-star giant envelope before the common envelope forms (Webbink 1984) or that of the common envelope itself on the assumption that it has swollen up to the size of the orbit (Iben & Tutukov 1984). The value of $\alpha_{\rm CE}$ is expected

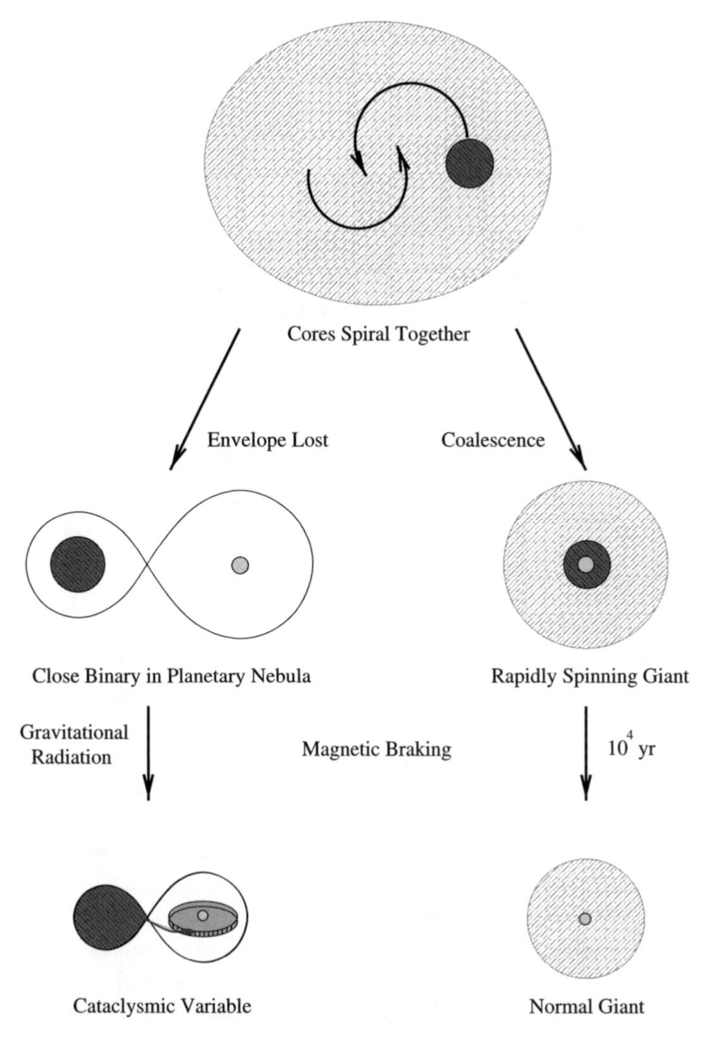

Fig. 11.9. Common-envelope evolution. After dynamical mass transfer from a giant, a common envelope enshrouds the relatively dense companion and the core of the original giant. These two spiral together as their orbital energy is transferred to the envelope until either the entire envelope is lost or they coalesce. In the former case a close white-dwarf and main-sequence binary is left, initially as the core of a planetary nebula. Magnetic braking or gravitational radiation may shrink the orbit and create a cataclysmic variable. Coalescence results in a rapidly rotating giant, which will very quickly spin down by magnetic braking

to be less than one because at least part of the released energy should be radiated away. However, population synthesis models that recreate sufficient numbers of cataclysmic variables and other close systems, such as X-ray binaries and the progenitors of SNe Ia, indicate that large values of α_{CE} are required. Typically, about three times the energy released seems to be needed (Hurley, Tout & Pols 2002).

Sources of energy other than the orbital energy are available but it is not yet established exactly how they might be tapped. There is always ongoing nuclear burning around the giant's core and indeed this energy is important if it is assumed that the common envelope expands to fill the orbit as it forms and so is included surreptitiously in the formalism of Iben & Tutukov but not in that of Webbink. In general, this requires that the timescale for common-envelope evolution be comparable with or longer than the thermal timescale of the envelope so that the nuclearly generated energy is comparable with the envelope binding energy. It also requires an efficient means of converting this nuclear luminosity to the kinetic energy of mass loss and avoid radiation. Han, Podsiadlowski & Eggleton (1994) include the ionization energy in the binding energy of the envelope. This greatly reduces what is required but to such an extent that the envelopes of many normal AGB star models are unbound. It is also difficult to see how this energy can be tapped in an envelope that is hot enough to remain fully ionised. Yet another source has been identified by Ivanova & Podsiadlowski (2001). During the formation of a common envelope a stream of hydrogen-rich material can penetrate to hot hydrogen-exhausted regions where rapid non-equilibrium burning takes place. Indeed, in their models, often enough energy is released to destroy the envelope before any spiralling of the cores has begun.

11.5.5 Type Ia Supernovae

Luminous SNe Ia are amongst the brightest objects in the Universe and their use as standard candles by cosmologists has elevated the need to understand their progenitors. The major energy source of SNe Ia is the decay of ^{56}Ni to ^{56}Fe and the total energy released in a SN Ia is consistent with the decay of approximately a solar mass of ^{56}Ni. These facts strongly implicate the thermonuclear explosion of a white dwarf though the actual explosion mechanism is not fully understood (Hillebrandt & Niemeyer 2000). White dwarfs may be divided into three major types: (i) helium white dwarfs, composed almost entirely of helium, form as the degenerate cores of low-mass red giants, which lose their hydrogen envelope before helium can ignite; (ii) carbon/oxygen white dwarfs, composed of about 20% carbon and 80% oxygen, form as the cores of asymptotic giant branch stars or naked helium burning stars that lose their envelopes before carbon ignition; and (iii) oxygen/neon white dwarfs, composed of heavier combinations of elements, form from giants that ignite carbon in their cores but still lose their envelopes before the degenerate centre collapses to a neutron star.

In binary systems, mass transfer can increase the mass of a white dwarf. Close to the Chandrasekhar mass ($M_{\rm Ch} \approx 1.44\,M_\odot$), degeneracy pressure can no longer support the star that collapses releasing its gravitational energy. The ONe white dwarfs lose enough energy in neutrinos and collapse sufficiently, before oxygen ignites, to avoid explosion (accretion induced collapse, AIC). The CO white dwarfs, on the other hand, reach temperatures early enough during collapse (at a mass of $1.38\,M_\odot$) for carbon fusion to set off a thermonuclear runaway under degenerate conditions and release enough energy to create a SN Ia. Accreting He white dwarfs reach sufficiently high temperatures to ignite helium at $M \approx 0.7\,M_\odot \ll M_{\rm Ch}$ (Woosley, Taam & Weaver 1986). An explosion under these conditions is expected to be quite unlike a SN Ia.

The process is further complicated by the nature of the accreting material. If it is hydrogen-rich, accumulation of a layer of only $10^{-4}\,M_\odot$ or so leads to ignition of hydrogen burning sufficiently violent to eject most, if not all of or more than, the accreted layer in the novae outbursts of cataclysmic variables. The white dwarf mass does not significantly increase and ignition of its interior is avoided. However, if the accretion rate is high $\dot{M} > 10^{-7}\,M_\odot\,{\rm yr}^{-1}$, hydrogen can burn as it is accreted, bypassing novae explosions (Paczyński & Żytkow 1978) and allowing the white dwarf mass to grow. Though if it is not much larger than this, $\dot{M} > 3 \times 10^{-7}\,M_\odot\,{\rm yr}^{-1}$, hydrogen cannot burn fast enough and accreted material builds up a giant-like envelope around the core and burning shell that rapidly leads to more drastic interaction with the companion and the end of the mass transfer episode. Rates in the narrow range for steady burning are found only when the companion is in the short-lived phase of thermal-timescale expansion as it evolves from the end of the main sequence to the base of the giant branch. Super-soft X-ray sources (Kahabka & van den Heuvel 1997) are probably in such a state but, without invoking some special feedback mechanism, such as disc winds (Hachisu, Kato & Nomoto 1996), cannot be expected to remain in it for very long and white dwarf masses very rarely increase sufficiently to explode as SNe Ia.

At first sight, a more promising scenario might be mass transfer from one white dwarf to another. In a very close binary orbit gravitational radiation can drive two white dwarfs together until the less massive fills its Roche lobe. If both white dwarfs are CO and their combined mass exceeds $M_{\rm Ch}$, enough mass could be transferred to set off a SN Ia. However, if the mass ratio $M_{\rm donor}/M_{\rm accretor}$ exceeds 0.628, mass transfer is dynamically unstable because a white dwarf expands as it loses mass. Based on the calculations at somewhat lower, steady accretion rates, Nomoto & Iben (1985) have claimed that the ensuing rapid accretion of material allows carbon to burn in mild shell flashes, converts the white dwarf to ONe and ultimately leads to AIC and not a SN Ia. They found a limit of one fifth of the Eddington accretion rate was necessary to avoid igniting carbon non-degenerately. The Eddington accretion rate is that rate at which the outward radiation pressure that results from the energy released as the material falls into the potential well of the star balances the gravitational attraction on an atom. Even for stable mass transfer driven by gravitational radiation, this is exceeded. Recently, Martin, Tout & Lesaffre (2005) have found that the accretion limit for steady accretion is more like two-fifths of the Eddington rate and further that short periods of accretion at much higher rates can be tolerated. They showed that a 1.1 $M_\odot$ white dwarf could accrete all the material from a companion white dwarf of 0.3 $M_\odot$ at the full rate driven by gravitational radiation and still ignite degenerately at the centre. However, there is no simple way to create a 0.3 $M_\odot$ CO white dwarf and accretion of helium rich material can lead to similar to but more extreme explosions than novae. We are still searching for the progenitors of SNe Ia from among the diverse binary systems in the stellar zoo.

References

Allen R. H., 1899, Star Names and Their Meanings. Stechert, New York
Crawford J. A., 1955, ApJ, 121, 71
Eggleton P. P., 1983, ApJ, 268, 368
Eggleton P. P., 1985, in Pringle J. E., Wade R. A., eds, Interacting Binary Stars. Cambridge Univ. Press, Cambridge, p. 21
Eggleton P. P., 2006, Evolutionary Processes in Binary and Multiple Stars. Cambridge Univ. Press, Cambridge
Fabian A. C., Pringle J. E., Rees M. J., 1975, MNRAS, 172, 15
Goodricke J. J., 1783, Phil. Trans. R. Soc. London, 73, 474
Hachisu I., Kato M., Nomoto K., 1996, ApJ, 470, L97
Han Z., Podsiadlowski P., Eggleton P. P., 1994, MNRAS, 270, 121
Herschel W., 1803, Phil. Trans. R. Soc. London, 93, 339
Hillebrandt W., Niemeyer J. C., 2000, ARA&A, 38, 191
Hoyle F., 1955, Frontiers of Astronomy. Heinemann, London
Hurley J. R., Tout C. A., Pols O. R., 2002, MNRAS, 329, 897
Hut P., 1980, A&A, 92, 167
Hut P., 1981, A&A, 99, 126
Iben I. Jr., Tutukov A. V. 1984, ApJS, 54, 335
Ivanova N., Podsiadlowski P., 2001, in Podsiadlowski P., Rappaport S., King A. R., D'Antona F., Burderi L., eds, ASP Conf. Ser. Vol. 229, Evolution of Binary and Multiple Star Systems. Astron. Soc. Pac., San Fransisco, p. 261
Kahabka P., van den Heuvel E. P. J., 1997, ARA&A, 35, 69
Kippenhahn R., Wiegert A., 1967, Z. Astrophys., 65, 251
Kopal Z., 1959, Close Binary Systems. Chapman and Hall, London
Martin R. G., Tout C. A., Lesaffre P., 2005, MNRAS, 373, 263
Mestel L., Spruit H. C., 1987, MNRAS, 226, 57
Michell J., 1767, Phil. Trans. R. Soc. London, 57, 234
Montanari 1671, Prose di Signori Academici Gelati di Bologna. (see Kopal 1959, p. 12)
Morton D. C., 1960, ApJ, 132, 146
Nomoto K., Iben I. Jr., 1985, ApJ, 297, 531
Paczyński B., 1966, A&A, 16, 231
Paczyński B., 1971, ARA&A, 9, 183
Paczyński B., 1976, in Eggleton P. P., Mitton S., Whelan J., eds, Proc. IAU Symp. 73, Structure and Evolution of Close Binary Systems. Reidel, Dordrecht, p. 75
Paczyński B., Ziółkowski J., 1967, A&A, 17, 7

Paczyński B., Żytkow A. N., 1978, ApJ, 222, 604
Parenago P. P., 1950, Astron. Zh., 27, 41
Peters P. C., Mathews J., 1963, Phys. Rev., 131, 435
Plavec M., Kříž S., Harmenec P., Horn J., 1968, Bull. Astr. Inst. Czech., 19, 24
Rasio F. A., Tout C. A., Livio M., 1996, MNRAS, 470, 1187
Sandage A. R., 1953, AJ, 58, 61
Struve O., 1948, Ann. Astrophys., 11, 117
Struve O., 1949, MNRAS, 109, 487
Tout C. A., Eggleton P. P., 1988, MNRAS, 231, 823
Vogel N. C., 1890, Astron. Nachr., 123, 289
Walter K., 1931, Königsberg Veröff., 2, (see Kopal 1959, p. 545)
Warner B., 1995, Cataclysmic Variable Stars. Cambridge Univ. Press, Cambridge
Webbink R. F., 1984, ApJ, 277, 355
Webbink R. F., 1985, in Pringle J. E., Wade R. A., eds, Interacting Binary Stars. Cambridge Univ. Press, Cambridge, p. 39
Wood F. B., 1950, ApJ, 112, 196
Woosley S. E., Taam R. E., Weaver T. A., 1986, ApJ, 301, 601

12

N-Body Binary Evolution

Jarrod R. Hurley

Centre for Astrophysics and Supercomputing, Swinburne University of Technology, P.O. Box 218, VIC 3122, Australia
jhurley@swin.edu.au

12.1 Introduction

It has long been recognized that binary stars represent a significant and important population within a star cluster and are present from the time of formation (Hut et al. 1992). As such, binary stars have been included in N-body models of star cluster evolution for quite some time (Heggie & Aarseth 1992, for example). However, these early models focused only on the dynamical evolution of binaries – orbital changes resulting from encounters with other cluster stars. It was not until the emergence of rapid binary evolution algorithms (also called population synthesis codes: Tout et al. 1997; Yungelson et al. 1995) that facets of internal binary evolution such as mass-transfer were followed in N-body codes. This chapter provides a description of how binary evolution is treated in NBODY4 and NBODY6 and what is included in the algorithm. It follows closely on from the overview of N-body stellar evolution given in Chap. 10 and the theory of binary stars presented in Chap. 11, so it is strongly suggested that these are read beforehand. The material in this chapter does not deal with dynamical considerations, such as the transformation of the two-body orbital elements to regularized variables for a more accurate treatment of close encounters, the integration of hierarchical subsystems, and gravitational perturbations of binary orbits. These are covered in Chaps. 1 and 3 as well as comprehensively in Aarseth (2003).

12.2 The BSE Package

The modelling of binary evolution in NBODY4 and NBODY6 follows closely the Binary Star Evolution (BSE) algorithm presented in Hurley, Tout & Pols (2002). Before discussing the implementation of this algorithm in the N-body codes, it will first be useful to give an overview of what it entails. This will also serve to give the reader some insight into how a prescription-based binary evolution code operates. BSE is the binary evolution analogue of the

Hurley, J.R.: *N-Body Binary Evolution*. Lect. Notes Phys. **760**, 321–332 (2008)
DOI 10.1007/978-1-4020-8431-7_12

Single Star Evolution (SSE) package described in Chap. 10. The SSE package is fully incorporated within BSE and provides the underlying stellar evolution of the binary stars as the orbital characteristics are evolved. Throughout the description of the binary evolution algorithm given below, references will be made to SSE subroutines as listed in Table 10.2 of Chap. 10.

The first step in the evolution algorithm is to initialize the binary. This requires setting the masses of the two stars (which we will call M_1 and M_2), an orbital separation (or equivalently an orbital period), and an eccentricity. In the next section there will be some discussion of how these parameters can be chosen from appropriate distribution functions, but for now it is assumed they are simply set to arbitrary values. For the purposes of stellar evolution the metallicity, Z, is also required, and it is generally assumed that this is the same for the two stars. Normally, the evolution begins with both stars on the zero-age main-sequence (ZAMS) and a separation such that the binary is detached. However, beginning with evolved stars and/or a semi-detached state is possible. A final consideration for the initialization phase is the spins, or rotation rates, of the stars. Unless otherwise specified, each star begins with a ZAMS spin set by SSE according to the ZAMS stellar mass (this is based on a fit to rotational data of observed main-sequence stars as described in Hurley, Pols & Tout 2000). Other options, such as starting the stars in co-rotation with the orbit, i.e. tidally locked, are available.

For the purposes of the algorithm, the evolution of a binary is separated into two distinct phases:

1. *detached* evolution if neither star is filling its Roche lobe;
2. *roche* evolution if one or both of the stars are filling their Roche lobes.

The Roche-lobe radius is calculated using the expression given by Eggleton (1983), which depends on the mass-ratio of the stars and the orbital separation. If the radius of a star exceeds its Roche-lobe radius, it is deemed to be filling its Roche lobe. In its most basic form the algorithm can be seen as moving the binary forward in time within the *detached* phase (according to some chosen timestep) until one of the stars fills its Roche lobe and is therefore starting to transfer mass to the companion star. The evolution then switches to the *roche* phase, which deals with all facets of the evolution associated with mass transfer, including contact and common-envelope evolution. This may once again involve moving the binary forward through a series of timesteps or the outcome may be decided immediately. Switching between the *detached* and *roche* phases is permitted, as is the possibility of following the evolution of a single star after a merger event.

Each iteration during a timestep Δt within the detached phase includes the following steps (taken in turn):

- calculate the stellar wind mass-loss rate from each star (via a call to `mlwind.f`) and determine if any of this material is accreted by the companion;

- calculate the rate of change of the orbital angular momentum and eccentricity owing to stellar wind mass loss and accretion;
- calculate the rate of change of orbital angular momentum and eccentricity owing to gravitational radiation (only effective for separations less than $10\,R_{\odot}$);
- calculate the change in the intrinsic spin of each star owing to mass changes and magnetic braking;
- calculate the rate of change of the spin of each star and the orbital eccentricity owing to tidal interactions between the stars and the orbital motion (spin-orbit coupling);
- restrict Δt if necessary to ensure that the relative changes in stellar mass, spin angular momentum owing to magnetic braking, and orbital angular momentum owing to tides are less than 1%, 3% and 2%, respectively;
- update the mass of each star and, for main-sequence (MS) and sub-giant stars, adjust the `epoch` parameter if necessary (see Chap. 10 and Hurley, Tout & Pols 2002 for usage);
- update the intrinsic spin of each star with a check to ensure that the star does not exceed its break-up speed;
- update the orbital parameters (angular momentum, separation, period and eccentricity);
- advance the time by Δt;
- evolve each star to the current time using calls to `star.f` and `hrdiag.f` in order to update the stellar parameters (stellar type, radius, core-mass, etc.);
- if a supernova has occurred, call `kick.f` and adjust the orbital parameters accordingly, including a check that the orbit is still bound (the next iteration is done with $\Delta t = 0$);
- check if either star now fills its Roche lobe and switch to the `roche` phase if this is true (if the Roche-lobe radius exceeds the stellar radius by more than 1%, the algorithm interpolates backwards until this condition is met before switching);
- for an eccentric binary, check if a collision is expected at periastron and switch to the `roche` phase if this is true;
- choose a new Δt from the minimum of the current recommended stellar evolution timestep for each star (based on the stellar type and a requirement that the radius changes by less than 10%: see Chap. 10);
- start the next iteration.

Note that if a single star emerges from the *roche* phase after a coalescence/merger of the binary stars, this new star will be evolved within the *detached* phase. Likewise, if the binary becomes unbound, the evolution of two single stars can be followed in the *detached* phase with the irrelevant steps, such as tidal evolution and mass accretion, skipped.

The general steps involved with each iteration of the *roche* phase are as follows:

- calculate the dynamical timescale for the primary star (the star filling its Roche lobe);
- determine if mass-transfer occurs on a dynamical timescale (dependent on the stellar types and the mass-ratio), and if this is true determine the instantaneous outcome – either a single star or a post-common-envelope binary – and switch back to the *detached* phase;
- otherwise, the mass-transfer occurs on a nuclear or thermal timescale and the algorithm proceeds by first calculating the amount of mass transferred from the primary per orbital period;
- determine what fraction of the mass transferred from the primary will be accreted by the companion star – this depends on the nature of the companion star as well as the mass-transfer rate and includes intricacies such as novae eruptions;
- set Δt (based on a relative mass loss from the primary of 0.5%);
- calculate the change in orbital angular momentum owing to mass loss from the system during the mass-transfer (any mass not accreted by the companion) and adjust the spin angular momentum of each star owing to mass-transfer;
- calculate mass loss and accretion owing to stellar winds as for the *detached* phase;
- calculate any changes to the orbital angular momentum and stellar spins owing to stellar-wind mass changes, magnetic braking, gravitational radiation and/or tidal interaction as for the *detached* phase;
- update the stellar spins;
- update the mass of each star and for the companion check for special cases (such as the mass of a carbon–oxygen white dwarf reaching the Chandrasekhar mass, which results in a type Ia supernova and a return to the *detached* phase with only the primary remaining to evolve);
- update the orbital parameters;
- advance the time by Δt and evolve both stars to the current time;
- if a supernova has occurred, call `kick.f`; and if the binary has become unbound, return to the *detached* phase;
- test whether or not the primary still fills its Roche lobe (return to the *detached* phase if it does not);
- test if the companion fills its Roche lobe, i.e. a contact binary (merge the two stars and return to the *detached* phase to evolve the merger product if true);
- start the next iteration of the *roche* phase.

Details of the calculations and decision-making involved in each step of the algorithm can be found in Hurley, Tout & Pols (2002). In most cases these are based on expressions and theory sourced from the literature. For example, the equations that parameterize tidal evolution are taken from Hut (1981) with additions from Zahn (1977) and Campbell (1984) for tides raised on radiative and degenerate stars, respectively. Prescribed outcomes are derived from the most accepted theory or models available at the time. For example, models suggest that white dwarfs (WDs), composed primarily of oxygen and neon that reach the Chandrasekhar mass by accreting oxygen-rich material, will collapse to form a neutron star (Nomoto & Kondo 1991). Therefore, this is the outcome currently adopted in `BSE`. If the theory changes or new models emerge suggesting a different outcome, the algorithm is updated to reflect this. Updates to the `BSE` algorithm since its publication in Hurley, Tout & Pols (2002) include the addition of an expression to calculate if an accretion disk is present during Roche-lobe overflow (as given by Ulrich & Burger 1976). The disk itself is not modelled within `BSE` but its presence is accounted for when making changes to the orbital angular momentum. Future updates might include an extension of the Roche-lobe treatment to include non-circular theory, along the lines of Sepinsky et al. (2007).

As with the `SSE` package, `BSE` can be obtained by downloading it from `http://astronomy.swin.edu.au/jhurley/bsedload.html` or by contacting the author. Within this package, the steps describing the *detached* and *roche* phases are contained in the `evolv2.f` subroutine. The package also contains a subroutine `comenv.f` to deal with common-envelope evolution: this is called from `evolv2.f` during the *roche* phase if the mass-transfer is deemed to be dynamical and the primary is a giant-like star. If the binary evolves into contact (both stars filling their Roche lobes), the two stars are merged and the subroutine `mix.f` is called to determine the outcome after complete mixing. An additional routine `gntage.f` is included to calculate the parameters of the new star that results from such a merger or from coalescence during common-envelope evolution.

Parameterized binary evolution naturally involves a number of input parameters that reflect uncertainties in the underlying theory. These can affect the evolution and outcomes. An example in `BSE` is the common-envelope parameter α, which determines the efficiency with which energy is transferred from the orbit to the envelope surrounding the two stellar cores as they spiral towards each other. Other parameters affect aspects of the evolution such as mass accretion from a stellar wind, mass ejected in a nova explosion, and the change in orbital angular momentum when mass is lost from the binary system during mass-transfer. These features will be returned to in the next section and full descriptions can be found in Hurley, Tout & Pols (2002).

12.3 N-Body Implementation

To evolve a population of binaries using the `BSE` population synthesis algorithm is a straightforward process. It simply involves taking each binary in turn, evolving it to the desired physical time (such as the age of the Galaxy), and recording the outcome. Thus, only one call to `evolv2.f` is required for each binary. In an N-body code it is not so straightforward as the binary evolution must be performed in step with the dynamical evolution of the star cluster. If the mass of a binary changes owing to mass transfer, this must be communicated to the dynamical interface of the code with minimal delay so that the gravitational force calculations remain accurate. Conversely, dynamical interactions between a binary and cluster stars can lead to perturbations that alter the orbital parameters of the binary, including disassociation, with consequences for the binary evolution outcomes. Binary evolution within three- and four-body sub-systems must also be accounted for (see Chap. 3), as well as the possible existence of non-primordial binaries that form during the cluster evolution. The binary evolution treatment must also interface with the regularization methods that are used to follow accurately the dynamical evolution of binaries, sub-systems and close encounters (see Aarseth 2003).

In NBODY4/6 the tasks performed in the BSE subroutine `evolv2.f` are split with the *detached* phases implemented in `mdot.f` and the *roche* phases contained in the `roche.f` subroutine. Stars in a binary have their individual `tev` values (time of next stellar evolution update) set equal (to the minimum of the two) so that they will be evolved together within `mdot.f`. This allows corrections to the spin and orbital angular momentum owing to stellar wind mass changes to be performed as the stars are evolved. Gravitational radiation for short-period detached binaries is taken care of by the subroutine `grrad.f` from `mdot.f`. Similarly, tidal interactions within circular binaries are accounted for by `bsetid.f` – tidal circularization of eccentric binaries is dealt with elsewhere as part of the two-body regularization process (see below). The subroutine `brake.f` is then used by `mdot.f` in order to update the binding energy of the binary and re-scale the associated two-body regularization variables after any orbital changes.

Decision-making for binaries is aided by assigning the centre-of-mass particle for each binary its own `tev0` and `tev` values. Here `tev` is the expected time of the next mass-transfer update: the next call to `roche.f` for the binary. For detached binaries this will be the time when one of the component stars has evolved to fill its Roche lobe and is estimated by the subroutine `trflow.f` (called from `mdot.f` each time a stellar evolution update is performed for the component stars). For a semi-detached binary in an ongoing Roche-lobe overflow phase, this will be set in `roche.f` (see below). The binary `tev` values are included in setting `TMDOT` (the smallest `tev`) and if `mdot.f` is called owing to `tev(i)` being less than the current time where i represents a centre-of-mass particle[1], the evolution update switches to `roche.f` (called from `mdot.f`).

The subroutine `roche.f` includes all of the processes outlined in the *roche* phase of the BSE algorithm with a few N-body related additions. First, as mentioned above, a steady mass-transfer phase must now be dealt with in a piece-wise fashion so that the binary evolution time does not get too far ahead of the dynamical time. This is put into place using the `tev` and `tev0`

[1] For a system of N stars and `NBIN` binaries the centre-of-mass particle for binary `j` sits at position $\mathtt{i} = N + \mathtt{j}$ in the various arrays. The component stars sit at $(2 \times \mathtt{j}) - 1$ and $2 \times \mathtt{j}$ while the single stars occupy the $(2 \times \mathtt{NBIN}) + 1$ to N positions.

variables: each call to `roche.f` evolves the binary from `tev0(i)` $\rightarrow$ `tev(i)`, unless something happens within the interval, such as a merger. Before exiting `roche.f`, the routine sets `tev0(i)` = `tev(i)` and updates `tev(i)`. If `roche.f` signals termination because the primary star no longer fills its Roche lobe, this is done with a call to `trflow.f`. Otherwise, `tev(i)` is set to the current time plus some multiple of the current mass-transfer timestep (as described in the previous section). This multiplication factor is in the range of 10–50 depending on whether or not the binary has a nearby perturber. The update of `tev(i)` also takes into account any major stellar evolution changes on the horizon for the component stars, such as an impending supernova explosion.

Analogous to the stellar type index used to describe the evolution state of individual stars, there is also a `kstar` index for the binary centre-of-mass particle that describes the current state of each binary. This takes on values such as 0 for a standard eccentric binary, -2 for a circularizing binary and 10 for a circular binary. The first time that a binary enters `roche.f` the `kstar` index is set to 11, and when the binary next becomes detached it is set to 12. Subsequently, `kstar` is increased by one each time a binary switches from a *detached* to a *roche* phase, and vice-versa, such that `kstar(i)` = 16 would indicate that binary $i - N$ is currently *detached* but has previously evolved through three distinct `roche` phases.

Another addition to the N-body version of the *roche* process is the subroutine `coal.f`, which is called from `roche.f` when mass-transfer has ended in coalescence of the two stars. This routine takes care of the associated N-body book-keeping such as removing the second star and the centre-of-mass particle from the relevant arrays and performing the necessary force corrections.

Unlike isolated binary evolution, the cluster environment provides for the formation of non-standard binary configurations through dynamical interactions. An example would be an eccentric binary that emerges from a four-body hierarchy with one of the stars filling its Roche lobe. If such a binary enters `roche.f`, it is currently dealt with by first calculating the tidal circularization timescale and, if this is less than 10 Myr, calling `bsetid.f` to circularize the binary before proceeding with the mass-transfer process.

Some of the subroutines associated with the *roche* phase are also utilized via an NBODY4/6 subroutine `cmbody.f`. This is called from various parts of the N-body code when a hyperbolic collision or a collision at periastron in an eccentric (and non-Roche-lobe filling) binary is detected. If one or both of the stars involved in the collision is a sub-giant or giant, `cmbody.f` calls `expel.f` which in turn calls `comenv.f` to determine the outcome via common-envelope evolution. Otherwise, the two stars are merged directly with `mix.f`, which determines the outcome. If this results in the formation of a new giant star, the BSE routine `gntage.f` is used to set the appropriate age and initial mass to match the core-mass and mass of the star (this routine is also used by `comenv.f` and `roche.f` when needed).

The main difference between the treatment of binary evolution within BSE and that of the N-body codes relates to how tidal interactions for eccentric binaries are dealt with. Mardling & Aarseth (2001) have developed algorithms that combine tidal circularization neatly with the two-body regularization method for following the orbital evolution of binaries. These algorithms also cope with N-body complications such as the orbit of an eccentric binary becoming chaotic owing to perturbations. The subroutines involved are `tcirc.f` and `spiral.f` (as well as some subsidiary routines). There is also a related subroutine `synch.f`, which models tidal synchronization. The underlying theory for tides in the Mardling & Aarseth (2001) algorithm is Hut (1981), as it is in BSE, so the two treatments are consistent. However, the option to model tidal circularization within NBODY4/6 using the BSE algorithm may be added in the future for the sake of completeness.

Subroutines in NBODY4/6 that are directly related to binary evolution are summarized in Table 12.1. The only one not yet mentioned above is `rl.f`, which contains the Eggleton (1983) function for calculating the Roche-lobe radius of a star.

An important facet of binary evolution is setting the initial parameters – for a population of binaries this is critical in determining the range of outcomes that are possible. In the case of a star cluster, the relative number of tightly bound binaries is an important factor in how the cluster itself will evolve. The first step towards initializing a population of primordial binaries in NBODY4/6 is to decide how many are to be included. This is set by the parameter NBIN0 read from the input file in the `data.f` subroutine. If NBIN0 is non-zero, the subroutine `binpop.f` generates the parameters of the NBIN0 binaries. This involves a number of choices that are controlled by a line of input variables read from the input file in `binpop.f`. These include SEMI0, ECC0, RATIO, RANGE and ICIRC. Both SEMI0 and RANGE affect the semi-major axes of the binaries: if RANGE is negative, the log-normal distribution from Eggleton, Fitchett & Tout (1989) is used with a peak at SEMI0 (in AU); if RANGE is positive, a uniform logarithmic distribution is used with a maximum of SEMI0 (in N-body units) and covering RANGE orders of magnitude; and if RANGE = 0, SEMI0 is the semi-major axis of all binaries. The input variable ECC0 determines the eccentricity distribution (constant or thermal distribution) and RATIO controls

Table 12.1. Subroutines in NBODY4 and NBODY6 associated with binary evolution

BSE-related	Other
`bsetid.f`	`brake.f`
`comenv.f`	`cmbody.f`
`gntage.f`	`coal.f`
`grrad.f`	`expel.f`
`mdot.f`	`tcirc.f`
`mix.f`	`trflow.f`
`rl.f`	`spiral.f`
`roche.f`	`synch.f`

how the masses of the two stars are assigned from the binary mass (see also `imf.f`). If the variable `ICIRC` is non-zero, pre-MS eigen-evolution of the orbital parameters is invoked (Kroupa 1995).

There are also a number of input options that affect binary evolution and related diagnostic output. The option `kz(34)` must be set non-zero for binary evolution (Roche-lobe mass-transfer and tides) to occur. If `kz(34)` = 1 tidal synchronization of circular binaries is performed using `synch.f`, otherwise it is performed using `bsetid.f`. The option `kz(6)` controls the level of diagnostic output for regularized binaries and `kz(8)` affects output relating to primordial binaries. To date, input parameters in `BSE` that affect particular aspects of the binary evolution algorithm are not included as input variables in NBODY4/6. Instead, they are hardwired into the various subroutines where they are used. For example, the common-envelope efficiency parameter mentioned in the previous section is set in the header of `comenv.f` while a number of parameters are set in `roche.f` – the fraction of accreted mass that is ejected from the surface of a WD in a nova explosion (`EPSNOV`), the Eddington-luminosity factor (`EDDFAC`) and the stellar-wind velocity factor (`BETA`), to name a few.

This completes the overview of how binary evolution is treated in NBODY4 and NBODY6. It is by no means a comprehensive description, but should give the interested user enough information to get started. More details can be found in Aarseth (2003) and Hurley et al. (2001).

12.4 Binary Evolution Results

The colour-magnitude diagram (CMD) of a binary-rich NBODY4 simulation is shown in Fig. 12.1. This simulation started with 28 000 stars and a 40% primordial binary fraction. The initial separations (or equivalently, orbital periods) of the binaries were drawn from the Eggleton, Fitchett & Tout (1989) distribution, with a peak at 10 AU and a maximum of 100 AU. The model shown is at an age of 4 000 Myr when the binary fraction is still at about 40% – preservation of the primordial binary fraction is a common feature of star cluster evolution noted in Hurley, Aarseth & Shara (2007). However, as the cluster evolution progresses, it becomes increasingly likely that a significant component of the binary population will be non-primordial. For the model in Fig. 12.1 about 20% of the binaries are non-primordial and these are primarily the result of exchange interactions. The exact proportion of binaries formed by dynamical processes depends on factors such as the fraction of binaries in relatively wide orbits, the cluster density and the stage of evolution. Figure 12.1 can be compared to the CMD at 4 000 Myr shown in Fig. 10.1 of Chap. 10 for a simulation starting with 30 000 stars and 0% binaries. The effects of binary evolution on the locus of points in the CMD is clearly seen and the result is much closer to the reality presented by the observations of open clusters (Fan et al. 1996, for example).

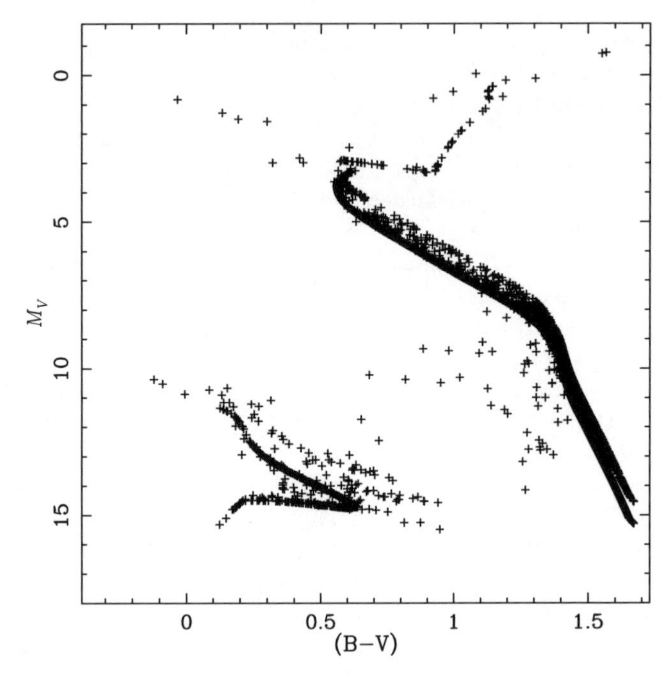

Fig. 12.1. Colour-magnitude diagram after 4 000 Myr of evolution for a $Z = 0.02$ NBODY4 simulation that started with 12 000 single stars and 8 000 binaries. At 4 000 Myr, there are 3 382 single stars and 2 360 binaries in the model cluster. Each binary is shown as a single point, i.e. unresolved. The luminosity and effective temperature provided for each star by SSE/BSE have been converted to magnitude and colour using the bolometric corrections given by the models of Kurucz (1992) and, in the case of white dwarfs, Bergeron, Wesemael & Beauchamp (1995)

Some features to note in Fig. 12.1 include the broadening of the MS owing to the presence of MS–MS binaries, with the upper edge defined by the equal-mass binaries. Similar behaviour can be seen for the WD sequence owing to WD–WD binaries. Points below the MS but distinct from the WD sequence are MS–WD binaries. These evolve away from the WD sequence and towards the MS as the WD cools and the MS star comes to dominate the colour. The points that form an extension of the MS, hotter and *bluer* than the MS turn-off, represent blue stragglers (BSs). These are MS stars that have longer central hydrogen-burning lifetimes than expected for their mass. That is to say, if these stars were born in the cluster with their current mass (or higher), they would already have evolved away from the MS to become giants or WDs. Their presence is explained by obtaining their current mass either through steady mass-transfer in a short-period binary or as the result of a merger of two MS stars. Either way they are a product of binary evolution. In Hurley et al. (2005), NBODY4 models were used to demonstrate how the combination of the cluster environment and close binary evolution could explain the number and

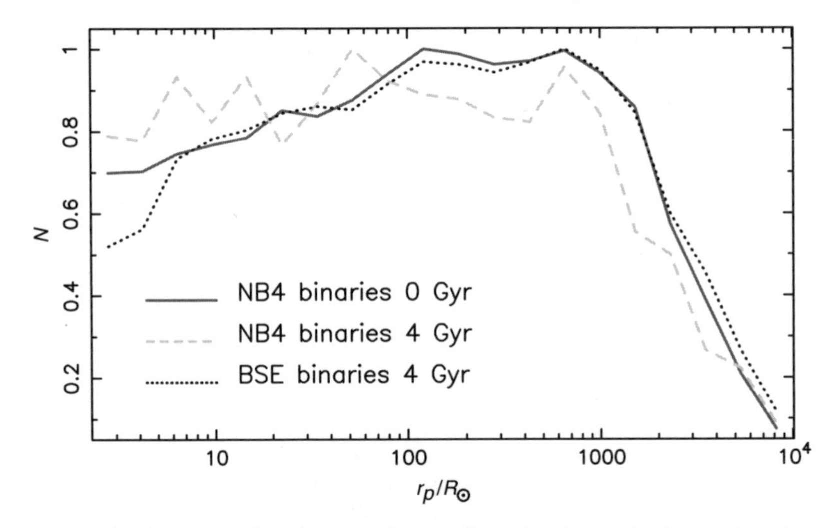

Fig. 12.2. Distribution of periastron, $R_{\rm p} = a\,(1-e)$, where a is the semi-major axis and e the eccentricity, for: the 8 000 primordial binaries in the NBODY4 simulation described in Fig. 12.1 (*solid line*); the binaries remaining in this simulation after 4 000 Myr (*dashed line*); and the primordial binaries evolved to the same age using BSE only (*dotted line*). Each distribution is normalized to a maximum of unity

nature of the BSs observed in the old open cluster M67. This included the production of BSs in eccentric binaries, which cannot be explained by binary evolution alone.

Figure 12.2 shows the periastron distribution for the binaries in the NBODY4 simulation of Fig. 12.1, and compares this to the primordial distribution as well as the distribution obtained when the binaries are evolved to the same age using BSE only. We see from comparing the latter two distributions that binary evolution steadily removes binaries with short periastron distances. However, the NBODY4 distribution shows that a star cluster is effective in replenishing the relative numbers of interacting binaries. This is done at the expense of the wide binaries, which are broken up in encounters with other cluster members. In closing it is noted that binary evolution is important for proper accounting of the orbital properties of the binary populations of star clusters, especially, as the presence of binaries, and in particular tightly bound binaries, can critically affect properties such as the structure and lifetime of a star cluster.

References

Aarseth S. J., 2003, Gravitational N-Body Simulations. Cambridge Univ. Press, Cambridge

Bergeron P., Wesemael F., Beauchamp A., 1995, PASP, 107, 1047

Campbell C. G., 1984, MNRAS, 207, 433
Eggleton P. P., 1983, ApJ, 268, 368
Eggleton P. P., Fitchett M., Tout C. A., 1989, ApJ, 347, 998
Fan X., et al., 1996, AJ, 112, 628
Heggie D. C., Aarseth S. J., 1992, MNRAS, 257, 513
Hurley J. R., Pols O. R., Tout C. A., 2000, MNRAS, 315, 543
Hurley J. R., Tout C. A., Aarseth S. J., Pols O. R., 2001, MNRAS, 323, 630
Hurley J. R., Tout C. A., Pols O. R., 2002, MNRAS, 329, 897
Hurley J. R., Pols O. R., Aarseth S. J., Tout C. A., 2005, MNRAS, 363, 293
Hurley J. R., Aarseth S. J., Shara M. M., 2007, ApJ, 665, 707
Hut P., 1981, A&A, 99, 126
Hut P., McMillan S., Goodman J., Mateo M., Phinney E. S., Pryor C., Richer H. B., Verbunt F., Weinberg M., 1992, PASP, 104, 981
Kroupa P., 1995, MNRAS, 277, 1507
Kurucz R. L., 1992, in Barbuy B., Renzini A., eds, Proc. IAU Symp. 149, The Stellar Populations of Galaxies. Kluwer, Dordrecht, p. 225
Mardling R. A., Aarseth S. J., 2001, MNRAS, 321, 398
Nomoto K., Kondo Y., 1991, ApJ, 367, L19
Sepinsky J. F., Willems B., Kalogera V., Rasio F. A., 2007, ApJ, 667, 1170
Tout C. A., Aarseth S. J., Pols O. R., Eggleton P. P., 1997, MNRAS, 291, 732
Ulrich R. K., Burger H. L., 1976, ApJ, 206, 509
Yungelson L., Livio M., Tutukov A., Kenyon S. J., 1995, ApJ, 447, 656
Zahn J.-P., 1977, A&A, 57, 383

13

The Workings of a Stellar Evolution Code

Ross Church[1,2]

[1] University of Cambridge, Institute of Astronomy, Madingley Road, Cambridge, CB3 0HA, UK
rpc25@srcf.ucam.org
[2] Centre for Stellar and Planetary Astrophysics, Monash University, PO Box 28M, Clayton, Victoria 3800, Australia

13.1 Introduction

Models of stellar clusters link the theoretical gravitational N-body problem to the study of real astrophysical systems. Such models require a description of the stars contained within the cluster. Stars are interesting objects in their own right, and the study of stellar evolution is important across astronomy, from the formation of exotic objects such as X-ray binaries and gamma-ray bursts to measuring the ages of galaxies.

The physical processes important for stellar evolution theory as well as qualitative results are discussed elsewhere in this book. Here the technical problem of computing the structure and evolution of the stars is considered. How can we solve the set of differential equations that describe the interior of a star to obtain a model of its physical properties? A brief mention will be made of some of the uncertainties in stellar physics and how they affect the results obtained.

The stellar evolution code used as an example in this text is stars, the Cambridge Stellar Evolution Code. Written originally by Peter Eggleton (1971), it is widely used by astronomers working in the field of stellar evolution. It has the advantage of being relatively concise and simple in its construction, owing mainly to the elegant treatment of meshpoint placement and convective mixing. The code itself can be downloaded from `http://www.ast.cam.ac.uk/research/stars`.

13.2 Equations

In order that a star can be modelled efficiently for its entire lifetime, which greatly exceeds its dynamical timescale, simplifying physical assumptions must be made. The star is usually taken to be spherically symmetric and in hydrostatic equilibrium. This reduces the problem to a single spatial dimension, but necessitates that the process of convection be treated empirically. These

Church, R.: *The Workings of a Stellar Evolution Code*. Lect. Notes Phys. **760**, 333–345 (2008)
DOI 10.1007/978-1-4020-8431-7_13

assumptions lead to the four equations of stellar structure. A detailed derivation of these equations can be found in any standard text on stellar structure and evolution, for example, Schwarzschild (1965); Cox & Giuli (1968); Kippenhahn & Weigert (1994); Prialnik (2000), as well as Chap. 9 of this book. In summary, the equations are

$$\frac{\mathrm{d}m}{\mathrm{d}r} = 4\pi r^2 \rho, \tag{13.1}$$

$$\frac{\mathrm{d}P}{\mathrm{d}r} = -\frac{Gm\rho}{r^2}, \tag{13.2}$$

$$\frac{\mathrm{d}T}{\mathrm{d}r} = \begin{cases} -\dfrac{3\kappa\rho L}{16\pi a c r^2 T^3} & \text{(radiative regions)} \\ \nabla_{\mathrm{a}} \dfrac{T}{P}\dfrac{\mathrm{dP}}{\mathrm{dr}} + \Delta\nabla T & \text{(convective regions)} \end{cases}, \tag{13.3}$$

$$\frac{\mathrm{d}L}{\mathrm{d}r} = 4\pi r^2 \rho\epsilon, \tag{13.4}$$

where m is the mass within radius r of the centre of the star, P the pressure, ρ the density, L the luminosity, T the temperature and κ the Rosseland mean opacity. The adiabatic temperature gradient ∇_{a} is calculated from the equation of state of the star, whilst the superadiabatic temperature gradient $\Delta\nabla T$ is obtained from mixing length theory. The energy liberation rate per unit mass, ϵ, contains contributions from gravitational expansion and contraction, nuclear reactions and neutrino emission.

In addition to the equations of stellar structure, it is necessary to take into account composition changes owing to nuclear burning and mixing. The process of mixing can be modelled as diffusion with an appropriate coefficient. This leads to a set of equations for the evolution of the chemical composition:

$$\frac{\partial X_i}{\partial t} = \frac{m_i}{\rho}\left(\sum_j R_{ji} - \sum_k R_{ik}\right) - \frac{\partial}{\partial r}\left(\sigma^2 \frac{\partial X_i}{\partial r}\right), \tag{13.5}$$

where R_{ij} is the rate of conversion of element i into element j per unit volume, m_i is the atomic mass of element i and σ is a diffusion coefficient, usually obtained from mixing length theory.

13.2.1 Boundary Conditions

The central boundary conditions of a stellar model are straightforward; at the centre $m = 0$, $r = 0$ and $L = 0$, although in practice STARS does not use a central meshpoint. The surface of the star is placed where the temperature equals the effective temperature, given by

$$L = 4\pi R^2 \sigma T_{\mathrm{eff}}^4 \tag{13.6}$$

for a star of luminosity L and radius R.

The Eddington closure approximation, together with a thin grey atmosphere, is used to obtain the gas pressure at the surface,

$$P_{\mathrm{g}} = \frac{2}{3}\frac{g}{\kappa}\left(1 - \frac{L}{L_{\mathrm{Edd}}}\right), \tag{13.7}$$

where L_{Edd} is the limiting Eddington luminosity and g the surface gravity. The total mass of the star, equal to the value of m at the outermost meshpoint, changes according to

$$\frac{\mathrm{d}M}{\mathrm{d}t} = -W, \tag{13.8}$$

where W is the stellar wind. There is no general theory of stellar winds and a number of empirically determined formulae are commonly used. During the main-sequence phase all but the most massive stars are assumed to lose no mass, the solar wind being evolutionarily negligible. For the red-giant phase the formula of Kudritzki & Reimers (1978) is commonly used, whereas on the asymptotic giant branch (AGB) the formulae of Blöcker (1995) and Vassiliadis & Wood (1993) are popular.

13.3 Variables and Functions

A stellar model is defined in terms of a set of independent variables.[1] The physical variables used in STARS are $\log T$, $\log m$, $\log r$, L, $\log f$, $X_{^1\mathrm{H}}$, $X_{^4\mathrm{He}}$, $X_{^{12}\mathrm{C}}$, $X_{^{16}\mathrm{O}}$ and $X_{^{20}\mathrm{Ne}}$. The first four are standard physical quantities defined above. Note that the luminosity can be negative and hence its logarithm cannot be used. The quantity f is a function of the electron degeneracy parameter ψ and is explained in Sect. 13.3.2. The composition of the star is measured by the mass fractions X_i of various isotopes. Because the mass fractions must sum to unity, these numbers also determine the mass fraction of another isotope, ^{14}N. All other compositions are assumed to either be constant, for example, iron, or zero.

13.3.1 The Mesh

Whilst a real star is continuous, a computer can only hold a finite quantity of data and hence the star must be discretised on to a mesh of points. The placement of these points is crucial to the functionality of the code. Areas of interest in a star must be sufficiently resolved; in particular the burning

[1]Speaking in a strictly mathematical sense, there are only two independent variables in the problem, m and t. All the other variables are dependent on these implicitly through the equations listed in Sect. 13.2. To explicitly define a model of a star, however, one needs values of all the 11 variables and it is possible to vary these variables independently; the resulting model may not, however, represent a physical star. Hence, it is reasonable to refer to them as independent variables.

shells in giants and ionisation zones in the envelope. The use of too many meshpoints, however, increases the memory requirements and slows the code down. An Eulerian mesh of points at constant radii performs poorly because the stellar radius can change by several orders of magnitude over the star's lifetime. Meshpoints placed at constant mass co-ordinates to form a Lagrangian mesh work better, but then the points must be moved as the evolution proceeds to keep interesting parts of the star well resolved. A unique feature of STARS is that the mesh is positioned automatically by the equation solving package. A further equation is solved by the code to make the gradient with respect to the meshpoint number of a function Q constant throughout the star. The function is chosen to cause points to be placed in regions of physical significance. The form usually adopted is

$$Q = c_4 \log(P) + c_5 \log\left(\frac{P + c_9}{P + c_1}\right) + c_2 \log\left(\frac{P + c_{10}}{P + c_1}\right) + c_7 \log\left(\frac{T}{T + c_{11}}\right) + \log\left(\frac{c_6 M^{2/3}}{c_6 M^{2/3} + m^{2/3}}\right) + c_3 \log\left(\frac{r^2}{c_8} + 1\right), \quad (13.9)$$

where the constants c_i are chosen by the user. Because

$$C = \frac{\mathrm{d}Q}{\mathrm{d}k} = \frac{\mathrm{d}Q}{\mathrm{d}m}\frac{\mathrm{d}m}{\mathrm{d}k} \quad (13.10)$$

is constant, the mass resolution, which is inversely proportional to $\mathrm{d}m/\mathrm{d}k$, is largest where Q varies most quickly with mass. Given appropriate values of the coefficients, the second and third terms have the effect of driving meshpoints into the hydrogen and helium burning shells. This substantially improves numerical stability during thermal pulses on the AGB.

13.3.2 The Equation of State

It is necessary to have an equation of state for the material that makes up a star. A common approach is to use a set of tables for different temperatures, densities, etc. STARS, conversely, utilises the semi-analytic equation of state described by Pols et al. (1995). Contributions to the Helmholtz free energy from radiation, ions and electrons are considered, along with some non-ideal effects. The Fermi-Dirac integral over the momentum states of the electron is simplified by working with the quantities f and g chosen so that a power series therein has the correct asymptotic form for limiting values of ψ and T. The quantities f and g are defined by

$$\psi = 2\sqrt{1+f} + \log\frac{\sqrt{1+f}-1}{\sqrt{1+f}+1} \quad (13.11)$$

and

$$g = \frac{kT}{m_{\mathrm{e}}c^2}\sqrt{1+f}. \quad (13.12)$$

Full details of the series can be found in Eggleton, Faulkner & Flannery (1973).

Although most of the equation of state is calculated in real time, there are still a few tabulated quantities. The opacities are too complicated to be calculated analytically, likewise the nuclear reaction and neutrino loss rates. These are included as tables of numerical values; bicubic spline interpolation is used within the opacity tables.

13.4 Method of Solution

The Henyey, Forbes & Gould (1964) relaxation method solves the equations of stellar structure and evolution by making small changes to the structure obtained at the previous timestep, and adjusting the resulting model until it solves the equations. This use of information from a previous timestep greatly improves the speed of calculations over a simple shooting method and is used in almost every modern stellar evolution code.

If the subscript i is allowed to run over the set of N_{e} equations at N_{p} meshpoints, and the subscript j over the N_{v} variables at N_{p} meshpoints, by bringing all the terms on to one side of the equations of the solved code can be written implicitly as

$$E_i(v_j) = 0. \tag{13.13}$$

Then for a complete stellar model v_j the degree to which it does not satisfy the equations is

$$\delta E_i = E_i(\boldsymbol{v}). \tag{13.14}$$

The model from the previous timestep is used as an initial guess for $\boldsymbol{v}$. By numerical differentiation of each equation with respect to each variable, one can obtain

$$A_{ij} = \frac{\partial E_i}{\partial v_j}. \tag{13.15}$$

Most of the entries in $\mathbf{A}$ vanish. Because the equations are either first or second order, spatially an element in $\mathbf{A}$ depends only on values within the adjacent one or two meshpoints; hence $\mathbf{A}$ is block-diagonal. This enables it to be economically inverted and corrections to the variables are calculated as

$$\delta v_j = A_{ji}^{-1} \delta E_i. \tag{13.16}$$

This process is iterated in a manner analogous to the Newton-Raphson method until the convergence criterion is met. It is required that the average change in δv_j in a single iteration is less than a user-supplied constant. In practice, this procedure is sometimes slightly modified to improve stability of the solution method. Only part of the correction is applied under some circumstances to prevent the solution being overshot. This is equivalent to reducing the magnitudes of the eigenvalues of the iteration matrix. It is also usually better to use $v_j + \delta v_j$ from the previous timestep as a first guess rather than v_j; that is, to start the iteration with the changes applied at the previous timestep.

13.4.1 Timesteps

The timestep $\delta\tau_i$ that the code uses is determined by an ad-hoc formula:

$$\delta\tau_i = \delta\tau_{i-1} \times \frac{\Delta}{\sum_{jk} |\delta X_{jk}|}, \tag{13.17}$$

where $\delta\tau_{i-1}$ is the previous timestep, δX_{jk} the change in variable j at meshpoint k and Δ is a user-supplied constant. The sum is evaluated over the variables omitting the luminosity, because this fluctuates too much to be useful. A larger value of Δ allows the variables to change more in a single timestep and hence larger timesteps to be taken. Because the change at a single meshpoint is independent of the number of meshpoints, it is necessary to scale Δ linearly with the number of meshpoints; different values are appropriate to different phases of evolution. In the standard case of 199 meshpoints $\Delta = 5$ provides adequate results.

If the iterative process fails to find a set of values for the variables that satisfy the equations with sufficient accuracy, a model is deemed to have not converged. The code reverts to the previous model and the timestep is reduced by a factor of 0.8. Multiple reductions in timestep are possible for a system that is failing to converge, but when the timestep has fallen below 1% of its first tried value the code stops attempting to converge.

A graph of the variation of the timestep with model number during the evolution of a $1\,M_\odot$ star is shown in Fig. 13.1. One can see that it has a

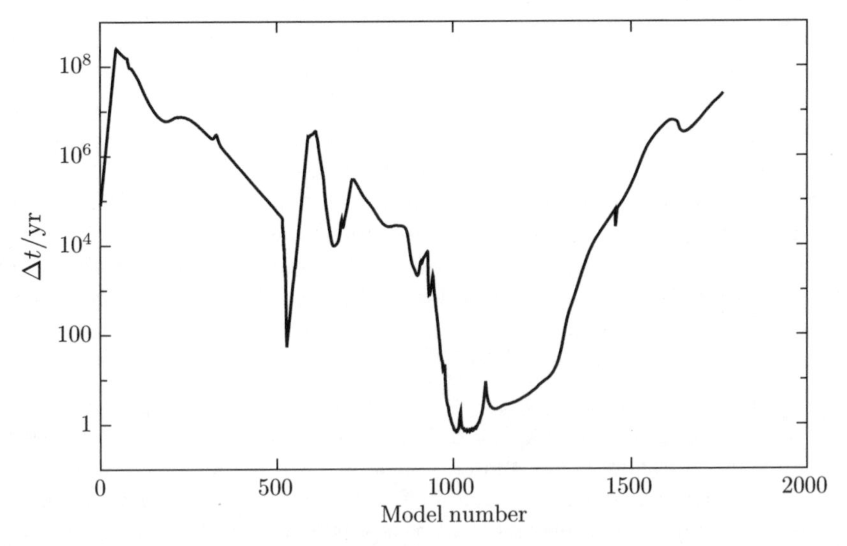

Fig. 13.1. Variation of the timestep during the evolution of a $1\,M_\odot$ star. The model number is plotted on the abscissa; this increments by unity for each converged stellar model. The ordinate shows the timestep in years. This model was run with 199 meshpoints and $\Delta = 5$ throughout

very large dynamic range; there is a difference of approximately 10^9 between the shortest and the longest timesteps. The initial peak in the timestep and that around model number 600 are the main sequence and horizontal branch, respectively. The timestep increases substantially again towards the end of the run as the star descends the white dwarf cooling track. The discontinuity around model 500 represents pseudo-evolution through the helium flash (see Sect. 13.6.1) and the period of short timesteps from model 1000 onward on the post-AGB.

13.5 The Structure of STARS

STARS comprises 20 subprocedures, which can be divided up into four groups, the solution package, physics package, the flow control routines and the initial setup routines, as well as a few vestigial routines. The *solution package* consists of the following procedures:

- `solver`, which solves the implicit matrix equation (see (13.13)),
- `difrns`, which differentiates the equations to be solved,
- `elimin8`, which carries out some matrix manipulations and
- `divide`, which implements matrix inversion.

The *physics package* contains

- `equns1`, which calculates the values of the difference equations and their boundary conditions,
- `funcs1`, which calculates various quantities from the principal variables, mostly for use in `equns1`,
- `statef`, which evaluates the equation of state at a given meshpoint,
- `statel`, which decides whether it is necessary to call `statef`,
- `fdirac`, which evaluates Fermi-Dirac integrals,
- `pressi`, which approximates pressure ionisation,
- `opacty`, which does spline interpolation within the opacity tables and
- `nucrat`, which calculates nuclear reaction rates.

The *flow control routines* are

- `main`, which provides the main integration loop and basic flow control,
- `printa`, which determines the next timestep, updates the matrix, controls input and output and does sundry minor tasks for which there is no obvious alternative location and
- `printb`, which writes most of the output files.

Finally, the *initial setup routines* are

- `opspln`, which sets up the opacity tables,
- `spline`, which calculates spline coefficients,
- `remesh`, which attempts to remesh the model to a different grid,

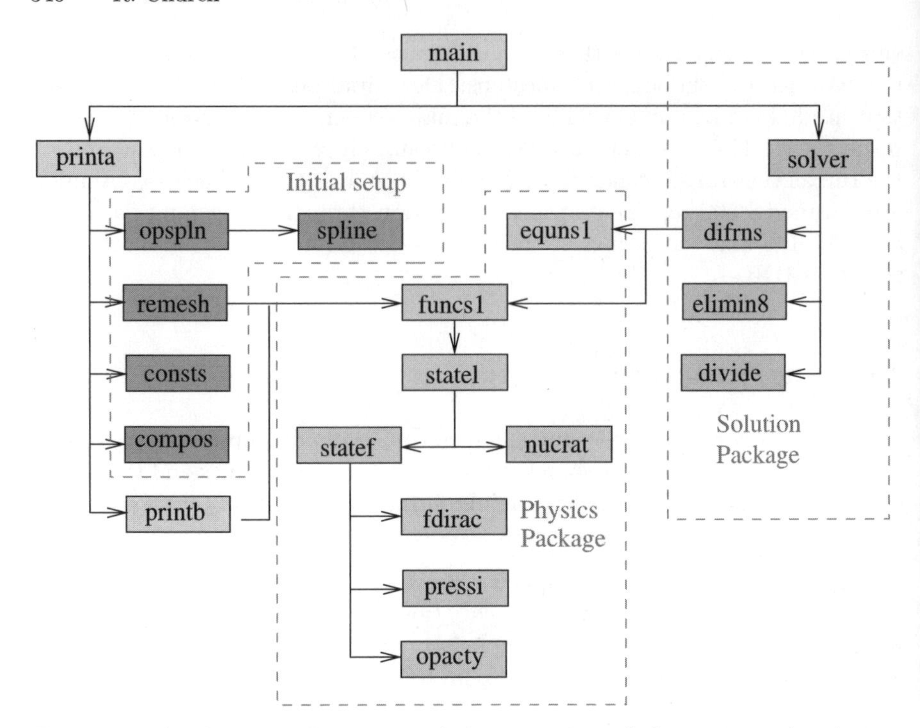

Fig. 13.2. A schematic illustration of the operation of the STARS code. *Arrows* indicate the direction in which one subroutine calls another. The division of the code into sections with different functionality is shown

- `consts`, which sets up physical constants and
- `compos`, which sets small or negative compositions to zero.

The interaction of the first three groups of these subroutines can be seen in Fig. 13.2. Note that the physics package is called via `funcs1`, from several places in the rest of the code.

13.6 Problematic Phases of Evolution

The iterative procedure that STARS uses to converge a model is not guaranteed to arrive at a solution. Usually, the desired solution is sufficiently close to the starting model that it does so, but in some situations this is not the case. Problematic phases of evolution are mostly those where the structure of the star is changing quickly. As well as requiring small timesteps, such phases of evolution often cause the mesh to move rapidly through the model. The advection terms in the equations that are included to deal with movement of the mesh are then large in magnitude but opposite in sign. This causes numerical problems.

Phases of evolution that routinely cause problems are the helium flash, thermal pulses on the AGB, the post-AGB, degenerate carbon ignition in super-AGB stars, heavy element burning subsequent to neon ignition and the very late stages of white dwarf evolution. Brief notes on how these problems can be tackled are given below.

13.6.1 The Helium Flash

In stars of $M \lesssim 2.3\, M_\odot$ the core is degenerated at the time of helium ignition. The increased temperature owing to helium burning does not cause expansion and thermonuclear runaway occurs (Schwarzschild & Härm 1962). This is the helium flash. To circumvent these problems, one can use an empirical procedure to construct approximate post-flash models with stable core helium burning. A star of mass $M \simeq 3\, M_\odot$ that has evolved successfully through non-degenerate core helium ignition is taken and matter removed from the envelope until the desired mass is reached. The hydrogen burning shell is allowed to burn outwards with helium consumption disabled in order to obtain the correct core mass. The envelope compositions are reset to their pre-flash values and normal evolution is resumed. Whilst not physically rigorous, this process provides models that can be used to study subsequent evolution.

13.6.2 The AGB

Evolution through thermal pulses on the AGB using STARS is possible, but only with a modified version of the code and considerable effort (Stancliffe, Tout & Pols 2004). An easier, though less accurate, approach is to avoid modelling the pulses. A relatively low resolution of 199 meshpoints per model and a comparatively large value of the timestep control parameter, $\Delta = 5$, suppress thermal pulses on the AGB. Their exclusion changes the composition of material ejected in stellar winds and, for the more massive AGB stars, the mass of the core and hence the final white dwarf mass.

13.6.3 Late Stages of Intermediate-Mass and High-Mass Stars

The problems in the late stages of the lives of intermediate-mass and high-mass stars are more tricky to deal with. Degenerate carbon ignition in lower-mass super-AGB stars and the post-AGB cannot be avoided as thermal pulses and the helium flash can. Stars that ignite carbon mildly degenerately, probably go on to form oxygen-neon white dwarfs, although the most massive amongst them may end their lives as neutron stars. The post-AGB is the final stage of evolution of AGB stars, and it is reasonable to assume that once a star reaches this point it forms a white dwarf.

Heavy element burning is only really of interest for the calculation of pre-supernova models. Very little stellar evolution significant for N-body calculations takes place after the ignition of neon and it is reasonable to terminate a star's evolution at this point. Likewise, problems in the evolution of white dwarfs mostly occur at times comparable with the Hubble time. In any case, the bulk properties of the star change very little after this point.

13.7 Robustness of Results

The theory of stellar structure and evolution contains substantial uncertainty. In particular, some of the input physics is not well determined. Convection is a three-dimensional process and the one-dimensional mixing length theory used to approximate it cannot be entirely accurate. Mixing length theory contains a free parameter, α, related to the length scale of convective plumes. Its value is usually obtained by fitting a solar model, but there is no reason why it should not vary between stars of different masses or in different evolutionary phases. There is substantial evidence that for many stars the amount of mixing predicted by the Schwarzschild criterion is insufficient and that processes that cause extra mixing occur in stars. Some candidates for these are stellar rotation, convective overshooting and internal gravity waves. Nuclear reaction rates, even some of those most important to the structure of a stellar model, are substantially uncertain. For example, the rate of the $^{14}\mathrm{N}(\mathrm{p},\gamma)$ reaction that forms the slowest step in the CNO cycle is uncertain to approximately a factor of 2 (Herwig, Austin & Lattanzio 2006). There is no general theory of stellar mass loss, so it is necessary to use empirically measured values of questionable accuracy. There are also uncertainties in the opacity of stellar material and in models of stellar atmospheres.

To illustrate briefly the effects of two of these uncertainties, a set of stellar models with varying input physics are presented here. Models of masses $1\,M_\odot$, $2\,M_\odot$, $4\,M_\odot$, $8\,M_\odot$ and $16\,M_\odot$ have been calculated varying two uncertain physical parameters. In one set of models extra mixing was added according to the prescription of Schröder, Pols & Eggleton (1997). In the the other the rate of the $^{14}\mathrm{N}(\mathrm{p},\gamma)$ reaction was doubled. This is the slowest step in the CNO cycle and hence determines how fast hydrogen burning occurs according to that process.

13.7.1 HR Diagrams

The effects on the HR diagram of changing the input physics are largest in the case of the $4\,M_\odot$ and $8\,M_\odot$ stars. HR diagrams for these two stars are presented in Fig. 13.3. It can be seen that changing the degree of extra mixing has a dramatic effect on the position of the blue loop (horizontal branch) in the HR diagram. The increased mixing draws more hydrogen into the core, increasing its size and hence the luminosity of the star. There is also a slight, but much less pronounced, difference when the CNO burning rate is changed.

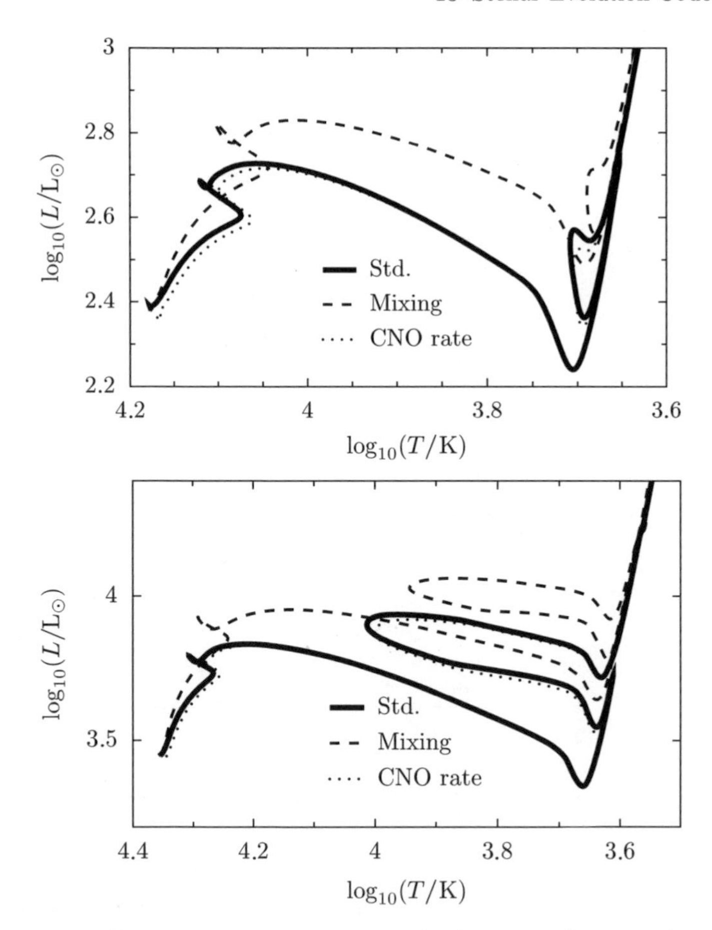

Fig. 13.3. HR diagrams for stellar models of mass $4\,M_\odot$ (*top panel*) and $8\,M_\odot$ (*bottom panel*). The *thick solid line* is the standard model, the *dashed line* the model with extra mixing and the *dotted line* the model with the enhanced CNO burning rate

13.7.2 Stellar Lifetimes

The effect of increased mixing and the enhanced CNO rate on main-sequence lifetimes is shown in Fig. 13.4. Stars spend the majority of their lives on the main sequence and hence this time is a useful measure. It also has the advantage of being better defined than the total stellar lifetime.

The main effect that can be seen is that models more massive than the Sun with extra mixing have substantially increased lifetimes. This is because their convective cores are enlarged by the extra mixing. The cores have more fuel to burn and hence the main sequence is prolonged. As the $1\,M_\odot$ model has a radiative core, it is unaffected by changing the degree of convective mixing.

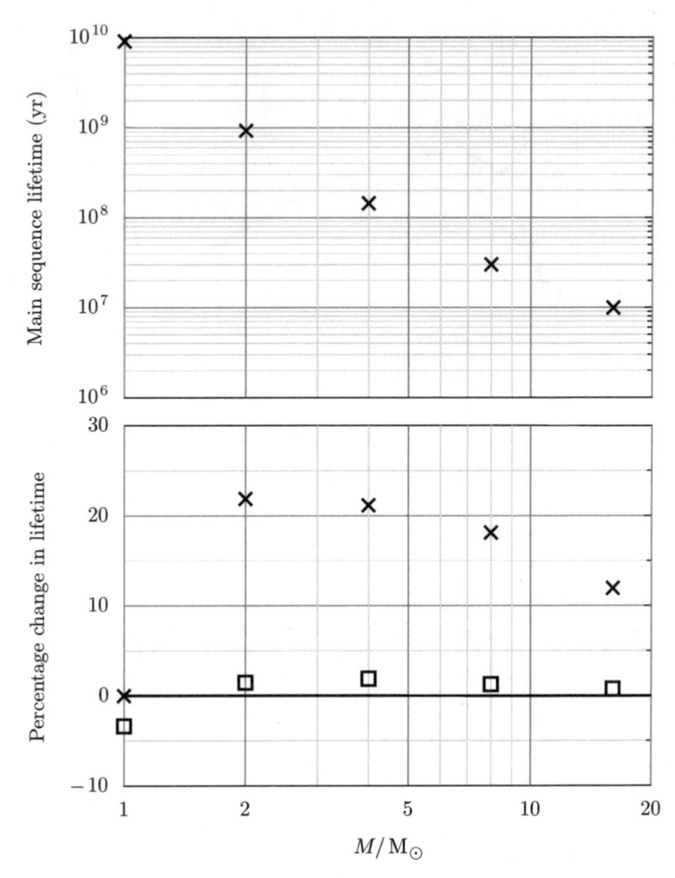

Fig. 13.4. The effect of enhanced mixing and increased CNO reaction rate on the main-sequence lifetimes of stellar models. The *top panel* shows the lifetimes of the standard stellar models as a function of their masses. The *lower panel* shows the percentage change in the main-sequence lifetime with respect to the standard model when the input physics is changed. The *crosses* represent the calculations with extra mixing, the *squares* those with an enhanced CNO reaction rate

The effect of increasing the CNO rate on the main-sequence lifetime is considerably counter-intuitive. For the stars in which the CNO cycle is the dominant reaction on the main sequence the lifetime increases slightly, whereas for the 1 $M_{\odot}$ model where it is not the dominant reaction it decreases slightly. The reason for the increase in lifetime is that the structure of the model depends on the conditions in the core. If the CNO rate is doubled from the standard value, too much energy is generated in the core of the star for the structure that it supports. As a result, the star expands and the core becomes cooler and less dense until equilibrium is regained. At the new equilibrium point the structure is such that a lower energy flux is needed to support the star. Hence, hydrogen burns more slowly and the star lives longer. In the

$1\,M_\odot$ model the dominant reaction rate is the pp chain and hence the change in the CNO rate does not have the same structural effect on the model. The small amount of CNO burning that does take place, however, is increased and hence the main-sequence lifetime is reduced. This effect demonstrates another important point about stellar evolution: it is a highly non-linear process, and simple assumptions about the behaviour of stars that are not supported by detailed calculations often turn out to be incorrect.

References

Blöcker T., 1995, A&A, 297, 727

Cox J. P., Giuli R. T., 1968, Principles of Stellar Structure. Gordon and Breach

Eggleton P. P., 1971, MNRAS, 151, 351

Eggleton P. P., Faulkner J., Flannery B. P., 1973, A&A, 23, 325

Henyey L. G., Forbes J. E., Gould N. L., 1964, ApJ, 139, 306

Herwig F., Austin S. M., Lattanzio J. C., 2006, Phys. Rev. C, 73, 025802

Kippenhahn R., Weigert A., 1994, Stellar Structure and Evolution. Springer-Verlag

Kudritzki R. P., Reimers D., 1978, A&A, 70, 227

Pols O. R., Tout C. A., Eggleton P. P., Han Z., 1995, MNRAS, 274, 964

Prialnik D., 2000, An Introduction to the Theory of Stellar Structure and Evolution. Cambridge Univ. Press, Cambridge

Schröder K.-P., Pols O. R., Eggleton P. P., 1997, MNRAS, 285, 696

Schwarzschild M., 1965, Structure and Evolution of the Stars. Dover Publication

Schwarzschild M., Härm R., 1962, ApJ, 136, 158

Stancliffe R. J., Tout C. A., Pols O. R., 2004, MNRAS, 352, 984

Vassiliadis E., Wood P. R., 1993, ApJ, 413, 641

14

Realistic N-Body Simulations of Globular Clusters

A. Dougal Mackey

Institute for Astronomy, University of Edinburgh, Royal Observatory, Blackford Hill, Edinburgh, EH9 3HJ, UK
dmy@roe.ac.uk

14.1 Introduction

This chapter is an introduction to realistic N-body modelling of globular clusters – specifically, why it might be desired to conduct such models and what constitutes their key ingredients. Detailed consideration is also given to the analysis of data from such simulations, and how it is increasingly becoming more important to perform simulated observations in order to derive quantities that are directly comparable with real-world measurements. The most salient points from this general discussion are illustrated via an extensive case study concerning N-body modelling of massive stellar clusters in the Large and Small Magellanic Clouds.

14.2 Realistic N-Body Modelling – Why and How?

N-body modelling has long been an important tool for exploring the evolution of star clusters. All major phases of cluster evolution, from early mass loss through to core collapse, gravothermal oscillations, and tidal disruption, have been investigated with N-body simulations (as well as other types of modelling), and they have played a large part in forming our current understanding of cluster evolutionary processes (see e.g. the review by Meylan & Heggie 1997). Even so, due to the massive computational workload involved with the direct, accurate integration of a large number of particles over very long time-scales, historically N has been restricted to relatively small values (a few thousand, or with major effort, a few tens of thousand). In addition, much of the complexity of real clusters (such as the processes involved with stellar evolution, binary star evolution, stellar collisions, time-varying tidal fields, and so on) has often, by necessity, been neglected. These two factors have meant that the investigation of globular cluster evolution with N-body modelling has generally involved the extrapolation of results to larger N, and

Mackey, A.D.: *Realistic N-Body Simulations of Globular Clusters*. Lect. Notes Phys. **760**, 347–376 (2008)
DOI 10.1007/978-1-4020-8431-7_14

approximations due to incomplete implementation of the complicated interplay between various internal and external evolutionary processes.

In the last decade, however, and particularly within the last few years, there have been two major advances that have propelled the field of cluster N-body modelling into a new era. The first of these is the advent of special purpose hardware, most recently the GRAPE-6 machines (Makino et al. 2003; Fukushige, Makino & Kawai 2005), to accelerate the direct N^2 summation of gravitational forces. These have greatly reduced the computational bottleneck associated with large N, and simulations covering a Hubble time of evolution with $N \sim 10^5$ – that is, at the lower end of the globular cluster mass function – are now within reach.

The second advance concerns the sophistication of the N-body codes themselves. Several of the major codes, such as Aarseth's NBODY4 (Chap. 1; see also Aarseth 2003)[1] and the STARLAB software environment[2], have now progressed to the stage where most, if not all, of the major internal and external evolutionary processes in a star cluster have successfully been incorporated. Such processes include single-star and binary star evolution, stellar collisions and the formation and destruction of hierarchical systems and, arbitrary external tidal fields. The sophistication of available N-body codes, combined with the integrating power of special purpose hardware, means that direct, realistic simulations of massive stellar clusters are now possible.

This aim of this chapter is to present an overview of realistic N-body modelling of globular clusters. In particular, we will discuss in what situations it is desirable to invest the time and effort to run and analyse a realistic N-body model, and examine the most important aspects of the N-body codes, which allow such realism. Since many (if not all) of the latter have been covered in significant detail elsewhere in this series of lectures, we will spend most of our time examining the processes involved with reducing the large amounts of data that come out of a realistic simulation, and in particular discuss the concept of "simulated observations" which is becoming increasingly prominent. Since this constitutes some very general discussion, much of it from an observer's perspective, the best way to illustrate the most important points is via a specific case study – we examine recent direct, realistic N-body modelling of the evolution of massive stellar clusters in the Magellanic Clouds.

14.2.1 Why Run a Realistic N-Body Model?

There are a number of advantages to running large-scale, realistic N-body models. First, unlike with many methods used to model star cluster evolution, a sophisticated N-body code includes all the important physics with a minimum of simplifying assumptions. Hence, for example, if one is interested in investigating the long-term evolution of hierarchical systems within a stellar cluster, in a realistic N-body simulation it is possible to integrate directly

[1] Available for download from `http://www.ast.cam.ac.uk/research/nbody`

[2] See `http://www.ids.ias.edu/~starlab/`

the orbits of all stars – no gravitational softening, or similar modifications, are required.

Similarly, because all the important physics is being included in a self-consistent manner (e.g. the stars and binaries are evolving in step with the cluster evolution) one can be reasonably confident that the complex interplay between various evolutionary processes in a cluster is being accounted for. Even though star clusters are generally considered to be relatively simple astrophysical systems, in that they are often approximately spherically symmetric, and consist of stars with a uniform single age and metallicity, they are in fact complicated objects and it is often extremely difficult to isolate (or predict) the effects of individual physical processes in a cluster.

For example, consider the production of blue stragglers in a globular cluster. It is generally accepted that there are a number of channels leading to the formation of such objects – for example, Roche-lobe mass transfer in a binary star, or the coalescence of a highly eccentric binary star after a strong interaction. It is complicated to determine the relative importance of formation channels in a star cluster, and the resulting properties of the blue stragglers, because much interplay between competing processes occurs. For example, the structural and dynamical state of the cluster plays an integral role in defining the collision (strong interaction) rate between individual members. However, the state of the cluster is strongly affected by the stellar evolution within the cluster, by related parameters such as the initial mass function, the metallicity, and so on, and by the properties of the external tidal field. In addition, the properties of any binary stars in the cluster are strongly affected by both the structural and the dynamical state of the cluster, as well as the stellar evolution of the individual members of the binary (especially if processes such as mass transfer occur). In certain cases (such as during deep core collapse), the binaries themselves can in turn affect the cluster structure and dynamics. Given all this, if one wishes to investigate the production and properties of blue stragglers in a cluster, a realistic N-body simulation offers a very powerful means of accounting for (and following) this complicated interplay.

A third advantage to running realistic N-body simulations is that with present technology one is now able to directly compare simulations with real clusters for realistic N up to that corresponding to low-mass globular clusters. Even for higher-mass clusters, it is almost always possible to choose an N which corresponds within an order of magnitude. We are therefore now moving into the regime where many of the scaling-with-N issues which have been necessary to account for in the past when applying the results of N-body simulations to the evolution of real clusters (e.g., Aarseth & Heggie 1998), are circumvented. In addition, with such large N, fluctuations in the global evolution of the N-body model are reduced to the point where they are not significant. For small-N models, it has been standard practice to average the results of a number of simulations to reduce such fluctuations, the amplitudes of which increase with decreasing N (e.g., Giersz & Heggie 1994; Wilkinson et al. 2003). For large-N models, it is becoming increasingly clear that this process is not necessary (e.g., Hurley et al. 2005; Mackey et al. 2007, 2008a).

Finally, given both the fact that processes such as stellar evolution are modelled along with the gravitational interactions between particles, and that we often do not have to worry about extrapolating our results to larger N, it is possible to apply sophisticated techniques to the analysis of realistic N-body simulations. More specifically, it is possible to realistically *simulate observations* of N-body models. This aspect is especially important if one is trying to compare an N-body simulation with a real system (which will inevitably have properties defined through observation), or if one is trying to make predictions about the properties of a real system (which will have to be tested observationally). This concept is discussed in more detail below, in Sect. 14.2.3, and examples are given in Sect. 14.3.

Even taking into account the above advantages, it is important to understand that it will not always be necessary to invest the time and effort in running a large-scale, realistic N-body model. One should always consider carefully what question is under investigation and how best to answer it. If the physics can be sufficiently well modelled with small-N clusters, or without needing to include degrees of sophistication such as stellar evolution or simulated observations, then running less complicated models will naturally be preferable (and almost certainly far quicker and more efficient) than investing in a direct, realistic N-body simulation.

14.2.2 Key Ingredients in a Realistic N-Body Model

There are two main ingredients in setting up and running a realistic N-body model – the N-body code itself, and the generation of initial conditions.

N-Body Codes

It is worth considering briefly the major components of a realistic N-body code. As noted earlier, there are a number of such codes publicly available. Prominent examples are NBODY4 (for use with the GRAPE-6 special purpose hardware), NBODY6 (for use without GRAPE-6) and NBODY6++ (a parallelised version of NBODY6), and the STARLAB environment. Here we will consider the code NBODY4 and note that much of the discussion also applies to the other codes. Since most of the following is covered in great detail by other contributions to this lecture series, we will not delve too deeply into the computational details. Nonetheless, it is important to understand what primary ingredients make up a realistic N-body code.

These main components can be divided into three different groups: the integration routines, the stellar evolution routines, and the binary evolution routines. Let us consider these in order. In NBODY4, the equations of motion are integrated using the fourth-order Hermite scheme (Makino 1991), in combination with a GRAPE-6. An external tidal field is incorporated by integrating the equations of motion in an accelerating but non-rotating reference frame, centred on the cluster's centre-of-mass (see e.g. Wilkinson et al. 2003, and references therein for more details). The integration proceeds using the N-body units of Heggie & Mathieu (1986), which are converted to physical units for output using a length scale generally set at the beginning of a run via comparison to a real cluster (see Sect. 14.3.2). A close multiple system (such as a hard binary) is treated as a combined centre-of-mass object in the Hermite integration, while the detailed orbits of the individual components of the multiple system are integrated separately using state-of-the-art two-body or chain regularization schemes, as applicable (Mikkola & Aarseth 1993, 1998). The point of two-body regularization is that binary star orbits, and particularly perturbed binary motion, can be followed at high accuracy without resorting to the introduction of gravitational softening. Chain regularization extends this possibility to close encounters between more than two stars (such as in a binary–binary interaction).

Stellar evolution in NBODY4 is incorporated by means of the analytical formulae of Hurley, Pols & Tout (2000), who derived them from detailed stellar evolution models, following stars from the zero-age main sequence through to remnant phases (such as white dwarfs, neutron stars and black holes). Each star is initially assigned a mass (the formulae cover the mass range 0.1–$100\,\mathrm{M}_{\odot}$), and a single metallicity for the cluster may be selected in the range $Z = 0.0001$–0.03. The stellar evolution is calculated in step with the dynamical integration, and includes a mass-loss prescription such that evolving stars lose gas through winds and supernova explosions. This gas is instantaneously removed from the cluster, which is a reasonable approximation since outflow speeds are generally large compared to the cluster escape velocity. An important consequence of the introduction of stellar evolution is that each star possesses a finite radius (as opposed to being a point mass), which varies as its evolution progresses. This is vital when considering close encounters between stars, including effects such as tidal capture. Furthermore, the stellar evolution parameters calculated in the routines in NBODY4 (such as luminosity and effective temperature) may be used to derive absolute magnitudes and colours, although this is not done within the code itself. This allows simulated observations of the model cluster to be made if necessary.

Binary star evolution is calculated in a similar manner to single-star evolution, following the analytical prescription of Hurley, Tout & Pols (2002) and allowing for such phases as the tidal circularization of orbits, mass transfer, common-envelope evolution, and mergers. Algorithms such as stability tests, which allow the consideration of triples and higher-order hierarchical systems, are also implemented within the code. Details of the tidal evolution and stability routines are discussed in Chap. 3, and Mardling & Aarseth (2001). As with the single-star evolution, binary star evolution is calculated in step with the overall dynamical integration.

Initial Conditions

Generating high-quality initial conditions is of paramount importance when running a realistic N-body model. Generally, the reason for wanting to run a realistic N-body simulation will be to directly model one or more real clusters. In such cases, the initial conditions are defined naturally by the clusters under consideration, although it may be necessary to infer them (for example, if the real clusters are dynamically evolved). In addition, since the initial conditions for the real clusters are almost certainly defined (or at least constrained) by observational measurements, it may well be necessary to implement simulated observations in order to confirm the generated initial conditions in the N-body model are as accurate as possible (see e.g., Sect. 14.3.7).

There is a significant number of variables to consider when setting initial conditions and the parameter space can therefore be very large. For example, consider the following (non-exhaustive) list:

- What is the initial cluster structure? The central density, core and/or half-mass radius, tidal limit, and the radial density profile all need to be appropriate to the problem under consideration.
- What is the initial dynamical state? Should the cluster be starting in virial equilibrium or is some other state more appropriate?
- What is the most appropriate initial mass function (IMF)?
- What is the most appropriate range of stellar masses?
- What is the total cluster mass $M_{\rm tot}$?
- $M_{\rm tot}$, the IMF, and the stellar mass range allow N to be calculated. Is this number realistic to model in a reasonable time-frame?
- What is the cluster metallicity?
- Should there be any primordial mass segregation in the cluster?
- Are there any primordial binaries in the cluster? If so, then what should the overall binary fraction be, and how should they be distributed spatially?
- What properties do any primordial binaries have? What are the distributions for the mass ratio, semi-major axis, and orbital ellipticity?
- What is the external tidal field?
- Are any special modifications to the code required? For example, to incorporate specific stellar evolution, or a new external tidal field, etc.

It is also important to consider practicalities for a given simulation, like its required duration (this will be constrained by the real systems being modelled), how frequently data should be produced during the run (this will be constrained by the temporal resolution required to investigate properly all questions under consideration) and whether the resulting disk space requirements can be met.

14.2.3 Data Analysis: Simulated Observations

There is a number of reasons why one may be running a large-scale, realistic N-body simulation. For example, the aim may be to directly model one specific cluster (see e.g., Hurley et al. 2005), to try and understand the global properties of a system of clusters (see e.g., Mackey et al. 2007), or to investigate a more general question like the effect of cluster metallicity on structural evolution (see e.g., Hurley et al. 2004). In most (if not all) such cases, the problems under investigation will be defined by the observations of real systems. Furthermore, any results from the simulations may lead to predictions for real systems that will require observational verification. For these reasons, it is necessary to treat the analysis of data from a realistic N-body simulation with some degree of sophistication. Specifically, the most useful results are likely to be obtained by simulating observations of the model cluster(s).

This will not constitute *all* of the data analysis for a given simulation. It is still necessary to perform more traditional analysis to understand the aspects of the global or specific evolution of a model cluster. Nonetheless, if one wishes to obtain measurements from an N-body simulation, which are to be compared directly with observational measurements of real systems, considerable care must be taken that the derived quantities are indeed directly comparable. If this is not the case, significant error can result, as highlighted in Sect. 14.3.6. The most straightforward means by which it can be ensured that directly comparable quantities are obtained is by closely reproducing the original observational analysis on the N-body model.

In undertaking such a process, the most important thing is to adopt an observer's perspective. In particular, it is vital to be aware of the circumstances and limitations of the genuine observations, and make sure that these are applied to the simulated observations. It should be clearly understood exactly what was observed in a cluster (e.g. maybe just red giant branch stars), what quantities were actually measured, and what processes were used to obtain these measured parameters. Detailed examples of this methodology are set out in Sects. 14.3.4, 14.3.6 and 14.3.7. For a theoretician or N-body modeller, accustomed to being able to consider any aspect of a simulated cluster at will, it is often surprising how crude many genuine observations are. Detailed observing in a globular cluster can be a very difficult feat, which has only recently become fairly routine due to the arrival of extremely high-quality telescopes and instruments, such as the Hubble Space Telescope (HST), particularly its associated cameras (WFPC2, ACS, etc.); and the Very Large Telescope (VLT), particularly its spectrographs (UVES, FLAMES) and adaptive optics instruments (e.g., NACO). Even so, the process of obtaining simulated observational measurements from a realistic N-body run will invariably involve degrading the data significantly, because star cluster observations generally only measure a small fraction of the stars in a cluster.

Simulated observations serve a number of functions in addition to their use in the primary analysis of the results from an N-body simulation. As discussed above in Sect. 14.2.2, in many situations the initial conditions for a realistic N-body model will be defined or constrained by the observations of a genuine system or systems. In such cases, simulated observations of the initial state of the model N-body cluster can be used to verify the validity of the adopted initial conditions, and can often be used to fine-tune these initial conditions. Examples of this are provided in Sects. 14.3.3 and 14.3.7.

Furthermore, simulated observations of an N-body cluster can provide important information about the quality of the real set of observations they are designed to reproduce. Since it is possible to do "perfect" observations on an N-body model and thus gauge the true state of the model at any particular time, by then degrading the observational quality to that of the real measurements, one can investigate how accurately those real measurements quantify that state and search for any biases that may have been introduced. Subsequently, it may be possible to use further simulated observations to examine the modifications that could be made to the real observations or data reduction procedure in order to improve their quality. An example of such a process is presented in Sect. 14.3.6.

Similarly, if one has calculated a realistic N-body model that makes some kind of prediction about a quantity which can potentially be observed in a globular cluster, it is important to examine whether it is feasible to search for that signature with presently available facilities. Simulated observations, in which the capabilities of a given telescope and/or instrument are incorporated, can provide such information, and also allow one to assess the complexity of such observations along with the time allocation requirements for them to be carried out.

Conducted with due care and attention, simulated observations of realistic N-body models can be an extremely powerful tool for both modellers and observers.

14.3 Case Study: Massive Star Clusters in the Magellanic Clouds

The above discussion is quite general, and many of the points are best illustrated via a specific case study. For the remainder of this chapter we will therefore examine recent work concerning the evolution of globular clusters in the Large and Small Magellanic Clouds (LMC and SMC, respectively) (Mackey et al. 2007, 2008a).

Before proceeding to this, however, it is worth noting that another excellent example of realistic N-body modelling, with a different focus to the case study considered below, is the recent 'work concerning' the old Galactic open cluster M67 by Hurley et al. (2005), in which they investigate the evolution of the cluster structure and mass loss, along with formation mechanisms and properties of blue stragglers, evolution of the cluster colour–magnitude diagram and various stellar populations, and modification of the cluster luminosity function due to external tidal forces. Some aspects of this work are discussed in Chap. 12.

14.3.1 Observational Background: The Radius–Age Trend

The star cluster systems belonging to the LMC and SMC (which are two close companion galaxies of the Milky Way) are of fundamental importance in star cluster astronomy, particularly the field of star cluster evolution. While the Galactic system provides the nearest globular cluster ensemble, from an observational point of view these objects are not ideal for studying cluster evolution because of their uniform ancient nature (ages $\sim 10 - 13$ Gyr). Therefore, we can determine very well the end-points of massive star cluster evolution, but must infer the complete long-term development that brought them to these observed states.

In contrast, the LMC and SMC possess extensive systems of star clusters with masses comparable to the Galactic globulars, but crucially *of all ages:* $10^6 \leq \tau \leq 10^{10}$ yr. These systems are hence the nearest places we can observe direct snapshots of cluster development over the last Hubble time.

Elson and her collaborators were among the first to consider the structural evolution of massive star clusters in the LMC (Elson, Fall & Freeman 1987; Elson, Freeman & Lauer 1989; Elson 1991, 1992). They measured radial brightness profiles and derived structural parameters for a sample of clusters covering a wide range of ages, to search for evolutionary trends. The most striking relationship they discovered concerns the sizes of the cluster cores.[3] The spread in core radius was observed to be a strongly increasing function of age, in that the youngest clusters possessed compact cores with $r_c \sim 1 - 2$ pc, while the oldest clusters exhibited a range $0 \leq r_c \leq 6$ pc (cf. Fig. 14.1). They did not observe any significant trend between cluster mass and radius. The radius-age trend provided intriguing evidence that our understanding of massive star cluster evolution may be incomplete, since quasi-equilibrium models of star cluster evolution do not predict large-scale core expansion over the cluster lifetime (see e.g., Meylan & Heggie 1997).

The advent of the Hubble Space Telescope has allowed this problem to be re-addressed observationally in significantly more detail than was possible with ground-based facilities. HST imaging can resolve LMC and SMC star clusters (at distances of ~ 50 and ~ 60 kpc, respectively) even in their inner cores, so that star counts may be conducted to very small projected radii and very accurate surface density/brightness profiles constructed. Work with HST observations, using the Wide Field Planetary Camera 2 (WFPC2) and Advanced Camera for Surveys (ACS), has recently been conducted (Mackey & Gilmore 2003a,b; Mackey et al. 2008b). These authors have a combined sample consisting of 84 LMC and 23 SMC clusters, covering the full age range and with masses generally comparable to those of the Galactic globular clusters. For the interested reader, full details of the data reduction, construction of surface brightness profiles, and measurement of structural parameters may be found in Mackey & Gilmore (2003a) and Mackey et al. (2008b).

[3] As parametrised by the observational core radius, r_c, defined in this case as the radius at which the surface brightness is half its central value.

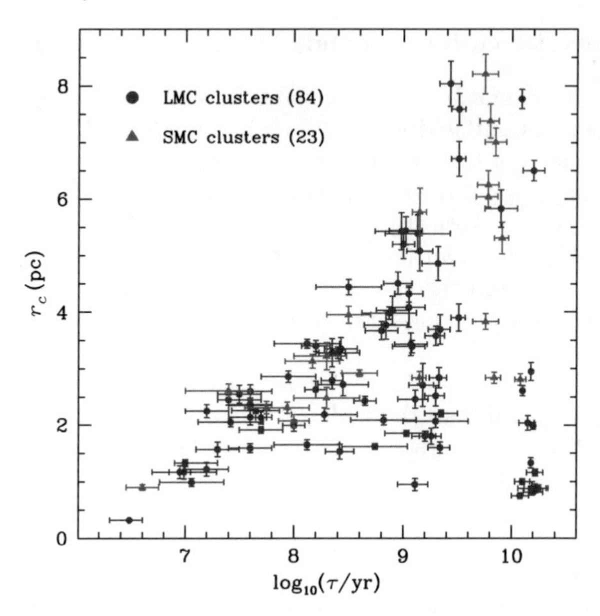

Fig. 14.1. Core-radius versus age for massive stellar clusters in the Large and Small Magellanic Clouds. This figure includes all clusters from the HST/WFPC2 measurements of Mackey & Gilmore (2003a,b) as well as the HST/ACS measurements of Mackey et al. (2008b)

The resulting core-radius versus age diagram is shown in Fig. 14.1. This represents the most up-to-date information available regarding the radius-age trend in the LMC and SMC cluster systems. The upper envelope is very well defined for all ages up to a few Gyr. At older times than this, the full range of core radii observed for massive stellar clusters is allowed. In fact, the situation is even more dramatic than appreciated in earlier studies. Several of the oldest clusters in the sample lie off the top of the diagram: the Reticulum cluster in the LMC, with age $\tau \sim 12-13$ Gyr and $r_c \sim 14.8$ pc; and Lindsay 1 and 113 in the SMC, with $\tau \sim 9$ Gyr and $r_c \sim 16.4$ pc, and $\tau \sim 5$ Gyr and $r_c \sim 11$ pc, respectively. Hence the range for the oldest clusters is $0 \leq r_c \leq 17\,\mathrm{pc}$.

It is interesting to note that the observed distribution of core radii for the oldest clusters is quite consistent with that observed for Galactic globular clusters. Indeed, if only globular clusters in the remote outer Milky Way halo are considered (where destructive tidal processes, particularly affecting diffuse clusters, are minimized), the distributions match very closely indeed (Mackey et al. 2008a). It is worth emphasizing, however, that the radius-age relationship cannot be inferred solely from the observations of the Galactic globular clusters – the full trend is only evident when the age spectrum present in the LMC and SMC cluster systems is exploited.

14.3.2 Realistic N-Body Modelling of Magellanic Cloud Clusters

The key question resulting from these observations concerns the origin of the radius-age trend. This is important for our understanding of star cluster evolution – since standard models never predict an order-of-magnitude expansion of the cluster core radius over the cluster lifetime, these models are possibly incomplete.

There exist a number of interpretations of the radius-age diagram. The most straightforward (which we consider here) postulates that massive star clusters (or at least the long-lived variety) are always formed as compact objects, and that some, for an as-yet unidentified reason, expand for the duration of their lives while the remainder do not. In this case we are searching for a dynamical explanation of the trend – a problem ideally suited to large-scale realistic N-body modelling.

A number of possible dynamical mechanisms for the radius-age trend have previously been proposed and investigated; however, none can fully explain the observed distribution of clusters. For example, a strongly varying intra-cluster IMF (Elson et al. 1989) or binary star fraction (Wilkinson et al. 2003) have been ruled out as viable explanations, as have the effects of a temporally varying tidal field, such as that which a cluster on a highly elliptical orbit might feel (Wilkinson et al. 2003). In the present case study, we consider the effects of a population of stellar-mass black holes (BHs). Usually, such objects are assumed to receive a large velocity kick at formation in a supernova explosion, which means they rapidly escape from their cluster. Therefore, we consider here the effects if a star cluster can somehow retain a fraction of these BHs. Large-scale realistic N-body modelling has been conducted to investigate this question, using the NBODY4 code (Mackey et al. 2007, 2008a).

As discussed in more general terms earlier in this chapter, there are two key aspects to conducting realistic N-body simulations. The first is to develop model clusters that have properties as similar as possible to those observed for the real LMC and SMC clusters. The second concerns the data analysis. Since we are trying to reproduce an observationally defined trend, we must obtain measurements from the simulations that are directly comparable to the measurements which were determined for the real clusters. The most logical way to do this is to perform simulated observations of the simulated clusters, in just the manner that the genuine observations were conducted. This will be discussed in more detail in Sect. 14.3.4, below.

Returning then to the question of setting up realistic models, we must first identify the key characteristics of the youngest LMC and SMC clusters. These are summarized in Fig. 14.2. All the observed young LMC and SMC clusters have profiles with cores (rather than cusps) – even the ultra-compact cluster R136 exhibits a small core (see e.g. the detailed discussion in Mackey & Gilmore 2003a, and the references therein). The radial brightness profiles of the youngest clusters are well fit by models of the form (Elson, Fall & Freeman 1987; EFF models hereinafter):

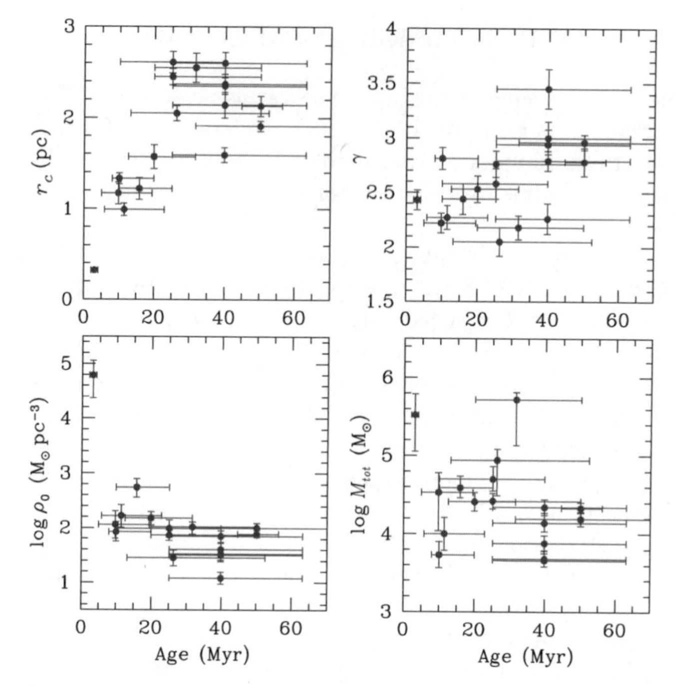

Fig. 14.2. Properties of the youngest massive clusters observed in the LMC and SMC. Structural data are taken from Mackey & Gilmore (2003a,b), while the central density and total mass estimates are taken from McLaughlin & van der Marel (2005)

$$\mu(r) = \mu_0 \left(1 + \frac{r_p^2}{a^2}\right)^{-\gamma/2} , \tag{14.1}$$

where r_p is the projected radius (i.e. the radius on the sky), μ_0 is the central surface brightness, γ determines the power-law slope of the fall-off in surface brightness at large radii, and a is the scale length. It is straightforward to show that this latter parameter is related to the core-radius by:

$$r_c = a(2^{2/\gamma} - 1)^{1/2} . \tag{14.2}$$

Typical values for these structural parameters in young LMC and SMC clusters are $r_c \leq 2$ pc and $\gamma \sim 2.6$. Excluding R136, the young LMC and SMC clusters generally have central densities in the range $1.6 \leq \log \rho_0 \leq 3.0$, and total masses in the range $4 \leq \log M_{\text{tot}} \leq 5$. R136 is the youngest cluster in the sample, ~ 3 Myr, and also has the greatest central density with $\log \rho_0 \approx 4.8$.

Given these observational constraints, we generate model clusters in virial equilibrium according to an EFF profile with $\gamma = 3$ – this is the member of the EFF family of models closest to $\gamma \sim 2.6$, which possesses analytic expressions for the radial dependence of the enclosed mass and isotropic velocity dispersion. Full details of the generation procedure may be found in Mackey et al. (2008a).

Using the IMF of Kroupa (2001), we assign a range of masses to the stars in a model cluster according to the multiple-part power law

$$\xi(m) \propto m^{-\alpha_i}, \tag{14.3}$$

where $\xi(m)\mathrm{d}m$ is the number of single stars falling in the mass interval m to $m + \mathrm{d}m$, and the exponents α_i are:

$$\begin{aligned} \alpha_0 &= +0.3 \pm 0.7, \quad 0.01 \leq m/\mathrm{M}_\odot < 0.08 \\ \alpha_1 &= +1.3 \pm 0.5, \quad 0.08 \leq m/\mathrm{M}_\odot < 0.50 \\ \alpha_2 &= +2.3 \pm 0.3, \quad 0.50 \leq m/\mathrm{M}_\odot < 1.00 \\ \alpha_3 &= +2.3 \pm 0.7, \quad 1.00 \leq m/\mathrm{M}_\odot. \end{aligned} \tag{14.4}$$

Kroupa (2001) derived his IMF from a large compilation of measurements from young stellar clusters, including many in the LMC. This is in contrast with many other widely used IMFs – the Kroupa (2001) IMF is therefore the most suitable for the present N-body modelling.

We impose a stellar mass range 0.1–100 $\mathrm{M}_\odot$ for our model clusters. The lower mass limit is set by the lowest mass stars for which stellar evolution routines are incorporated in NBODY4, while the upper limit is consistent with the observations of very young massive star clusters. Note that the lower mass limit means that in practice only the exponents α_1–α_3 in the IMF described above are utilized.

Selection of the IMF described above, along with the requirement that our model clusters have masses typical of those of young LMC and SMC clusters (Fig. 14.2), allows the total number of stars in each given model to be assigned. For all present simulations, $N \sim 10^5$ stars, which gives typical initial total cluster masses of $M_{\mathrm{tot}} \sim 56\,000\mathrm{M}_\odot$ (i.e., $\log M_{\mathrm{tot}} \sim 4.75$).

In the interest of maintaining a high degree of realism in the simulations, model clusters are evolved in a weak external tidal field rather than in isolation. This external field is incorporated by imposing the gravitational potential of a point-mass LMC with $M_{\mathrm{g}} = 9 \times 10^9\mathrm{M}_\odot$, and placing the clusters on circular orbits of galactocentric radius $R_{\mathrm{g}} = 6$ kpc. Adopting a point-mass LMC is a significant over-simplification; however, as described by Wilkinson et al. (2003), the gradient of this potential is within a factor of 2 of that in the LMC mass model of van der Marel et al. (2002) at the assigned orbital radius. In any case, the relatively weak tidal field of the LMC does not significantly affect the core-radius evolution of its massive stellar clusters (Wilkinson et al. 2003).

Incorporating a tidal field in the N-body modelling serves two important purposes. First, it allows the gradual evaporation of stars from a simulated cluster to be modelled in a self-consistent fashion, so that the rates of evaporation between different models with the same external potential and escape criterion may be easily compared. Second, it lets us impose a natural scaling between N-body units, in which the integration is computed, and physical units, which we use to compare the model cluster to observational results. In particular, the length scaling controls the physical density of the cluster and hence the physical time-scale on which internal dynamical processes occur. The tidal radius, r_t, of a star cluster (mass $M_{\rm cl}$) on a circular orbit of radius $R_{\rm g}$ in the external point-mass potential of a point-mass galaxy (mass $M_{\rm g}$) may be estimated from the relationship (King 1962):

$$r_t = R_{\rm g} \left(\frac{M_{\rm cl}}{3M_{\rm g}} \right)^{\frac{1}{3}} . \tag{14.5}$$

The initial tidal radius of the cluster, estimated via (14.5), is used to determine the length-scale conversion. It is important to check that this results in cluster densities consistent with those observed for young LMC and SMC clusters – we quantify this more carefully below.

Since we wish to examine the dynamical effects of populations of stellar-mass black holes on star cluster evolution, it is important to consider how such objects may be incorporated naturally into our N-body simulations. The most unambiguous method is to generate black holes from the supernova explosions of the most massive stars in the cluster. NBODY4 includes such formation in its stellar evolution routines; however, we added small modifications so that the progenitors, masses, and natal kicks of the generated BHs could be controlled. To ensure a sizeable population of BHs, we form one whenever a star with an initial mass greater than $20\,{\rm M}_\odot$ explodes. For a cluster with $N = 10^5$ stars and a Kroupa (2001) IMF with an upper mass limit of $100\,{\rm M}_\odot$, this results in $N_{\rm BH} = 198$ BHs. When a BH is formed, we assign it a mass randomly selected from a uniform distribution in the range $7 < M_{\rm BH} < 13\,{\rm M}_\odot$, so that the mean mass is $10\,{\rm M}_\odot$. This process is again undoubtably a simplification; however, the mass characteristics of the progenitors and BHs are reasonably consistent with theoretical expectations (see e.g., Zhang, Woosley & Heger 2007) as well as observational evidence (see e.g., Casares 2006).

The natal kicks which the BHs are given are very important. A large kick (a few hundred km s^{-1}) is usually used for both black holes and neutron stars. This generally means no BHs are retained in a typical cluster, which might have an escape velocity of 10–20 km s^{-1}. In order to control the retention fraction we modified NBODY4 so that the natal kicks given to generated BHs could be easily controlled, and varied from run to run.

It is also important to specify the metallicity of the model clusters, since this parameter strongly affects the stellar evolution and hence the mass loss at early times in the N-body simulations (see e.g. Hurley et al. 2004). In the present example we select solar metallicity ($Z = 0.02$) to be consistent with observations of young clusters in the Magellanic Clouds. However, it is important to be aware that since there is a strong age-metallicity relation in both Clouds, there is a metallicity gradient across the radius-age diagram (i.e. the oldest clusters are also very metal poor). In any ensemble of N-body runs seeking to explain the radius-age trend, the significance of this fact should be investigated (although we do not consider it any further in the present example).

One additional key aspect of young LMC and SMC clusters is that those which have been observed in detail generally exhibit some degree of mass segregation – that is, the most massive stars in a given cluster are preferentially located near the centre of that cluster. For example, mass segregation has been observed in the LMC clusters NGC 1805 and NGC 1818 (de Grijs et al. 2002a,b) and R136 (Malumuth & Heap 1994; Brandl et al. 1996; Hunter et al. 1995, 1996), as well as the SMC cluster NGC 330 (Sirianni et al. 2002). It does not necessarily follow from these observations that mass segregation occurs in *all* young LMC and SMC clusters, and nor is it clear whether the segregation is primordial or dynamical in the clusters where it has been found; however, mass segregation is clearly an important factor, which we must consider in our models.

In order to produce mass-segregated clusters in a self-consistent fashion (i.e. close to virial equilibrium, with all members having appropriate velocities) a cluster is first generated as described above (with no mass segregation). We then implement a mass-truncation, setting all stars in the cluster with masses greater than $8\,M_\odot$ to have mass $8\,M_\odot$. Next, the cluster is evolved dynamically using NBODY4 but with the stellar evolution routines turned off. Hence the cluster begins to dynamically relax and mass segregate. The degree of primordial mass segregation is controlled by the length of time for which the cluster is "pre-evolved". The truncation limit of $8\,M_\odot$ is selected so that the pre-evolution can extend for a reasonable period (a few hundred Myr) without the most massive stars sinking to the cluster centre, forming a collapsed core, and ejecting each other through close interactions. Once the desired pre-evolution time is reached, the simulation is halted, the mass-truncated stars replaced with their original masses, and the resulting cluster taken as the input for the simulation proper.

The truncation and replacement process introduces some small inconsistencies in the velocities of some stars, once the simulation proper is started. However, these are small, and are erased by dynamical processes within a few crossing times. In addition, during the pre-evolution phase, some stars escape from the cluster. This process is very gradual, however, and even clusters with long pre-evolution times (several hundred Myr) only lose a few per cent of their mass. Since the scaling of all models is set by (14.5), which varies as the cube root of the cluster mass, the differences in scaling between non-segregated and primordially segregated clusters are tiny.

It is important to check whether this artificial mass segregation process produces clusters that have properties comparable to the observed mass-segregated young LMC and SMC clusters. We do this by comparing simulated observations of the model clusters with the genuine cluster observations. This is considered in more detail in the next section, and in Sect. 14.3.7.

14.3.3 Summary of N-Body Runs

With the initial conditions specified as described above, four N-body simulations are required to address the question under consideration – namely the dynamical effects of a population of stellar-mass black hole remnants on massive star cluster evolution – at a basic level. The parameter space of interest is spanned by two types of clusters – those with no primordial mass segregation and those with a strong degree of primordial mass segregation. In each of these types, we consider evolution with no black holes (that is, where the natal kick is large so the retention fraction is zero) and a significant population of black holes (that is, where the natal kick is zero so the retention fraction is unity).

These four runs cover the extreme limits of the parameter space we aim to investigate, and hence are expected to cover the extreme limits of cluster evolutionary behaviour. Subsequent to their completion, it is sensible to check this is indeed the case, by adding further runs which sample intermediate regions of the parameter space (e.g. a cluster with only moderate mass segregation, or a black hole retention fraction around 0.5). Although such runs have been carried out, we will not consider them in any detail here.

The properties of the four N-body runs are listed in Table 14.1. Note that for Runs 3 and 4, "strong mass segregation" is rather difficult to define numerically; however, a pre-evolution duration of ~ 450 Myr is adequate to reproduce observational results of mass segregation in young Magellanic cloud clusters. This aspect is discussed in more detail in Sect. 14.3.7 below. Each model is run until late times ($T_{\rm max} > 10$ Gyr), which match the ages of the oldest Magellanic Cloud globular clusters. Each such run took approximately 2 weeks of full-time calculation on the GRAPE-6 at the Institute of Astronomy in Cambridge. The first week takes any given run to an age of ~ 1.5 Gyr after which time the computation becomes rather swifter, mainly due to decreasing particle number and much less demanding stellar evolution calculations.

We selected data for output every 1.5 Myr at ages less than 100 Myr, and every 15 Myr thereafter. This allowed close examination of the early phases

Table 14.1. Details of N-body runs and initial conditions. Each cluster begins with N_0 stars with masses summing to $M_{\rm tot}$, and initial central density ρ_0. Initial cluster structure is "observed" to obtain r_c and γ. Each model is evolved until $T_{\rm max}$

Name	N_0	$\log M_{\rm tot}$ ($M_\odot$)	$\log \rho_0$ ($M_\odot\,{\rm pc}^{-3}$)	r_c (pc)	γ	Initial mass segregation	Black hole kicks	$T_{\rm max}$ (Myr)
Run 1	100 881	4.746	2.31	1.90	2.96	None	Large	16 996
Run 2	100 881	4.746	2.31	1.90	2.96	None	Zero	10 668
Run 3	95 315	4.728	4.58	0.25	2.33	452 Myr	Large	11 274
Run 4	95 315	4.728	4.58	0.25	2.33	452 Myr	Zero	10 000

of cluster evolution and suitable resolution at all times to consider in detail the development and evolution of any black hole populations. Typically, each $\sim 10\,\mathrm{Gyr}$, $N \sim 10^5$ star run takes up $\sim 10\,\mathrm{Gb}$ of space on disk. This can be reduced considerably by compressing the output for storage and backup.

For each run, we measured the initial cluster mass, central density, and the structural parameters r_c and γ – these are all listed in Table 14.1. The structural parameters were derived from simulated observations, as discussed in Sect. 14.3.4, below. It is worth re-emphasizing how closely these correspond to the observed quantities for the youngest massive clusters in the Magellanic clouds. This can be seen explicitly by comparing the values listed in Table 14.1 with the plots in Fig. 14.2. In addition, the evolution of the central density (ρ_0) over the first tens of Myr for Runs 1 and 3 is plotted in Fig. 14.3.

The model clusters with no primordial mass segregation have $r_c \sim 1.9\,\mathrm{pc}$, $\gamma \sim 3$, and $\log \rho_0 \sim 2.3$. These clusters therefore appear very similar to a number of Magellanic Cloud clusters with ages of $\sim 20\,\mathrm{Myr}$. In contrast, the heavily mass-segregated model clusters have much smaller cores and higher central densities, with $r_c \sim 0.3$ pc and $\log \rho_0 \sim 4.8$. They also have flatter power-law fall-offs, with $\gamma \sim 2.3$. In this respect, they look very similar to the very compact massive young LMC cluster R136, which has an age of $\sim 3 - 4\,\mathrm{Myr}$.

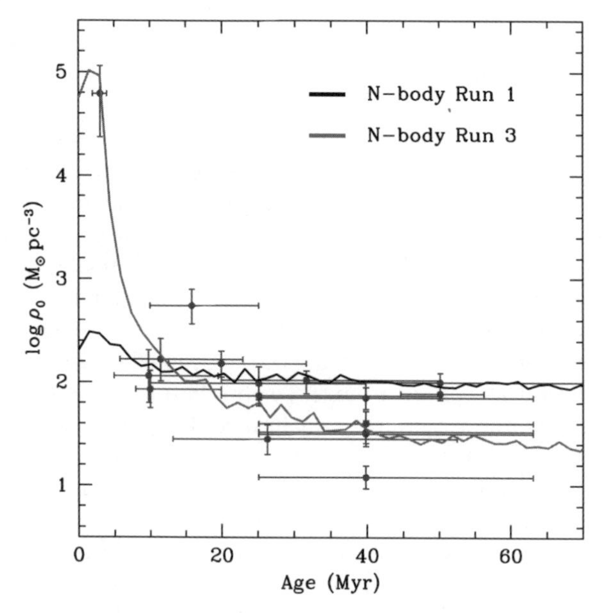

Fig. 14.3. Early evolution of the central density ρ_0 for Runs 1 and 3 (*solid lines*), compared with the observations for young LMC clusters (points). Run 1 has no primordial mass segregation while Run 3 is heavily segregated. Run 3 looks very similar to R136 at early times but by a few tens of Myr looks more like other observed young LMC and SMC clusters, and indeed rather similar to Run 1

14.3.4 Simulated Observations of Core Radius Evolution

As described in Sect. 14.2, a key advantage of running realistic N-body simulations is that they allow the opportunity to conduct simulated observations on the models. In particular, this is a vital ingredient if the problem under investigation is defined observationally. If this is the case, it is essential to ensure that whatever measurements obtained from the N-body modelling are directly comparable to those determined observationally.

In our present case study, we are investigating the origin of the radius-age trend in the LMC and SMC star cluster systems. This trend is defined observationally, through measurements of cluster core radii. To determine whether our N-body simulations have been successful in reproducing the trend or not, a directly comparable parameter must be obtained from them. The most unambiguous method of achieving this is by passing the N-body data through as similar a process as possible to that which generated the observed measurements.

The first step is to identify and account for the limitations of the cluster observations. In any given LMC or SMC cluster in the sample displayed in Fig. 14.1, only a fraction of the stars in the cluster were imaged and used to produce the brightness profiles from which core-radius measurements were made. There are two primary reasons for this. First, the HST field of view (whether it be with WFPC2 or ACS) is not large enough to cover the full spatial extent of an LMC or SMC cluster. The core is imaged but the radial profile is cut off typically at ~ 20 pc, much less than the nominal tidal radius of roughly ~ 40–50 pc .

Second, the exposure times are too short to see the faintest stars in the cluster, and too long to allow accurate measurement of the brightest stars. This point is illustrated in Fig. 14.4. The displayed colour-magnitude diagram (CMD) is from ACS imaging of 47 Tuc, a bright Galactic globular cluster. The main sequence is clearly visible, as is the turn-off. The image exposure times were not long enough to measure stars fainter than $\sim$6 mag below the turn-off. A large fraction of the stars in 47 Tuc are fainter than this (for example, no white dwarfs were observed), but would not be included in any star counts used to construct a brightness profile from these observations. At the bright end, the data are cut off just above the sub-giant branch. Brighter stars (i.e. all the red giant branch and horizontal branch stars) do appear on the images; however, the exposure times were long enough that these objects were saturated on the CCD. That is, the pixels imaging these stars have received too many photons and the signal has overflowed into neighbouring areas. Accurate photometry cannot be done above a certain level of saturation, hence the bright cut-off limit on the CMD in Fig. 14.4. None of the saturated stars would be counted in a radial brightness profile either.

Exactly similar processes apply to the LMC and SMC clusters we are trying to model. Each has a bright and faint cut-off determined by the exposure times of the imaging. These are illustrated in Fig. 14.4, for the complete

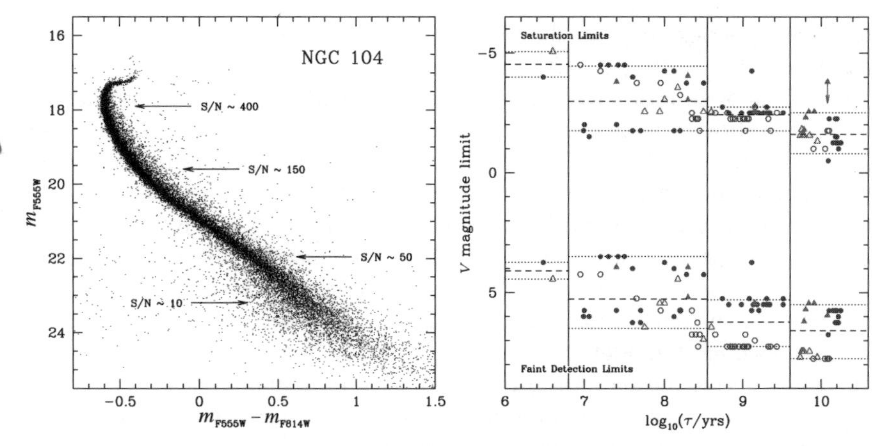

Fig. 14.4. *Left:* Colour-magnitude diagram of the Galactic globular cluster 47 Tuc from HST/ACS imaging. The measured signal-to-noise ratios for the detected stars are indicated in several places. The bright and faint cut-offs are evident. *Right:* Bright and faint stellar detection limits on the HST/WFPC2 and ACS images of LMC and SMC clusters used for the measurements presented in Fig. 14.1. LMC clusters are blue circles, while SMC objects are magenta triangles. Filled symbols represent the WFPC2 imaging described in Mackey & Gilmore (2003a,b), while open symbols are the ACS imaging from Mackey et al. (2008b). Clusters are split into four age bins, shown with *solid vertical lines.* Within each bin, the mean bright and faint detection limits are marked by *dashed lines*, while the approximate maximum scatter about each mean is marked by a pair of *dotted lines*

sample. The clusters are split into four age bins, delineated on the plot with solid vertical lines. Within each of these, the mean bright and faint detection limits are marked with dashed lines, and the approximate maximum scatter about these means with dotted lines. From this figure it is clear that the bright and faint limits, and hence the portion of the mass function sampled by the observations, vary systematically with cluster age. This is due to the fact that observations of star clusters in the LMC and SMC are commonly aimed at targeting stars near the main-sequence turn-off. Consequently, the required exposure time increases with cluster age, meaning that both the brighter and the fainter detection limits decrease with age.

To observe our model clusters, we pass the N-body data at each output time through a measurement pipeline essentially identical to that used to obtain structural quantities for the real LMC and SMC cluster sample (full details of the observational pipeline may be found in Mackey & Gilmore 2003a). At a given output time, the luminosity and effective temperature of each star in the cluster is first converted to magnitude and colour, using the bolometric corrections of Kurucz (1992) (see also e.g., Hurley et al. 2005). We also convert the position and velocity of each star to physical units using the appropriate length-scale and velocity factor (see Sect. 14.3.2). With this completed, we next impose the bright and faint detection limits appropriate to the output time (these are the dashed mean limits in Fig. 14.4). This leaves an ensemble of stars with which to construct a surface brightness profile. We project the three-dimensional position of each star onto a plane (to mimic the observation of a cluster projected onto the sky), construct annuli of a given width about the cluster centre, and calculate the surface brightness in each annulus. For consistency with the observational pipeline, we use a variety of annulus widths so that both the bright inner core and the fainter outer regions of the cluster are well measured. Measurements are truncated at a radius commensurate with that imposed by the HST field of view, as discussed above. We next fit an EFF model to the resulting surface brightness profile, and from this model derive the structural parameters, in particular the core radius. To reduce noise we repeat this process for each of the three orthogonal planar projections at each output time and average the results.

14.3.5 Results from the Simulations

In this chapter, we are primarily concerned with investigating the processes involved in running realistic N-body simulations and analysing the resulting data, illustrated through the examination of a case study. Therefore, we will not delve deeply into the results of the four N-body runs themselves (the interested reader is referred to Mackey et al. (2008a) for full details). Nonetheless, it is interesting to take a moment to consider these results in the context of the radius-age trend described in Sect. 14.3.1.

Because we have taken care to construct models where N is sufficiently large that no scaling with N is necessary to interpret the output, and because we have taken care to obtain measurements closely mimicking the real observations, it is legitimate to directly plot the core-radius evolution of our N-body models over Fig. 14.1. This is shown in Fig. 14.5 for Runs 1 and 2, and Fig. 14.6 for Runs 3 and 4.

The simplest model is Run 1, which is not primordially mass-segregated, and in which black holes formed in supernova explosions receive a large natal kick, ejecting them almost immediately from the cluster. The retention fraction is thus zero. As could be expected, the evolution follows the standard path expected for an ordinary globular cluster (see e.g., Meylan & Heggie 1997). There is an initial phase of violent relaxation and mass loss due to stellar evolution, which lasts for the first $\sim$100 Myr. This phase is hardly reflected in the core-radius evolution, because as there is no primordial mass segregation, the mass loss is distributed widely over the cluster. The remainder of the cluster evolution consists of a slow contraction of the core as dynamical mass segregation is established, and the cluster moves towards core collapse, which happens near the end of the run at $\sim$15 Gyr.

Run 2 is identical, except for the fact that natal black hole kicks are set to be zero, so that the retention fraction is one. This results in a population of 198 stellar mass black holes within the cluster. Initially, the core radius

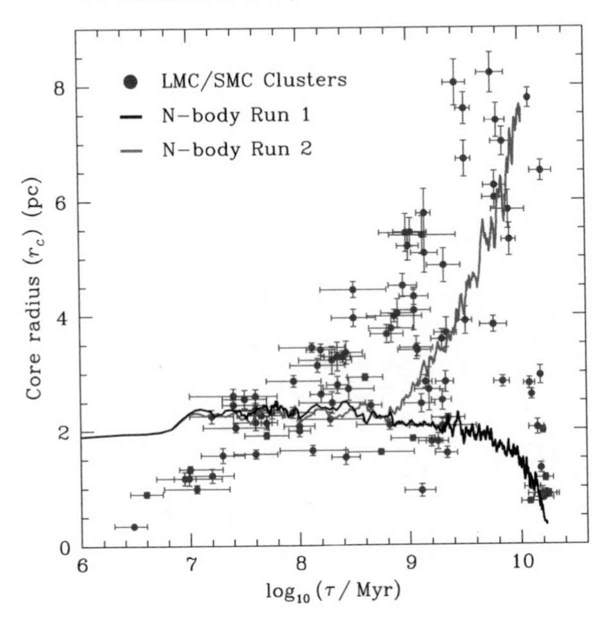

Fig. 14.5. Core-radius evolution of N-body Runs 1 and 2. Both runs have no primordial mass segregation and start from identical initial conditions. The only difference between them is the retention fraction of stellar-mass black holes (zero and one, respectively). Run 1 evolves exactly as expected, with the main trend being a slow contraction in r_c as the cluster relaxes and moves towards core collapse. In stark contrast, Run 2 evolves very similarly up to a point, after which strong expansion in the core radius is observed. The presence of 198 stellar-mass black holes in this cluster thus leads to strikingly different core radius evolution

evolution appears identical to that of Run 1. The mass loss phase passes and relaxation processes set in. However, starting at about ∼500 Myr, the core radius of Run 2 begins to expand dramatically. This is due to the dynamical influence of the black holes. These objects, because they are dark, are not included in the core-radius measurements (they fall far below the faint cut-off on the CMD). All we can see is how the stars which are included in the profile calculations are affected. After their formation, and a few tens of Myr of stellar evolution within the cluster, the black holes are by far the most massive cluster members. They therefore sink rapidly to the cluster centre via dynamical mass segregation, and, after a few hundred Myr, form a compact black hole core. The densities within this core are such that close encounters between BHs are frequent, and soon black hole binaries are formed. Encounters between binary BHs and single BHs, and between binary BHs and other binary BHs scatter single BHs out of the core, which then sink back in again via mass segregation. Since an individual BH may undergo this process a number of times, significant energy is transferred to the core stars through the repeated mass segregation. In addition, in very strong encounters, BHs are ejected from

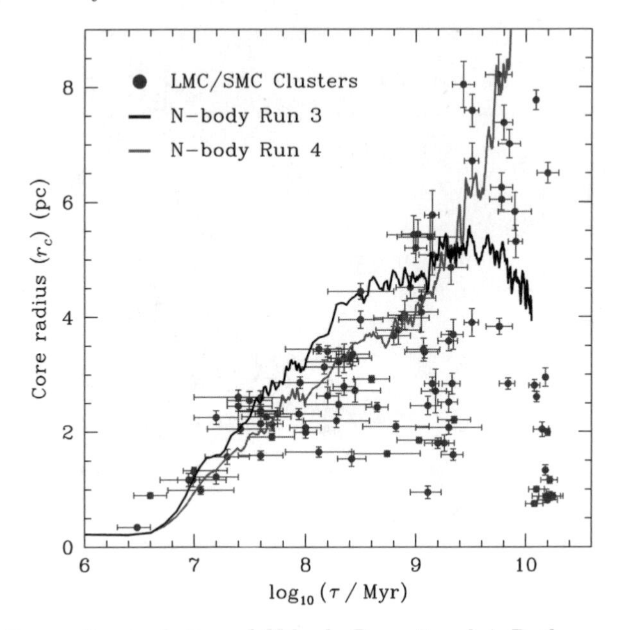

Fig. 14.6. Core-radius evolution of N-body Runs 3 and 4. Both runs have strong primordial mass segregation and start from identical initial conditions. The only difference between them is the retention fraction of stellar-mass black holes (zero and one, respectively). Compared with Runs 1 and 2, there is strong early expansion due to the concentrated central mass loss. Subsequently, Run 3, without black holes, begins to mass segregate and contract, whereas Run 4 undergoes continued expansion due to the dynamical effect of its black hole population

the cluster. By the end of the run, only about 30% of the original population remains. This ejection process serves as an additional heating mechanism.

In contrast to Runs 1 and 2, the primordially mass-segregated Runs 3 and 4 expand dramatically at early times. Given that these two runs have the same IMF as Runs 1 and 2, this early expansion must be a direct result of their different initial structure. Unlike in Runs 1 and 2, where the early mass loss from stellar evolution is spread throughout the cluster, in Runs 3 and 4 it is heavily concentrated in the core, which in turn reacts with strong expansion. This expansion lasts for the first ~250 Myr, by which time the highest mass stars in the cluster have completed their evolution, and the stellar mass-loss rate has been significantly reduced. After this point, the evolution follows very similar paths to those for Runs 1 and 2. The model with no black hole retention (Run 3) gradually begins to dynamically relax, and mass segregation sets in causing slow contraction. Because this cluster expanded at early times, it is less dense than Run 1, and hence its relaxation time is longer. Thus, it does not reach a state of core collapse by the end of the simulation. In contrast, Run 4, where the black hole retention fraction is unity, undergoes core expansion for the full duration of its evolution. As in Run 2, the BHs segregate to the centre of the cluster and form a compact core by $\sim$ 500 Myr. This initiates BH–BH interactions, inducing the expansion. Because of its additional early expansion, Run 4 reaches larger core radii than Run 2 at late times, evolving off the top of the figure to $r_c \sim 12$ pc.

This set of four runs hence demonstrates that we can cover all regions of the observed cluster distribution in the radius-age plane simply by varying two basic parameters within the ranges constrained by observation – the degree of initial mass segregation in a cluster, and the retention fraction of stellar mass black holes. Additional runs have been performed, which demonstrate that, as should be expected, models with intermediate degrees of mass segregation, or an intermediate BH retention fraction, evolve somewhere between the four extremes modelled in the present example.

14.3.6 More Detail on Simulated Observations of r_c

As well as directly addressing the question of the origin of the radius-age trend, conducting simulated observations of the four N-body models described above also allows us to investigate the quality of the data reduction carried out on the original observational data.

For example, when we examined the bright and faint saturation limits present in the imaging and constructed Fig. 14.4, it became clear that these limits vary systematically with cluster age. We were able to implement this variation in our simulated observations of the N-body clusters, and hence account for any systematic effect on the measurement of the core radius. However, it also became clear that at any given age there is considerable scatter in the bright and faint limits between clusters – something we did not account for in the simulated observations. This raises the question as to whether this cluster-to-cluster variation at similar ages introduces significant scatter into the observed distribution of clusters on the radius-age diagram. Furthermore, if it does, is it possible to reduce this scatter by re-analysing the observational data and artificially imposing uniform bright and faint limits at a given age.

To investigate these questions, we re-calculated the core-radius evolution of the N-body clusters using simulated observations with new bright and faint limits implemented in place of the mean limits previously adopted. In these new calculations, we used the “maximum scatter” limits marked in Fig. 14.4 – in one set we used the brightest pair of limits at any given age and in a second set we used the faintest pair of limits at any given age. The resulting evolution is plotted in Fig. 14.7, along with the evolution derived using the mean bright and faint limits.

It can be seen from this figure that in all four runs for the majority of the evolution the selected bright and faint limits make little difference to the calculated core radius, at least at the level of the cluster-to-cluster scatter determined to be present in the Magellanic Cloud cluster observations. However, in the case where a cluster is heavily mass-segregated *and* where it still

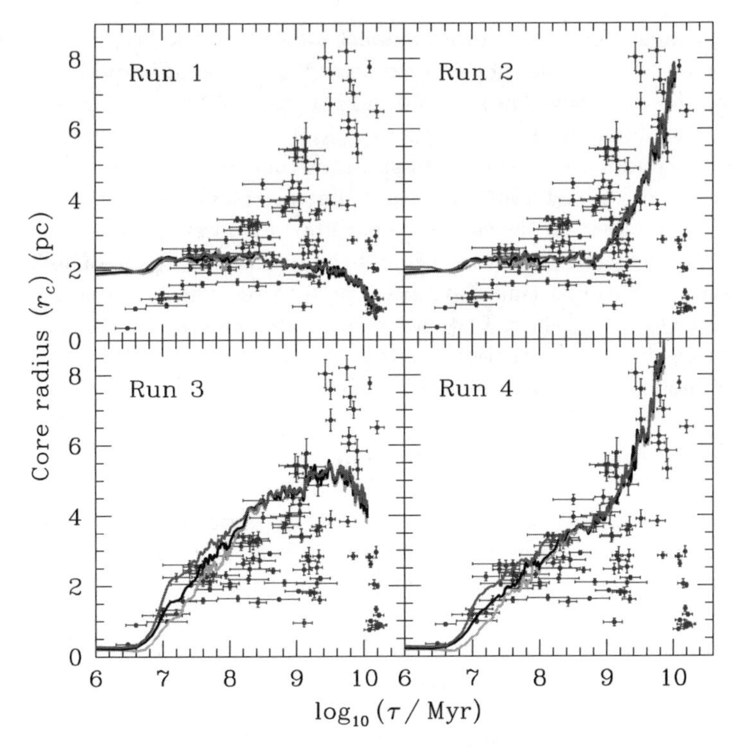

Fig. 14.7. Core-radius evolution derived from the simulated observations with three different sets of saturation and faint limits implemented, as indicated in Fig. 14.4. The *black lines* represent r_c calculated using the mean limits, as in Figs. 14.5 and 14.6, while the *magenta lines* represent r_c calculated using the brightest maximum scatter limits, and the *green lines* represent r_c calculated using the faintest maximum scatter limits. Agreement between the three is excellent, except in the case where a cluster is mass-segregated *and* young (so that it still possesses massive luminous stars)

possesses massive, luminous stars, the adopted bright and faint limits make a significant difference to the measured core radius.

This result is readily understood. In any given cluster, since we construct brightness profiles rather than simple stellar density profiles, the presence of any luminous stars strongly weights the resulting structural calculations. In particular, when mass segregation is present, the most luminous stars are predisposed to lie near the cluster centre, resulting in a small core radius. Hence, if the saturation limit is varied in the observations of such a cluster, different numbers of luminous stars will be included in the calculation, resulting in a strong variation in the measured r_c. This is clearly evident for Runs 3 and 4 at early times in Fig. 14.7, and suggests that the cluster-to-cluster scatter in saturation limits present in the observational data for the youngest clusters may have introduced significant scatter in the positions of clusters in the radius-age diagram for ages up to ~200 Myr. It would therefore be worthwhile re-reducing the observational data for clusters younger than this limit, artificially imposing uniform bright and faint detection limits. With this done, a major source of scatter in the positions of the youngest clusters on the radius-age diagram would be removed.

This example shows that while genuine cluster observations define simulated observations to be carried out on any N-body modelling of these clusters, additional simulated observations of the N-body models can lead to improvements in the genuine cluster observations, in an iterative process. This illustrates one of the key advantages to running direct, realistic N-body simulations and implementing a sophisticated data reduction procedure.

One additional aspect worth a brief investigation is a comparison between the measured core-radius (now using the mean bright and faint limits again) and the core-radius computed internally by NBODY4, which one might be tempted to use rather than proceeding down the more complicated and time-consuming path of implementing simulated observations.

The core-radius calculated by NBODY4 is more correctly termed the density radius ($r_{\rm d}$) and is based on a quantity described by Casertano & Hut (1985), so that $r_{\rm d}$ is defined as the density-weighted average of the distance of each star from the density centre of the cluster (see e.g., Aarseth 2003). The local density at each star is computed from the mass within the sphere containing the six nearest neighbours. This parameter was designed to behave in a similar manner to the observational core radius; however, as we will see, it can be strongly biased by particles that would not be included in any genuine observation aimed at deriving the structural parameters of a cluster.

In Fig. 14.8, comparison between the observational core radius, as calculated above in Sect. 14.3.4, and the density radius computed by NBODY4 is presented for each of the four runs. For Runs 1 and 3, where black holes are not retained, the agreement between the two radii is generally satisfactory, although there is a significant tendency for the density radius to be larger than the observational core radius. In comparison for Runs 2 and 4 where black holes are retained, the agreement is very poor indeed, with no correlation between the behaviour of the two radii. The reason for this is simple – black holes are included in the computation of r_d, but not included in the computation of r_c (since they are dark particles). Hence, for Runs 2 and 4 r_d is effectively tracing only the evolution of the black hole sub-system rather than the distribution of the luminous matter.

Based on this result, it is clear why one should be very careful about selecting measurements that are directly comparable to any observations being modelled. If two disparate quantities are compared, the potential for serious mistakes exists. In the above example, if the density radius from NBODY4 had been taken as a proxy for the observational core-radius instead of making use of the simulated observations method, the dramatic expansion evident in Figs. 14.5 and 14.6 may not have been noticed, and an ultimately successful explanation for the radius-age trend possibly not investigated any further.

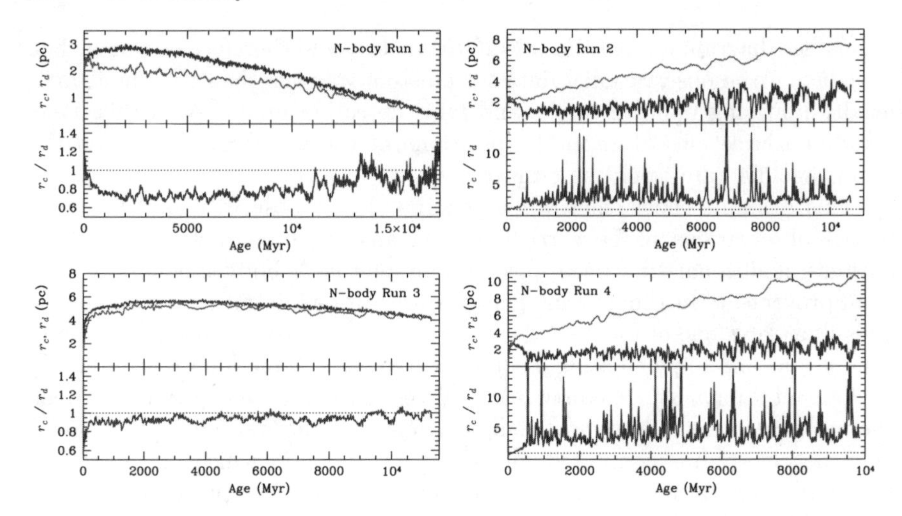

Fig. 14.8. Comparison between the evolution of the core radius r_c, derived from simulated observations, and the density radius r_d implemented in NBODY4, for each of the four N-body runs. In each plot, the *upper panel* shows the evolution of the two radii (r_c in *magenta*; r_d in *blue*), while the *lower panel* shows the evolution of the ratio r_c/r_d. A ratio of unity is marked with a *dashed line*. In runs with black hole populations, the density radius is a poor match to the observational core radius

14.3.7 Simulated Observations of the Initial Mass Segregation

As a final example, it is worth investigating the fact that we can use detailed simulated observations to examine the quality of the initial conditions we constructed in Sect. 14.3.2, especially for the primordially mass-segregated models. We have already demonstrated that these model clusters closely resemble the youngest massive LMC and SMC clusters in terms of their basic structural parameters, central densities and masses. However, we would like to verify that the method used to primordially segregate these clusters produces mass segregation similar to that observed in genuine objects. Ideally, we would also like to integrate stellar velocities into the initial conditions (so that we can see whether the assumption of virial equilibrium is valid); however, unfortunately, suitably detailed internal velocity measurements for young, massive Magellanic Cloud clusters do not yet exist.

Nonetheless, detailed observations of the radial dependence of the mass function in such clusters do exist. In particular, there are three studies that are very useful to us – that of Hunter and collaborators for R136 (Hunter et al. 1995, 1996); that of de Grijs and collaborators for NGC 1805 and NGC 1818 (de Grijs et al. 2002a,b); and that of Sirianni and collaborators for NGC 330 (Sirianni et al. 2002). R136, in the LMC, is the youngest of these four clusters (~3 Myr), followed by NGC 1805 (~10 Myr) and NGC 1818 (~20 Myr), both also in the LMC, and finally NGC 330 (~30 Myr) in the SMC. This age range allows us to closely trace the evolution of the primordially mass-segregated models, by comparing simulated observations to genuine observations reported in the relevant papers.

Consider first R136, and the work of Hunter et al. (1995, 1996) who used HST/WFPC2 observations of this cluster to measure the slope of the mass function as a function of projected radius. Their results are reproduced in Fig. 14.9. Note that in their work the mass function is represented by a function $\zeta(m)$, which is the number of single stars per *logarithmic* mass interval, as opposed to the mass function $\xi(m)$ defined in (14.3). It is straightforward to

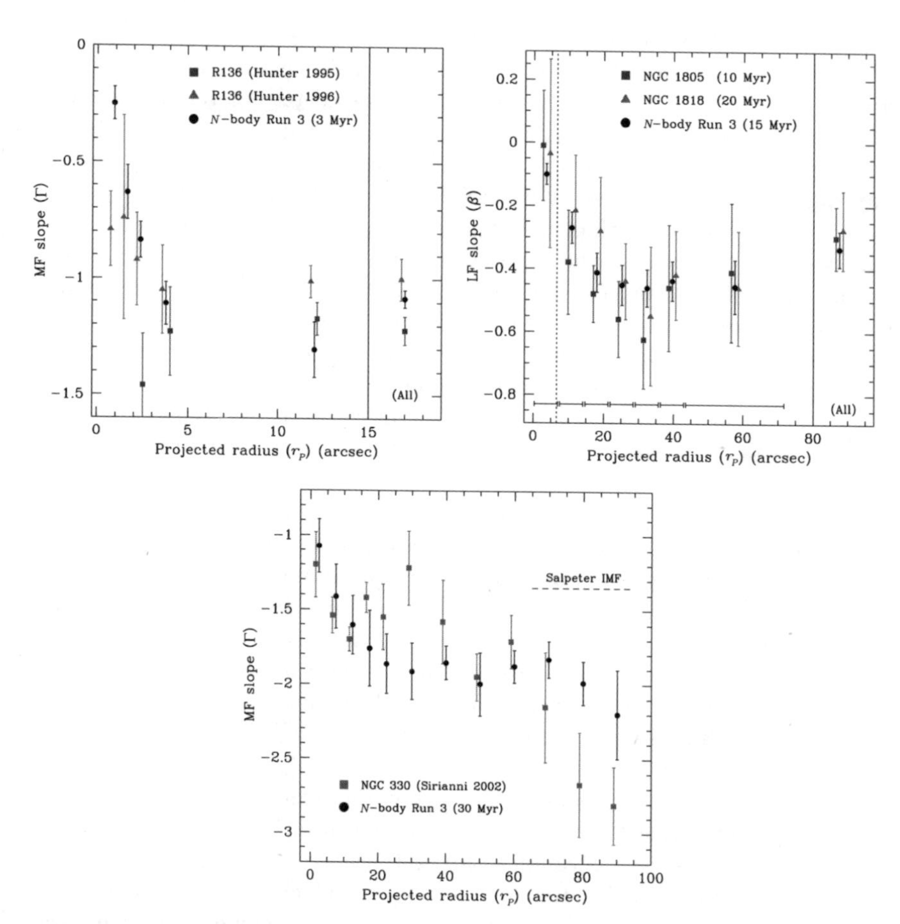

Fig. 14.9. Mass and luminosity function slopes as a function of projected radius for various young LMC and SMC clusters, compared with the results from simulated observations of N-body Run 3. *Upper left:* Mass function slope Γ as a function of radius in R136 in the LMC, from Hunter et al. (1995, 1996). *Upper right:* Luminosity function slopes β as a function of projected radius for NGC 1805 and NGC 1818, in the LMC, from de Grijs et al. (2002b). *Lower:* Mass function slope Γ for NGC 330, in the SMC, from Sirianni et al. (2002)

demonstrate that if a function $\xi(m)$ has an exponent $-\alpha$, then the function $\zeta(m)$, also a power law, has exponent $\Gamma = -\alpha + 1$. Hence, the exponent $\alpha_3 = 2.3$ in the Kroupa (2001) IMF in (14.5) becomes $\Gamma = 1.3$ if the mass function is represented by $\zeta(m)$ rather than $\xi(m)$.

Hunter et al. (1996) found some flattening of the mass function slope with increasingly small radius in R136. Using their annulus widths, together with the specific bright and faint detection limits they list for each annulus, we directly simulated their measurements on N-body Run 3, at an output time of 3 Myr. As usual, it is vital to this process that the annulus widths and bright and faint limits per annulus are exactly reproduced so that directly comparable mass function slopes are derived. Radii in arcseconds were obtained by applying an LMC distance modulus of 18.5, which defines a scale of 4.116 arcsec per parsec. The N-body results are plotted on the relevant panel in Fig. 14.9, and clearly closely match the results of Hunter et al. (1996). The greatest deviation occurs in the innermost part of the cluster, where severe crowding prevented Hunter et al. (1996) from obtaining a secure measurement. It is also worth noting that the overall mass function slope agrees well. This value is flatter than the input value (i.e. flatter than $\Gamma = -1.3$, which is the slope in the mass ranges under consideration), because we are only considering the innermost 15 arcsec of Run 3 to match the radial extent of the genuine R136 measurements. In the outer regions of the N-body cluster the mass function slope is somewhat steeper than the input slope, so that in the entire cluster we obtain $\Gamma = -1.3$. Observations of R136 extending to large projected radii would presumably also find a steeper mass function slope in its outer regions.

We followed a similar procedure to reproduce the observations of de Grijs et al. (2002b) for NGC 1805 and NGC 1818 (in this case we used an intermediate output time from Run 3 of 15 Myr), and the observations of Sirianni et al. (2002) for NGC 330 (we used an output time from Run 3 of 30 Myr). In each case we adopted the annulus widths and annulus-specific detection limits listed by the authors. Note that in the case of NGC 1805 and NGC 1818, the slope β of the luminosity function (rather than the mass function) is measured. This is easily reproduced by using the brightnesses of the N-body stars rather than their masses.

Our N-body measurements are plotted on the relevant panels in Fig. 14.9. In all cases, agreement is close. The largest deviation comes in the outer regions of NGC 330, where Sirianni et al. (2002) note that their measurements are uncertain due to field star contamination (which is not present in the N-body models, and which is not straightforward to include in simulated observations). The fact that this more detailed testing of our initial conditions matches well the best available observations of young LMC and SMC clusters suggests we have managed to set up sufficiently realistic clusters, and validates the procedure we used to generate primordial mass segregation in the N-body models. Once even more detailed observations of young Magellanic cloud clusters are available (say, velocity profiles, for example) these will be able to be incorporated into the initial conditions in a very similar manner.

14.4 Summary

Realistic large-scale N-body modelling of low-mass globular clusters, such as those found in the LMC and SMC, is now feasible and routinely carried out. This is mainly due to the advent of special purpose hardware combined with the ever-increasing sophistication of leading N-body codes, which now incorporate all the major physical processes that occur in star clusters. Direct modelling of typical mass globular clusters is still an order of magnitude out of reach (this is the so-called million body problem); however, within a few years this goal should be reached. The next generation GRAPE machine will shortly be in production (GRAPE-DR), and it is expected that this will provide the required order of magnitude leap. Furthermore, exciting new code developments are taking place. For example, Church (PhD dissertation, University of Cambridge) includes live stellar evolution in an N-body code (as opposed to stellar evolution calculated from analytic formulae). Borch, Spurzem & Hurley (2007) are associating spectral libraries with evolving stars in N-body clusters. These will allow new levels of sophistication and realism in both the models themselves and the types of simulated observations it will be possible to carry out.

This chapter has provided an introduction to what is presently possible within the field of realistic N-body simulations and a general description of various aspects of the philosophy and methodology required for successful simulations and data analysis. A detailed example has demonstrated how the interaction between observation and modelling is essential throughout the process of applying realistic large-scale N-body simulations to real systems.

References

Aarseth S. J., Heggie D. C., 1998, MNRAS, 297, 794
Aarseth S. J., 2003, Gravitational N-Body Simulations. Cambridge Univ. Press, Cambridge
Brandl B., et al., 1996, ApJ, 466, 254
Borch A., Spurzem R., Hurley J., 2007, 328, 662
Casares J., 2006, in Karas V., Matt G., eds, Proc. IAU Symp. 238, Black Holes: From Stars to Galaxies. Cambridge Univ. Press, Cambridge, p. 3
Casertano S., Hut P., 1985, ApJ, 298, 80
de Grijs R., Johnson R. A., Gilmore G. F., Frayn C. M., 2002a, MNRAS, 331, 228
de Grijs R., Gilmore G. F., Johnson R. A., Mackey A.D., 2002b, MNRAS, 331, 245
Elson R. A. W., 1991, ApJS, 76, 185
Elson R. A. W., 1992, MNRAS, 256, 515
Elson R. A. W., Fall S. M., Freeman K. C., 1987, ApJ, 323, 54

Elson R. A. W., Freeman K. C., Lauer T. R., 1989, ApJ, 347, L69
Fukushige T., Makino J., Kawai A., 2005, PASJ, 57, 1009
Giersz M., Heggie D. C., 1994, MNRAS, 268, 257
Heggie D. C., Mathieu R. D., 1986, in Hut P., McMillan S., eds, Lecture Notes in Physics Vol. 267, The Use of Supercomputers in Stellar Dynamics. Springer-Verlag, Berlin, p. 233
Hunter D. A., Shaya E. J., Holtzman J. A., Light R. M., 1995, ApJ, 448, 179
Hunter D. A., O'Neil Jr. E. J., Lynds R., Shaya E. J., Groth E. J., Holtzman J. A., 1996, 459, L27
Hurley J. R., Pols O. R., Tout C. A., 2000, MNRAS, 315 543
Hurley J. R., Tout C. A., Pols O. R., 2002, MNRAS, 329, 897
Hurley J. R., Tout C. A., Aarseth S. J., Pols O. R., 2004, MNRAS, 355, 1207
Hurley J. R., Pols O. R., Aarseth S. J., Tout C. A., 2005, MNRAS, 363, 293
King I. R., 1962, AJ, 67, 471
Kroupa P., 2001, MNRAS, 322, 231
Kurucz R. L., 1992, in Barbuy B., Renzini A., eds, Proc. IAU Symp. 149, The Stellar Populations of Galaxies. Kluwer, Dordrecht, p. 225
Mackey A. D., Gilmore G. F., 2003a, MNRAS, 338, 85
Mackey A. D., Gilmore G. F., 2003b, MNRAS, 338, 120
Mackey A. D., Wilkinson M. I., Davies M. B., Gilmore G. F., 2007, MNRAS, 379, L40
Mackey A. D., Wilkinson M. I., Davies M. B., Gilmore G. F., 2008a, MNRAS, in press
Mackey A. D., et al., 2008b, in prep.
Makino J., 1991, ApJ, 369, 200
Makino J., Fukushige T., Koga M., Namura K., 2003, PASJ, 55 ,1163
Malumuth E. M., Heap S. R., 1994, AJ, 107, 1054
Mardling R. A., Aarseth S. J., 2001, MNRAS, 321, 398
McLaughlin D. E., van der Marel R. P., 2005, ApJS, 161, 304
Meylan G., Heggie D. C., 1997, A&AR, 8, 1
Mikkola S., Aarseth S. J., 1993, Celest. Mech. Dyn. Astron., 57, 439
Mikkola S., Aarseth S. J., 1998, New Astron., 3, 309
Sirianni M., Nota A., De Marchi G., Leitherer C., Clampin M., 2002, ApJ, 579, 275
van der Marel R. P., Alves D. R., Hardy E., Suntzeff N. B., 2002, AJ, 124, 2639
Wilkinson M. I., Hurley J. R., Mackey A. D., Gilmore G. F., Tout C. A., 2003, MNRAS, 343, 1025
Zhang W., Woosley S. E., Heger A., 2008, ApJ, 679, 639

15

Parallelization, Special Hardware and Post-Newtonian Dynamics in Direct N-Body Simulations

Rainer Spurzem[1,5], Ingo Berentzen[1,5], Peter Berczik[1,5], David Merritt[2], Pau Amaro-Seoane[3], Stefan Harfst[4,2,5] and Alessia Gualandris[2,4]

[1]Astronomisches Rechen-Institut, Zentr. Astron. Univ. Heidelberg (ZAH), Mönchhofstrasse 12-14, 69120 Heidelberg, Germany
[2]College of Science, Dept. of Physics, Rochester Instute of Technology, 85 Lomb Memorial Drive, Rochester, NY 14623-5603, USA
[3]Max-Planck Institut für Gravitationsphysik (Albert-Einstein-Institut), Am Mühlenberg 1, D-14476 Potsdam, Germany
[4]Astronomical Institute *Anton Pannekoek* and Section Computational Science, University of Amsterdam, The Netherlands
[5]The Rhine Stellar Dynamical Network
`spurzem@ari.uni-heidelberg.de`

15.1 Introduction

The formation and evolution of supermassive black hole (SMBH) binaries during and after galaxy mergers is an important ingredient for our understanding of galaxy formation and evolution in a cosmological context, e.g. for predictions of cosmic star formation histories or of SMBH demographics (to predict events that emit gravitational waves). If galaxies merge in the course of their evolution, there should be either many binary or even multiple black holes, or we have to find out what happens to black hole multiples in galactic nuclei, e.g. whether they come sufficiently close to merge resulting from emission of gravitational waves, or whether they eject each other in gravitational slingshot interactions.

According to the standard theory, the subsequent evolution of the black holes is divided in three successive stages (Begelman, Blandford & Rees 1980). 1. Dynamical friction causes a transfer of the black holes' kinetic energy to the surrounding field stars, and the black holes spiral to the centre where they form a binary. 2. While hardening, the effect of dynamical friction reduces and the evolution is dominated by superelastic scattering processes, that is, the interaction with field stars closely encountering or intersecting the binaries' orbit, thereby increasing the binding energy. 3. Finally, the black holes coalesce through the emission of gravitational radiation, potentially detectable by the planned space-based gravitational wave antennae LISA. For a more detailed

Spurzem, R. et al.: *Parallelization, Special Hardware and Post-Newtonian Dynamics in Direct N-Body Simulations*. Lect. Notes Phys. **760**, 377–389 (2008)
DOI 10.1007/978-1-4020-8431-7_15

account of the state of research in this field, see Milosavljević & Merritt (2001, 2003); Makino & Funato (2004); Berczik, Merritt & Spurzem (2005). In our context the problem will be used as an example, where relativistic dynamics becomes important during the evolution of an otherwise classical Newtonian N-body system.

15.2 Relativistic Dynamics of Black Holes in Galactic Nuclei

Relativistic stellar dynamics is of paramount importance for the study of a number of subjects. For instance, if we want to have a better understanding of what the constraints on alternatives to supermassive black holes are, in order to explore the possibility of ruling out stellar clusters, one must do detailed analysis of the dynamics of relativistic clusters. Furthermore, the dynamics of compact objects around an SMBH or multiple SMBHs in galactic nuclei requires the inclusion of relativistic effects. Our current work deals with the evolution of two SMBHs, in bound orbit, and looks at the phase when they get close enough to each other that relativistic corrections to Newtonian dynamics become important, which ultimately leads to gravitational radiation losses and coalescence.

Efforts to understand the dynamical evolution of a stellar cluster in which relativistic effects may be important have already been made by Lee (1987), Quinlan & Shapiro (1989, 1990) and Lee (1993). In the earlier work, $1\mathcal{PN}$ and $2\mathcal{PN}$ terms were neglected (Lee 1993) and the orbit-averaged formalism (Peters 1964) used. We describe here a method to deal with deviations from Newtonian dynamics more rigorously than in most existing literature (but compare Mikkola & Merritt (2007); Aarseth (2007), which are on the same level of $\mathcal{PN}$ accuracy). We modified the NBODY6++ code to allow for post-Newtonian ($\mathcal{PN}$) effects of two particles getting very close to each other, implementing the $1\mathcal{PN}$, $2\mathcal{PN}$ and $2.5\mathcal{PN}$ corrections fully from Soffel (1989) and Kupi, Amaro-Seoane & Spurzem (2006).

Relativistic corrections to the Newtonian forces are expressed by expanding the relative acceleration between two bodies in a power series of $1/c$ in the following way (Damour & Dereulle 1987; Soffel 1989),

$$\underline{a} = \underbrace{\underline{a}_0}_{\text{Newt.}} + \underbrace{\underbrace{c^{-2}\underline{a}_2}_{1\mathcal{PN}} + \underbrace{c^{-4}\underline{a}_4}_{2\mathcal{PN}}}_{\text{periastron shift}} + \underbrace{\underbrace{c^{-5}\underline{a}_5}_{2.5\mathcal{PN}}}_{\text{grav. rad.}} + \mathcal{O}(c^{-6}), \tag{15.1}$$

where $\underline{a}$ is the acceleration of particle 1, $\underline{a}_0 = -Gm_2\underline{n}/r^2$ is the Newtonian acceleration, G is the gravitation constant, m_1 and m_2 are the masses of the two particles, r is the distance of the particles, $\underline{n}$ is the unit vector pointing from particle 2 to particle 1, and the $1\mathcal{PN}$, $2\mathcal{PN}$ and $2.5\mathcal{PN}$ are post-Newtonian corrections to the standard acceleration, responsible for the pericentre shift

($1\mathcal{PN}$, $2\mathcal{PN}$) and the quadrupole gravitational radiation ($2.5\mathcal{PN}$), correspondingly, as shown in (15.1). The expressions for the accelerations are

$$\underline{a}_2 = \frac{Gm_2}{r^2} \cdot \left\{ \underline{n} \left[-v_1^2 - 2v_2^2 + 4v_1v_2 + \frac{3}{2}(nv_2)^2 + 5\left(\frac{Gm_1}{r}\right) + 4\left(\frac{Gm_2}{r}\right) \right] \right.$$
$$\left. + (\underline{v}_1 - \underline{v}_2)\left[4nv_1 - 3nv_2\right] \right\}, \qquad (15.2)$$

$$\underline{a}_4 = \frac{Gm_2}{r^2} \cdot \left\{ \underline{n} \left[-2v_2^4 + 4v_2^2(v_1v_2) - 2(v_1v_2)^2 + \frac{3}{2}v_1^2(nv_2)^2 \right. \right.$$
$$+ \frac{9}{2}v_2^2(nv_2)^2 - 6(v_1v_2)(nv_2)^2 - \frac{15}{8}(nv_2)^4$$
$$+ \frac{Gm_2}{r} \cdot \left(4v_2^2 - 8v_1v_2 + 2(nv_1)^2 - 4(nv_1)(nv_2) - 6(nv_2)^2 \right)$$
$$\left. + \frac{Gm_1}{r} \cdot \left(-\frac{15}{4}v_1^2 + \frac{5}{4}v_2^2 - \frac{5}{2}v_1v_2 + \frac{39}{2}(nv_1)^2 - 39(nv_1)(nv_2) + \frac{17}{2}(nv_2)^2 \right) \right]$$
$$+ (\underline{v}_1 - \underline{v}_2)\left[v_1^2(nv_2) + 4v_2^2(nv_1) - 5v_2^2(nv_2) - 4(v_1v_2)(nv_1) \right.$$
$$+ 4(v_1v_2)(nv_2) - 6(nv_1)(nv_2)^2 + \frac{9}{2}(nv_2)^3$$
$$\left. \left. + \frac{Gm_1}{r} \cdot \left(-\frac{63}{4}nv_1 + \frac{55}{4}nv_2 \right) + \frac{Gm_2}{r} \cdot \left(-2nv_1 - 2nv_2 \right) \right] \right\}$$
$$+ \frac{G^3m_2}{r^4} \cdot \underline{n} \left[-\frac{57}{4}m_1^2 - 9m_2^2 - \frac{69}{2}m_1m_2 \right], \qquad (15.3)$$

$$\underline{a}_5 = \frac{4}{5}\frac{G^2m_1m_2}{r^3} \left\{ (\underline{v}_1 - \underline{v}_2)\left[-(\underline{v}_1 - \underline{v}_2)^2 + 2\left(\frac{Gm_1}{r}\right) - 8\left(\frac{Gm_2}{r}\right) \right] \right.$$
$$\left. + \underline{n}(nv_1 - nv_2)\left[3(\underline{v}_1 - \underline{v}_2)^2 - 6\left(\frac{Gm_1}{r}\right) + \frac{52}{3}\left(\frac{Gm_2}{r}\right) \right] \right\}. \qquad (15.4)$$

In the last expressions $\underline{v}_1$ and $\underline{v}_2$ are the velocities of the particles. For simplification, we have denoted the vector product of two vectors, $\underline{x}_1$ and $\underline{x}_2$, as x_1x_2. The basis of direct NBODY4 and NBODY6++ codes relies on an improved Hermite integration scheme (Makino & Aarseth 1992; Aarseth 1999) for which we need not only the accelerations but also their time derivatives. These derivatives are not included here for succinctness. We include our correction terms in the KS *regularisation* scheme (Kustanheimo & Stiefel 1965) as perturbations, similarly to what is done to account for passing stars influencing a KS pair. Note that formally the perturbing force in the KS equations does not need to be small compared to the two-body force (Mikkola 1997). If the internal KS time step is properly adjusted, the method works even for relativistic terms becoming comparable to the Newtonian force component.

15.3 Example of Application to Galactic Nuclei

In Fig. 15.1 the importance of relativistic, post-Newtonian dynamics for the separation of the binary black holes in our simulations is seen. The curve deviates from the Newtonian results when gravitational radiation losses set in and causes a sudden coalescence ($1/a \to \infty$) at a finite time. Gravitational radiation losses are enhanced by the high eccentricity of the SMBH binary. It is interesting to note that the inclusion or exclusion of the conservative $1\mathcal{PN}$ and $2\mathcal{PN}$ terms changes the coalescence time considerably. Details of these results will be published in a larger parameter study (Berentzen et al. 2008, in preparation). Note that Aarseth (2003a) presents two models very similar to those discussed here, which agree qualitatively with our work regarding the relativistic merger time and the eccentricity of the SMBH binary.

Once the SMBH binary starts to lose binding energy dramatically due to gravitational radiation, its orbital period drops from a few thousand years to less than a year very quickly (time-scale much shorter than the dynamical time-scale in the galactic centre, which defines our time unit). Then the SMBH binary will enter the LISA band, i.e. its gravitational radiation will be detectable by LISA. The Laser Interferometer Space Antenna is a system of three space probes with laser interferometers to measure gravitational waves, see e.g. `http://lisa.esa.int/`. Once the SMBH binary decouples from the rest of the system we just follow its relativistic two-body evolution, starting

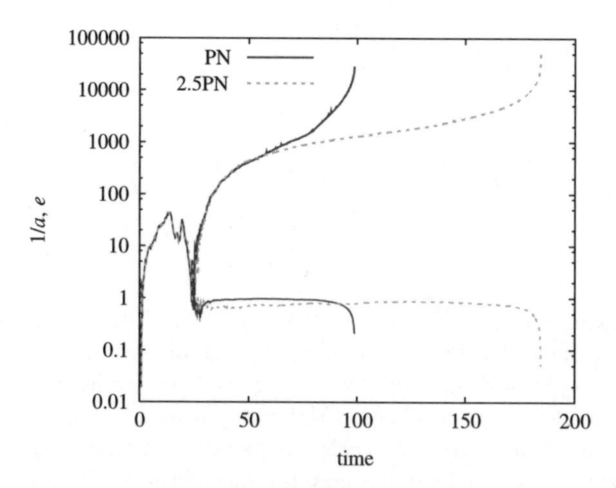

Fig. 15.1. Effect of post-Newtonian (PN) relativistic corrections on the dynamics of black hole binaries in galactic nuclei. Plotted are inverse semi-major axis and eccentricity as a function of time. The solid line uses the full set of PN corrections, while the dashed line has been obtained by artificially only using the dissipative 2.5PN terms. Note that the coalescence time in the latter case has changed significantly. Further details will be published elsewhere (Berentzen et al. 2008, in preparation)

with exactly the orbital parameters (including eccentricity) as they were extracted from the N-body model. It is then possible to predict the gravitational radiation of the SMBH binary relative to the LISA sensitivity curve (Preto et al. 2008, in preparation). For some values of the eccentricity our simulated SMBH binaries indeed enter the LISA sensitivity regime; for a circular orbit the $n = 2$ harmonic of the gravitational radiation is dominant, while for eccentric orbits higher harmonics are stronger (Peters & Mathews 1963; Peters 1964).

15.4 N-Body Algorithms and Parallelization

Numerical algorithms for solving the gravitational N-body problem (Aarseth 2003) have evolved along two main lines in recent years. Direct-summation codes compute the complete set of N^2 interparticle forces at each time step. These codes are designed for systems in which the finite-N graininess of the potential is important or in which binary- or multiple-star systems form, and until recently, were limited by their $\mathcal{O}(N^2)$ scaling to moderate ($N < 10^5$) particle numbers. The best-known examples are the `NBODY` series of codes (Aarseth 1999) and the `Starlab` environment developed by McMillan, Hut and collaborators (e.g. Portegies Zwart et al. 2001).

A second class of N-body algorithms replaces the direct summation of forces from distant particles by an approximation scheme. Examples are the Barnes–Hut tree code (Barnes & Hut 1986), which reduces the number of force calculations by subdividing particles into an oct-tree, and fast multipole algorithms that represent the large-scale potential via a truncated basis-set expansion (van Albada & van Gorkom 1977; Greengard & Rokhlin 1987). Such algorithms have a milder $\mathcal{O}(N \log N)$ or even $\mathcal{O}(N)$ scaling for the force calculations and can handle much larger particle numbers, although their accuracy are substantially lower than that of the direct-summation codes (Spurzem 1999). The efficiency of both sorts of algorithm can be considerably increased by the use of individual time steps for advancing particle positions (Aarseth 2003).

A natural way to increase both the speed and the particle number in an N-body simulation is to parallelize (Dubinski 1996; Pearce & Couchman 1997). Parallelization on general-purpose supercomputers is difficult, however, because the calculation cost is often dominated by a small number of particles in a single dense region, e.g. the nucleus of a simulated galaxy. Communication latency becomes the bottleneck; the time to communicate particle positions between processors can exceed the time spent computing the forces. The best such schemes use systolic algorithms (in which the particles are successively passed around a ring of processors) coupled with non-blocking communication between the processors to reduce the latency (Makino 2002; Dorband, Hemsendorf & Merritt 2003).

A major breakthrough in direct-summation N-body simulations came in the late 1990s with the development of the GRAPE series of special-purpose computers (Makino & Taiji 1998), which achieve spectacular speed-ups by implementing the entire force calculation in hardware and placing many force pipelines on a single chip. The GRAPE-6, in its standard implementation (32 chips, 192 pipelines), can achieve sustained speeds of about 1 Tflops at a cost of just $\sim$ \$50 K. In a standard setup, the GRAPE-6 is attached to a single host workstation, in much the same way that a floating-point or graphics accelerator card is used. Advancement of particle positions $[\mathcal{O}(N)]$ is carried out on the host computer, while coordinate and velocity predictions and inter-particle forces $[\mathcal{O}(N^2)]$ are computed on the GRAPE. More recently, "mini-GRAPEs" (GRAPE-6A) (Fukushige, Makino & Kawai 2005) have become available, which are designed to be incorporated into the nodes of a parallel computer. The mini-GRAPEs have four processor chips on a single PCI card and deliver a theoretical peak performance of $\sim$ 131 Gflops for systems of up to 128 K particles, at a cost of about \$6 K. By incorporating mini-GRAPEs into a cluster, both large (10^6) particle numbers and high (1 Tflops) speeds can be achieved.

In the following we describe the performance of direct-summation N-body algorithms on two computer clusters that incorporate GRAPE hardware.

15.5 Special Hardware, GRAPE and GRACE Cluster

The GRAPE-6A board (Fig. 15.2, top panel) is a standard PCI short card on which a processor, an interface unit and a power supply are integrated. The processor is a module consisting of four GRAPE-6 processor chips, eight SSRAM chips and one FPGA chip. The processor chips each contain six force calculation pipelines, a predictor pipeline, a memory interface, a control unit and I/O ports (Makino et al. 2003). The SSRAM chips store the particle data. The four GRAPE chips can calculate forces, their time derivatives and the scalar gravitational potential simultaneously for a maximum of 48 particles at a time; this limit is set by the number of pipelines (six force calculation pipelines each of which serves as eight virtual multiple pipelines). There is also a facility to calculate neighbour lists from predefined neighbour search radii; this feature is not used in the algorithms presented below. The forces computed by the processor chips are summed in an FPGA chip and sent to the host computer. A maximum of 131 072 (2^{17}) particles can be held in the GRAPE-6A memory. The peak speed of the GRAPE-6A is 131.3 Gflops (when computing forces and their derivatives) and 87.5 Gflops (forces only), assuming 57 and 38 floating-point operations, respectively, per force calculation (Fukushige, Makino & Kawai 2005). The interface to the host computer is via a standard 32-bit/33 MHz PCI bus. The FPGA chip (Altera EP1K100FC256) realizes a 4-input, 1-output reduction when transferring data from the GRAPE-6 processor chip to the host computer. The complete

Fig. 15.2. *Top*: interior of a node showing a GRAPE-6A card (note the large black fan) and an Infiniband card. *Bottom*: the GRACE cluster at ARI. The head node and the 14 Tbyte raid array are visible on the central rack. The other four racks hold a total of 32 compute nodes, each equipped with a GRAPE-6A card and MPRACE cards

GRAPE-6A unit is roughly 11 cm × 19 cm × 7 cm in size. Note that 5.8 cm of the height is taken up by a rather bulky combination of cooling body and fan, which may block other slots on the main board. Possible ways to deal with this include the use of even taller boxes for the nodes (e.g. 5U) together with a PCI riser of up to 6 cm, which would allow the use of slots for interface cards beneath the GRAPE fan, or the adoption of the more recent, flatter designs such as that of the GRAPE6-BL series. The reader interested in more technical details should seek information from the GRAPE (`http://astrogrape.org`) and Hamamatsu Metrix (`http:/www.metrix.co.jp`) websites.

A computer cluster incorporating GRAPE-6A boards became fully operational at the Rochester Institute of Technology (RIT) in February 2005. This cluster, named "gravitySimulator," consists of 32 compute nodes plus one head node, each containing dual 3 GHz-Xeon processors. In addition to a standard Gbit-ethernet, the nodes are connected via a low-latency Infiniband network with a transfer rate of 10 Gbits. The typical latency for an Infiniband network is of the order of 10^{-6} seconds, or a factor ~ 100 better than the Gbit-Ethernet. A total of 14 Tbyte of disc space is available on a level 5 RAID array. The disc space is equivalent to 2.5×10^5 N-body data sets each with 10^6 particles. The discs are accessed via a fast Ultra320 SCSI host adapter from the head node or via NFS from the compute nodes, which in addition are each fitted with an 80 Gbyte hard disc. Each compute node also contains a GRAPE-6A PCI card (Fig. 15.2, top panel). The total, theoretical peak performance is approximately 4 Tflops if the GRAPE boards are fully utilized. Total cost was about $ 450 000, roughly half of which was used to purchase the GRAPE boards.

Some special considerations were required in order to incorporate the GRAPE cards into the cluster. Since our GRAPE-6A's use the relatively old PCI interface standard (32 bit/33 MHz), only one motherboard was available, the SuperMicro X5DPL-iGM, that could accept both the GRAPE-6A and the Infiniband card. (A newer version of the GRAPE-6A which uses the faster PCI-X technology is now available.) The PC case itself has to be tall enough (4U) to accept the GRAPE-6A card and must also allow good air flow for cooling since the GRAPE card is a substantial heat source. The cluster has a total power consumption of 17 kW when the GRAPEs are fully loaded. Cluster cooling was achieved at minimal cost by redirecting the air conditioning from a large room toward the air-intake side of the cluster. Temperatures measured in the PC case and at the two CPUs remain below 30°C and 50°C, respectively.

A similar cluster, called "GRACE" (GRAPE + MPRACE), has been installed in the Astronomisches Rechen-Institut (ARI) at the University of Heidelberg (Fig. 15.2, bottom panel). There are two major differences between the RIT and ARI clusters. (1) Each node of the ARI cluster incorporates a reconfigurable FPGA card (called "MPRACE") in addition to to the GRAPE board. MPRACE is optimized to compute neighbour forces and other non-Newtonian forces between particles, in order to accelerate calculations of molecular dynamics, smoothed-particle hydrodynamics, etc. (2) The newer main board SuperMicro X6DAE-G2 was used, which supports Pentium Xeon chips with 64-bit technology (EM64T) and the PCIe (PCI express) bus. This made it possible to use dual-port Infiniband interconnects via the PCI express Infiniband ×8 host interface card, used in the ×16 Infiniband slot of the board (it has another ×4 Infiniband slot, which is reserved for the MPRACE-2 Infiniband card). As discussed below, the use of the PCIe bus substantially reduces communication overhead. The benchmark results presented here for the ARI cluster were obtained from algorithms that do not access the FPGA cards.

15.6 Performance Tests

Initial conditions for the performance tests were produced by generating Monte-Carlo positions and velocities from self-consistent models of stellar systems. Each of these systems is spherical and is completely described by a steady-state phase-space distribution function $f(E)$ and its self-consistent potential $\Psi(r)$, where $E = v^2/2 + \Psi$ is the particle energy and r is the distance from the centre. The models were a Plummer sphere, two King models with different concentrations and two Dehnen models (Dehnen 1993) with different central density slopes. The Plummer model has a low central concentration and a finite central density; it does not represent any class of stellar system accurately, but is a common test case. King models are defined by a single dimensionless parameter W_0 characterizing the central concentration (e.g. ratio of central to mean density); we used $W_0 = 9$ and $W_0 = 12$, which are appropriate for globular star clusters. Dehnen models have a divergent inner density profile, $\rho \propto r^{-\gamma}$. We took $\gamma = 0.5$ and $\gamma = 1.5$, which correspond approximately to the inner density profiles of bright and faint elliptical galaxies.

In what follows we adopt standard N-body units $G = M = -4E = 1$, where G is the gravitational constant, M the total mass and E the total energy of the system. In some of the models, the initial time step for some particles was smaller than the minimum time step $t_{\rm min}$ set to 2^{-23}. These models were then rescaled to change the minimum time step to a large enough value. Since the rescaling does not influence the performance results, we will present all results in the standard N-body units.

We realized each of the five models with 11 different particle numbers, $N = 2^k$, $k = [10, 11, \ldots, 20]$, i.e. $N = [1\,\mathrm{K}, 2\,\mathrm{K}, \ldots, 1\,\mathrm{M}]$.[1] We also tested Plummer models with $N = 2\,\mathrm{M}$ and $N = 4\,\mathrm{M}$; the latter value is the maximum N-value allowed by filling the memory of all 32 GRAPE cards. Thus, a total of 57 test models were used in the timing runs.

Two-body relaxation, i.e. exchange of energy between particles due to gravitational scattering, induces a slow change in the characteristics of the

[1] Henceforth, we use K to denote a factor of $2^{10} = 1024$ and M to denote a factor of $2^{20} = 1,048,576$.

models. In order to minimize the effects of these changes on the timing runs, we integrated the models for only one time unit. The standard softening ϵ was set to zero for the Plummer models and to 10^{-4} for the Dehnen and King models. For the time step parameters used see Harfst et al. (2007).

We analyzed the performance of the hybrid scheme as a function of particle number and also as a function of number of nodes, using $p = 1, 2, 4, 8, 16,$ and 32 nodes. The compute time w for a total of almost 350 test runs was measured using `MPI_Wtime()`. The timing was started after all particles had finished their initial time step and ended when the model had been evolved for one time unit. No data evaluation was made during the timing interval.

The top panel of Fig. 15.3 shows wallclock times $w_{N,p}$ from all integrations on the ARI cluster. For any p, the clock time increases with N, roughly as N^2 for large N. However, when N is small, communication dominates the total clock time, and w *increases* with increasing number of processors. This behaviour changes as N is increased; for $N > 10\,\mathrm{K}$ (the precise value depends on the model), the clock time is found to be a decreasing function of p, indicating that the total time is dominated by force computations.

The speedup for selected test runs is shown in the bottom panel of Fig. 15.3. The speedup s is defined as

$$s_{N,p} = \frac{w_{N,1}}{w_{N,p}}. \tag{15.5}$$

The ideal speedup (optimal load distribution, zero communication and latency) is $s_{N,p} = p$. For particle numbers $N \geq 128\,\mathrm{K}$ the wallclock time $w_{N,1}$ on one processor is undefined as N exceeds the memory of the GRAPE card. In that case we used $w_{N,1} = w_{128\,\mathrm{K},1}(N/128\,\mathrm{K})^2$, assuming a simple N^2-scaling. In general, the speedup for any given particle number is roughly proportional to p for small p, then reaches a maximum before reducing at large p. The number of processors at maximum speedup is "optimum" in the sense that it provides the fastest possible integration of a given problem. The optimum p is roughly the value at which the sum of the communication and latency times equals the force computation time; in the zero-latency case, $p_{opt} \propto N$ (Dorband, Hemsendorf & Merritt 2003). Figure 15.3 (bottom panel) shows that for $N \geq 128\,\mathrm{K}$, $p_{opt} \geq 32$ for all the tested models. The reader interested in more details is referred to Harfst et al. (2007).

15.7 Outlook and Ahmad–Cohen Neighbour Scheme

At present there exist only the relatively simple parallel N-body code described above and in Harfst et al. (2007), which uses GRAPE special hardware in parallel, but always computes full forces for every particle at every step. This code, sometimes dubbed p-GRAPE (sources are freely available, see link in the cited paper) also does not include any special few-body treatments (regularisations), as in the N-body codes of Aarseth (1999, 2003).

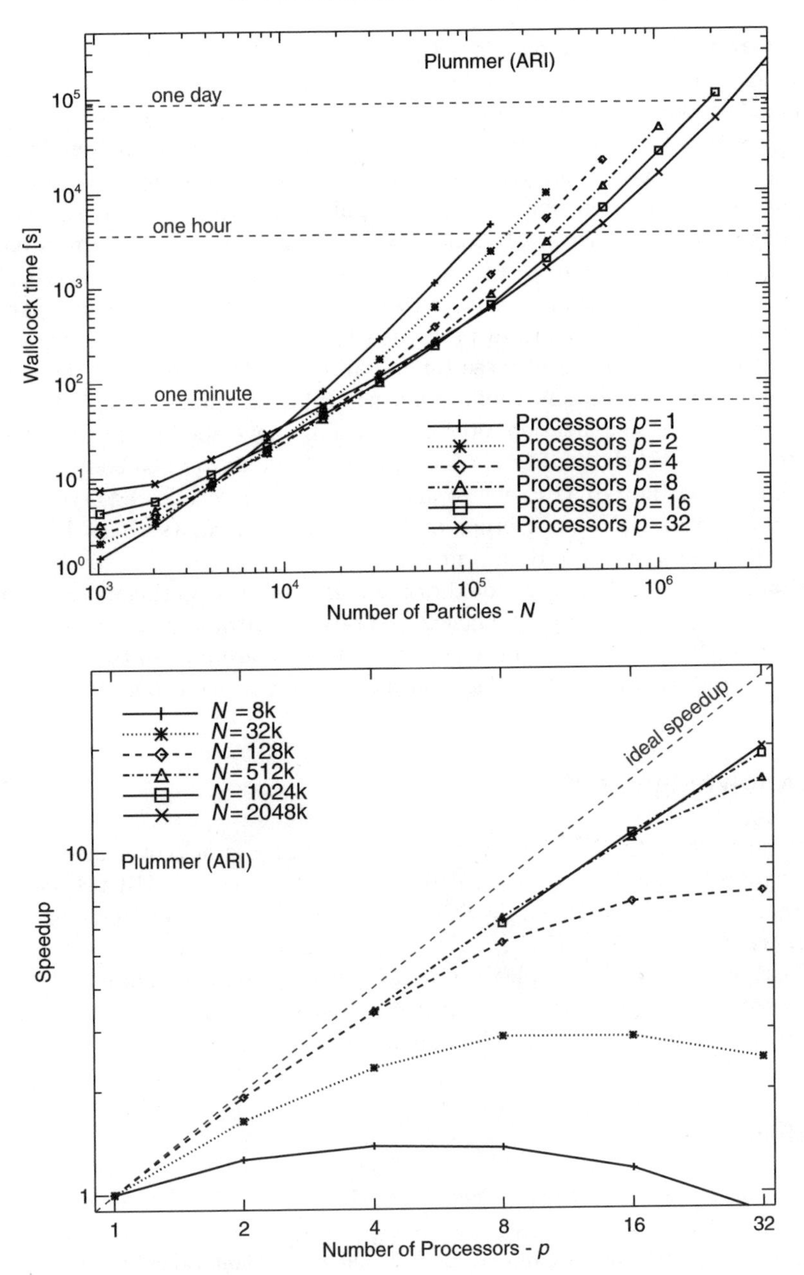

Fig. 15.3. *Top*: wallclock time w versus particle number N for different numbers of processors p. *Bottom*: speedup s versus processor number p for different N. Both the plots show the results obtained for a Plummer model on the ARI cluster

There is the already mentioned parallel N-body code NBODY6++, which includes all regularizations and the use of the Ahmad-Cohen neighbour scheme (Ahmad & Cohen 1973) as in the standard NBODY6 code. However, the publicly provided source code (`ftp://ftp.ari.uni-heidelberg.de/pub/staff/spurzem/nb6mpi/`) is not yet able to make parallel use of special hardware. It parallelizes very efficiently over the regular *and* irregular force loops (cf. Spurzem 1999; Khalisi et al. 2003), but current work is in progress on an implementation of NBODY6++ for special-purpose hardware (such as GRAPE, MPRACE or graphical processing units GPU) as well as on an efficient parallel treatment of many regularized perturbed binaries (see first results in Maalej et al. 2005). New results in these topics will be published early at the wiki of NBODY6++ developers and users at `http://nb6mpi.pbwiki.com/`. Last but not least, a nice visualization interface, specially developed for NBODY6++, is hosted by FZ Jülich, see `http://www.fz-juelich.de/jsc/xnbody/`.

Similar to the GRAPE development nearly two decades ago, the recent introduction of GPUs and other new hardware devices (such as FPGA or MPRACE cards in the GRACE project,

`http://www.ari.uni-heidelberg.de/grace/`) is inspiring a new interest in improving and developing efficient N-body algorithms. It is expected that very soon the use of most advanced special hardware and software (such as NBODY6 and NBODY6++) will not mutually exclude each other any more.

Acknowledgement

Computing time at NIC Jülich on the IBM Jump is acknowledged. Financial support comes partly from Volkswagenstiftung (I/80 041-043), German Science Foundation (DFG) via SFB439 at the University of Heidelberg and Schwerpunktprogramm 1177 (Project ID Sp 345/17-1) 'Black Holes Witnesses of Cosmic History'. It is a pleasure to acknowledge many enlightening discussions with and support by Sverre Aarseth, and very useful interactions about relativistic dynamics with A. Gopakumar and G. Schäfer.

References

Aarseth S. J., 1999, PASP, 111, 1333

Aarseth S. J., 2003a, ApSS, 285, 367

Aarseth S. J., 2003, Gravitational N-Body Simulations. Cambridge University Press, Cambridge

Aarseth S. J., 2007, MNRAS, 378, 285

Ahmad A., Cohen L., 1973, J. Comput. Phys., 12, 349

Barnes J., Hut P., 1986, Nature, 324, 446

Begelman M. C., Blandford R. D., Rees M. J., 1980, Nature, 287, 307

Berczik P., Merritt D., Spurzem R., 2005, ApJ, 633, 680

Berczik P., Merritt D., Spurzem R., Bischof H.-P., 2006, ApJ, 642, L21
Berentzen I., Preto M., Berczik P., Merritt D., Spurzem R., 2008, to be submitted
Damour T., Dereulle N., 1987, Phys. Lett., 87, 81
Dehnen W., 1993, MNRAS, 265, 250
Dorband E. N., Hemsendorf, M., Merritt, D., 2003, J. Comput. Phys., 185, 484
Dubinski J., 1996, New Astron., 1, 133
Fukushige T., Makino J., Kawai A., 2005, PASJ, 57, 1009
Greengard L., Rokhlin V., 1987, J. Comput. Phys., 73, 325
Harfst S., Gualandris A., Merritt D., Spurzem R., Portegies Zwart S., Berczik P., 2007, New Astron., 12, 357
Khalisi E., Omarov C. T., Spurzem R., Giersz M., Lin D. N. C., 2003, in Krause E., Jaeger W., Resch M., eds, Performance Computing in Science and Engineering. Springer Verlag, p. 71
Kupi G., Amaro-Seoane P., Spurzem R., 2006, MNRAS, 371, L45
Kustaanheimo P., Stiefel E., Journ. für die reine und angew. Math., 1965, 218, 204
Lagoute C., Longaretti P. -Y., 1996, A&A, 308, 441
Lee H. M., 1987, ApJ, 319, 801
Lee M. H., 1993, ApJ, 418, 147
Maalej K. P., Boily C., David R., Spurzem R., 2005, in Casoli F., Contini T., Hameury J. M., Pagani L., eds, SF2A-2005: Semaine de l'Astrophysique Francaise. EdP-Sciences, Conference Series, p. 629
Makino J., 2002, New Astron., 7, 373
Makino J., Aarseth S. J., 1992, PASJ, 44, 141
Makino J., Fukushige T., Koga M., Namura K., 2003, PASJ, 55, 1163
Makino J., Funato Y., 2004, ApJ, 602, 93
Makino J., Taiji M., 1998, Scientific Simulations with Special-Purpose Computers — the GRAPE systems. Wiley
Mikkola S., 1997, Celes. Mech. Dyn. Ast., 68, 87
Mikkola S., Merritt D., 2007, ArXiv e-prints 709, arXiv:0709.3367
Milosavljević M., Merritt D., 2001, ApJ, 563, 34
Milosavljević M., Merritt D., 2003, ApJ, 596, 860
Pearce F. R., Couchman H. M. P., 1997, New Astron., 2, 411
Peters P. C., 1964, Phys. Rev., 136, B1224
Peters P. C., Mathews J., 1963, Phys. Rev., 131, 435
Portegies Zwart S. F., McMillan S. L. W., Hut P., Makino J., 2001, MNRAS, 321, 199
Preto M., Berentzen I., Berczik P., Spurzem R., 2008, in preparation
Quinlan G. D., Shapiro S. L., 1989, ApJ, 343, 725
Quinlan G. D., Shapiro S. L., 1990, ApJ, 356, 483
Soffel M. H., 1989, Relativity in Astrometry, Celestial Mechanics and Geodesy. Springer-Verlag
Spurzem R., 1999, J. Comput. Applied Math., 109, 407
van Albada T. S., van Gorkom J. H., 1977, A&A, 54, 121

A

Educational N-Body Websites

Francesco Cancelliere[1], Vicki Johnson[2], and Sverre Aarseth[3]

[1] Free University, Brussels, Pleinlaan 2, B-1050 Brussels, Belgium
fcancell@vub.ac.be
[2] Interconnect Technologies LLC, POB 1517, Placitas, NM, 87043, USA
vlj@interconnect.com
[3] University of Cambridge, Institute of Astronomy, Madingley Road, Cambridge CB3 0HA, UK
sverre@ast.cam.ac.uk

A.1 Introduction

The 2006 Cambridge N-body School introduced participants to educational websites for N-body simulations, *www.Sverre.com* and *www.NBodyLab.org*. These websites run versions of the freely available, open-source `NBODY4`, `TRIPLE` and `CHAIN` codes (Aarseth 2003) that have been adapted for the web. The websites provide guidance and documentation. They support simulations of small N (3 and 4 bodies) on both sites and higher N (up to 15,000) on NBodyLab.org. Numerical results, graphics and animations are displayed. NBodyLab.org supports `NBODY4`, running on a GRAPE-6A hardware accelerator, and demonstrates its accuracy and speed. The websites were developed with different approaches; NBodyLab.org runs N-body codes on the server side, and Sverre.com uses Java to run locally.

The websites were recommended as homework before the N-body School and practical demonstrations were given during the School. Use of these sites by participants also continued afterwards. Such web-based tools can be a useful and convenient part of the curriculum for teaching N-body simulations and also serve as test-beds for prospective buyers of GRAPE hardware accelerators for large simulations. This Appendix describes the websites and their educational utility.

A.2 www.NBodyLab.org

NBodyLab.org (Johnson & Aarseth 2006) is a laboratory where we can experiment with small N-body simulations with a desktop GRAPE-6A supercomputer (Fukushige, Makino & Kawai 2005, Makino & Taiji 1998). The `NBODY4`, `TRIPLE` and `CHAIN` codes are adapted for the web from the current versions

Cancelliere, F. et al.: *Educational N-Body Websites*. Lect. Notes Phys. **760**, 391–396 (2008)
DOI 10.1007/978-1-4020-8431-7_16

of the Unix/FORTRAN codes [1] and simulations are run on the server side. Plots and 3D animations are created from the simulation output.

NBodyLab was initially developed in 2002 to augment an undergraduate astrophysics course. Prior to upgrading to NBODY4, NBodyLab was used for homework assignments, an undergraduate senior thesis on tidal shocking of globular clusters, small system studies of Ursa Major, Hyades, Collinder 70, the solar system, Halley's comet and a master's thesis on N-body simulations and HR diagrams of nearby stars (Johnson & Ates 2004). Incorporating NBODY4 has significantly improved the site's N-body simulation capabilities.

Examples of NBODY4 simulations that can be run on NBodyLab.org include

- single Plummer sphere cluster model ($N = 1000$),
- single Plummer sphere cluster with 200 additional primordial binaries,
- two Plummer models in orbit,
- massive perturber and planetesimal disk,
- evolution of a dominant binary and
- upload specialized initial conditions.

Input parameters are entered via forms (NBODY4 concise style, or simplified):

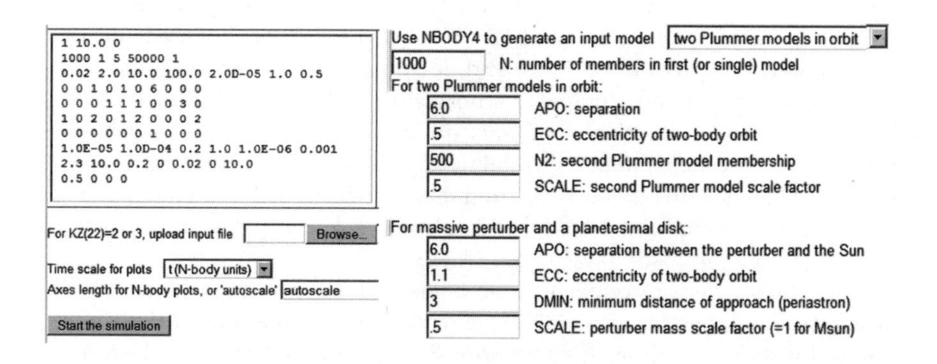

The presentation of NBodyLab.org at the N-body School included discussion of the site's goals, parameter limits, and an overview of the main features of NBODY4, such as GRAPE acceleration for direct integration, regularization of close encounters and stellar evolution with mass loss and collisions. The main NBODY4/6 input parameters were introduced, including model options, choices for binaries, stellar evolution and mass loss, initial mass function, scaling and chain regularization. NBODY4 and NBODY6 were compared. It should be noted that NBODY6 uses a neighbour scheme to speed up the integration. Output data analysis and output quantities were discussed, along with plots and stellar evolution features, such as the time

[1] downloads at http://www.ast.cam.ac.uk/research/nbody

dependence of the half-mass radius and core radius in N-body units, as well as the HR diagram for the initial and final population of single stars (see next figures).

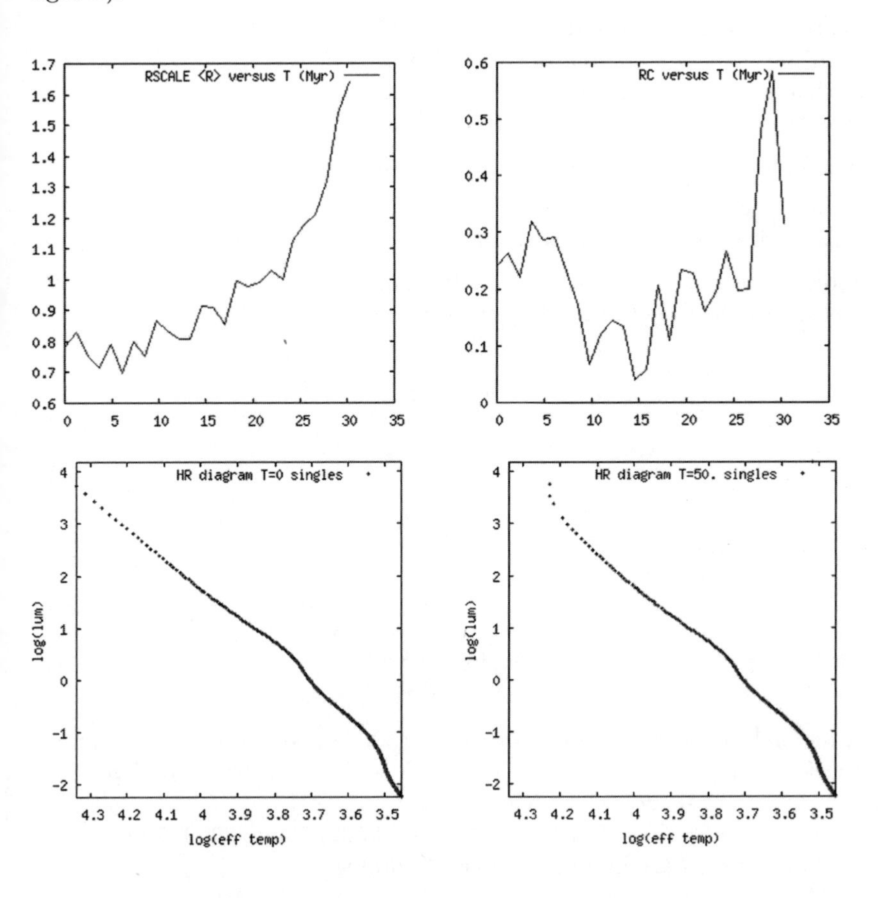

Animations of model evolution can be viewed in 3D with a Java applet:

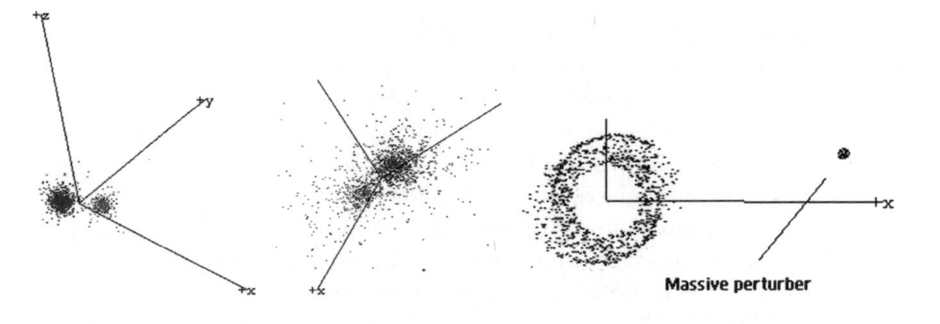

`TRIPLE` and `CHAIN` with regularization are used for small N simulations on NBodyLab.org. Examples of three-body simulations with 3D animations include

- figure-8 periodic orbit and perturbations (Heggie 2000),
- idealized triple system and perturbations,
- Pythagorean problem and perturbations and
- criss-cross periodic orbit and perturbations (Moore 1993)

and examples of four-body simulations with 3D animations include

- great circle unstable orbit and
- symmetrical exchange for two binaries.

Examples of graphics for the three-body figure-8 stable orbit and with perturbations are displayed in the following figures.

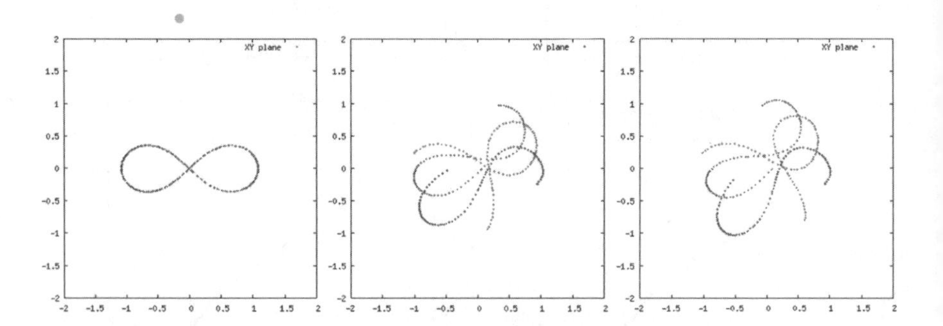

A manual for running simulations with `NBODY4` and `NBODY6` was prepared for the Cambridge N-body School (Aarseth & Johnson 2006). It covers parameter selection, suggested simulations, astrophysical and N-body units, integration methods, the relationship between `NBODY4` and `NBODY6` and other topics. Sample runs are interpreted and annotated.

A.3 www.Sverre.com

This interactive website was made available in 2005 to support movies of the three-body problem, where the initial conditions are specified online. In the summer of 2006 a second similar presentation was implemented for the four-body problem. The main technical difference is that a three-body regularization method (Aarseth & Zare 1974) is used for the former while $N = 4$ is handled by chain regularization (Mikkola & Aarseth 1993), which can also deal with $N = 3$ after one body escapes. The calculations are done in real time by a Java applet or Java application that can be downloaded. In spite of considerable loss in programming efficiency, owing to the use of Java instead of FORTRAN, the viewing time is sufficiently short even at the highest time-step resolution.

Online simulations can be instructive and also great fun. For practical convenience, only 2D calculations are performed. A number of useful features are available, such as a scale factor for magnification, smoothness index to vary the viewing time, maximum run time (otherwise until escape), a facility for play, pause or reset and also for displaying the orbits at the end. The screen shots show initial and final configurations for the two movie versions, with the interactive initial conditions specified in appropriate boxes. The basic FORTRAN codes without the interactive part, as well as `TRIPLE` and `CHAIN`, can be downloaded from the URL specified above.

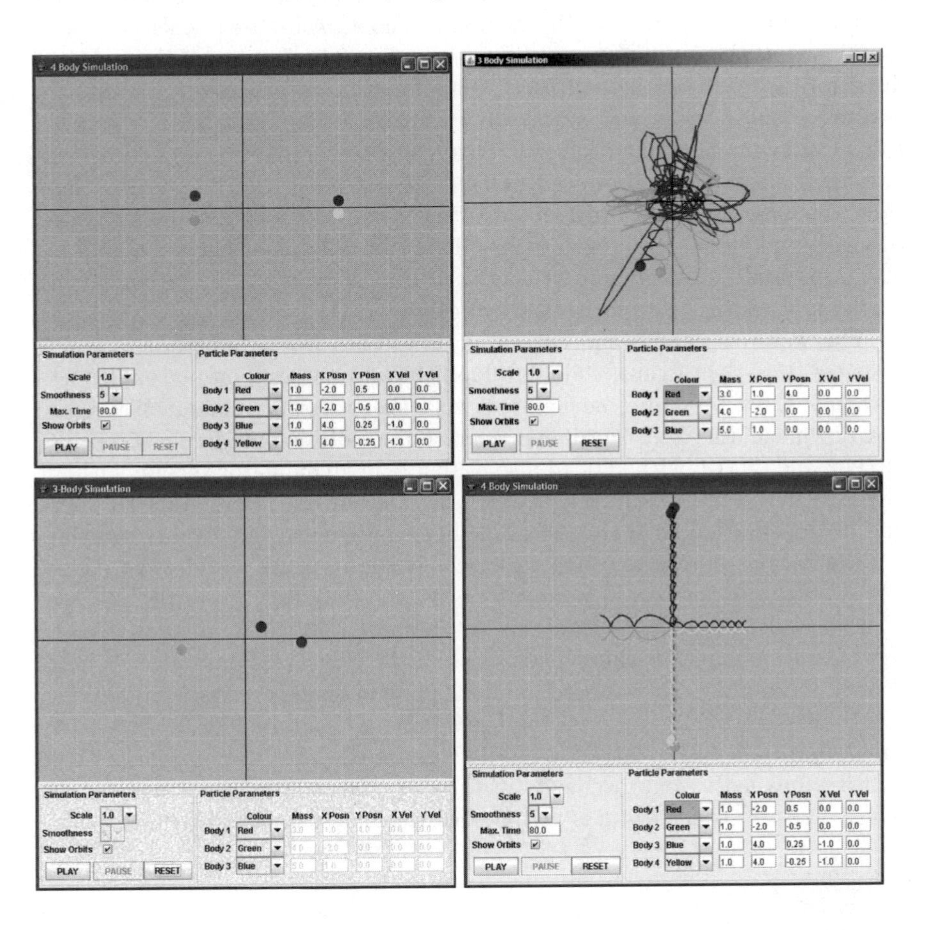

Some examples of interesting initial conditions are provided as templates and shown above, together with the final orbits. Users are encouraged to experiment by exploring the large parameter space. It can be seen that very small changes in the initial conditions may produce widely different behaviour owing to the chaotic nature of the problem. Although most solutions should be accurate, complex interplays of long duration are notoriously difficult and even small errors are subject to exponential growth, which may lead to the wrong outcome. However, since close two-body encounters are treated very accurately with regularization, the result of the strong interactions themselves is reliable.

A.4 Educational Utility

For undergraduate and graduate astronomy and physics courses and special advanced programs such as the N-body School, web-based tools can be a useful part of the curriculum. The primary educational utility of the websites discussed here is their ease-of-use. Documentation is available for beginners and experienced users and initial values are given for interesting examples. Runs can be made with a click of a button and no compilation and additional graphical displays are produced, which are not supported in the standard code versions. Specially constructed initial conditions can also be uploaded to satisfy individual requirements for GRAPE simulations. The websites have also been used by researchers writing their own N-body codes, for comparing results and testing (e.g., for stellar evolution).

The websites enable and encourage migration from simulations via the websites to in-depth runs, code development and research on personal workstations. After becoming acquainted with the program functionality, users can download the freely available open-source software and run `NBODY4/6`, `TRIPLE` and `CHAIN`, with `NBODY4` also available in an emulator version without GRAPE hardware. Discussion of the programs in the book (Aarseth 2003) and documentation on the websites facilitate online use and local computing.

Simulations on the websites have been made by users world-wide. About 300 simulations per month were run on www.NBodyLab.org in the last half of 2006 and the guide *Introduction to Running* `NBODY4/6` *Simulations* was downloaded about 100 times per month. Following the Cambridge N-body School, NBodyLab.org was used in late 2006 in assigned exercises for students of a Stellar Dynamics course at the University of Bonn. In 2007, a three-body simulation code with relativistic effects was added. The development of these websites has led to improvements in the N-body codes and documentation. Suggestions for other features and new educational uses are welcomed.

References

Aarseth S. J., 2003, Gravitational N-Body Simulations, Cambridge Univ. Press, Cambridge
Aarseth S. J., Johnson, V. L., 2006, posted on NBodyLab.org
Aarseth S. J., Zare, K., 1974, Celes. Mech., 10, 185
Fukushige T., Makino, J., Kawai, A., 2005, PASJ, 57, 1009
Heggie D. C., 2000, MNRAS, 318, L61
Johnson V. L., Aarseth, S. J., 2006, in C. Gabriel, C. Arviset, D. Ponz, E. Solano, eds, ADASS XV, ASP Conf. Ser., 351, 165
Johnson V. L., Ates, A., 2004, in P. Shopbell, M. Britton, R. Ebert, eds, ADASS XIV, ASP Conf. Ser., 347, 524
Makino J., Taiji M., 1998, Scientific Simulations with Special-Purpose Computers, the GRAPE System, John Wiley & Sons
Mikkola S., Aarseth, S. J., 1993, Celes. Mech. Dyn. Ast., 57, 439
Moore C., 1993, Phys. Rev. Lett. 70, 3675

Index